Diagram Geometries

ANTONIO PASINI
Professor of Geometry
University of Siena

CLARENDON PRESS • OXFORD
1994

Oxford University Press, Walton Street, Oxford OX2 6DP
Oxford New York
Athens Auckland Bangkok Bombay
Calcutta Cape Town Dar es Salaam Delhi
Florence Hong Kong Istanbul Karachi
Kuala Lumpur Madras Madrid Melbourne
Mexico City Nairobi Paris Singapore
Taipei Tokyo Toronto
and associated companies in
Berlin Ibadan

Oxford is a trade mark of Oxford University Press

Published in the United States
by Oxford University Press Inc., New York

A catalogue record for this book is available from the British Library

Library of Congress Cataloging in Publication Data

ISBN 0 19 853497 3

Typeset by the author
Printed in Great Britain on acid-free paper by
Biddles Ltd., Guildford & King's Lynn

Preface

I view diagram geometry as a craft that gets information on a multi-dimensional geometric object (a geometry of rank ≥ 3) from information on its bi-dimensional properties. Diagrams are a way of packing up the local information we start from. Borrowing an image from chemistry, geometries of rank 2 are like atoms, but I am interested in the molecules that we can build by connecting two or more of those atoms. This being my point of view, I will discuss rank 2 geometries only to the extent that I need in my 'molecular' perspective. Thus, I will leave out the point–line approach to multi-dimensional geometries, where geometries of rank ≥ 3 are produced as systems of subspaces of certain bi-dimensional geometries. Moreover, an excellent book already exists and will soon appear, where the potentiality of the point–line approach is fully exploited (I mean the book by F. Buekenhout and A. Cohen [25]).

Good books on buildings are also available (K. Brown [14] and M. Ronan [181]; also Tits [222], of course). Thus, I have not insisted too much on the general theory of buildings, even if many of the geometries considered in this book are in fact buildings. I will only recall what I need of that theory in Chapter 13.

Perhaps some readers would like to see a list of all classification theorems so far obtained in diagram geometry, something like an 'Atlas of Geometries and Theorems', but they will not find this in my book. A fairly complete survey of classification theorems, updated in 1992, is given in [27] and I do not think it is worthwhile to duplicate that work here. Moreover, my main purpose in this book is not to exhibit what people have been able to do in diagram geometry, but to enable the reader to master the tools of our craft. In fact we use many tools, and each requiring a particular skill. Thus, I am afraid that some readers will find this book a bit difficult. I have done my best to make it as easy as possible, but I don't want to cheat the reader by presenting only the easiest and most amusing things, hiding the rest.

The book is organized as follows. I recall some prerequisites in Chapter 0. In Chapter 1 I state the first definitions and describe a number of examples, continuing that survey of examples in Chapter 2. The language of diagrams is explained in Chapter 3. Chapters 4, 5 and 6 are devoted to the elementary theory of diagram geometry (the direct sum theorem, geometries with string diagrams, Grassmann geometries and shadows, intersection property). My purpose in Chapter 7 is to show how the tools prepared in Chapters 4, 5 and 6 can be used when we want to classify geometries with prescribed diagrams. I turn to morphisms and automorphisms in Chapters 8, 9, 10 and 11. Morphisms and quotients are discussed in Chapter 8. Automorphism groups are studied in Chapters 9 and 10. I turn back to quotients in the first section of Chapter 11, considering quotients by groups. The rest of Chapter 11 might be viewed as an anticipation of the theory of universal covers, developed in Chapter 12. In the last Chapters (13, 14 and 15) I want to show how the instruments set up in Chapters 8, 9, 10, 11 and 12 can be used to carry on the analysis begun in Chapter 7, improving some of the results obtained there. I will focus on Coxeter diagrams of spherical type (Chapters 13 and 14), in particular on C_n and F_4 (Chapter 14), and on C_n-like diagrams (Chapter 15).

I should thank many friends for remarks, suggestions and encouragement. Particular thanks are due to two of them: Daniel Hughes and William Kantor. Daniel has generously collaborated with me on the first three chapters almost as a coauthor. William has patiently read the first draft of my book, making a lot of valuable remarks, suggesting plenty of improvements, correcting mistakes, doing all he could to make this book more readable.

Siena A.P.
April 1994

Contents

Chapter 0

Prerequisites

We presume that the reader is familiar with linear algebra and with the basic concepts of group theory: cosets, Lagrange's theorem, Sylow subgroups, inner and outer automorphisms, normal subgroups, factor groups, ... We also assume he is acquainted with some elementary notions of projective and affine geometry, such as the definition of projective and affine planes, of Desarguesian projective planes (which we also call *classical* projective planes) and the construction of the projective geometry of linear subspaces of a vector space (anyway, we will recall this construction in Section 1.3.1).

We will also use a few elementary notions from graph theory and from the theory of simplicial complexes. The concepts we need will be defined in Sections 0.4 and 0.5, even if most of them are probably well known to the reader.

In this chapter we state some notation and terminology, to fix them in those cases where alternative conventions appear elsewhere. This is especially necessary for graph-theoretic notions, since even the meaning of the word 'graph' is not the same for all authors.

0.1 A few conventions for mappings and sets

We write compositions of mappings from right to left. Thus, given mappings $f : A \longrightarrow B$ and $g : B \longrightarrow C$, the composition of f and g will be denoted by gf. Given a mapping $f : A \longrightarrow B$, the preimage of an element $b \in f(A)$ is denoted by $f^{-1}(b)$ and it is called a *fibre* of f.

Given sets A and B, we write $A - B$ for the complement of B in A, even if B is not a subset of A.

We will seldom distinguish between an element a and its singleton $\{a\}$.

Thus, we will often use expressions such as $A \cup a$, $A - a$, $A \cap B = a$ etc. even though the correct set-theoretic notation is $A \cup \{a\}$, $A - \{a\}$, $A \cap B = \{a\}$, etc.

We denote the cardinality of a set A by $|A|$. When $|A| = n < \infty$ we also say that A is an *n-set.*

0.2 Notation for equivalence relations

Given a relation Φ on a set S and elements $x, y \in S$, we write $x\Phi y$ for $(x, y) \in \Phi$. If Φ is an equivalence relation, then we also write $x \equiv y\ (\Phi)$ for $x\Phi y$ and we denote by $[x]\Phi$ the class of all elements of S equivalent to x in Φ. Thus, $\{[x]\Phi \mid x \in S\}$ is the partition of S defined by Φ. We will often make no distinction between an equivalence relation and the partition defined by it.

Given two equivalence relations Φ and Ψ on the same set S, the intersection of Φ and Ψ will be denoted by $\Phi \cap \Psi$. The *join* $\Phi \vee \Psi$ of Φ and Φ is the intersection of all equivalence relations on S containing both Φ and Ψ. The symbol $\Phi\Psi$ denotes the relation defined as follows: x and y are equivalent in $\Phi\Psi$ iff $x \equiv z\ (\Phi)$ and $z \equiv y\ (\Psi)$ for some $z \in S$.

We write $\Phi \leq \Psi$ to mean that Φ is a refinement of Ψ. We denote by Ω the *trivial* equivalence relation on S; that is, $\Omega = S \times S$. The *identity* relation $=$ will be denoted by $\mho$.

0.3 Notation for partially ordered sets

We write 'poset' for 'partially ordered set', for short. Given a poset $\mathcal{P} = (P, \leq)$ and elements $x, y \in P$ with $x \leq y$, we define *intervals* as follows:

$$
\begin{aligned}
&[x, +[= \{z \mid x \leq z \in P\},\\
&]x, +[= \{z \mid x < z \in P\},\\
&]-, y] = \{z \mid y \geq z \in P\},\\
&]-, y[= \{z \mid y > z \in P\},\\
&[x, y] = \{z \mid x \leq z \leq y\},\\
&[x, y[= \{z \mid x \leq z < y\},\\
&]x, y] = \{z \mid x < z \leq y\},\\
&]x, y[= \{z \mid x < z < y\}.
\end{aligned}
$$

Note that some of the above intervals might be empty for some choices of x and $y \geq x$. The intervals $]-, y[$, $]x, +[$, $]x, y[$ are said to be *open.* The interval $[x, y]$ is *closed.* We also write $]-, +[$ for P.

The *length* of a chain $x_0 < x_1 < ... < x_m$ of $\mathcal{P}$ is the number m. We say that $\mathcal{P}$ is *graded* of *rank* n if all maximal chains of $\mathcal{P}$ are finite of the same length $n-1$ and, for every element $x \in P$, all maximal chains ending at x have the same length $t(x)$. We call $t(x)$ the *dimension* of x. The function $t : P \longrightarrow \mathbf{N}$ mapping every element $x \in P$ of a graded poset $\mathcal{P}$ onto its dimension $t(x)$ is the *dimension function* of $\mathcal{P}$. Elements of $\mathcal{P}$ of dimension 0 are called *atoms.*

Given posets $\mathcal{P}_1 = (P_1, \leq_1)$, $\mathcal{P}_2 = (P_2, \leq_2)$, ..., $\mathcal{P}_n = (P_n, \leq_n)$, their *direct product* $\prod_{i=1}^{n} \mathcal{P}_i$ is the poset with $P_1 \times P_2 \times ... \times P_n$ as set of vertices and with order relation $\leq$ defined as follows: $(x_1, x_2, ..., x_n) \leq (y_1, y_2, ..., y_n)$ iff $x_i \leq_i y_i$ for all $i = 1, 2, ..., n$.

0.4 Some terminology for graphs

We define graphs as follows: a *graph* is a pair $\Gamma = (V, E)$ where V is a nonempty set and E is a (possibly empty) set of unordered pairs of distinct elements of V. The elements of V (of E) are called *vertices* (*edges*, respectively). Given two distinct vertices $x, y \in V$, if $\{x, y\}$ is an edge, then we write $x \sim y$ and we say that x and y are *adjacent.* We will often write $\Gamma = (V, \sim)$ instead of $\Gamma = (V, E)$, putting emphasis on the adjacency relation $\sim$.

We say that a graph is *trivial* if it has no edges. A graph is said to be *complete* if every pair of distinct vertices is an edge). A *clique* of a graph Γ is a (possibly empty) set of pairwise adjacent vertices of Γ. Henceforth we will not make any distinction between a vertex x and the clique $\{x\}$. A *coclique* of Γ is a set of pairwise non-adjacent vertices of Γ.

Given a positive integer n and a graph $\Gamma = (V, \sim)$, an *n-partition* of Γ is a partition of V into n cocliques. If Γ admits an n-partition, then it said to be *n-partite.* If Γ admits an n-partition Θ such that any two vertices belonging to distinct classes of Θ are always adjacent, then we say that Γ is a *complete n-partite* graph.

We say that a graph $\mathcal{X}$ with set of vertices X is a *subgraph* of a graph $\Gamma = (V, \sim)$ if $X \subseteq V$ and the adjacency relation of $\mathcal{X}$ is a refinement of the relation induced by $\sim$ on X. If the adjacency relation of $\mathcal{X}$ is the relation induced by $\sim$ on X, then $\mathcal{X}$ is called the graph *induced* by Γ on X.

Given an n-partition Θ of an n-partite graph $\Gamma = (V, \sim)$, let X be a nonempty subset of V meeting just m classes of Θ and let $\mathcal{X}$ be the graph induced by Γ on X. Then the partition induced by Θ on X is an m-partition of $\mathcal{X}$. We call it the m-partition *induced* by Θ on $\mathcal{X}$.

Given a vertex x of a graph $\Gamma = (V, \sim)$, the *neighbourhood* of x is the set of vertices of Γ distinct from x and adjacent to x. We denote this set by

Γ_x. If X is a nonempty clique of Γ, then the set $\Gamma_X = \bigcap_{x\in X}\Gamma_x$ is called the *neighbourhood* of X. We also set $\Gamma_\emptyset = V$ and we say that V is the *neighbourhood* of $\emptyset$. Note that for a clique X we have $\Gamma_X = \emptyset$ if and only if the clique X is maximal. We have defined the neighbourhood of a clique X as a set of vertices of Γ. However, Γ_X can naturally be viewed as a graph, equipped with the graph structure inherited from Γ. We will often choose this point of view, identifying Γ_X with the graph induced by Γ on it.

A *path* of *length* m *from* a vertex a *to* a vertex b (also, *joining* a with b) is a sequence $\alpha = (a_0, a_1, ..., a_m)$ of $m+1$ vertices such that $a_0 = a$, $a_m = b$ and $a_{i-1} \sim a_i$ for all $i = 1, 2, ..., m$. Note that adjacent vertices are distinct, by the definition of $\sim$. Hence $a_{i-1} \neq a_i$ for every $i = 1, 2, ..., m$. If $a_0 = a_m$ (hence $m \neq 1$), then we say that α is *closed* and *based at* a_0. If $m = 0$ (that is, $\alpha = (a_0)$), then α is a *null* path.

We say that a path is *simple* if its vertices are mutually distinct (note that non-null closed paths are not simple in this meaning). A closed path $\alpha = (a_0, a_1, ..., a_m)$ of length $m \geq 3$ is called a *circuit* if the path $(a_0, a_1, ..., a_{m-1})$ is simple.

Given two paths $\alpha = (a_0, a_1, ..., a_m)$ and $\beta = (b_0, b_1, ..., b_s)$ with $a_m = b_0$, the *product* (also *composition*) of α and β is the path

$$\alpha\beta = (a_0, a_1, ..., a_{m-1}, a_m = b_0, b_1, ..., b_s).$$

The *inverse* of a path $\alpha = (a_0, a_1, ..., a_m)$ is the path

$$\alpha^{-1} = (a_m, a_{m-1}, ..., a_0).$$

The relation 'being joined by a path' is evidently an equivalence relation on the set of vertices of Γ. The equivalence classes of this relation are called the *connected components* of Γ. The graph Γ is said to be *connected* if it has just one connected component; that is, if any two distinct vertices of Γ can always be joined by a path.

A *tree* is a connected graph with no circuits. A finite tree where every vertex is adjacent to at most two other vertices is called a *string.* A finite connected graph where every vertex is adjacent to just two other vertices is called a *cycle.* Thus, a graph is a string (a cycle) if it can be viewed as a simple path (as a circuit).

Given vertices a and b of a connected graph $\Gamma = (V, \sim)$, the minimal length of a path from a to b is the *distance* of a from b and it is denoted by $d(a, b)$. If there is a positive integer k such that $d(a, b) \leq k$ for all $a, b \in V$, then the positive integer $d = max(d(a, b) \mid a, b \in V)$ is called the *diameter* of Γ. Otherwise, we say that Γ has diameter $d = \infty$.

Given graphs Γ_1 and Γ_2, a *morphism* from Γ_1 to Γ_2 is a mapping f from the set of vertices of Γ_1 to the set of vertices of Γ_2 such that, for any two adjacent vertices x, y of Γ_1, either $f(x)$ and $f(y)$ are adjacent in Γ_2 or $f(x) = f(y)$.

A *coloured graph* is a pair (Γ, t) where $\Gamma = (V, E)$ is a graph with edge set E and $t : E \longrightarrow I$ is a surjective mapping from E to a finite set I, called the set of *colours*. However, in Chapter 12 we will also consider coloured graphs with two colouring mappings $t_V : V \longrightarrow I_V$ and $t_E : E \longrightarrow I_E$, one for vertices and one for edges.

0.5 Simplices and simplicial complexes

0.5.1 Simplices

An *n-dimensional simplex* ($0 \leq n < \infty$) is the set $\mathcal{S}$ of all subsets of a finite set S of size $n+1$. The elements of S are called the *vertices* of the simplex $\mathcal{S}$, the unordered pairs of distinct vertices are the *edges* of $\mathcal{S}$ and the proper nonempty subsets of S are the *faces* of $\mathcal{S}$. If X is a face of $\mathcal{S}$ and i is the number of elements of X, then $i-1$ is called the *dimension* of X. Note that, if $n > 0$, then the vertices of $\mathcal{S}$ are the 0-dimensional faces. If $n > 1$, then the edges of $\mathcal{S}$ are the 1-dimensional faces of $\mathcal{S}$.

0.5.2 Simplicial complexes

Let $\{S_i\}_{i \in I}$ be a nonempty collection of finite nonempty sets with $S_i \not\subseteq S_j$ for all $i, j \in I$, $i \neq j$. For every $i \in I$, let $\mathcal{S}_i$ be the simplex with S_i as the set of points. The join $\mathcal{K} = \bigcup_{i \in I} \mathcal{S}_i$ of all those simplices is called a *simplicial complex*. The simplices $\mathcal{S}_i$ ($i \in I$) are the *simplices* of $\mathcal{K}$ and their vertices, edges and faces are called *vertices*, *edges* and *faces* of $\mathcal{K}$, respectively. The *dimension* of a face X of $\mathcal{K}$ is the number of elements of X diminished by 1, as above. When $|I| > 1$, the sets S_i ($i \in I$) are also said to be *faces* of $\mathcal{K}$ (they are the maximal faces). If $|I| = 1$, then $\mathcal{K}$ is a simplex; we do not consider its set of vertices as a face of $\mathcal{K}$.

If there is a positive integer k such that all simplices of $\mathcal{K}$ have dimension $\leq k$, then the maximal dimension of a simplex of $\mathcal{K}$ is called the *dimension* of $\mathcal{K}$. Otherwise, we say that $\mathcal{K}$ has dimension ∞.

The *star* $\mathcal{K}_x$ of a vertex x of a simplicial complex $\mathcal{K}$ is the set of faces of $\mathcal{K}$ properly containing x. Clearly, if $\mathcal{K}_x \neq \emptyset$, then $\mathcal{K}_x$ is a simplicial complex, with vertices the edges of $\mathcal{K}$ containing x.

The vertices and the edges of a simplicial complex $\mathcal{K}$ form a graph, called the *skeleton* of $\mathcal{K}$. We say that $\mathcal{K}$ is *connected* if its skeleton is connected.

A few more definitions on simplicial complexes will be given later in this book, when we shall need them.

0.6 Division rings and vector spaces

0.6.1 Division rings

We use the word *division ring* to mean any ring where the non-zero elements form a (possibly non-abelian) group. Given a division ring K, we denote its multiplicative group by K^*. The centre of the group K^* is the *centre* of K. A division ring K is said to be *commutative* if K^* is abelian. Commutative division rings will be called *fields*.

The *dual* of a division ring K is the division ring K^{op} defined as follows: K^{op} has the same elements and the same sum operation as K, but its multiplication $\cdot^{op}$ is defined as follows: $x \cdot^{op} y = yx$ $(x, y \in K)$, where yx is the product of y and x in K. An *anti-automorphism* of a division ring K is an isomorphism between K and K^{op}. Trivially, K is a field if and only if $K = K^{op}$ (if and only if the anti-automorphisms of K are automorphisms of K).

0.6.2 Vector spaces

Given a division ring K and a positive integer n, the n-dimensional left vector space over K will be denoted by $V(n, K)$. It is well known that the linear mappings from $V(n, K)$ to K form an n-dimensional right vector space over K, called the *dual* of $V(n, K)$. Every left (right) vector space over a division ring K can also be viewed as right (left) vector space over the dual K^{op} of K. Thus, the dual of $V(n, K)$ can also be identified with $V(n, K^{op})$. In particular, if K is field, then $V(n, K)$ is (isomorphic to) its own dual.

Given a prime power $q > 1$, we denote the field of order q by $GF(q)$. We write $V(n, q)$ for $V(n, GF(q))$.

0.7 Notation for groups

Given a group G, we write $A \leq G$ (respectively, $A \trianglelefteq G$) to mean that A is a subgroup of G (a normal subgroup of G). By abuse of notation, we will sometimes write $A \leq G$ and $A \trianglelefteq G$ to mean that A is isomorphic to a subgroup of G or to a normal subgroup of G, respectively. However, we will do so only when the context will avoid any possible confusions.

Given a set X of elements of a group G, the subgroup of G generated by X will be denoted by $\langle X \rangle$. Given a family $(X_i)_{i \in I}$ of subsets of G, we write $\langle X_i \rangle_{i \in I}$ for $\langle \bigcup_{i \in I} X_i \rangle$, for short. Given elements $a_1, a_2, ..., a_n \in G$, we write $\langle a_1, a_2, ..., a_n \rangle$ for $\langle \{a_1, a_2, ..., a_n\} \rangle$.

Given a subgroup A of G, we denote the normalizer (the centralizer) of A in G by $N_G(A)$ (by $C_G(A)$, respectively). Given another subgroup B of G, we write $N_B(A)$ and $C_B(A)$ for $N_G(A) \cap B$ and $C_G(A) \cap B$, respectively. The *centre* $Z(G)$ of G is $Z(G) = C_G(G)$.

Given $A \leq G$ and $g \in G$, the coset gA (respectively, Ag) is called a *right* coset (a *left* coset).

The factor group of G over a normal subgroup A of G will be denoted by G/A, as usual.

Given a normal subgroup A of G and a group B, we say that G is an *extension* of A by B and we write $G \approx A \cdot B$, if $G/A \cong B$. If furthermore $A \leq Z(G)$, then we say that G is a *central* extension of A by B. By abuse of notation, we often write $G \approx A \cdot B$ even if A is not a subgroup of G, to mean that $G \approx A_1 \cdot B$ for some $A_1 \leq G$ isomorphic to A.

We use the symbol $\approx$ instead of $=$ in this context to remind the reader that the structures of A and B and the information $A \trianglelefteq G$ and $G/A \cong B$ might be insuffïcent to uniquely determine the isomorphism type of G. More generally, if H is an expression such as $A \cdot (B \cdot C)$, $(A \cdot B) \times C$ etc., when we write $G \approx H$ we want to make it clear that H only gives us perhaps rich but possibly incomplete information on the structure of G. We write $G \cong H$ (also $G = H$) only when H determines G up to isomorphism.

If $G \approx A \cdot B$ and G contains a normal subgroup $A_1 \cong A$ and a subgroup $B_1 \cong B$ such that $A_1 \cap B_1 = 1$ and $G = A_1 B_1$, then we write $G \approx A : B$ and we say that G is a *split* extension of A by B. Note that in this case G is a semidirect product of A_1 and B_1.

Given groups G and B, we say that B is *involved* in G if B is isomorphic to a subgroup of a factor group of G.

Given a nonempty set S and a group G, an *action* of G on S is a homomorphism φ from G into the group of all bijections of S onto S. The *kernel* of the action φ is the kernel $Ker(\varphi)$ of the homomorphism φ. The action φ is said to be *faithful* if its kernel is trivial. We often identify the action φ with its image $\varphi(G)$, or even with the factor group $G/Ker(\varphi)$, saying that $\varphi(G)$ (or $G/Ker(\varphi)$) is the *action* of G on S, although this is a terminological abuse.

Given an action of a group G on a set S and an element x of S, the stabilizer of x in (that action of) G and the orbit of x under (that action of) G will be denoted by G_x and $G(x)$, respectively. A similar notation is used for subsets of S (needless to say, an action of G on S also determines an action of G on the set of subsets of S).

Note that, if G is a semidirect product $A : B$ of two groups A and B, then B acts on A as a group of automorphisms of A. The structure of G is uniquely determined by the action of B on A. When that action is known (for instance, if it is clear from the context), then we write $G = A : B$, using the symbol $=$ instead of $\approx$.

The symbol $\wr$ will denote wreath products. We denote direct products by $\times$, as usual. We write G^n for $G \times G \times ... \times G$ (n times).

Given a positive integer $s \geq 2$, we denote the cyclic group of order s by the boldface figure $\mathbf{s}$ or by Z_s, according to which notation is more convenient at the time. Thus, if p is prime, $\mathbf{p}^n$ $(= Z_p^n)$ is the elementary abelian p-group of order p^n. The symbol $\mathbf{p}^{n+m}$ will denote an extension of $\mathbf{p}^n$ by $\mathbf{p}^m$. The additive group of integers will be denoted by $\mathbf{Z}$.

A *Frobenius group* is a semidirect product $G = F : K$ of a finite group F and a non-trivial subgroup H of $Aut(F)$ acting fixed-point-freely on the set of non-identity elements of F. The subgroup F is called the *Frobenius kernel* of G. The symbol $Frob_t^s$ will denote a Frobenius group of order st with Frobenius kernel of order t. Note that the symbol $Frob_t^s$ denotes a class of groups and the groups in that class might not be mutually isomorphic, in general. However, for certain values of s and t all groups named $Frob_t^s$ are isomorphic. This happens when t is prime, for instance.

Apart from the above conventions for cyclic, elementary abelian and Frobenius groups, we closely follow the Atlas of Finite Groups [41] for the names of particular finite groups.

The group of inner automorphisms of a group G is the action of G onto itself by conjugation. We denote it by $Inn(G)$. The *outer* automorphism group of G is the factor group $Aut(G)/Inn(G)$. It is denoted by $Out(G)$.

0.8 Presentations of groups

0.8.1 Words and relations

Given a nonemtpy set X of elements of a group G with $1 \notin X$, a *word* of *length* $m > 0$ in G on the *set of letters* X is a sequence $x_1x_2...x_m$ where, for every $i = 1, 2, ..., m$, either $x_i \in X$ or $x_i^{-1} \in X$. The identity element $1 \in G$ is called the *null word*, of *length* 0. A subsequence $x_ix_{i-1}...x_{j-1}x_j$ of a word $x = x_1x_2...x_m$ is called a *subword* of x. The null word is a subword of every word. Needless to say, every word represents an element of the subgroup $\langle X \rangle$ of G. On the other hand, every element of $\langle X \rangle$ can be represented as a word in infinitely many ways. Thus, one should carefully distinguish between words and the elements they represent, also using an appropriate notation. However, a sloppy notation is often harmless. Our notation will actually be sloppy: we will often use the same symbol for a word and for the element it represents.

When G is the free group $\mathcal{F}(X)$ on the set of generators X, then we call the words in G on X *abstract words* on X.

Given a set X of letters, a set of *relations* on the set X of *generators* is a set $\mathbf{R}$ of formulas $x = y$, with x, y abstract words on X. A relation $x = y \in \mathbf{R}$ is uniquely determined by its two terms x and y. Thus, $\mathbf{R}$ can also be viewed as a set of pairs (x, y) of abstract words on x; or even as a set of abstract words on X, rewriting every relation $x = y$ in the form $xy^{-1} = 1$. This is precisely the point of view we choose in what follows: $\mathbf{R}$ is just a set of words.

0.8.2 Presentations

Given a mapping φ from X to a group G, we say that $\mathbf{R}$ holds in G for the *interpretation* φ of the generators if the homomorphism from the free group $\mathcal{F}(X)$ to G extending φ maps all words of $\mathbf{R}$ onto the identity element of G. We say that G is *presented* by the set $\mathbf{R}$ of relations if $\mathbf{R}$ holds in G for a given interpretation φ of the generators and, for every group Y and every mapping ψ from X to Y such that $\mathbf{R}$ holds in Y with respect to ψ, there is just one homomorphism $f : G \longrightarrow Y$ such that $f\varphi = \psi$.

Let G be presented by $\mathbf{R}$ for some interpretation $\varphi : X \longrightarrow G$ of the generators. Then $G = \langle \varphi(X) \rangle$ and the interpretations of X in G for which $\mathbf{R}$ presents G are precisely those of the form $\alpha\varphi$, with $\alpha \in Aut(G)$. That is, the interpretation of X in G for which $\mathbf{R}$ presents G is uniquely determined up to automorphisms of G. Thus, we will often omit mentioning it. The following is also evident: $\mathbf{R}$ is a presentation of G if and only if $G \cong \mathcal{F}(X)/N_{\mathbf{R}}$, where $N_{\mathbf{R}}$ is the normal subgroup of $\mathcal{F}(X)$ generated by $\mathbf{R}$.

0.8.3 Amalgamated products

Given a group G, a nonempty family $\mathcal{X} = (X_i)_{i \in I}$ of subgroups of G and a subset Y of $\bigcup_{i \in I} X_i - 1$, let us define a set of symbols X and a set of relations $\mathbf{R}_{\mathcal{X},Y}$ as follows:

(i) X contains a symbol x_i for every $i \in I$ and every element $x \neq 1$ of X_i (we may identify x_i with the pair (x, i), if we like so); furthermore, all elements of Y are symbols of X;

(ii) for every $i \in I$ and every word w on letters x_i with $x \in X_i$, if $w = 1$ then $w \in \mathbf{R}_{\mathcal{X},Y}$; for every $i \in I$ and every element $x \in X_i$, if $x = y$ for some $y \in Y$, then $x_i y^{-1} \in \mathbf{R}_{\mathcal{X},Y}$.

We say that G is the *amalgamated product* of the family $\mathcal{X}$ over Y if G is presented by $\mathbf{R}_{\mathcal{X},Y}$, with the natural interpretation of the generators, mapping a symbol x_i representing an element x of a member X_i of $\mathcal{X}$ onto that element x and mapping every element of Y onto itself. Trivially, if G is the amalgamated product of $\mathcal{X}$ over Y, then $G = \langle X_i \rangle_{i \in I}$. Furthermore, G is the amalgamated product of $\mathcal{X}$ over Y if and only if, for every group A and every choice of homomorphisms $f_i : X_i \longrightarrow A$ $(i \in I)$ and of a mapping $g : Y \longrightarrow A$ such that $f_i(y) = g(y)$ for every $i \in I$ and every $y \in Y \cap X_i$, there is just one homomorphism $f : G \longrightarrow A$ such that f induces f_i on X_i, for every $i \in I$.

If G is the amalgamated product of $\mathcal{X}$ over Y and Y is the set of non-identity elements of the members of another family $\mathcal{Y}$ of subgroups of G, then we say that G is the amalgamated product of $\mathcal{X}$ with *amalgamation* of (the members of) $\mathcal{Y}$.

The reader might feel a bit unhappy with our definition of abstract words based on the notion of free groups, since in the standard definition of free groups it is presumed that words have previously been defined (as abstract polynomial symbols, not as sequences of elements of a group). However, we are not suggesting any unusual way to define free groups here. We only needed to fix a few conventions on the word 'word' and we have chosen the most economical way to do that. The reader may use a more accurate definition of words if he does not like the one we have given, rephrasing for it the things we have said above.

Chapter 1

Basic Concepts and Examples

1.1 Definitions

We define geometries inductively, as follows:

(i) a *geometry of rank* 1 is just a set with at least two elements (namely, a trivial graph with at least two vertices);

(ii) a *geometry of rank* $n \geq 2$ is a pair (Γ, Θ) where Γ is a connected graph and Θ is an n-partition of Γ such that, for every vertex x of Γ, the neighbourhood Γ_x of x is a geometry of rank $n-1$ with respect to the partition induced by Θ on Γ_x.

In particular, a rank 2 geometry is just a connected bipartite graph where every vertex is adjacent to at least two vertices. A geometry of rank 3 is a connected 3-partite graph where the neighbourhood of every vertex is connected and every edge is contained in at least 2 cliques of size 3.

We warn the reader that the definition we have chosen is more restrictive than others which can be found in the literature (see [19] or [227], for instance). In particular, properties as the firmness (see [19]) and the residual connectedness (see [227]) or the strong connectedness (see [19]) can be obtained as consequences from our definition (see Section 1.4, Lemma 1.8 and Section 1.5, Theorems 1.16 and 1.18), whereas many authors do not include those properties in the definition of geometries, preferring to assume them as additional hypotheses. However, our definition is general enough to cover almost all structures that are currently considered in geometry.

The vertices of Γ are called *elements* of the geometry (Γ,Θ). Two elements are said to be *incident* if they are adjacent in Γ, but we also state that every element is incident to itself, by convention. The graph Γ and the n-partition Θ are called the *incidence graph* and the *type partition* of the geometry (Γ,Θ). The classes of Θ are the *types* of the geometry. The incidence relation is denoted by $*$.

We will often write Γ instead of (Γ,Θ) for short, identifying a geometry (Γ, Θ) with its incidence graph Γ. This might look like an abuse, but we will see later (Theorem 1.24) that it is not really so. Indeed, we will prove that the type partition Θ is uniquely determined by the incidence graph Γ.

Given an element x of the geometry Γ, the neighbourhood Γ_x of x in the graph Γ is called the *residue* of x.

The *flags* of the geometry Γ are the cliques of the graph Γ. We write $F * G$ to mean that two flags F and G are *incident*, namely that $F \cup G$ is a flag. In particular, if x and F are an element and a flag respectively, then $x * F$ (x is *incident* to F) means that $F \cup x$ is a flag.

The *rank* (the *corank*) of a flag F is its size $|F|$ (the number $n-|F|$, respectively). Clearly, every flag picks up at most one element from each of the types (classes of Θ). Hence, every flag has rank at most n.

Given a flag F, the set of types meeting F (not meeting F) is called the *type* of F (*cotype* of F, respectively). Clearly, the type of an element is the type which that element belongs to.

1.2 Examples of rank 2

Throughout this section Γ is a rank 2 geometry with types P and L (the two classes of Θ), where the letters P and L have been chosen to remind us of the words 'point' and 'line'.

1.2.1 Partial planes

The geometry Γ is a *partial plane* [61] if no circuit of length 4 (i.e. quadrangle) occurs in the graph Γ. This can be rephrased in a more familiar way as follows. Choose one of the two types, say P, as the set of *points*. The elements of the other type L are the *lines*. Then assuming that no 4-circuit exists in Γ is the same as assuming that, given any two distinct points, there is at most one line incident to both of them. Equivalently, given any two distinct lines, there is at most one point incident to both.

Let Γ be a partial plane. Then the lines of Γ can be identified with the sets of points incident to them. Indeed, as distinct points are incident to at most one common line, distinct lines are incident to distinct sets of points. Furthermore, every point is incident to at least two lines and every line is incident to at least two points, as every element of a geometry of rank 2 is incident to at least two elements different from it. Thus, every line of a partial plane can be viewed as a proper subset of the set P of points, of size at least 2.

Hence, given a point a and a line x, we may write $a \in x$ to mean that $a * x$ and phrases such as 'the point a is in the line x', 'the line x passes through the point a', 'the lines x, y meet at the point a', etc. may be freely used. In particular, two distinct points a, b are said to be *collinear* (*non-collinear*) if there is a line through both of them (if there are no lines incident to both of them). We write $a \perp b$ (respectively, $a \not\perp b$) to mean that a and b are collinear (non-collinear). The graph $(P,\perp)$, with P as set of vertices and the collinearity relation $\perp$ as adjacency relation, is called the *collinearity graph* of the partial plane Γ. Trivially, the connectedness of (the incidence graph of) Γ implies the connectedness of the collinearity graph.

Given a point a, we denote by $a^{\perp}$ the set of points collinear with a, including a among them by convention. Given $X \subseteq P$, $X^{\perp}$ is defined as $X^{\perp} = \bigcap(a^{\perp} \mid a \in X)$.

The previous conventions create an asymmetry between the roles of the types P and L. If we interchange their roles, taking the elements of L as points and those of P as lines, then we may say that we are considering Γ *in the dual way*; also, that we consider the *dual* of Γ, for short.

1.2.2 Linear spaces

We say that a partial plane Γ is a *linear space* if its collinearity graph is complete; that is, if any two distinct points are joined by a (unique) line. This property is equivalent to the following in the incidence graph of Γ: every point has distance 2 from any other point and distance 1 or 3 from any line (whence, every line has distance 2 or 4 from any other line).

A *dual linear space* is the dual of a linear space: namely, a partial plane where any two distinct lines intersect in a point. A (possibly degenerate) projective plane is a linear space that is also a dual linear space ([90], III.2); that is, a projective plane is a partial plane where the incidence graph has diameter 3.

Affine planes ([90], III.4) and systems of points and lines of projective or affine geometries ([61], Section 1.4; [90], Chapter II) are other well-known examples of linear spaces.

Circular spaces are finite linear spaces where every line has precisely two points. This means that a circular space is a finite complete graph with at least three vertices, vertices and edges of the graph playing the role of points and lines, respectively.

1.2.3 Gonality and diameters

We need a bit of general theory of rank 2 geometries before describing further examples. Let (Γ, Θ) be a geometry of rank 2, with $\Theta = \{P, L\}$. Every circuit of the graph Γ has even length, because Γ is bipartite. Let Γ admit circuits and let $2g$ be the minimal length of a circuit of Γ. Then g is called the *gonality* of Γ. If there are no circuits in Γ (that is, if Γ is a tree), then we say that Γ has gonality $g = \infty$.

According to this definition, partial planes are precisely the rank 2 geometries of gonality $g \geq 3$.

Given two elements u,v of Γ, let $d(u, v)$ be their distance in the graph Γ. The *diameter* d of the geometry Γ is the diameter of the graph Γ. We may also define a *P-diameter* d_P and an *L-diameter* d_L as follows (recall that P and L are the two types of Γ):

$$d_P = max(d(a, v) \mid a \in P, v \in P \cup L)$$

$$d_L = max(d(v, x) \mid x \in L, v \in P \cup L)$$

For instance, linear spaces are characterized by the following relations: $g = d_P = 3 \leq d_L \leq 4$. Projective planes are characterized by the relations $g = d_P = d_L = 3$.

Note that d_P and d_L might be infinite. This is the case when $g = \infty$, for instance. However, there are many examples of geometries of rank 2 with $d_P = d_L = \infty > g$. For instance, the Euclidean plane can be tessellated into squares; the vertices and the edges of those squares can be taken as points and lines of a geometry of rank 2, with the natural incidence relation (a vertex a and an edge x are said to be incident when $a \in x$). This geometry has gonality $g = 4$ and diameters $d_P = d_L = \infty$.

Some relations hold in general between the gonality g and the diameters d, d_P and d_L of a geometry Γ of rank 2. Trivially, $d = max(d_P, d_L)$. Furthermore:

Proposition 1.1 *We have $g \leq min(d_P, d_L)$ and $d \leq 1 + min(d_P, d_L)$. Moreover, if d is an odd integer or $d = \infty$, then $d = d_P = d_L$.*

Proof. If $g = \infty$ then Γ is a tree, whence $g = d_P = d_L = \infty$. Let $g < \infty$. We first prove the inequality $g \leq min(d_P, d_L)$. Let $v_0, v_1, ..., v_{2g} = v_0$ be a circuit of Γ of minimal length $2g$. Interchanging the roles of P and L if necessary, we may always assume that $d_P \leq d_L$ and that $v_0 \in P$. Let $g > d_P$, if possible. Then a path $v_0, v'_1, ..., v'_k = v_g$ can be found, with $k \leq d_P$. This gives us a closed path $v_0, v_1, ..., v_g, v'_{k-1}, ..., v'_1, v_0$ of length less than $2g$: contradiction.

Let us turn to the inequality $d \leq 1 + min(d_P, d_L)$. Given a line $x \in L$ and an element v at maximal distance d_L from x, let $x = v_0, v_1, ..., v_m$ be a path of minimal length $m = d_L$ from x to v. Then $v_1, v_2, ..., v_m$ is a path of minimal length from the point v_1 to v. Therefore $d_P \geq d_L - 1$. Similarly, $d_L \geq d_P - 1$, interchanging the roles of P and L. Therefore $d \leq 1 + min(d_P, d_L)$. In particular, $d = d_P = d_L = \infty$ if one of d_L or d_P is infinite.

Finally, let d be an odd integer and let x, y be elements of Γ at distance d. Since Γ is a bipartite graph, elements of Γ at odd distance belong to distinct types. Therefore, one of x, y is a point and the other one is a line. Hence $d_P = d_L = d$. □

Choosing P and L as sets of *points* and *lines*, respectively, a *collinearity graph* $(P,\perp)$ can be defined for arbitrary rank 2 geometries as for partial planes: two distinct points $a, b \in P$ are said to be *collinear* ($a \perp b$) if there is a line incident to both of them.

Proposition 1.2 *Let a, b be points of Γ and let m be their distance in the collinearity graph of Γ. Then a and b have distance $2m$ in the incidence graph of Γ.*

Proof. Let k be the distance from a to b in the (incidence) graph (of) Γ. Note that k is even, as a and b belong to the same type. Given a path $a_0, a_1, ..., a_m = b$ of length m in $(P, \perp)$ from a $(= a_0)$ to b $(= a_m)$, let u_i be a line incident to a_{i-1} and a_i, for $i = 1, 2, ..., m$. Then $a_0, u_1, a_1, ..., a_{m-1}, u_m, a_m$ is a path of length $2m$ from a to b in the graph Γ. Therefore $k \leq 2m$.

Conversely, let $b_0, v_1, b_1, ..., v_{k/2}, b_{k/2}$ be a path of length k from a $(= b_0)$ to b $(= b_{k/2})$ in the graph Γ. Then $b_0, b_1, ..., b_{k/2}$ is a path of length $k/2$ from a to b in $(P, \perp)$. Therefore $m \leq k/2$. Hence $2m = k$. □

By this proposition, if $d_P < \infty$, then $(P, \perp)$ has diameter $d_P/2$ or $(d_P - 1)/2$, according to whether d_P is even or odd. If $d_P = \infty$, then $(P, \perp)$ has infinite diameter.

Remark. We have used the word 'line' till now. Some authors prefer to use the word 'block' instead of 'line' when they deal with rank 2 geometries of gonality 2. We will sometimes follow this habit, but only in those geometries for which the word 'line' is almost never used in the literature.

1.2.4 Generalized m-gons

A rank 2 geometry Γ is a *generalized m-gon* ($m \geq 2$) if $g = d_P = d_L = m$.

Generalized 2-gons (*generalized digons*) are complete bipartite graphs with at least two vertices in each of the two classes of the bipartition; these two classes are the two types of elements of the generalized digon. It is easily seen that any of the two relations $d_P = 2$ or $d_L = 2$ is sufficent to obtain the full set of equations $g = d_P = d_L = 2$. Therefore, generalized digons can also be characterized as rank 2 geometries where one of the two diameters d_P or d_L is 2.

Generalized digons are quite common. For instance, the points and planes on a given line of a projective or affine geometry of dimension ≥ 3 ([61], [90]) form a generalized digon.

Generalized 3-gons (also called *generalized triangles*) are precisely (possibly degenerate) projective planes.

Generalized 4-gons (*generalized quadrangles*) can be characterized as partial planes satisfying the following property:

($*$) for every line $x \in L$ and every point $a \in P$ not on x, we have $|a^{\perp} \cap x| = 1$.

Indeed, a partial plane Γ has gonality $g \geq 4$ if and only if every clique of its collinearity graph is contained in some line, and this latter property holds if and only if $|a^{\perp} \cap x| \leq 1$ for every line x and every point $a \notin x$. On the other hand, given a point a and a line x of Γ, we have $a^{\perp} \cap x = \emptyset$ if and only if $d(a, x) \geq 5$ in the incidence graph of Γ (compare Proposition 1.2). Therefore, if $d(a, x) \leq 4$, then $a^{\perp} \cap x \neq \emptyset$. It is now clear that ($*$) holds in a partial plane Γ if and only if Γ has gonality $g \geq 4$ and diameter $d \leq 4$. By Proposition 1.1, ($*$) holds if and only if $g = d_P = d_L = 4$.

Generalized quadrangles where every point is on just two lines are called *grids*. Let Γ be a grid. Given lines x, y of Γ, if $x \cap y = \emptyset$ or $x = y$, then we say that x and y are *parallel* and we write $x \parallel y$. Given a line x and a point $a \notin x$, just one of the two lines of Γ on a meets x, by ($*$). Therefore the parallelism relation $\parallel$ is an equivalence relation on the set of lines of Γ, with just two equivalence classes. It is now evident that the dual of Γ can also be described as a complete bipartite graph with classes of size ≥ 2 (the vertices and the edges of the graph are respectively points and lines in the dual of Γ).

Generalized m-gons with $m = 5, 6, 7, 8$ are called *generalized pentagons*, *generalized hexagons*, *generalized heptagons* and *generalized octagons*, respectively). The reader may see [106] for some properties and constructions of generalized hexagons and octagons. As for generalized m-gons with $m = 5$ or 7 or $m \geq 9$, we will see later (Chapter 3, Theorem 3.6) that the finite ones are necessarily 'degenerate', i.e. they admit lines with precisely two points or points belonging to just two lines.

Generalized m-gons where every element is incident to precisely two other elements are usually called *ordinary m-gons*, or just *m-gons*. Ordinary m-gons with $2 < m < \infty$ are also called *ordinary polygons*, or just *polygons*. We do not include the extremal cases $m = 2$ and $m = \infty$ among ordinary polygons.

Finally, a few remarks on the case of $m = \infty$: the ordinary ∞-gon can be realized taking the set of integer numbers as the set of points and the pairs of consecutive integers as lines. Thus, the ∞-gon is nothing but a two-sided infinite chain. The incidence graph of a generalized ∞-gon is an infinite tree with no terminal nodes and every such tree gives rise to a generalized ∞-gon.

1.2.5 Classical generalized quadrangles

Given a left vector space V over a division ring K, an anti-automorphism σ of K and an element ε of K^* as in Appendix I of this chapter, let f be a non-degenerate trace valued (σ, ε)-sesquilinear form of Witt index 2 in V (see Appendix I; note that the dimension of V might be infinite, even if f has Witt index 2).

The non-trivial subspaces of V totally isotropic for f form a generalized quadrangle, which we denote by $Q(f)$. The incidence relation $*$ of $Q(f)$ is defined by means of the inclusion relation: if X, Y are totally isotropic subspaces of dimension 1 and 2 respectively, then $X * Y$ means that $X \subset Y$. As these subspaces are points and lines of the projective geometry of the linear subspaces of V, the incidence relation $*$ can also be read as $\in$.

Let us check that $Q(f)$ is indeed a generalized quadrangle. Given a 'point' Y of $Q(f)$ (i.e., a totally isotropic 1-dimensional subspace of V) and a 'line' X of $Q(f)$ not on Y (a totally isotropic 2-dimensional subspace of V not containing Y), we have $dim(Y^{\perp} \cap X) = 1$ by (v) of Appendix I (the symbol $\perp$ has now the meaning stated in Appendix I). Furthermore, $Y^{\perp} \cap X$ is the unique 'point' on the 'line' X collinear with Y and the 2-dimensional totally isotropic subspace $Y + (Y^{\perp} \cap X)$ is the unique 'line' of $Q(f)$ joining Y with $Y^{\perp} \cap X$. Trivially, every 'line' of $Q(f)$ has $1+|K|$ 'points', whereas every 'point' is in at least two 'lines', by (vi) of Appendix I. Therefore, $Q(f)$ is a partial plane satisfying $(*)$ of Section 1.2.4. That is, $Q(f)$ is a generalized quadrangle.

We now state some notation for generalized quadrangles defined by alternating or symmetric bilinear forms of Witt index 2 or Hermitian forms of Witt index 2 (see Appendix I, 1.8.2).

(1) Let f be an alternating bilinear form of Witt index 2 in $V(4, K)$ (with K a field). The generalized quadrangle $Q(f)$ is called a *symplectic variety* of rank 2 and it is denoted by the symbol $\mathcal{S}_3(K)$ or $W(K)$. The subscript 3 in $\mathcal{S}_3(K)$ should remind us that the elements of $\mathcal{S}_3(K)$ are points and lines of the 3-dimensional projective geometry $PG(3, K)$ of linear subspaces of $V(4, K)$. When $K = GF(q)$, the notation $\mathcal{S}_3(q)$ or $W(q)$ is normally used instead of $\mathcal{S}_3(K)$ or $W(K)$.

(2) Let f be a symmetric bilinear form of Witt index 2 in $V(4, K)$, with K a field and $char(K) \neq 2$. Then $Q(f)$ is called a *hyperbolic quadric* of rank 2 and it is denoted by $\mathcal{Q}_3^+(K)$ (by $\mathcal{Q}_3^+(q)$ when $K = GF(q)$). The subscript 3 in $\mathcal{Q}_3^+(K)$ has the same meaning as in $\mathcal{S}_3(K)$.
Remark. $\mathcal{Q}_3^+(K)$ is a grid (the proof of this claim is an exercise in linear algebra; we leave it for the reader).

(3) Let f be a symmetric bilinear form of Witt index 2 in $V(5, K)$, with K a field and $char(K) \neq 2$. Then $Q(f)$ is called a *quadric* of rank 2 and it is denoted by $\mathcal{Q}_4(K)$ (by $\mathcal{Q}_4(q)$ when $K = GF(q)$).
Remark. $\mathcal{Q}_4(q)$ and $\mathcal{S}_3(q)$ are dually isomorphic (see [166], Chapter 3).

(4) Let f be a symmetric bilinear form of Witt index 2 in $V(6, q)$, q odd. Then $Q(f)$ is called an *elliptic quadric* of rank 2 and it is denoted by $\mathcal{Q}_5^-(q)$.

(5) Let f be a Hermitian form of Witt index 2 in $V(m, q^2)$, $m = 4$ or 5. Then $Q(f)$ is called a *Hermitian variety* of rank 2 and it is denoted by $\mathcal{H}_3(q^2)$ or $\mathcal{H}_4(q^2)$, according to whether $m = 4$ or 5.
Remark. If q is odd, then we can consider $Q_5^-(q)$ (see above); the generalized quadrangles $\mathcal{H}_3(q^2)$ and $\mathcal{Q}_5^-(q)$ are dually isomorphic (see [166], Chapter 3).

The definitions stated in (2), (3) and (4) must be slightly modified in the characteristic 2 case, using non-singular quadratic forms instead of symmetric bilinear forms (Appendix I, 1.8.4, (2),(3),(4)). Indeed, when $char(K) = 2$ the expressions we have given for alternating bilinear forms and symmetric bilinear forms of Witt index 2 in $V(4, K)$ define the same forms and the expression we have used to define symmetric bilinear forms of Witt index 2 in $V(5,K)$ gives us a degenerate form (Appendix I, 1.8.2); the expression for symmetric bilinear forms of Witt index 2 in $V(6,q)$ is undefined if q is even (Appendix I, 1.8.2).

Thus, instead of isotropic vectors we must now take vectors singular for a non-singular quadratic form g; instead of totally isotropic subspaces we take linear subspaces totally singular for g (see Appendix I, 1.8.3). As in the case of sesquilinear forms, the system of non-trivial totally singular subspaces of a non-singular quadratic form of Witt index 2 is a generalized quadrangle (indeed the analogues of (v) and (vi) of Section 1.8.1 hold for totally singular subspaces of non-singular quadratic forms; Appendix I, 1.8.3).

The generalized quadrangles arising from non-singular quadratic forms of Witt index 2 in $V(2n,K)$ ($char(K) = 2$; Appendix I, 1.8.4(2)), in $V(2n+1,K)$ ($char(K) = 2$; Appendix I, 1.8.4(3)) and in $V(2n+2,q)$ (q even; Appendix I, 1.8.4(4)) are denoted by $\mathcal{Q}_3^+(K)$, $\mathcal{Q}_4(K)$ and $\mathcal{Q}_5^-(q)$ respectively, as in the case of characteristic $\neq 2$.

It is worth mentioning that, if q is even, then $\mathcal{S}_3(q)$ and $\mathcal{Q}_4(q)$ are isomorphic and dually somorphic at the same time, hence they are self-dual (see [166], Chapter 3). We have remarked above that $\mathcal{Q}_5^-(q)$ (q odd) and $\mathcal{H}_3(q^2)$ are dually isomorphic; the same is true when q is even (see [166], Chapter 3).

We were above thinking of the quadratic forms defined in (2), (3) and (4) of Appendix I, 1.8.4, but it is clear that a generalized quadrangle can be obtained in that way from any non-singular (σ,ε)-quadratic form of Witt index 2 in a vector space V over a division ring K with $char(K) = 2$ (Appendix I, 1.8.3). On the other hand, there is no need to use quadratic forms when $char(K) \neq 2$. Indeed when $char(K) \neq 2$ every (σ,ε)-quadratic form g is uniquely determined by its sesquilinearization f (Appendix I, 1.8.3) and a linear subspace of V is totally singular for g if and only if it is totally isotropic for f (Appendix I, 1.8.3).

The generalized quadrangles obtained from sesquilinear or quadratic forms as above are called *classical*, except for $\mathcal{Q}_3^+(K)$ which, being a grid, is considered as a degenerate case and is not normaly included among classical examples. It follows from Theorems 1.25 and 1.26 that $\mathcal{S}_3(q)$, $\mathcal{Q}_4(q)$, $\mathcal{Q}_5^-(q)$, $\mathcal{H}_3(q^2)$ and $\mathcal{H}_4(q^2)$ are the only examples of finite classical generalized quadrangles. In some sense, classical generalized quadrangles are the analogues of Desarguesian projective planes. A number of non-classical generalized quadrangles also exist, apart from grids, but we will not discuss them here (we will mention only one example in Section 2.3). The reader can find rich information on finite non-classical generalized quadrangles in

[166] and [210].

1.2.6 More examples

(1) **Nets.** A partial plane Γ is a *net* if an equivalence relation $\|$ is given between the lines of Γ (called a *parallelism relation*) such that every point is in precisely one line of each of the equivalence classes of $\|$ (*parallelism classes*) and two lines have no point in common precisely when they belong to the same parallelism class. A net has diameters $d_P = 3$ or 4, $d_L = 4$ and gonality $g = 3$ or 4.

Affine planes and grids (see Section 1.2.4) are instances of nets. A net is an affine plane if and only if it has P-diameter $d_P = 3$. It is a grid if and only if it has gonality $g = 4$. A net which is neither an affine plane nor a grid is said to be *proper*. That is, a net is proper if and only if it has gonality $g = 3$ and P-diameter $d_P = 4$.

There are many examples of proper nets. For instance, we can take all points of an affine plane of order ≥ 3 and at least 3 bundles of parallel lines of that plane, discarding at least one bundle of parallel lines. Or we can take all points but one of an affine plane of order ≥ 3, and at least 3 bundles of parallel lines, deleting all lines through the discarded point.

Dual nets are defined interchanging the roles of points and lines in the definition of nets. Instead of the parallelism relation $\|$ we now have an equivalence relation T on the set of points, called a *transversality relation*.

(2) **Mőbius, Laguerre and Minkowski planes.** A geometry Γ of rank 2 is a *Mőbius plane* (also, an *inversive plane*) if any two distinct points of Γ are incident to some common line and, for every point a, the points different from a and the lines incident to a form an affine plane with the incidence relation inherited from Γ. A Mőbius plane has gonality $g = 2$ and diameters $d_P = 3$ and $d_L = 4$.

Given a geometry Γ of rank 2 and points x, y of Γ, write xTy to mean that x and y are non-collinear or equal. We say that Γ is a *Laguerre plane* if T is an equivalence relation, every line is incident to precisely one point in each of the equivalence classes of T and, for every point a, the points collinear with a and the lines incident to a form a net and a dual net at the same time, with respect to the incidence relation inherited from Γ. The gonality and the diameters now are as follows: $g = 2$ and $d_P = d_L = 4$.

A geometry Γ of rank 2 is a *Minkowski plane* if the following hold in it:

(i) two equivalence relations T_1, T_2 can be defined on the set P of points of Γ such that $T_1 \cap T_2 = \mho$ (where $\mho$ denotes the identity relation on P) and $T_1 \cup T_2 = T$ (where aTb means that a and b are non-collinear or equal, as above);

(ii) for every point a, the points collinear with a and the lines incident to a form a net with the following property: if a point b is not incident to a line x, then b is collinear with all but two points of x.

The gonality and the diameters of a Minkowski plane are as in Laguerre planes. The reader may see [20] for other systems of axioms for Mőbius, Laguerre and Minkowski planes, equivalent to those given above.

Classical examples of Mőbius and Minkowski planes are obtained from elliptic quadrics and hyperbolic quadrics, respectively, in the 3-dimensional projective geometry $PG(3, K)$ of linear subspaces of $V(4, K)$. As points we take the points of the quadric and the lines are the intersections of the quadric with secant planes (namely, planes of $PG(3, K)$ intersecting the quadric in non-degenerate conics). The elliptic quadric in $PG(3, K)$ (denoted by $\mathcal{Q}_3^-(K)$) is defined by a quadratic form of Witt index 1 in $V(4, K)$ (see Appendix I). The hyperbolic quadric is the grid $\mathcal{Q}_3^+(K)$ (see Section 1.2.5). The two bundles of parallel lines of this grid give us the two equivalence relations T_1 and T_2 mentioned in the definition of Minkowski planes.

Classical Laguerre planes are obtained from a cone in $PG(3, K)$, namely a degenerate quadric represented by the following equation with respect to some basis of the 4-dimensional vector space over K: $x_1x_2 + x_3^2 = 0$. As points we take the points of the cone other than the vertex and the lines are the intersections of the cone with the planes of $PG(3, K)$ not passing through the vertex. The lines of the cone are the equivalence classes of the relation T in the Laguerre plane.

General constructions for all finite Minkowski and Laguerre planes (including non-classical ones) will be given in Chapter 2 (Exercises 2.12 and 2.14).

Remark. The lines of Mőbius, Laguerre and Minkowski planes are usually called *circles* in the literature (also *blocks*, in the finite case).

(3) **Partial geometries.** A *partial geometry* of *order* (s,t) and *index* α ($1 \leq \alpha \leq min(s+1, t+1)$, $s, t < \infty$) is a partial plane where all lines have $s+1$ points, every point is in $t+1$ lines and, given a non-incident point–line pair (a,x), we have $|a^\perp \cap x| = \alpha$ (where $\perp$ has the meaning stated in Section 1.2.1).

In particular, a partial geometry is a linear space (a dual linear space) if $\alpha = s+1$ ($\alpha = t+1$, respectively). Trivially, the relations $\alpha = s+1 = t+1$ characterize projective planes of order s. The relations $\alpha = s+1 = t$

characterize affine planes of order t.

A partial geometry with $\alpha = 1$ is a finite generalized quadrangle (compare (*) of Section 1.2.4). A partial geometry with $\alpha = t$ ($\alpha = s$) is a finite net (a finite dual net). In particular, if $\alpha = t = 1$ (if $\alpha = s = 1$), then we obtain finite grids (finite dual grids). Note that finite grids with lines of two different sizes are not partial geometries.

The dual of a partial geometry of order (s,t) and index α is a partial geometry of order (t, s), with the same index α.

A partial geometry is said to be *proper* if $1 < \alpha < min(s+1, t+1)$. Proper partial geometries have gonality $g = 3$ and diameters $d = d_P = d_L = 4$.

Several examples of proper partial geometries are known, often produced inside some finite projective or affine geometries. The reader wanting to know more on this topic may see [51].

(4) **Some restrictions on α, s and t.** The order (s, t) and the index α of a partial geometry Γ must satisfy some conditions, besides the inequalities $1 \leq \alpha \leq min(s+1, t+1)$.

For instance, let $\alpha = s + 1$ (hence Γ is a linear space). Then $s \leq t$ (because $s+1 = \alpha \leq t+1$). Furthermore, Γ has $(t+1)s+1$ points. Indeed, given a point a, all remaining points are reached taking the $t + 1$ lines on a and the s points on each of these lines other than a. As every point is on precisely $t + 1$ lines and every line has $s + 1$ points, the total number of lines of the linear space Γ is $((t+1)s+1)(t+1)/(s+1)$. Hence $s + 1$ divides $t(t+1)$.

A similar divisibility condition is obtained when $\alpha = t+1$ (that is, when Γ is a dual linear space), interchanging s and t.

Let $\alpha < s+1, t+1$. Then the collinearity graph $(P, \perp)$ of Γ is regular [13] with diameter 2 and valency $(t+1)s$. Furthermore, given two distinct points a, b of Γ, we have $\mid a^{\perp} \cap b^{\perp} \mid = t(\alpha - 1) + s + 1$ or $(t+1)\alpha$, according to whether $a \perp b$ or $a \not\perp b$. Therefore $(P, \perp)$ is strongly regular ([13]; see also Appendix II) and Γ has $(1+s)(1+st/\alpha)$ points. Since every point is on $t+1$ lines and every line has $s+1$ points, Γ has $(1+t)(1+st/\alpha)$ lines. Hence α divides $st(s+1, t+1)$ (by (x, y) we mean the greatest common divisor of the integers x and y).

This condition says nothing when Γ is a generalized quadrangle ($\alpha = 1$), a net ($\alpha = t$) or a dual net ($\alpha = s$). When $\alpha = 1$, Γ has $(1+s)(1+st)$ points and $(1+t)(1+st)$ lines. When $\alpha = t$, Γ has $(1+s)^2$ points and $(1+t)(1+s)$ lines. When $\alpha = s$, Γ has $(1+s)(1+t)$ points and $(1+t)^2$ lines.

More restrictions on s, t and α can be obtained considering eigenvalues of the adjacency matrix of $(P, \perp)$ (see Appendix II).

(5) **Designs.** Let Γ be a finite rank 2 geometry with types P and L. We shall keep calling the elements of P points, as we have done till now, but we now call the elements of L *blocks* instead of lines, to be consistent with the terminology commonly used in design theory.

The geometry Γ is said to be a *t-(v, k, λ)-design* (with t, v, k, λ positive integers and $2 \leq t \leq k < v$) if $|P| = v$, every block is incident to precisely k points, t distinct points are incident to precisely λ blocks and distinct blocks are incident to distinct sets of points (that is, blocks are subsets of P).

Every design with $t = 2$ and $\lambda = 1$ is a linear space. In particular, a 2-$(q^2 + q + 1, q + 1, 1)$-design is a projective plane and a 2-$(q^2, q, 1)$-design is an affine plane.

When $\lambda > 1$ or $t > 2$, then Γ has gonality $g = 2$, P-diameter $d_P = 3$ and L-diameter $d_L = 3$ or 4. We have $d_L = 4$ if and only if there are non-intersecting blocks in Γ.

t-$(v, k, 1)$-designs are also called *Steiner systems* and denoted by the symbol $S(t, k, v)$. Five important examples of Steiner systems are related to the five Mathieu groups. For instance, the Steiner system $S(5, 6, 12)$ for the Mathieu group M_{12} arises from the 5-transitive action of M_{12} on a set S of 12 objects. The elements of S are the points of that Steiner system. The stabilizer in M_{12} of 5 points of S fixes a unique additional point. These 6 points form a block and all blocks are obtained in this way. The Steiner system $S(4, 5, 11)$ for the Mathieu group M_{11} can be described in a similar way, by the 4-transitive action of M_{11} on a set of 11 objects. The reader can see [91] (Chapter 8) for a description of the Steiner systems $S(5, 8, 24)$, $S(4, 7, 23)$ and $S(3, 6, 22)$ associated with the Mathieu groups M_{24}, M_{23} and M_{22}, respectively.

Note that Steiner systems $S(3, n + 1, n^2 + 1)$ are inversive planes (see Section 1.2.6(2)).

If both Γ and its dual are t-(v, k, λ)-designs for given positive integers t, v, k, λ, then Γ is called a *symmetric* t-(v, k, λ)-design. For instance, a finite projective plane of order q is a symmetric 2-$(q^2 + q + 1, q + 1, 1)$-design.

All the examples mentioned above are Steiner systems ($\lambda = 1$). However, there are many designs with $\lambda > 1$. For instance, the system of points and planes of the projective geometry $PG(3, q)$ of linear subspaces of $V(4, q)$ is a symmetric design with $t = 2$ and $\lambda = q + 1$ (compare Section 1.6.1, (4) and (6)). Given a finite projective plane Π of order $q > 1$, we can define a symmetric 2-$(q^2 + q + 1, q^2, q + 1)$-design with the same points as Π, but taking as blocks the complements of the lines of Π in the set of points of Π. Given an integer $n \geq 4$ and an n-set S, the elements and the $(n-1)$-subsets of S form a t-$(n, n-1, n-t)$-design for every $t = 2, 3, ..., n-2$ (when $t = n-1$, we have a Steiner system). Every t-(v, k, λ)-design with $t \geq 3$ is also an s-(v, k, μ)-design, with $s = t-1$ and $\mu = \lambda(v-t+1)/(k-t+1)$.

The reader wanting to know more on designs can see [91].

Remark. The previous definition of designs is more restrictive than other definitions frequently used in the literature (see [91] or [61]). For instance, it does not cover designs with just one block (note that these designs are not rank 2 geometries in the meaning stated in Section 1.1). It does not even cover 1-designs (however, many 1-designs are not geometries in our meaning).

1.3 Examples of higher rank

1.3.1 Projective and affine geometries

Classical projective and affine geometries are well-known ([61], [90], [234]). In fact, we took it for granted in previous sections (1.2.5 and 1.2.6(2)) that the reader is acquainted with them, when we mentioned projective geometries of linear subspaces of vector spaces. Nevertheless, we will now discuss finite dimensional classical projective and affine geometries in some detail, and also fix some notation and terminology which we will often use later.

The projective geometry $PG(n, K)$. Given a division ring K and a positive integer n, let us write V for $V(n+1, K)$. The *n-dimensional projective geometry* over K is the system of all linear subspaces of V. It is usually denoted as $PG(n, K)$ ($PG(n, q)$ when $K = GF(q)$). The i-dimensional linear subspaces of V are called the $(i-1)$-*dimensional subspaces* of $PG(n, K)$. In particular, linear subspaces of V of dimension 1, 2, 3 or n (of dimension 0, 1, 2 or $n-1$ in $PG(n, K)$) are called *points*, *lines*, *planes*, *hyperplanes* of $PG(n, K)$, respectively.

The trivial subspace $\{0\}$ of V has dimension -1 in $PG(n, K)$ and a nontrivial subspace of $PG(n, K)$ is uniquely determined by its set of points. Therefore, the subspaces of $PG(n, K)$ can be identified with their sets of points. In particular, the trivial subspace $\{0\}$ is identified with the empty set.

The proper nonempty subspaces of $PG(n, K)$ form a geometry Γ of rank n with type partition $\Theta = \{T_i\}_{i=0}^{n-1}$ defined as follows: T_i is the set of all i-dimensional subspaces of $PG(n, K)$, for $i = 0, 1, ..., n-1$. The incidence relation $*$ is defined as symmetrized inclusion, namely $x * y$ means that either $x \subseteq y$ or $x \supseteq y$. As usual, points are identified with their singletons, so that writing $x \in y$ or $x \subseteq y$ means the same when x is a point.

Flags are chains of nonempty proper subspaces, the rank of a flag being the number of subspaces it contains. Types can naturally be identified with dimensions. Hence the type of a flag is the set of dimensions of the subspaces in that flag.

Of course, when $n = 1$ the geometry Γ is just a set and our combinatorial view misses the projective line structure $PG(1, K)$.

Assume $n > 1$. Given a point or a hyperplane x, the residue Γ_x of x is isomorphic to $PG(n-1, K)$. Given a subspace x of dimension $i = 1, 2, ..., n-2$, the residue of x splits in two parts, say Γ_x^+ and Γ_x^-, where $\Gamma_x^- \cong PG(i, K)$ (respectively, $\Gamma_x^+ \cong PG(n-i-1, K)$) and consists of subspaces of dimension less than i (greater than i). Every element of Γ_x^- is incident to all elements of Γ_x^+.

Actually, we should have checked the connectedness properties of the graph Γ before claiming that Γ is indeed a geometry. We do this now. Any two points of a projective geometry are joined by a line and all subspaces contain points. Therefore Γ is connected. The previous analysis of residues shows that an inductive argument can be applied, so that residues are geometries by the induction hypothesis. Hence Γ is a geometry.

The affine geometry $AG(n, K)$. Given a division ring K, let n be an integer greater than 1 and let V denote $V(n, K)$. The cosets in the additive group of V of the linear subspaces of V are called the *affine subspaces* of V. The system of all affine subspaces of V is the *n-dimensional affine geometry* over K, denoted by $AG(n, K)$ (by $AG(n, q)$ when $K = GF(q)$). Trivially, given an affine subspace x of V (a *subspace* of $AG(n, K)$, for short), there is just one linear subspace x_0 of V such that x is a coset of x_0. This linear subspace x_0 is called the *direction* of x, and its dimension is the *dimension* of x. Subspaces of $AG(n, K)$ of dimension 0, 1, 2, $n-1$ are called *points*, *lines*, *planes*, *hyperplanes*, respectively, as in the projective case. The trivial subspace $\{0\}$ (namely, the empty subspace of $PG(n-1, K)$) is the common direction of all points. The empty set $\emptyset$ is also considered as a subspace of $AG(n, K)$, the (-1)-dimensional one.

Two subspaces x and y with the same direction are said to be *parallel*. The expression $x \parallel y$ means that x and y are parallel. Clearly, $\parallel$ is an equivalence relation on the set of subspaces of $AG(n, K)$ and the quotient of $AG(n, K)$ by $\parallel$ is isomorphic to the projective geometry $PG(n-1, K)$ of linear subspaces of V. The direction x_0 of a subspace x of $AG(n, K)$ can be chosen as canonical representative of the parallelism class of x and this choice of representatives preserves inclusions of subspaces. $PG(n-1, K)$ is called the *geometry at infinity* of $AG(n, K)$.

The affine geometry $AG(n, K)$ can be constructed in another way, too. Let us fix a hyperplane H of $PG(n, K)$ and discard H and all subspaces of $PG(n, K)$ contained in H. What is left of $PG(n, K)$ is a model of $AG(n, K)$. This can be seen as follows. Clearly, H inherits the vector space structure $V(n, K)$ from the vector space $V(n+1, K)$ giving rise to $PG(n, K)$. Let us fix a vector $e \notin H$ in $V(n+1, K)$. For every point x of $PG(n, K)$ not in H there is precisely one vector $\varepsilon(x)$ in H such that $e + \varepsilon(x) \in x$. This defines a mapping ε from the set of points of $PG(n, K)$ not in H onto the set of vectors of $V(n, K)$ and it is an easy exercise to check that ε maps subspaces of $PG(n, K)$ not in H onto affine subspaces of $V(n, K) = H$ and is bijective. On the other hand, ε^{-1} maps affine subspaces of $V(n, K)$ onto subspaces of $PG(n, K)$ not in H. Therefore, ε is an isomorphism and $PG(n, K) - H$ is a model of $AG(n, K)$.

A rank n geometry Γ is also associated with $AG(n, K)$. The elements of Γ are the nonempty proper subspaces of $AG(n, K)$. The incidence relation and the type partition are defined by means of inclusion and dimensions, as we have done for $PG(n, K)$. If x is an i-dimensional affine subspace of V with $i = 1, 2, ..., n-2$, then $\Gamma_x^- \cong AG(i, K)$ and $\Gamma_x^+ \cong PG(n-i, K)$. If x is a point then we have $\Gamma_x \cong PG(n, K)$ and, if x is a hyperplane, then $\Gamma_x \cong AG(n-1, K)$.

The proof that the structure Γ defined in this way is indeed a geometry of rank n is similar to that given in the projective case. We leave it for the reader.

1.3.2 Matroids

Projective and affine geometries are instances of matroids. A number of equivalent definitions can be given for matroids (see [239] or [44]). We choose a definition in terms of closure operators.

Let S be a set with at least two elements and $P(S)$ be the family of all subsets of S. A *matroid* on the *set of points* S is a mapping $\langle ... \rangle : P(S) \longrightarrow S$ satisfying the following properties:

(i) $\langle \emptyset \rangle = \emptyset$ and $\langle x \rangle = x$ for all points $x \in S$;

(ii) $\langle \langle X \rangle \rangle = \langle X \rangle$ for all $X \subseteq S$;

(iii) for all $X, Y \subseteq S$, we have $\langle X \rangle \subseteq \langle Y \rangle$ if and only if $X \subseteq \langle Y \rangle$;

(iv) for every $X \subseteq S$, there is a finite subset Y of X such that $\langle Y \rangle = \langle X \rangle$;

(v) (*Exchange axiom*). For all $X \subseteq S$ and $x, y \in S$, if $y \in \langle X \cup x \rangle$ and $y \notin \langle X \rangle$, then $x \in \langle X \cup y \rangle$.

The following are easy consequences of (iii):

(vi) $X \subseteq \langle X \rangle$ for all $X \subseteq S$;

(vii) for all $X, Y \subseteq S$, if $X \subseteq Y$, then $\langle X \rangle \subseteq \langle Y \rangle$.

A subset X of S is said to be *closed* if $X = \langle X \rangle$. It is an easy exercise to check that the family $\mathcal{M}$ of closed subsets of S is closed under arbitrary intersections. Hence, as S itself is closed, $\mathcal{M}$ is a complete lattice. We may identify a matroid with its lattice of closed subsets.

A *generating* (or *spanning*) set for a closed subset X is a subset Y of X such that $\langle Y \rangle = X$. A subset Y of S is said to be *independent* if $x \notin \langle Y - x \rangle$ for all $x \in Y$. Using the exchange axiom (v), it is not difficult to prove that, given a closed subset X, a subset Y of S is a minimal generating set for X if and only if it is a maximal independent subset in X and that all maximal independent subsets of X have the same size (necessarily finite, by (iv)). The maximal independent subsets of a given closed subset X are called *bases* of X and, if $i+1$ is their common size, then i is called *dimension* of X and denoted by $dim(X)$. Trivially, we have $dim(\emptyset) = -1$ and $dim(x) = 0$ for all points $x \in S$. The number $n = dim(S)$ is called the *dimension* of the matroid $\mathcal{M}$.

The exchange axiom (v) can now be rephrased as follows:

(v') given a closed subset X of dimension $i < n-1$ and a point $x \notin X$, there is just one closed subset of dimension $i+1$ containing both X and x.

The word *subspace* is sometimes used instead of 'closed subset'; we adopt this terminology. The subspaces of dimension 1, 2, $n-1$ are called *lines*, *planes* and *hyperplanes*, respectively. Note that the system of points and lines of a matroid of dimension $n > 1$ is a linear space and that linear spaces are precisely the 2-dimensional matroids.

As in $PG(n, K)$ and $AG(n, K)$, the nonempty proper subspaces of an n-dimensional matroid $\mathcal{M}$ form a geometry of rank n, where the elements are partitioned in types according to their dimensions as subspaces of $\mathcal{M}$ and the symmetrized inclusion provides the incidence relation. Let us denote this geometry by the same symbol $\mathcal{M}$ used to denote the matroid.

If X is a hyperplane of the matroid $\mathcal{M}$, then its residue $\mathcal{M}_X$ in the geometry $\mathcal{M}$ is an $(n-1)$-dimensional matroid. If x is a point, then $\mathcal{M}_x$ is an $(n-1)$-dimensional matroid (the points of this matroid are the lines of $\mathcal{M}$ on x). The matroid $\mathcal{M}_x$ is also called the *star* of x. If X is a subspace of dimension $i = 1, 2, ..., n-2$, then its residue $\mathcal{M}_X$ consists of two parts $\mathcal{M}_X^-$ and $\mathcal{M}_X^+$, as in the case of projective and affine geometries. $\mathcal{M}_X^-$ is an i-dimensional matroid with X as set of points whereas $\mathcal{M}_X^+$ (the *star* of X) is an $(n-i-1)$-dimensional matroid with the $(i+1)$-dimensional subspaces containing X as lines.

1.3.3 Back to projective and affine geometries

We can now give abstract definitions of projective and affine geometries in the language of matroids.

Projective geometries. Following Jónsson [102], we say that a matroid $\mathcal{M}$ of dimension n is an *n-dimensional projective geometry* if, for every subspace X of $\mathcal{M}$ and every point $x \notin X$, all lines on x contained in $\langle x \cup X \rangle$ meet X in some point. The uniqueness of this point follows.

If, furthermore, all lines of $\mathcal{M}$ have at least three points, then the projective geometry $\mathcal{M}$ is said to be *non-degenerate*, or *ordinary*; otherwise, we say that $\mathcal{M}$ is *degenerate*, or *weak*.

It is quite evident that, when $n = 2$, the above definition just gives us projective planes. As usual, we have nothing to say on the case of $n = 1$. Let us turn to the case of $n > 2$. Obviously, $PG(n, K)$ is also a projective geometry in the previous abstract sense. The next well-known theorem states that the converse holds when $n > 2$.

Theorem 1.3 (Veblen–Young [234]) *Let $\mathcal{M}$ be an ordinary projective geometry of dimension $n > 2$. Then $\mathcal{M} = PG(n, K)$ for some division ring K.*

We warn the reader that the definition of projective geometries given by Veblen and Young in [234] is not quite the same as above: they define a projective geometry as the system of all subspaces of a linear space which satisfies the Pasch axiom (Section 7.1.4, property (P)) and where all lines have at least three points (see Section 1.6.2 for a definition of subspaces of partial planes). In Section 7.1.4 we will slightly change our approach of projective geometries, coming closer to the point of view of Veblen and Young [234]. For the moment, we only remark that an ordinary projective geometry as in our previous definition is also a projective geometry in the meaning of Veblen and Young (see [102]).

Let $\mathcal{M}_1$, $\mathcal{M}_2$, ..., $\mathcal{M}_k$ be projective geometries. Let S_i and n_i be the set of points and the dimension of $\mathcal{M}_i$ $(i = 1, 2, ..., k)$ and assume that

$S_i \cap S_j = \emptyset$ for $1 \leq i < j \leq k$. Let us set $S = \bigcup_{i=1}^{k} S_i$ and let $\mathcal{M}$ be the family of the subsets X of S such that $X \cap S_i$ is a subspace of $\mathcal{M}_i$ for every $i = 1, 2, ..., m$. It is not difficult to check that $\mathcal{M}$ is a degenerate projective geometry of dimension $n = -1 + \sum_{i=1}^{k}(n_i + 1)$. It is called the *direct product* of the projective geometries $\mathcal{M}_1$, $\mathcal{M}_2$, ..., $\mathcal{M}_k$.

We can also modify the previous definition allowing some or all factors of the product to be singletons instead of projective geometries: if $\mathcal{M}_i$ is a singleton, then we may agree to say that $\mathcal{M}_i$ has dimension 0 and that it has just two subspaces, namely $\emptyset$ and itself. With these conventions, products can be defined just as above and, as above, the product of a family $\mathcal{M}_1$, $\mathcal{M}_2$, ..., $\mathcal{M}_k$ of projective geometries or singletons is a degenerate projective geometry of dimension $n = -1 + \sum_{i=1}^{k}(n_i + 1)$.

Let $\mathcal{M}$ be a degenerate projective geometry. Define a relation $\equiv$ on the set of points of $\mathcal{M}$ by the following clause: $a \equiv b$ if and only if either $a = b$ or the line $\langle a, b\rangle$ has at least 3 points. A degenerate projective plane contains at most one line with more than two points. Hence the relation $\equiv$ is an equivalence relation and its equivalence classes either inherit from $\mathcal{M}$ the structure of non-degenerate projective geometries or are singletons. Clearly, $\mathcal{M}$ is a direct product of non-degenerate projective geometries and/or singletons, corresponding to the equivalence classes of $\equiv$.

The direct product of a set of $n+1$ points is a degenerate n-dimensional projective geometry — in fact an n-dimensional simplex (a tetrahedron when $n = 3$; a triangle when $n = 2$).

Affine geometries. An *n-dimensional affine geometry* with $n \geq 2$ is an n-dimensional matroid $\mathcal{M}$ where all planes are affine planes and, for every point x, the residue $\mathcal{M}_x$ of x is a $(n-1)$-dimensional projective geometry.

When $n = 2$, this definition just gives us affine planes. It is also clear that $AG(n, K)$ is an affine geometry in the above abstract sense. As for projective geometries, the converse holds when $n > 2$:

Theorem 1.4 (Jónsson [102]) *Let $\mathcal{M}$ be an affine geometry of dimension $n > 2$. Then $\mathcal{M} = AG(n, K)$ for some division ring K.*

We omit the details of the proof, giving only a sketch of it. However, we draw particular attention to its first (and main) step.

A binary relation $\|$ can be defined between the lines of $\mathcal{M}$, stating that $L \| M$ if L and M are non-intersecting coplanar lines or if $L = M$. This relation is evidently reflexive and symmetric. The first step of the proof is to show that $\|$ is also transitive, namely that it is an equivalence relation. This can be proved by way of contradiction.

Let L, M, N be lines with $L \parallel M \parallel N \not\parallel L$, if possible. Then L, M, N are pairwise distinct and $L \cup M \cup N$ spans a 3-dimensional subspace U of $\mathcal{M}$, as $\parallel$ is transitive when restricted to lines of a given plane, by a well-known property of affine planes. Let X, Y be the planes spanned by $L \cup M$ and by $M \cup N$ respectively. Given a point a of N, let Z be the plane spanned by $L \cup a$ and let N' be the line of Z parallel to L through a. We have $N \neq N'$. On the other hand, $Z \subseteq U$. The star of a in U is a projective plane. Hence $Z \cap Y$ is a line through a. Either $Z \cap Y \neq N$ or $Z \cap Y \neq N'$. Hence $Z \cap Y$ meets either M or L. In any case, $Z \cap Y \cap X \neq \emptyset$. This forces L to meet M: a contradiction. Therefore $L \parallel N$. The relation $\parallel$ is transitive.

The rest of the proof is a straightforward generalization of the well-known construction of the line at infinity of an affine plane. Since any three non-collinear points span a plane of $\mathcal{M}$, every point is in precisely one line of each of the equivalence classes of $\parallel$. These classes can be added to $\mathcal{M}$ as 'points at infinity'. Every plane of $\mathcal{M}$ defines a 'line at infinity' and the points and lines of $\mathcal{M}$ together with these new points and lines 'at infinity' form the point–line system of a projective geometry, say $\mathcal{M}'$. The points at infinity form a hyperplane of $\mathcal{M}'$, say $\mathcal{M}^{\infty}$, and $\mathcal{M}$ is obtained dropping $\mathcal{M}^{\infty}$ from $\mathcal{M}'$, as in Section 1.3.1. The proof is completed by applying Theorem 1.3 to $\mathcal{M}'$. □

1.3.4 Classical polar spaces

Let V be a left vector space over a division ring K and let f be a trace-valued non-degenerate reflexive (σ, ε)-sesquilinear form of finite Witt index $n \geq 2$ (Appendix I, 1.8.1).

We form a geometry Γ of rank n taking as elements the non-trivial linear subspaces of V that are totally isotropic for f. The incidence relation $*$ is defined as symmetrized inclusion, as we have done for projective and affine geometries and, more generally, for matroids. The elements of Γ are partitioned into n types according to their dimensions: the i-dimensional totally isotropic subspaces form the ith class of the type partition, for $i = 1, 2, ..., n$; we denote this class by i, for short.

We say that Γ is the *polar space* (of *rank* n) defined by f. Note that, when $n = 2$, Γ is just a classical generalized quadrangle (see Section 1.2.5).

We must still prove that Γ is a geometry, that is, that the connectedness properties involved in our definition of geometries hold.

We first prove that the graph Γ is connected. Let us fix an element X of Γ of type n (that is, a maximal totally isotropic subspace). Given any element Y of Γ of type $i < n$, $Z = Y + (Y^{\perp} \cap X)$ is a totally isotropic subspace. Furthermore, $Y^\perp \cap X \neq \{0\}$, by (v) of Appendix I. Therefore Z and $Y^\perp \cap X$ give us a path connecting Y to X. On the other hand, if Y had type n, we would only have to choose a non-trivial proper subspace Y' of Y and connect Y' to X as above. The connectedness of Γ is proved.

Before finishing the proof that Γ is a geometry, we need to examine the residues of the elements of Γ.

Given an element X of Γ of type $i < n-1$, let V_X be a complement of X inside the linear subspace $X^\perp$. Let f_X be the form induced by f on $V_X \times V_X$. Trivially, f_X is a trace valued reflexive (σ, ε)-sesquilinear form. Furthermore, f_X is non-degenerate and has Witt index $n-i$ (proving these two claims is simply an exercise in linear algebra). Therefore f_X defines a polar space of rank $n-i$, say Γ_X^+ (in particular, a generalized quadrangle when $i = n-2$). The isomorphism type of Γ_X^+ does not depend on the choice of the complement V_X of X in $X^\perp$. Indeed, given any totally isotropic subspace Y containing X (hence $Y \subseteq X^\perp$), let us set $\varphi_X(Y) = Y \cap V_X$. Then φ_X is an isomorphism from the system of totally isotropic subspaces properly containing X to the polar space Γ_X^+. Therefore, we may identify Γ_X^+ with the star of X, namely with the system of totally isotropic subspaces containing X. In particular, when $i = 1$, then the polar space Γ_X^+ is simply the residue Γ_X of X. When $i > 1$, then the residue of X splits into two pieces: the star Γ_X^+ of X and the projective geometry, say Γ_X^-, of all proper non-trivial linear subspaces of X. All elements of Γ_X^- are incident to all elements of Γ_X^+.

When X has type $n-1$, then its star Γ_X^+, consisting of the elements of Γ of type n containing X, contains at least two elements, by (vi) of Appendix I. Hence it is a geometry of rank 1. The other piece of the residue of X is the system Γ_X^- of all proper non-trivial subspaces of X, which is an $(n-2)$-dimensional projective geometry.

Finally, if X has type n, then its residue is an $(n-1)$-dimensional projective geometry.

An inductive argument can be applied now. If $n = 2$, then Γ is a generalized quadrangle (see Section 1.2.5). When $n > 2$, residues are geometries by the inductive hypothesis. Hence Γ is a geometry, as it is connected.

As we did in Section 1.2.5, we now give the notation currently used for polar spaces defined by alternating or symmetric bilinear forms or Hermitian forms.

(1) Alternating bilinear forms of Witt index n in $V(2n, K)$, with K a field. The polar spaces defined by these forms are called *symplectic varieties* of rank n and denoted by the symbol $\mathcal{S}_{2n-1}(K)$. When $K = GF(q)$, the

notation $\mathcal{S}_{2n-1}(q)$ is normally used instead of $\mathcal{S}_{2n-1}(K)$.

(2) Symmetric bilinear forms of Witt index n in $V(2n, K)$, with K a field and $char(K) \neq 2$. The polar spaces arising from these forms are called *hyperbolic quadrics* of rank n and denoted by $\mathcal{Q}^+_{2n-1}(K)$ (by $\mathcal{Q}^+_{2n-1}(q)$ when $K = GF(q)$).

(3) Symmetric bilinear forms of Witt index n in $V(2n+1, K)$, with K a field and $char(K) \neq 2$. The polar spaces defined by these forms are called *quadrics* of rank n and are denoted by $\mathcal{Q}_{2n}(K)$ (by $\mathcal{Q}_{2n}(q)$ when $K = GF(q)$).

(4) Symmetric bilinear forms of Witt index n in $V(2n+2, q)$, q odd. The polar spaces defined by these forms are called *elliptic quadrics* of rank n and denoted by $\mathcal{Q}^-_{2n+1}(q)$.

(5) Hermitian forms of Witt index n in $V(m, q^2)$, $m = 2n$ or $2n+1$. The polar spaces defined by these forms are called *Hermitian varieties* of rank n and denoted by $\mathcal{H}_{2n-1}(q^2)$ or $\mathcal{H}_{2n}(q^2)$, according to whether $m = 2n$ or $2n+1$.

When $char(K) = 2$ (when q is even), the three types of quadrics $\mathcal{Q}^+_{2n-1}(K)$, $\mathcal{Q}_{2n}(K)$ and $\mathcal{Q}^-_{2n+1}(q)$ must be defined using quadratic forms and totally singular subspaces.

We remark that $\mathcal{Q}_{2n}(q)$ and $\mathcal{S}_{2n-1}(q)$ are isomorphic when q is even (see [63]; also [62]).

Polar spaces arising as above from sesquilinear or quadratic forms are called *classical*. It follows from Theorems 1.26 and 1.27 of Appendix I that $\mathcal{S}_{2n-1}(q)$, $\mathcal{Q}^+_{2n-1}(q)$, $\mathcal{Q}_{2n}(q)$, $\mathcal{Q}^-_{2n+1}(q)$, $\mathcal{H}_{2n-1}(q^2)$ and $\mathcal{H}_{2n}(q^2)$ are the only examples of finite classical polar spaces of rank n (assuming $n > 2$ for $\mathcal{Q}^+_{2n-1}(q)$ to keep grids out).

1.3.5 Abstract polar spaces

We say that a family $\mathcal{P}$ of subsets of a nonempty set S is an (abstract) *polar space* of (finite) *rank* $n \geq 2$ if the following hold:

(i) we have $S = \bigcup_{X \in \mathcal{P}} X$ and $\emptyset \in \mathcal{P}$;

(ii) the family $\mathcal{P}$ is closed under arbitrary intersections;

(iii) every maximal element X of $\mathcal{P}$ together with all elements of $\mathcal{P}$ it contains forms an $(n-1)$-dimensional projective geometry with set of points X (degenerate projective geometries are allowed);

(iv) given a maximal element X of $\mathcal{P}$ and a non-maximal $Y \in \mathcal{P}$ such that $X \cap Y$ has codimension 1 in Y ($X \cap Y = \emptyset$ if $Y \in S$), there is precisely one maximal element Z of $\mathcal{P}$ containing Y and such that $Z \cap X$ is a hyperplane of the projective geometry X;

(v) every non-maximal element X of $\mathcal{P}$ can be obtained as the intersection of two maximal elements of $\mathcal{P}$.

These axioms are rather redundant (the reader may compare the axioms given in Chapter 7 of [222]). However, giving the best set of axioms for polar spaces is not our aim here.

When $n = 2$, the previous axioms are simply a way of rephrasing the definition of generalized quadrangles. Therefore, generalized quadrangles are precisely (abstract) polar spaces of rank 2.

Classical polar spaces are polar spaces in this abstract sense, too. The set S is the set of all 1-dimensional totally isotropic (or totally singular) linear subspaces of the vector space V with respect to the sequilinear (or quadratic) form f which we consider. The family $\mathcal{P}$ consists of all totally isotropic (or totally singular) subspaces of V (including $\{0\}$ among them), viewed as sets of 1-dimensional linear subspaces of V (thus, the trivial subspace $\{0\}$ of V is viewed as the empty set $\emptyset$). Checking that (i)–(v) hold is not difficult: (i) and (ii) are quite evident; as for (iii), the dimension of the projective geometry X is the Witt index of the form f, diminished by 1 unit. (iv) and (v) correspond to (v) and (vii) of Appendix I.

Turning to the general abstract case, let $\mathcal{P}$ be a polar space of rank n. By (iii), every element of $\mathcal{P}$ is a projective geometry. Therefore a dimension function dim can de defined on $\mathcal{P}$, stating that $dim(X)$ is the dimension of the projective geometry X, for $X \in \mathcal{P}$. In particular, $dim(\emptyset) = -1$ and $dim(x) = 0$ for all $x \in S$. Elements of dimension 0, 1 or 2 are called *points*, *lines* and *planes*, respectively; the elements of $\mathcal{P}$ are called *subspaces*, those of dimension $n-1$ being the *maximal subspaces*.

Given a subspace X of dimension $i = 0, 1, ..., n-2$, the *star* $\mathcal{P}_X^+$ of X is the set of subspaces properly containing X.

If we take the symmetrized inclusion as incidence relation, then $\mathcal{P} - \{\emptyset\}$ acquires the structure of an n-partite graph, the elements of $\mathcal{P} - \{\emptyset\}$ being partitioned in the n classes of this graph according to their dimensions, as in the classical case.

We still denote this graph by $\mathcal{P}$, by abuse of notation. As in the classical case, we can prove that the graph $\mathcal{P}$ is connected: we used (v) of Appendix I to do that in the classical case; we can now use the above axiom (iv), which is nothing but an abstract version of (v) of Appendix I.

In order to prove that $\mathcal{P}$ is in fact a geometry in the meaning of 1.1, we must examine the stars of subspaces of dimension 0, 1, ..., $n-2$ and show that they are geometries. When we have done this, the conclusion will follow from the fact that subspaces are projective geometries, as in the classical case.

Let $dim(X) = i = 1, 2, ..., n-2$. If $i = n-2$, then $\mathcal{P}_X^+$ is a rank 1 geometry, by (v). We already know that P is a geometry when $n = 2$, because rank 2 polar spaces are simply generalized quadrangles. We can work by induction now.

Assume $i < n-2$ (hence $n > 2$). It is easily seen that the axioms (i)–(v) are inherited by $\mathcal{P}_X^+ \cup \{\emptyset\}$, with the set of all $(i+1)$-dimensional subspaces containing X in the role of S and n substituted with $n-i-1$. Therefore $\mathcal{P}_X^+$ is a geometry, by the inductive hypothesis. We are done: $\mathcal{P}$ is a geometry.

We say that a polar space $\mathcal{P}$ is *thick-lined* if all lines of $\mathcal{P}$ have at least 3 points. Otherwise, $\mathcal{P}$ is said to be *weak*.

The following theorem is perhaps the most important one in the theory of polar spaces. It is the analogue of Theorems 1.3 and 1.4 for projective and affine geometries.

Theorem 1.5 (Tits) *All thick-lined polar spaces of rank $n \geq 4$ are classical. All thick-lined polar spaces of rank 3 with Desarguesian planes and at least three planes on each line are classical.*

The earliest proof of this theorem is due to Tits ([222], Chapter 8). It is long and difficult. Two shorter proofs are given by Buekenhout and Cohen in [25] (Chapters 9, 8 and 11). One of these two proofs develops earlier ideas by Veldkamp [235]. The other one is based on the existence of sufficently many automorphisms of polar spaces, proved by Tits ([222], Chapter 4) in a more general context.

Tits has also proved that there are just two families of non-cassical thick-lined polar spaces of rank 3 ([222], Chapter 9). One of them is the family of Grassmann geometries of lines of 3-dimensional projective geometries defined over non-commutative division rings. We will describe these geometries at the very end of the next section. We now only mention some features of the exceptional polar spaces arising in this way. Planes are Desarguesian, but there are two distinct (mutually dual) isomorphism types for planes. Every line is in just two planes, one for each isomorphism type.

The other exceptional family was discovered by Tits. We are not going to describe it. The reader may find complete information on these polar spaces in [222], Chapter 9 (also in [25]). We only mention that, if Γ is one of these exceptional polar spaces, then the lines of Γ are in infinitely many planes and the planes of Γ are Moufang, but not Desarguesian.

As a by-product of the above we immediately obtain the following:

Theorem 1.6 (Tits) *All finite thick-lined polar spaces of rank 3 are classical.*

Indeed it is well known that all finite division rings are fields and all finite Moufang projective planes are Desarguesian ([90], Theorem 6.20).

Weak polar spaces can be obtained as 'products' of thick-lined polar spaces and/or lines and/or points, as for projective geometries. We do not describe this product construction here. The reader interested in it may see [29] (also Section 5.3.2). We will only describe the structure of those weak polar spaces where all lines have precisely two points. In this case the polar space is obtained from a complete n-partite graph where all classes of the partition have size ≥ 2; the cliques of the graph are the subspaces of the polar space and n is the rank. In particular, when $n = 2$ this construction just gives us dual grids.

When all classes of the n-partition have size 2, then the previous construction gives us just the dual of the n-dimensional cube (in particular, when $n = 3$ we obtain the octahedron). This can be seen by marking the two elements of each of the n classes by 0 and 1, ordering the n classes of the partition in some way and interpreting the maximal cliques of the graph as n-dimensional vectors over $GF(2)$: a maximal clique X has ith entry j if the element of X in the ith class is marked by j ($j = 0$ or 1).

1.3.6 Hyperbolic quadrics revisited

Let $\mathcal{Q}$ be the $(2n-1)$-dimensional hyperbolic quadric $\mathcal{Q}^+_{2n-1}(K)$ (see Section 1.3.4), that is, $\mathcal{Q}$ is the polar space of totally singular subspaces of a non-singular quadratic form of Witt index n in $V(2n, K)$ (see Appendix I).

Given two maximal subspaces X, Y of $\mathcal{Q}$, we write $X\Theta Y$ to mean that the number $n-1-dim(X\cap Y)$ is even (dim denotes projective dimension, as in Section 1.3.5). The relation Θ defined in this way is an equivalence relation with just two equivalence classes. We will prove this claim later in this book (Lemma 5.22). However, the reader who wants an immediate proof of it may try to obtain it algebraically, as a (non-trivial) exercise in linear algebra.

Let M^+ and M^- be the two equivalence classes of Θ. Trivially, every $(n-2)$-dimensional subspace of $\mathcal{Q}$ is in precisely two maximal subspaces, one from each of the two families M^+ and M^-.

We can now define a new geometry, say $\Gamma(\mathcal{Q})$, as follows. The elements of $\Gamma(\mathcal{Q})$ are the elements of $\mathcal{Q}$ of dimension $i = 0, 1, ..., n-3, n-1$, partitioned into n classes T_0, T_1, ..., T_{n-3}, M^+ and M^-, where T_i is the set of i-dimensional subspaces of $\mathcal{Q}$ (when $n = 2$, we only have two types of elements in $\Gamma(\mathcal{Q})$, namely M^+ and M^-). The incidence relation between elements of dimension $< n-2$ or between these elements and maximal subspaces of $\mathcal{Q}$ is defined in $\Gamma(\mathcal{Q})$ just as in $\mathcal{Q}$, whereas two maximal subspaces X and Y of $\mathcal{Q}$ are said to be incident in $\Gamma(\mathcal{Q})$ if $dim(X \cap Y) = n-2$.

We still must prove that this definition indeed gives us a geometry. We prove this by induction on n. When $n = 2$, then $\mathcal{Q}$ is simply a grid and the above construction amounts to building a generalized digon taking M^+ and M^- as two types of elements.

Assume $n > 2$. We show first that the graph $\Gamma(\mathcal{Q})$ is connected. Let X, Y be elements of $\Gamma(\mathcal{Q})$ and $X = X_0$, X_1, ..., $X_m = Y$ be a path in the incidence graph of $\mathcal{Q}$ from X to Y. A path of $\Gamma(\mathcal{Q})$ from X to Y can be constructed as follows. If $X_j \in \Gamma(\mathcal{Q})$, then we keep it. Otherwise, $dim(X_j) = n-2$ and X_j is contained in precisely two maximal subspaces, say X_j^+ and X_j^-, forming an edge in the graph $\Gamma(\mathcal{Q})$. In this case we substitute the vertex X_j with the edge $\{X_j^+, X_j^-\}$ and the connection from X_{j-1} to X_{j+1} is saved in $\Gamma(\mathcal{Q})$, possibly modulo repetitions of consecutive vertices (but this is not a problem at all: we can always clean the path discarding repetitions). The connectedness of $\Gamma(\mathcal{Q})$ is proved.

In order to finish the proof that $\Gamma(\mathcal{Q})$ is a geometry, we must show that the neighbourhood $\Gamma(\mathcal{Q})_X$ of any element X of $\Gamma(\mathcal{Q})$ is a geometry, exploiting the inductive hypothesis to do this.

Assume $X \in M^+$ first (the case of $X \in M^-$ is quite similar to this, of course). As the $(n-2)$-dimensional subspaces of $\mathcal{Q}$ are in precisely two maximal subspaces, the hyperplanes of the projective geometry $\mathcal{Q}_X$ bijectively correspond to the maximal subspaces in M^- incident to X in $\Gamma(\mathcal{Q})$. This shows that $\Gamma(\mathcal{Q})_X \cong \mathcal{Q}_X$. Therefore, $\Gamma(\mathcal{Q})_X$ is an $(n-1)$-dimensional projective geometry.

Now let $1 \leq dim(X) < n-2$. Then $\Gamma(\mathcal{Q})_X$ splits into two parts, say $\Gamma(\mathcal{Q})_X^-$ and $\Gamma(\mathcal{Q})_X^+$, consisting of subspaces contained in X and containing X, respectively. Trivially, all elements in any of these two parts are incident to all elements in the other part and $\Gamma(\mathcal{Q})_X^- = \mathcal{Q}_X^-$, where $\mathcal{Q}_X^-$ and $\mathcal{Q}_X^+$ are the two parts of the residue of X in $\mathcal{Q}$, $\mathcal{Q}_X^+$ being the star of X.

Furthermore, we have $\Gamma(\mathcal{Q})_X^+ = \Gamma(\mathcal{Q}_X^+)$ and $\mathcal{Q}_X^+$ is a hyperbolic quadric (as for the last claim, the reader may see the argument used in Section 1.3.4 to prove that stars of non-maximal subspaces of classical polar spaces are isomorphic to classical polar spaces). By the inductive hypothesis on $\mathcal{Q}_X^+$, $\Gamma(\mathcal{Q}_X^+)$ is a geometry. Therefore $\Gamma(\mathcal{Q})_X^+$ is a geometry. The very same argument shows that $\Gamma(\mathcal{Q})_x$ is a geometry when x is a point of $\mathcal{Q}$. We are done.

When $n > 3$, the standard name for $\Gamma(\mathcal{Q})$ is a *building of type D_n over K*. Motivations for this seemingly odd terminology will become evident in Chapter 13.

We have already remarked that, when $n = 2$, $\Gamma(\mathcal{Q})$ is simply a generalized digon. Actually, each of the two types of elements of this generalized digon is endowed with the projective line structure $PG(1, K)$, but this is lost in our combinatorial perspective.

When $n = 3$, $\mathcal{Q} = \mathcal{Q}_5^+(K)$ is called the *Klein quadric over K*. In this case, $\Gamma(\mathcal{Q})$ has rank 3. We call its elements points, lines and planes, taking the elements of M^+ as points, the points of $\mathcal{Q}$ as lines and the elements of M^- as planes. Using (i)–(v) of Section 1.3.5 in $\mathcal{Q}$, it is not difficult to check that $\Gamma(\mathcal{Q})$ is a 3-dimensional projective geometry, as defined Section 1.3.3. The two planes of $\mathcal{Q}$ on a given line of $\mathcal{Q}$ form a point–plane flag in $\Gamma(\mathcal{Q})$. Hence the lines of $\mathcal{Q}$ bijectively correspond to the point–plane flags of $\Gamma(\mathcal{Q})$. Thus, given $X \in M^+$, the lines of $\mathcal{Q}$ in X bijectively correspond to the points of $\Gamma(\mathcal{Q})$ in the plane X. On the other hand, the points of $\mathcal{Q}$ in X are the lines of $\Gamma(\mathcal{Q})$ belonging to the plane X. Therefore we have an isomorphism between the plane X of $\mathcal{Q}$ and the dual of X, viewed as a plane of $\Gamma(\mathcal{Q})$. It is now clear that $\Gamma(\mathcal{Q}) = PG(3, K)$.

The previous construction of $PG(3, K)$ from $\mathcal{Q}_5^+(K)$ can be inverted as follows. Let K be a division ring (we do not need to assume that K is a field, now). We take the lines of $PG(3, K)$ as 'points' and the point–plane flags of $PG(3, K)$ as 'lines'. Points and planes of $PG(3, K)$ are gathered together in one type and taken as 'planes'. The incidence relation is the one inherited from $PG(3, K)$. The reader can check that this in fact gives us a geometry, which is a rank 3 thick-lined polar space where every 'line' is in precisely two 'planes'. This polar space is called the *Grassmann geometry of lines* of $PG(3, K)$. Let us denote it by $\mathcal{G}_{3,K}$.

If K is a field, then $\mathcal{G}_{3,K} = \mathcal{Q}_5^+(K)$. Indeed in this case the above is just the inverse of the construction of $PG(3, K) = \Gamma(\mathcal{Q})$ from $\mathcal{Q} = \mathcal{Q}_5^+(K)$.

Let K be non-commutative. Then we have two distinct isomorphism types for 'planes' of $\mathcal{G}_{3,K}$, namely $PG(2, K)$ and its dual, representing points and planes of $PG(3, K)$, respectively. Therefore the polar space

$\mathcal{G}_{3,K}$ is non-classical.

1.4 Chambers and galleries

Let us turn back to the general theory. Throughout this section Γ is a given geometry of rank n and $\Theta = \{T_1, T_2, ..., T_n\}$ is its type partition.

Maximal flags are called *chambers*. Since every flag picks up at most one element from each type, every flag has rank at most n, every flag is contained in some chamber and all flags of rank n are chambers (they take precisely one element from each type).

Lemma 1.7 *All chambers have rank n.*

Proof. This is trivial when $n = 1$. We go on by induction. Assume $n > 1$. Given a chamber C, let x be an element of C. The residue Γ_x of x is a geometry of rank $n-1$, by the very definition of geometries, and $C - x$ is a chamber of it. Hence $|C - x| = n - 1$, by the inductive hypothesis, and we have $|C| = n$. □

Lemma 1.8 *Every non-maximal flag is contained in at least two chambers.*

Proof. The statement is trivial when $n = 1$. We proceed by induction; assume $n > 1$. Given a non-maximal flag $F \neq \emptyset$, choose an element $x \in F$. Then $F - x$ is a flag of the residue Γ_x of x. By the inductive hypothesis, $F - x$ is contained in at least two chambers C, D of Γ_x. Clearly, $x \cup C$ and $x \cup D$ are two chambers of Γ containing F.

The case of $F = \emptyset$ is trivial: choose any element and two chambers containing it (they exist, by the above). □

By Lemmas 1.7 and 1.8, the corank of F is the number of elements we should add to F in order to form a chamber. The flags of corank 1 are called *panels*. Two chambers C and D are said to be *adjacent* if $C \cap D$ is a panel. We write $C \sim D$ to mean that C and D are adjacent chambers.

The *chamber system* $\mathcal{C}(\Gamma)$ of Γ is the graph having the chambers of Γ as vertices, the adjacency relation being the one defined above. A path $\gamma = (C_0, C_1, ..., C_m)$ in $\mathcal{C}(\Gamma)$ from a chamber $C = C_0$ to a chamber $D = C_m$ is called a *gallery* of *length* m from C to D.

We warn the reader that galleries are sometimes defined in a slightly different way, allowing repetitions of consecutive chambers (see [222] and [227]).

Theorem 1.9 *The graph $\mathcal{C}(\Gamma)$ is connected.*

Proof. The proof is by induction on n. The statement is trivial when $n = 1$.

Assume $n > 1$. Given two chambers C and D, let $x \in C$, $y \in D$ and let $x = x_0, x_1, ..., x_m = y$ be a path from x to y in Γ. For every $i = 1, 2, ..., m$, we can choose a chamber C_i containing the flag $\{x_{i-1}, x_i\}$, by Lemma 1.8. Let $C_0 = C$ and $C_{m+1} = D$. For every $j = 0, 1, ..., m$, the residue Γ_{x_j} of x_j has rank $n-1$. Therefore a gallery, say γ_i, can be found from $C_i - x_i$ to $C_{i+1} - x_i$ in the chamber system $\mathcal{C}(\Gamma_{x_i})$, by the inductive hypothesis. Clearly, γ_i gives us a gallery δ_i from C_i to C_{i+1} in $\mathcal{C}(\Gamma)$ and $\delta = \delta_0\delta_1...\delta_m$ is a gallery from C to D in $\mathcal{C}(\Gamma)$. □

Given two distinct adjacent chambers C and D, the two elements of $(C \cup D) - (C \cap D)$ belong to the same type, say T_i. We write $C\Phi^i D$ to mean this and we say that C and D are *T_i-adjacent*, or *i-adjacent*, for short. We also state $C\Phi^i C$, by convention (hence, $C \sim C$, too). Clearly, Φ^i is an equivalence relation. Theorem 1.9 can now be rephrased as follows:

Corollary 1.10 $\bigvee_{i=1}^{n} \Phi^i = \Omega$

The system of equivalence relations $(\Phi)_{i=1}^{n}$ may be viewed as a coloured graph, obtained giving each edge of the graph $\mathcal{C}(\Gamma)$ its appropriate colour $i = 1, 2, ..., n$.

The reader may have noticed that, when shortening the symbol T_i to i, we have identified the set of types $\Theta = \{T_1, T_2, ..., T_m\}$ with the set of positive integers $\{1, 2, ..., n\}$. Keeping this convention, we also give types to galleries, as follows: if $C_0, C_1, ..., C_m$ form a gallery and $C_{h-1}\Phi^i C_h$ ($h = 1, 2, ..., m$), then $(i_1, i_2, ..., i_m)$ is the *type* of that gallery.

An $(n-1)$-dimensional simplicial complex can also be associated with every geometry Γ of rank n, taking the elements and the chambers of Γ as vertices and simplices of the complex, respectively. The flags of rank $i+1$ ($i = 0, 1, ..., n-1$) are the i-dimensional faces of the complex. The incidence graph of Γ is the skeleton of this complex. We denote this simplicial complex by $\mathcal{K}(\Gamma)$ and we call it the *flag complex* of Γ.

Exercise 1.1. Prove that $\mathcal{K}(\Gamma)$ is a geometry of rank n.

(Hint: use Theorem 1.9 to prove the connectedness of the incidence graph of $\mathcal{K}(\Gamma)$; this already settles the rank 2 case (see also the next section). Then work by induction, as we have done several times in Section 1.3.)

Exercise 1.2. Prove that $\mathcal{K}(\Gamma)$ is a (weak) polar space of rank n with all lines of size 2, if and only if the incidence graph of Γ is a complete n-partite graph.

1.4.1 The rank 2 case

In this section Γ is a geometry of rank 2, P and L are the two types of Γ and g, d_P, d_L, d $(= max(d_P, d_L))$ have the meaning stated in Section 1.2. Consistently with this notation, Φ^P and Φ^L are the P- and L-adjacency relation, respectively.

Since Γ has rank 2, the flag complex (the chamber system) of Γ is the (dual of) the incidence graph of Γ.

The Structure of $\mathcal{K}(\Gamma)$. The flag complex $\mathcal{K}(\Gamma)$ is a geometry of rank 2. Its incidence graph is obtained taking the vertices and the edges of the graph Γ as new vertices and all incident vertex–edge pairs of the graph Γ as new edges.

Let P' and L' denote the sets of elements and chambers of Γ, respectively, viewed as points and lines of $\mathcal{K}(\Gamma)$ and let g', $d_{P'}$ and $d_{L'}$ denote the gonality and the diameters of $\mathcal{K}(\Gamma)$. The next relations easily follow from the description given above for the incidence graph of $\mathcal{K}(\Gamma)$:

$$g' = 2g, \qquad d_{P'} = 2d, \qquad d_{L'} = 2 \cdot min(d_P, d_L)$$

Furthermore, the diameter of the collinearity graph of $\mathcal{K}(\Gamma)$ equals d.

Note that $\mathcal{K}(\Gamma)$ is a partial plane, since $g' = 2g \geq 4$.

It follows from the previous relations that, if Γ is a generalized m-gon, then $\mathcal{K}(\Gamma)$ is a generalized $2m$-gon. In particular, if Γ is a generalized digon, then $\mathcal{K}(\Gamma)$ is a dual grid; if Γ is a projective plane, then $\mathcal{K}(\Gamma)$ is a generalized hexagon with all lines of size 2. Of course, we can continue the process, considering $\mathcal{K}(\mathcal{K}(\Gamma))$, $\mathcal{K}(\mathcal{K}(\mathcal{K}(\Gamma)))$, etc.

For instance, an infinite series of generalized $2^k m$-gons is obtained in this way, starting from a generalized m-gon.

The reader may have noticed that an example of a flag complex was already implicit in Section 1.3.6: the grid $\mathcal{Q} = \mathcal{Q}_3^+(K)$ is just the dual of the flag complex of the generalized digon $\Gamma(\mathcal{Q})$.

Diameters and adjacencies. We have $\Omega = \Phi^P \vee \Phi^L$, by Corollary 1.10. Let us state the following definitions:

$$\Phi_1^{P,L} = \Phi^P, \qquad \Phi_1^{L,P} = \Phi^L,$$

$$\Phi_{m+1}^{P,L} = \Phi^P \Phi_m^{L,P} = \Phi^P \Phi^L \Phi^P ... \ (m+1 \text{ factors})$$

$$\Phi_{m+1}^{L,P} = \Phi^L \Phi_m^{P,L} = \Phi^L \Phi^P \Phi^L ... \ (m+1 \text{ factors})$$

Proposition 1.11 *We have* $m = d_P$ *if and only if* $\Phi^{L,P}_m = \Omega \neq \Phi^{L,P}_{m-1}$. *Similarly for* $\Phi^{P,L}_m$ *and* d_L.

Proof. Given two chambers C, D, let x be the element of C of type P and y the element of D at maximal distance from x. Let $x = x_0, x_1, ..., x_{m-1} = y$ be a path from x to y in the incidence graph of Γ and define $C_i = \{x_{i-1}, x_i\}$ $(i = 1, 2, ..., m-1)$, $C_0 = C$ and $C_m = D$. Then we have $C_0\Phi^L C_1 \Phi^P C_3....$ Note that $C_{i-1} = C_i$ is allowed here. Indeed, if $C_{i-1} = C_i$, then both $C_{i-1}\Phi^P C_i$ and $C_{i-1}\Phi^L C_i$ hold and we may choose the one we need of these relations, according to whether i is even or odd.

Trivially, $m \leq d_P$. This shows that $\Phi^{L,P}_m = \Omega$ if $m = d_P$. Conversely, let $\Phi^{L,P}_m = \Omega$. Then starting, from any element $x \in P$, we need a path of at most m chambers to pass from x to any other element y of Γ (of course, the first and last chambers of the path will contain x and y, respectively). This proves that $m \geq d_P$ and completes the proof. □

Corollary 1.12 *The graphs* $\mathcal{C}(\Gamma)$ *and* Γ *have the same diameter.*

Proof. Let m_P (respectively, m_L) be the minimal m such that $\Phi^{L,P}_m = \Omega$ (respectively, $\Phi^{P,L}_m = \Omega$). Trivially, $max(m_P, m_L)$ is the diameter of $\mathcal{C}(\Gamma)$. The conclusion follows from the previous proposition. □.

Corollary 1.13 *The equation* $\Phi^P\Phi^L = \Phi^L\Phi^P$ *characterizes generalized digons.*

Proof. Indeed, the equation in the statement is equivalent to the pair of equalities $d_P = d_L = 2$, by Proposition 1.11, and these in turn characterize generalized quadrangles, because $2 \leq g \leq min(d_P, d_L)$, by Proposition 1.1. □

By essentially the same argument we also obtain the following.

Corollary 1.14 *We have* $d_P = d_L = m$ *if and only if* $\Phi^{L,P}_m = \Phi^{P,L}_m$ *but* $\Phi^{L,P}_{m-1} \neq \Phi^{L,P}_m$ *and* $\Phi^{P,L}_{m-1} \neq \Phi^{P,L}_m$.

Exercise 1.3 (Gonalities). Let us state the following definitions:

$\bar{\Phi}^{P,L}_0 = \bar{\Phi}^{L,P}_0 = \mho$,

$\bar{\Phi}^{P,L}_{m+1} = \Phi^P\bar{\Phi}^{L,P}_m - \bar{\Phi}^{L,P}_m$,

$\bar{\Phi}^{L,P}_{m+1} = \Phi^L\bar{\Phi}^{P,L}_m - \bar{\Phi}^{P,L}_m$.

Prove that the gonality g of Γ is the minimal positive integer m such that $\bar{\Phi}^{P,L}_m \cap \bar{\Phi}^{L,P}_m \neq \emptyset$.

Exercise 1.4. Using Proposition 1.11 and the above, characterize projective planes and generalized quadrangles by means of relations involving Φ^P and Φ^L.

Exercise 1.5. Let F, G be chambers of a generalized m-gon Γ at distance m in $\mathcal{C}(\Gamma)$. Prove that there are precisely two galleries of length m from F to G. They form an ordinary m-gon and have types $PLPL...$ (m factors) and $LPLP...$ (m factors).

Exercise 1.6. Let F, G be as above, but now assume that F and G have distance $k < m$. Prove that there is just one gallery of length k from F to G.

1.4.2 One example of rank 3

We now give some information on the structure of $\mathcal{C}(\Gamma)$ when Γ is a 3-dimensional projective geometry. We denote the three types of Γ by 0, 1, 2, thinking of projective dimensions.

Certain subgeometries of Γ appear to play the role of ordinary m-gons in generalized m-gons (Exercise 1.5). These subgeometries are tetrahedra. They are in fact instances of those substructures called apartments in buildings (to be defined in Chapter 13). However, the reader need not know anything about buildings to understand what follows.

Let Γ be a 3-dimensional projective geometry. Then $\mathcal{C}(\Gamma)$ has diameter 6. Indeed, let $A = \{a_0, a_1, a_2\}$ and $B = \{b_0, b_1, b_2\}$ be two chambers of Γ, where a_i, b_i have dimension $i = 0, 1, 2$, and suppose that A, B have maximal distance in $\mathcal{C}(\Gamma)$. Then $a_0 \notin b_2$. Let c_1 be the line through a_0 and b_0, let c_2 be the plane on c_1 and b_1 and define $C = \{a_0, c_1, c_2\}$, $C' = \{b_0, c_1, c_2\}$ and $C'' = \{b_0, b_1, c_2\}$. Then C, C', C'', B form a gallery of length 3. Since A and B are supposed to have maximal distance, the chambers $A - a_0$ and $C - a_0$ must have maximal distance in the projective plane $\mathcal{C}(\Gamma_{a_0})$. Hence they have distance 3 (Corollary 1.12). Therefore $\mathcal{C}(\Gamma)$ has diameter 6.

The chambers A and B uniquely determine a tetrahedron in Γ with vertices a_0, b_0, $c_0 = b_1 \cap a_2$ and $d_0 = a_1 \cap b_2$. Let $\Gamma(A, B)$ be that tetrahedron. All galleries of length 6 from A to B are contained in $\Gamma(A, B)$. A formal proof of this last claim is a bit awkward; however, a picture is enough for the reader to understand why that claim is true. Exploiting the fact that all galleries of length 6 from A to B are contained in the tetrahedron $\Gamma(A, B)$, it is not hard to prove that distinct galleries of length 6 from A to B have distinct types (see also Section 2.1.3).

Now let A and B be chambers at distance $k < 6$. Then every gallery of length k from A to B is contained in all tetrahedra in Γ containing both A and B. Indeed, in each of those tetrahedra we can choose a chamber C at distance $6-k$ from B and 6 from A. Attaching a gallery of length $6-k$ from B to C to a gallery of length k from A to B we obtain a gallery of length 6 from A to C. The conclusion follows from the above.

Exercise 1.7. It is implicit in the above that the chamber system of a tetrahedron has diameter 6. Prove that the chamber system of an octahedron (dually, of a cube) has diameter 9.

Exercise 1.8. Given a tetrahedron, let us write 0, 1, 2 for 'vertex', 'edge' and 'face', respectively. Prove that the galleries joining pairs of chambers at distance 6 are precisely those whose types are obtained from 012010 applying the following substitution rules (possibly several times, or even no times):

$$010 \longleftrightarrow 101, \qquad 121 \longleftrightarrow 212, \qquad 02 \longleftrightarrow 20.$$

Exercise 1.9. As above, but with an octahedron instead of a tetrahedron. We can now choose 012012012 as type to start from and the substitution rules to use are the following ones:

$$010 \longleftrightarrow 101, \qquad 1212 \longleftrightarrow 2121, \qquad 02 \longleftrightarrow 20.$$

Exercise 1.10. Prove that the chamber system of an n-dimensional projective geometry has diameter $(n+1)n/2$.

(Hint: use an inductive argument, already implicit in the above discussion of the 3-dimensional case.)

Exercise 1.11. Prove that the chamber system of a polar space of rank 3 has diameter 9.

(Hint. If $A = \{a_0, a_1, a_2\}$ and $B = \{b_0, b_1, b_2\}$ are two chambers at maximal distance, then $a_0 \notin b_2$. Let c_2 be the plane on a_0 meeting b_2 in a line ((iv) of 1.3.4), let c_1 be this line, $c_0 = c_1 \cap b_1$ and d_1 be the line joining a_0 with c_0. Then A has distance 4 from $\{a_0, d_1, c_2\}$, whereas a gallery of length 5 from this latter chamber to B is obtained substituting a_0 with c_0, then d_1 with c_1, c_2 with b_2, c_1 with b_1 and, finally, c_0 with b_0.)

Exercise 1.12. Show that two chambers A, B at distance 9 in a polar space of rank 3 uniquely determine an octahedron, containing all galleries of length 9 from A to B.

(Hint. Given chambers $A = \{a_0, a_1, a_2\}$ and $B = \{b_0, b_1, b_2\}$ at distance 9, as above, define $a'_0 = a_1 \cap b_0^{\perp}$, $b'_0 = b_1 \cap a_0^{\perp}$, $a''_0 = a_2 \cap b_1^{\perp}$ and $b''_0 = b_2 \cap a_1^{\perp}$, where $\perp$ means the collinearity relation in the point–line system of the polar space. The points a_0, a'_0, a''_0, b_0, b'_0, b''_0 are the six vertices of the octahedron.)

Exercise 1.13. Let Γ be a classical polar space of rank 3, defined by a reflexive (σ, ε)-sesquilinear form f in a vector space V of dimension $m \geq 6$. Let v_1, v_2, ..., v_m be vectors of V forming a basis and such that the first 6 of them determine an octahedron in Γ as above. Let M be the matrix representing f with respect to that basis. What does M look like ?

1.5 Residues and cells

In this section Γ is still a geometry of rank n, as in Section 1.4, and $\Theta = \{T_1, T_2, ..., T_n\}$ is its type partition.

As in Section 1.4, Φ^i is the i-adjacency relation of $\mathcal{C}(\Gamma)$ (for $i = 1, 2, ..., n$, abbreviating T_i to i). Given $J \subseteq \{1, 2, ..., n\}$, we define $\Phi^J = \bigvee_{j \in J} \Phi^j$. In particular, $\Phi^{\emptyset} = \mho$.

In accord with the notation chosen for adjacency relations, the type partition Θ is identified with the set of indices $\{1, 2, ..., n\}$. Therefore, we can abbreviate our notation further, writing Θ to denote that set of indices.

We can now define $\Phi_i = \Phi^{\Theta - i}$ (for $i \in \Theta = \{1, 2, ..., n\}$) and $\Phi_J = \bigcap_{j \in J} \Phi_j$ (for $J \subseteq \Theta$). In particular, $\Phi_{\emptyset} = \Phi^{\Theta}$ ($= \Omega$, by Corollary 1.10).

Residues of elements have already been defined in Section 1.1. More generally, given a flag F, the neighbourhood Γ_F of F in Γ is called the *residue* of F. Clearly, we have $\Gamma_F = \Gamma$ if and only if $F = \emptyset$ and $\Gamma_F = \emptyset$ if F is a chamber. The corank of F is called the *rank* of Γ_F.

Lemma 1.15 *If F is a non-maximal flag of corank m, then Γ_F is a geometry of rank m with respect to the type partition induced by Θ on Γ_F.*

Proof. By induction on n. If $n = 1$ or $F = \emptyset$, the statement is trivial. Assume $n > 1$ and $F \neq \emptyset$. For a chosen element $x \in F$, the residue of F in Γ is the residue of $F - x$ in Γ_x. The statement follows by the inductive hypothesis. □

Given a non-maximal flag F, let $\mathcal{C}(\Gamma)_F$ be the set of chambers containing F. We call it the *cell* of F.

As Γ_F is a geometry, we may consider the chamber system $\mathcal{C}(\Gamma_F)$ of Γ_F. Clearly, $\mathcal{C}(\Gamma_F)$ can naturally be identified with $\mathcal{C}(\Gamma)_F$. Indeed the chambers of Γ_F are obtained dropping F from the chambers containing F.

1.5.1 Connectedness properties

The following theorem is a trivial consequence of Lemma 1.15:

Theorem 1.16 (Residual connectedness) *For every flag F of corank ≥ 2, the graph Γ_F is connected.*

We also have the following:

Lemma 1.17 *For any two distinct types $T_h, T_k \in \Theta$, the graph induced by Γ on $T_h \cup T_k$ is connected.*

Proof. By induction on n. Of course, we assume $n \geq 2$.

If $n = 2$, the statement is trivial. Assume $n > 2$ and let a,b be distinct elements in $T_h \cup T_k$. Let $a = x_0, x_1, ..., x_m = b$ be a path from a to b in Γ. We must prove that we can choose it inside $T_h \cup T_k$.

We may assume that, for every $i = 1, 2, ..., m$, some of x_{i-1} or x_i is in $T_h \cup T_k$. Otherwise, we can choose a chamber C containing the flag $\{x_{i-1}, x_i\}$ and C meets both T_h and T_k (Lemmas 1.8 and 1.7). If we insert any of the two elements of $C \cap (T_h \cup T_k)$ between x_{i-1} and x_i and we do this for any pair $\{x_{i-1}, x_i\}$ disjoint from $T_h \cup T_k$, then we obtain a path with the required property.

Now let $x_i \notin T_h \cup T_k$. By the previous assumption, both x_{i-1} and x_{i+1} are in $T_h \cup T_k$. By the inductive hypothesis in Γ_{x_i}, we can find a path γ_i in $\Gamma_{x_i} \cap (T_h \cup T_k)$ connecting x_{i-1} with x_{i+1}. Substituting x_i with γ_i, and doing this for all $x_i \notin T_h \cup T_k$, we obtain a path joining a with b inside $T_h \cup T_k$. □

The next theorem is a trivial consequence of this lemma and Lemma 1.15:

Theorem 1.18 (Strong connectedness) *For every flag F of corank at least 2 and for any two distinct types T_h, T_k with $F \cap (T_h \cup T_k) = \emptyset$, the graph induced by Γ on $\Gamma_F \cap (T_h \cup T_k)$ is connected.*

By Theorem 1.9 (and Lemma 1.15) we also have the following:

Theorem 1.19 (Cell connectedness) *For any non-maximal flag F, the cell $\mathcal{C}(\Gamma)_F$ of F is a connected subgraph of $\mathcal{C}(\Gamma)$.*

This theorem can be rephrased in terms of adjacency relations, as we have done for Theorem 1.9 in Corollary 1.10:

Corollary 1.20 *For every flag F, we have $\mathcal{C}(\Gamma)_F = [C]\Phi^J$, where C is any chamber containing F.*

Namely, the cells of the flags of Γ are precisely the equivalence classes of the relations Φ^J ($J \subseteq \Theta$). In particular, for every $i \in \Theta$, the equivalence classes of Φ_i are the cells of the elements of Γ in the type T_i.

The next corollary is another way to say the above, without mentioning flags of Γ explicitly.

Corollary 1.21 *For every $J \subseteq \Theta$, we have $\Phi_J = \Phi^{\Theta - J}$.*

Indeed, for every flag F, we have $\mathcal{C}(\Gamma)_F = \bigcap_{x \in F} \mathcal{C}(\Gamma)_x$. The conclusion follows from Corollary 1.20.

In particular, we have $\Phi_I = \Phi^{\emptyset} = \mho$.

1.5.2 Reconstructing Γ from $(\Phi^i)_{i \in \Theta}$

Let $\mathcal{K}_{\mathcal{C}}(\Gamma)$ be the set of equivalence classes of the relations Φ^J $(J \subset \Theta)$, ordered by reverse inclusion. By Corollary 1.20, $\mathcal{K}_{\mathcal{C}}(\Gamma)$ is the system of cells of (nonempty) flags of Γ. Let $\mathcal{K}(\Gamma)$ be the flag complex of Γ, as in Section 1.4. We have the following:

Lemma 1.22 $\mathcal{K}(\Gamma) \cong \mathcal{K}_{\mathcal{C}}(\Gamma)$.

Proof. Indeed, given two flags F, G, we have $F \subseteq G$ if and only if $\mathcal{C}(\Gamma)_F \supseteq \mathcal{C}(\Gamma)_G$. The 'only if' part of this claim is trivial. Let us prove the 'if' part. Let $F \not\subseteq G$ and $x \in F - G$.

If $F \cup G$ is not a flag, then $\mathcal{C}(\Gamma)_F \cap \mathcal{C}(\Gamma)_G = \emptyset$ and, trivially, $\mathcal{C}(\Gamma)_G \not\subseteq \mathcal{C}(\Gamma)_F$.

Otherwise, let $C \in \mathcal{C}(\Gamma)_F \cap \mathcal{C}(\Gamma)_G$. By Lemma 1.8, there is a chamber $D \neq C$ containing the panel $C - x$. We have $x \not\in D$, since $D \neq C$. Therefore, $D \not\in \mathcal{C}(\Gamma)_F$. However, $D \subseteq C - x$. Hence $D \in \mathcal{C}(\Gamma)_G$. Again, $\mathcal{C}(\Gamma)_G \not\subseteq \mathcal{C}(\Gamma)_F$. □

It follows from this lemma that the incidence graph of Γ is also implicit in $\mathcal{K}_{\mathcal{C}}(\Gamma)$. Indeed, let $\Gamma(\mathcal{C})$ be the graph having the cells of elements of Γ as vertices, two such cells being adjacent in $\Gamma(\mathcal{C})$ precisely when they have nonempty intersection. By Lemma 1.22, $\Gamma(\mathcal{C})$ is isomorphic to the vertex–edge system of the complex $\mathcal{K}(\Gamma)$. The latter is in turn nothing but the incidence graph of Γ.

However, it is not yet clear from the above if the type partition Θ of Γ can also be recovered from the complex $\mathcal{K}_{\mathcal{C}}(\Gamma)$. Actually it can, but we shall prove this only at the end of this chapter. For the moment, we only remark that Θ can anyway be reconstructed from the system $(\Phi^i)_{i \in \Theta}$ of adjacency relations of $\mathcal{C}(\Gamma)$. Indeed, the equivalence classes of Φ_i are just the cells of the elements of Γ of type T_i, by Corollary 1.20. We may summarize the above as follows:

Theorem 1.23 *The coloured graph $(\Phi^i)_{i \in \Theta}$ uniquely determines the geometry Γ.*

1.5.3 Examples

We now describe some examples of residues of flags. We only consider polar spaces and buildings of type D_n (1.3.6), but it will be clear that quite similar remarks could be made for other cases, too (for matroids, for instance).

Given a polar space $\mathcal{P}$ of rank $n > 2$ and a subspace X of $\mathcal{P}$ of projective dimension $h = 1, 2, ..., n-2$, let $\{X_0, X_1, ..., X_{n-1}\}$ be any chamber containing X, with X_i of dimension i ($i = 0, 1, ..., n-1$; in particular, $X_h = X$). Let us set $F^- = \{X_0, X_1, ..., X_h\}$ and $F^+ = \{X_h, X_{h+1}, ..., X_{n-1}\}$. The two parts $\mathcal{P}_X^-$ and $\mathcal{P}_X^+$ of the residue $\mathcal{P}_X$ of X (see Section 1.3.4) are the residues of the flags F^+ and F^-, respectively. In particular, the residue of F^+ is a projective geometry and the residue of F^- is a polar space.

Let X, Y be subspaces of $\mathcal{P}$ of dimensions h and k, respectively, with $h < k$ and $X \subset Y$. Let C be any chamber containing both X and Y and let $F = C - \{X, Y\}$. The residue of F is a generalized digon if $h + 1 < k$, a projective plane if $h + 1 = k < n - 1$ and a generalized quadrangle if $h + 1 = k = n - 1$. Let $F = \{Z \mid X \subseteq Z \subseteq Y, Z \in C\}$ and $G = C - F$. Then the residue $\mathcal{P}_G$ of G is a $(k-h+1)$-dimensional projective geometry, whereas, if $0 < h$ and $k < n - 1$, then $\mathcal{P}_F$ splits into two parts $\mathcal{P}_F^-$ and $\mathcal{P}_F^+$, just as for residues of single subspaces of $\mathcal{P}$ of dimension $\neq 0$ or $n-1$. Furthermore, we have $\mathcal{P}_F^- = \mathcal{P}_X^-$ and $\mathcal{P}_F^+ = \mathcal{P}_Y^+$.

Turning to buildings of type D_n, let Γ be a building of type D_n over a field K. Let $\{X_0, X_1, ..., X_{n-3}, X^+, X^-\}$ be a chamber, where $dim(X_i) = i$ ($i = 0, 1, ..., n-3$) and $X^\pm \in M^\pm$ and let $F_h = \{X_0, X_1, ..., X_h\}$ and $G = \{X^+, X^-\}$. The residue of the flag F_h is the D_{n-h-1}-building for K if $h < n - 4$; it is $PG(3, K)$ when $h = n - 4$; it is a complete bipartite graph if $h = n - 3$. The residue of G is $PG(n-2, K)$. The residue of $F_{n-3} - X_0$ is a complete 3-partite graph.

1.6 Truncations

Let Γ be a rank n geometry and $\Theta = \{T_1, T_2, ..., T_n\}$ its type partition, as in the previous section. Let J be a proper subset of Θ.

If we drop all elements of Γ of type $T_i \in J$, then what is left is still a geometry, by the strong connectedness of Γ (Theorem 1.18). We denote this geometry by $Tr_J^-(\Gamma)$ and we call it the *J_--truncation* of Γ. Clearly, $Tr_J^-(\Gamma)$ is a geometry of rank $n - m$, where $m = |J|$, and $\Theta - J$ is its type partition. The chamber system of $Tr_J^-(\Gamma)$ consists of the flags of Γ of type $\Theta - J$, two such flags F, G being adjacent as chambers of $Tr_J^-(\Gamma)$ precisely when $F \cap G$ has rank $n - m - 1$.

Of course, we have $Tr^-_{\emptyset}(\Gamma) = \Gamma$.

Let F be a flag of Γ having a proper subset of $\Theta - J$ as type. Then we may consider both the J_--truncation $Tr^-_J(\Gamma_F)$ of the residue of F in Γ, and the residue $(Tr^-_J(\Gamma))_F$ of F in the J_--truncation of Γ. It is easily seen that $Tr^-_J(\Gamma_F) = (Tr^-_J(\Gamma))_F$. This allows us to write $Tr^-_J(\Gamma)_F$ instead of $Tr^-_J(\Gamma_F)$ or $(Tr^-_J(\Gamma))_F$, without any danger of ambiguity.

We remark that the connectedness of $\mathcal{C}(Tr^-_J(\Gamma)_F)$ (Theorem 1.19 applied to $Tr^-_J(\Gamma)$) gives us a generalized version of strong connectedness. In particular, when F has rank $n-m-2$, the connectedness of $\mathcal{C}(Tr^-_J(\Gamma)_F)$ is just the strong connectedness, as stated in Theorem 1.18.

When J is a nonempty subset of Θ, then it is sometimes convenient to use a notation opposite to the above, defining the *J_+-truncation* $Tr^+_J(\Gamma)$ as follows: $Tr^+_J(\Gamma) = Tr^-_{\Theta-J}(\Gamma)$. Needless to say, $Tr^+_{\Theta}(\Gamma) = \Gamma$.

1.6.1 Examples

(1) Let $\mathcal{M}$ be a matroid of dimension $n \geq 3$. The point–line system of $\mathcal{M}$ is the $\{0,1\}_+$-truncation of $\mathcal{M}$, where dimensions are taken as types. We have already remarked in Section 1.3.2 that the point–line system of $\mathcal{M}$ is a 2-dimensional matroid (that is, a linear space). More generally, if $0 < i < n$ and we drop subspaces of $\mathcal{M}$ of dimension $\geq i$, then we make a truncation of $\mathcal{M}$, which is again a matroid (of dimension i).

(2) Let $\mathcal{M}$ be as above and X be an i-dimensional subspace of $\mathcal{M}$, with $0 < i < n-1$. We know that the residue $\mathcal{M}_X$ of X splits into two parts, $\mathcal{M}^-_X$ and $\mathcal{M}^-_X$ (see Section 1.3.2). Both these parts are truncations of $\mathcal{M}_X$, as well as residues of flags containing X of type $\{i, i+1, ..., n-1\}$ and $\{0, 1, ..., i\}$ respectively (compare with Section 1.5.3). Of course, similar remarks can be made for other geometries where residues of elements of an 'intermediate' type split in two parts, as above (for polar spaces, for instance).

(3) Let Γ be a n-dimensional projective geometry, with $n \geq 5$. Choose a type i with $1 < i < n-2$. The residue in the $\{i-1, i, i+1\}_+$-truncation of Γ of an $(i-1)$-dimensional subspace (of an $(i+1)$-dimensional subspace) is the point–line system of a $(n-i)$-dimensional projective geometry (the dual point–line system of a $(i+1)$-dimensional projective geometry).

(4) Let $\Gamma = PG(n,q)$, with $n \geq 3$. The $\{0, n-1\}_+$-truncation of Γ is a symmetric 2-(v,k,λ)-design, with $v = (q^{n+1}-1)/(q-1)$, $k = (q^n-1)/(q-1)$ and $\lambda = (q^{n-1}-1)/(q-1)$.

(5) Let Γ be the building of type D_n for a field K (Section 1.3.6) and let $J = \{M^+, T_{n-3}, M^-\}$, where T_{n-3} is the set of $(n-3)$-dimensional subspaces. Then the residues in $Tr_J^+(\Gamma)$ of elements of M_+ (of M_-) are isomorphic to the point–line system of $PG(n-1, K)$, where elements of M_- (of M_+) play the role of points.

(6) Let Γ be as above, but now assume $n = 4$ and $K = GF(q)$ and consider the $\{T_1\}_-$-truncation of Γ. Residues of elements of this truncation are symmetric 2-$((q^2+1)(q+1), q^2+q+1, q+1)$-designs.

1.6.2 Point-line systems and subspaces

It happens sometimes that certain truncations of rank 2 contain enough information for recovering the larger geometry Γ which they are truncations of, or at least for recognizing many properties of Γ. The elements of such a truncation are often called *points* and *lines* of Γ and the truncation itself is called the *point–line system* of Γ.

This is nothing but a linguistic convention, of course; however, in many cases certain choices of points and lines look most natural. For instance, if Γ is projective or affine geometry, or any other matroid, or a polar space, then the usual points and lines of Γ form the natural point–line system of Γ.

Let $\mathcal{L}(\Gamma)$ be a rank 2 truncation of a geometry Γ, chosen as the point–line system of Γ. Then all concepts and notations stated in Section 1.2.3 for rank 2 geometries may be referred to Γ via $\mathcal{L}(\Gamma)$. Thus, the collinearity relation $\perp$, the collinearity graph and the diameter of the collinearity graph of $\mathcal{L}(\Gamma)$ are taken as the *collinearity relation*, *collinearity graph* and *collinearity diameter* of Γ, respectively (relatively to that choice of points and lines).

When $\mathcal{L}(\Gamma)$ is a partial plane, then we can also define subspaces: a (proper) subset S of the set of points of $\mathcal{L}(\Gamma)$ is called a (*proper*) *subspace* of $\mathcal{L}(\Gamma)$ if it contains every line meeting S in more than one point. (Needless to say, this definition can be stated for arbitrary partial planes, even when they are not point–line systems of geometries of higher rank.)

We warn the reader that subspaces of $\mathcal{L}(\Gamma)$ need not represent elements of the geometry Γ, in general. For instance, if Γ is a matroid obtained by truncating a higher dimensional matroid Γ' as in Section 1.6.1(1) and $\mathcal{L}(\Gamma)$, $\mathcal{L}(\Gamma')$ are the natural point–line systems of Γ and Γ', then $\mathcal{L}(\Gamma) = \mathcal{L}(\Gamma')$ and all elements of Γ' are subspaces of $\mathcal{L}(\Gamma)$; however, not all of them are saved in Γ. If Γ is the enrichment of a Steiner system $S(t, k, v)$ (see Section 2.4) or an affine geometry with lines of size 2, then its natural point–line system $\mathcal{L}(\Gamma)$ is a complete graph having the points of Γ as vertices. All subsets of the set of points of Γ are subspaces for $\mathcal{L}(\Gamma)$, whereas not all of them are subspaces for the matroid Γ. On the other hand, the subspaces of a projective geometry are precisely the subspaces of its natural point–line system. The same holds for affine geometries with lines of size > 2.

If Γ is a polar space and $\mathcal{L}(\Gamma)$ is its natural point–line system, then the elements of the geometry Γ are precisely those subspaces of $\mathcal{L}(\Gamma)$ that consist of pairwise collinear points (see [28] and [191]). However, the partial plane $\mathcal{L}(\Gamma)$ admits many subspaces which are not subspaces (elements) of the polar space (geometry) Γ. For instance, every coclique of the collinearity graph of $\mathcal{L}(\Gamma)$ with at least two points is a subspace of $\mathcal{L}(\Gamma)$, but it is not a subspace of Γ.

When there is some danger of confusion between subspaces of a polar space Γ (namely, elements of the geometry Γ) and subspaces of its point–line system $\mathcal{L}(\Gamma)$, the former are called *singular* subspaces. However, we will seldom deal with subspaces of $\mathcal{L}(\Gamma)$. Indeed, we omit the point-line theory of polar spaces from the scope of this book (the reader interested in that theory may see [28], [191] or [25]). Thus, we may safely keep the terminology stated for polar spaces in Section 1.3.5.

Exercise 1.14. Let d be the collinearity diameter of Γ (with respect to some choice of points and lines). Prove that the incidence graph of Γ has diameter $\leq 2d+2$.

Exercise 1.15. Let $\perp$ denote the collinearity relation of a polar space $\mathcal{P}$, with respect to the natural point–line system of $\mathcal{P}$. Let X, Y, Z be as in (iv) of Section 1.3.5. Prove that $X \cap Z = X \cap Y^{\perp}$.

Exercise 1.16 (Buekenhout–Shult property). Let $\mathcal{P}$ and $\perp$ be as above and let (a, x) be a non-incident point–line pair of $\mathcal{P}$.

Prove that either a and x are coplanar, or $|a^{\perp} \cap x| = 1$.

(Hint: choose a maximal subspace X on x. If X can be chosen so as to contain a, then we are done. Otherwise, $a^{\perp} \cap X$ is a hyperplane in X (Exercise 1.15) and $(a^{\perp} \cap X) \cap x = a^{\perp} \cap x$ is a single point.)

This property of polar spaces is often called the *Buekenhout-Shult property*, after [28]. As a consequence of it, the collinearity diameter of a polar space is 2. Note that, in the rank 2 case, the Buekenhout–Shult property is just $(*)$ of Section 1.2.4.

Exercise 1.17. Let $\mathcal{P}$ be a polar space of rank n, but now take as points and lines the maximal subspaces and the subspaces of projective dimension $n-2$, respectively. Prove that the collinearity diameter is now n.

(Hint: use induction, exploiting (iv) to go back to case of rank $n-1$.)

1.7 Types

In this section Γ is a rank n geometry and $\Theta = \{T_1, T_2, ..., T_n\}$ is its type partition, as in the previous section.

1.7.1 Getting rid of types

Let T be the relation defined on the set of elements of Γ by the following clause: we have xTy if and only if there is a panel F such that $F \cup x$ and $F \cup y$ are adjacent chambers.

Let $\overline{T}$ be the transitive closure of T. Identifying the relation $\overline{T}$ with the partition defined by it on the set of elements of Γ, we have the following:

Lemma 1.24 $\overline{T} = \Theta$.

Proof. It is clear that $\overline{T} \leq \Theta$ (Lemma 1.7). Conversely, let x, y be elements of Γ in the same type, say T_1. Let C, D be chambers containing x and y, respectively, and let $C = C_0, C_1, ..., C_m = D$ be a gallery from C to D. Let x_i be the element of C_i in T_1 (Lemma 1.7); in particular, $x_0 = x$ and $x_m = y$. If we show that, for every $i = 1, 2, ..., m$, either $x_{i-1} = x_i$ or $x_{i-1}Tx_i$, then $x\overline{T}y$ and $\Theta \leq \overline{T}$ follows (hence $\Theta = \overline{T}$).

Let $x_{i-1} \neq x_i$. Then $\{x_{i-1}, x_i\} = (C_{i-1} \cup C_i) - (C_{i-1} \cap C_i)$, whence $x_{i-1}Tx_i$, as we wished to prove. □

Recalling that chambers are just maximal cliques of Γ and that the definition of $\overline{T}$ does not mention Θ at all, we see we have just proved the following:

Theorem 1.25 *The type partition of a geometry is uniquely determined by the incidence graph of that geometry.*

Therefore, we might define geometries without mentioning types at all (see Exercise 1.19 in the next section).

The 'ancestor' of the above theorem is the definition of types for buildings stated by Tits in [222] (n.3.8).

1.7.2 Reinstating types

However, in spite of the above, we will put even more emphasis on types, assuming that they were given independently since the beginning as a set I of indices (like $\{0, 1, ..., n\}$ or $\{1, 2, ..., n\}$), or of names (like 'point', 'line', 'plane'; or 'vertex', 'edge', 'face'), or in some other way. Actually, this is precisely what we have already tacitly done many times in the previous sections.

Thus we say that a finite set I is a *set of types* for a geometry (Γ, Θ) if a bijection t is given from the type partition Θ of Γ to I. If such a pair (I, t) is given, then we say that Γ is a *geometry over the set of types* I with *type function* t. By abuse of notation, the bijection $t : \Theta \longrightarrow I$ will always be identified with the function from the set of elements of Γ onto I mapping every element x of Γ onto the element $t([x]\Theta)$ of I, where $[x]\Theta$ denotes the class of Θ containing x (namely, the type which x belongs to). Consistently with this, we shall always write $t(x)$ instead of $t([x]\Theta)$ and we say that the element $t(x)$ of I is the *type* of x. Similarly, given a flag F, we say that the subset $t(F)$ of I is the *type* of F and that $I - t(F)$ is the *cotype* of F and the *type* of the residue Γ_F of F.

Chambers are precisely flags of type I (Lemma 1.7).

Of course, the notation stated for adjacencies of chambers, types of galleries and truncations will also be changed consistently with the above (actually, we really have nothing to change there, apart from substituting I for the symbol Θ: we have simply another way to interpret that notation).

Exercise 1.18. Let I be a finite set, $\Gamma = (V, \sim)$ a graph and $t : V \longrightarrow I$ a mapping from V to I. Prove that (Γ, t) is a geometry over the set of types I if and only if all the following hold:

(i) every non-maximal clique of Γ is contained in at least two maximal cliques;
(ii) a clique F of Γ is maximal if and only if t induces a bijection from F to I;
(iii) for every clique F of size $|F| \leq |I| - 2$, the neighbourhood of F in Γ is connected.

Exercise 1.19 (A type-free definition of geometries). Let Γ be a connected graph and define a relation T on the set of vertices of Γ stating that, for any two vertices x, y, we have xTy precisely when there is a clique F such that both $F \cup x$ and $F \cup y$ are maximal cliques. Let $\overline{T}$ be the transitive closure of T. For every non-maximal clique F, let $\overline{T}_F$ be the relation defined in the neighbourhood Γ_F of F in the same way as $\overline{T}$ has been defined in Γ, but only using maximal cliques containing F.

Prove that Γ is geometry of rank n if and only if all the following hold:

(i) every non-maximal clique of Γ is contained in at least two maximal cliques;
(ii) all maximal cliques of Γ have size n;
(iii) every maximal clique of Γ meets eevery equivalence classe of $\overline{T}$ in

precisely one vertex;
(iv) for every non-maximal clique F of Γ, $\overline{T}_F$ is the relation induced by $\overline{T}$ on the neighbourhood Γ_F of F.

(Hint. The graph Γ is connected by assumption. Conditions (i)–(iii) define an n-partition of Γ. Next, (iv) forces adjacencies of cliques of size $< n-1$ to inherit the connectedness of Γ.)
(Remark. This definition of geometries is implicit in [24].)

Problem. Does the graph structure of $\mathcal{C}(\Gamma)$ uniquely determine Γ ? Note that we are now considering $\mathcal{C}(\Gamma)$ just as a graph, without splitting its adjacency relation into the i-adjacencies Φ^i. We are forgetting the complex $\mathcal{K}_{\mathcal{C}}(\Gamma)$ and are forbidden to use the information that the vertices of $\mathcal{C}(\Gamma)$ are in fact the chambers of Γ (otherwise, we already know from Section 1.5.2 that the answer would be affirmative).

Note that, if n is the rank of Γ, then, for every chamber C, the neighbourhood of C in $\mathcal{C}(\Gamma)$ is the disjoint union of n complete graphs, corresponding to the n equivalence classes $[C]\Phi^i$. Therefore, every equivalence class of i-adjacency relations is recognizable in $\mathcal{C}(\Gamma)$. The problem is to show that there is just one consistent way of gathering those classes into n distinct partitions of the set of vertices of $\mathcal{C}(\Gamma)$.

The answer to our question is affirmative when $n = 2$. Indeed, in this case, every circuit of $\mathcal{C}(\Gamma)$ has even length, whence there is just one way of grouping the above classes into two partitions.

1.8 Appendix I

In this Appendix we collect some information on sesquilinear and quadratic forms, referring to [10] (Chapter 9) and [222] (Chapter 8).

Henceforth K will be a division ring, V a left K-vector space, σ an anti-automorphism of K (see Section 0.6) and ε an element of the multiplicative group K^* of K such that $\sigma(\varepsilon) = \varepsilon^{-1}$ and $\sigma^2(x) = \varepsilon^{-1}x\varepsilon$ for all $x \in K$.

1.8.1 Sesquilinear forms

A function $f : V \times V \longrightarrow K$ is a *reflexive* (σ, ε)-*sesquilinear form* if the following hold:

(i) (*additivity*) for all $x, y, z \in V$,

$$f(x + y, z) = f(x, z) + f(y, z)$$

(ii) (*homogeneity*) for all $x, y \in V$ and $t \in K$,

$$f(tx, y) = tf(x, y)$$

(iii) (*reflexivity*) for all $x, y \in V$,

$$f(x, y) = \varepsilon \cdot \sigma(f(y, x))$$

The form f is said to be *trace valued* if $f(x, x) \in \{t + \varepsilon\sigma(t) \mid t \in K\}$ for all $x \in V$. We remark that non-trace valued forms can occur only when $char(K) = 2$, with σ inducing the identity on the centre of K (see [10] or [222]).

Given two vectors $x, y \in V$, we write $x \perp y$ to mean that $f(x, y) = 0$ (equivalently, $f(y, x) = 0$, by the reflexivity of f). The reader may be disturbed to see a new meaning for $\perp$, different from the one stated in Sections 1.2.1 and 1.2.3. However, this ambiguity will be quite innocuous in practice, as the reader will soon realize.

A vector $x \in V$ is said to be *isotropic* for f if $x \perp x$. Given a vector x, $x^\perp$ will denote the linear subspace $\{y \mid y \perp x\}$ of V and, given a set of vectors X, we define $X^\perp = \bigcap(x^\perp \mid x \in X)$. A linear subspace X of V is said to be *totally isotropic* if $X \subseteq X^\perp$. Trivially, all linear subspaces of a totally isotropic subspace are totally isotropic. In particular, $\{0\}$ is always a totally isotropic subspace (the *trivial* one).

The maximal totally isotropic subspaces have the same dimension, called the *Witt index* of f ([10], Chapter 9; we warn that the Witt index of f need not be finite).

The linear subspace $V^\perp$ is called the *radical* of f and denoted by $Rad(f)$. Trivially, $Rad(f)$ is contained in the intersection of all maximal totally isotropic subspaces. The form f is said to be *non-degenerate* if $Rad(f) = 0$.

Let f be non-degenerate, trace valued and with Witt index $n > 0$ (that is, some non-zero vectors of V are isotropic for f). Then all the following hold ([10], Chapter 9):

(iv) the maximal totally isotropic subspaces generate V;

(v) given a maximal totally isotropic subspace X and a totally isotropic subspace Y, we have:

$$dim(Y) + dim(Y^\perp \cap X) = dim(X) + dim(X \cap Y)$$

(vi) for every non-maximal totally isotropic subspace X and every maximal totally isotropic subspace Y containing X, there is a maximal totally isotropic subspace Z such that $Y \cap Z = X$.

The following are easy consequences of (vi):

(vii) every non-maximal totally isotropic subspace is the intersection of two maximal totally isotropic subspaces;

(vii) for every maximal totally isotropic subspace X there is a maximal totally isotropic subspace Y such that $X \cap Y = \{0\}$;

(ix) we have $2n \leq dim(V)$, where n is the Witt index of f (note that $2n = n$ if n is infinite; the inequality $2n \leq dim(V)$ is trivial in this case).

1.8.2 Examples

We will now give a few examples of non-degenerate trace valued reflexive sesquilinear forms in a finite dimensional vector space $V(m, K)$ over a field K. The vectors x, y, etc. of $V(m, K)$ will be written as $x = (x_1, x_2, ..., x_m)$, $y = (y_1, y_2, ..., y_m)$, etc., as usual, with respect to a suitable basis of $V(m, K)$.

(1) *Alternating bilinear forms* in $V(2n, K)$, with K a field, $\varepsilon = -1$, $\sigma = id$ and Witt index n. The form f can be expressed as follows, with respect to a suitable basis of $V(2n, K)$:

$$f(x, y) = \sum_{i=1}^{n} (x_{2i-1}y_{2i} - x_{2i}y_{2i-1})$$

(2) *Symmetric bilinear forms* of Witt index n in $V(2n, K)$, with K a field of characteristic $\neq 2$, $\varepsilon = 1$ and $\sigma = id$. The expression for f, with respect to a suitable basis of $V(2n, K)$, is

$$f(x, y) = \sum_{i=1}^{n} (x_{2i-1}y_{2i} + x_{2i}y_{2i-1})$$

(3) *Symmetric bilinear forms* of Witt index n in $V(2n+1, K)$, with K a field of characteristic $\neq 2$, $\varepsilon = 1$ and $\sigma = id$. The form f can be expressed as follows, up to a proportionality constant $a \in K^*$ (of course, this proportionality constant has no effect on the orthogonality relation $\perp$):

$$f(x, y) = x_1y_1 + \sum_{i=1}^{n} (x_{2i}y_{2i+1} + x_{2i+1}y_{2i})$$

(4) *Symmetric bilinear forms* of Witt index n in $V(2n+2, q)$, q odd, $\varepsilon = 1$ and $\sigma = id$. The form f can be expressed as follows (up to a proportionality

constant):

$$f(x,y) = x_1y_1 - bx_2y_2 + \sum_{i=2}^{n+1}(x_{2i-1}y_{2i} + x_{2i}y_{2i-1})$$

where b is a non-square element of $GF(q)$. As we have assumed q odd, the group of non-zero square elements of $GF(q)$ has index 2 in the multiplicative group Z_{q-1} of $GF(q)$. Therefore, given any non-square element b of $GF(q)$, we can always find a basis of $V(2n+2,q)$ with respect to which f is (proportional to a form) expressed as above, with that element b as the coefficent of x_2y_2.

(5) *Hermitian forms* of Witt index n in $V(m,q^2)$, $m = 2n$ or $2n+1$, $\varepsilon = 1$ and σ acting as follows: $\sigma(t) = t^q$ for all $t \in K$. The expression for f is

$$f(x,y) = \sum_{i=1}^{m} x_i y_i^q$$

(when $m = 2n+1$, this expression is given up to a proportionality constant).

The next theorem is well known [10]:

Theorem 1.26 *The five types of forms listed above give us all possible non-degenerate reflexive (σ,ε)-sesquilinear forms in finite dimensional vector spaces over $GF(q)$, q odd. Alternating bilinear forms and Hermitian forms are the only non-degenerate trace valued (σ,ε)-sesquilinear forms in finite dimensional vector spaces over $GF(q)$, q even.*

Note that, if $char(K) = 2$, then the expressions we have used above to define symmetric bilinear forms of Witt index n in $V(2n,K)$ and alternating bilinear forms are the same. We have assumed $char(K) \neq 2$ in the definition of symmetric bilinear forms of Witt index n in $V(2n,K)$ to keep them distinct from alternating bilinear forms.

If $char(K) = 2$, then the expression we have used to define symmetric bilinear forms of Witt index n in $V(2n+1,K)$ and $V(2n+2,K)$ define degenerate forms. Furthermore, no non-square elements exist in $GF(q)$ when q is even. Hence the expression we have used for symmetric bilinear forms of Witt index n in $V(2n+2,q)$ is undefined when q is even.

1.8.3 Quadratic forms

Let $char(K) = 2$ and set $K_{\sigma,\varepsilon} = \{t - \varepsilon\sigma(t) \mid t \in K\}$. Of course, as we have assumed that $char(K) = 2$, we might write $+$ instead of $-$ in the above definition, but the choice of the $-$ sign will later allow us to extend the following to the case of $char(K) \neq 2$.

It is easily seen that $K_{\sigma,\varepsilon}$ is a subgroup of the additive group of K. We set $K^{\sigma,\varepsilon} = K/K_{\sigma,\varepsilon}$.

Let h be a (possibly non-reflexive) (σ,1)-*sesquilinear form*, namely a mapping $h : V \times V \longrightarrow K$ satisfying (i) and (ii) of Section 1.8.1, with 1 in place of ε. Let $g : V \longrightarrow K^{\sigma,\varepsilon}$ be the mapping defined as follows: $g(x) = h(x,x) + K_{\sigma,\varepsilon}$. We say that g is a (σ,ε)-*quadratic form*.

Let us set $f(x,y) = h(x,y) + \varepsilon h(x,y)$. Then f is a trace valued reflexive (σ,ε)-sesquilinear form and we have $g(x+y) - g(x) - g(y) = f(x,y) + K_{\sigma,\varepsilon}$ for every choice of vectors $x, y \in V$. The form f is uniquely determined by g ([10]; also [222], Chapter 8) and it is called the *reflexive sesquilinear form associated with* g, or the *sesquilinearization* of g.

A vector $x \in V$ is said to be *singular* for g if $g(x) = 0$. A linear subspace X of V is said to be *totally singular* for g if $g(x) = 0$ for all $x \in X$. Every vector singular for g is also isotropic for the sesquilinearization f of g and two vectors x, y singular for g are orthogonal with respect to f if and only if they span a subspace totally singular for g ([10]; also [222], Chapter 8). Hence every linear subspace of V totally singular for g is also totally isotropic for f.

The maximal totally singular subspaces have the same dimension ([10]); also [222], Chapter 8), called the *Witt index* of g. Trivially, the Witt index of g is less than or equal to the Witt index of f.

The set of vectors $\{x \in Rad(f) \mid g(x) = 0\}$ is a linear subspace of V. We call this subspace the *radical* of g and we denote it by $Rad(g)$. The quotient space $Rad(f)/Rad(g)$ and its dimension are called the *nucleus* and the *defect* of g, respectively. We denote the defect of g by the symbol $d(g)$. The (σ,ε)-quadratic form g is said to be *non-singular* if $Rad(g) = 0$.

The analogues of properties (v)–(ix) of Section 1.8.1 hold for a non-singular quadratic form g of Witt index $n > 0$, provided we substitute the word 'singular' for 'isotropic' in those statements, denoting by $\perp$ the orthogonality relation of the sesquilinearization of g.

When $\varepsilon = 1$, g need not be uniquely determined by its sesquilinearization f (see [10]; also [222], Chapter 8). It may also happen that some of the totally isotropic subspaces for f are not totally singular for g. For instance, it may happen that g is non-singular but f is degenerate. In this case, $Rad(f)$ is totally isotropic for f but not totally singular for g. It may also happen that even though f is non-degenerate (hence g is non-singular), not all totally isotropic subspaces for f are totally singular for g (see the first example of the Section 1.8.4).

These pathological phenomena do not occur when $\varepsilon \neq 1$. Indeed, in this case we have $h(x,y) = (1+\varepsilon)^{-1} f(x,y)$ for all $x, y \in V$. Hence $g(x) =$

$(1+\varepsilon)^{-1}f(x,x)+K_{\sigma,\varepsilon}$ for all $x \in V$. Thus, g and f are interchangeable in this case.

(σ,ε)-quadratic forms can be defined also when $char(K) \neq 2$, provided that $(\sigma,\varepsilon) \neq (id,-1)$. However, in that case every (σ,ε)-quadratic form g is uniquely determined by its sesquilinearization f and the totally singular subspaces for g are the totally singular subpaces for f (see [222]). Therefore, (σ,ε)-quadratic forms and trace valued reflexive (σ,ε)-sesquilinear forms are interchangeable when $char(K) \neq 2$.

The above claim is quite evident when $\varepsilon \neq -1$. Indeed, in this case we have $g(x) = (1+\varepsilon)^{-1}f(x,x)+K_{\sigma,\varepsilon}$, for all $x \in V$ (compare with the case of $\varepsilon \neq 1$ when $char(K) = 2$).

1.8.4 Examples

We now give some examples of non-singular (σ,ε)-quadratic forms in a finite dimensional vector space V over a field K of characteristic 2. The expressions we give for g are relative to a suitable choice of the basis of V. In the first three examples we have $\sigma = id$ and $\varepsilon = 1$. Hence $K_{\sigma,\varepsilon} = 0$ in these cases, g maps V into K and we have $f(x,y) = g(x+y)-g(x)-g(y)$.

(1) *Quadratic forms* of Witt index n in $V(2n,K)$, K a field of characteristic 2, $\varepsilon = 1$, $\sigma = id$, $d(g) = 0$. Sesquilinearization: the non-degenerate alternating bilinear form f of Witt index n (not all totally isotropic subspaces for f are totally singular for g). The expression for g is

$$g(x) = \sum_{i=1}^{n} x_{2i-1}x_{2i}$$

(2) *Quadratic forms* of Witt index n in $V(2n+1,K)$, K a field of characteristic 2, $\varepsilon = 1$, $\sigma = id$, $d(g) = 1$. Sesquilinearization: the degenerate alternating bilinear form with radical of dimension 1 and maximal totally isotropic subspaces of dimension $n+1$. Up to a proportionality constant $a \in K^*$, the expression for g is

$$g(x) = x_1^2 + \sum_{i=1}^{n} x_{2i}x_{2i+1}$$

(3) *Quadratic forms* of Witt index n in $V(2n+2,q)$, q even, $\varepsilon = 1$, $\sigma = id$, $d(g) = 2$. Sesquilinearization: the degenerate alternating bilinear form with radical of dimension 2 and maximal totally isotropic subspaces of dimension $n+2$. Up to a proportionality constant, g can be given the

following expression, where $b \in GF(q)$ is such that the polynomial $p_b(t) = t^2 + bt + 1 = 0$ is irreducible in $GF(q)$:

$$g(x) = x_1^2 + bx_1x_2 + x_2^2 + \sum_{i=2}^{n+1} x_{2i-1}x_{2i}$$

Given $b, u \in GF(q)$ such that $p_b(u) \neq 0$, set $v = p_b(u)^{q/2}$ and $c = bv^{-1}$. Then $p_c(t) = v^{-2}p_b(vt + u)$, hence $p_c(t)$ is irreducible in $GF(q)$ if and only if $p_b(t)$ is irreducible in $GF(q)$. It follows from this remark that just $q/2$ of the q polynomials $p_b(t)$ ($b \in GF(q)$) are irreducible and that, if $p_b(t)$ and $p_c(t)$ are irreducible, then $p_c(t) = v^{-2}p_b(vt + u)$ for two choices of $u \in GF(q)$ and with $v = p_b(u)^{q/2}$.

Therefore, given any element $b \in GF(q)$ such that $p_b(t)$ is irreducible in $GF(q)$, we can always find a basis of $V(2n+2, q)$ with respect to which the quadratic form g is (proportional to a form) expressed as above, with that element b as the coefficent of x_1x_2.

(4) *Hermitian forms* of Witt index n in $V(m, q^2)$, $m = 2n$ or $2n+1$, $\varepsilon = 1$, σ is the involutory automorphism of $GF(q^2)$ mapping t onto t^q and $d(g) = 0$. Sesquilinearization: a Hermitian sesquilinear form f of Witt index n in $V(m, q^2)$. The totally isotropic subspaces for f are precisely the totally singular subspaces for g. The expression for g is as follows (up to a proportionality constant when $m = 2n + 1$)

$$g(x) = \sum_{i=1}^{m} x_i^{1+q}$$

Theorem 1.27 *When K is a finite field of even order, then the above are the only possible non-singular (σ, ε)-quadratic forms.*

A proof of this theorem may be found in [10].

1.9 Appendix II

We recall some results from graph theory in this section, which can be used to obtain conditions necessary for the existence of certain finite 'nice' geometries of rank 2. We will turn back to this kind of condition in Chapter 3.

Let $\mathcal{G}$ be a finite graph with v vertices and assume the set of vertices of $\mathcal{G}$ to be ordered in some way: $p_1, p_2, ..., p_v$. The *adjacency matrix* of $\mathcal{G}$ is the $v \times v$ symmetric matrix $A = (a_{ij})_{i,j=1}^{v}$ defined as follows: $a_{ij} = 1$ or 0 according to whether p_i and p_j are adjacent or not in $\mathcal{G}$.

The graph $\mathcal{G}$ is said to be *regular* of *valency* k if every vertex of $\mathcal{G}$ is adjacent to precisely k vertices other than itself. Let $\mathcal{G}$ be regular of valency k. Then k is an eigenvalue of the adjacency matrix A of $\mathcal{G}$. It is called the *trivial* eigenvalue and it has multiplicity 1 (see [13]). Some information on $\mathcal{G}$ can be obtained computing the non-trivial eigenvalues of A and their multiplicities, when possible.

For instance, let $\mathcal{G}$ be *strongly regular*; this means that $\mathcal{G}$ is regular, it has diameter 2 and the number of common neighbours of two distinct vertices a, b of $\mathcal{G}$ depends only on whether a and b are adjacent or not. We denote this number by λ when a and b are adjacent; otherwise, we denote it by μ. Assume that $k \neq 2\mu$ or $\lambda \neq \mu - 1$ (where k is the valency of $\mathcal{G}$, as above). Then the non-trivial eigenvalues of the adjacency matrix of $\mathcal{G}$ (*non-trivial eigenvalues of* $\mathcal{G}$, for short) are the two roots x_1, x_2 of the following equation [13]:

$$\text{(i)} \qquad x^2 + (\mu - \lambda)x + (\mu - k) = 0.$$

Furthermore, they are integral and their multiplicities m_1, m_2 satisfy the following relations [13]:

$$\begin{array}{ll} \text{(ii.1)} & m_1 + m_2 + 1 = v, \\ \text{(ii.2)} & x_1 m_1 + x_2 m_2 + k = 0. \end{array}$$

Since multiplicities are integral, a divisibility condition is implicit in the above. For instance, if $\mathcal{G}$ is the collinearity graph of a partial geometry of order (s,t) and index α, with $\alpha < s+1$ and $(s,t) \neq (2\alpha, \alpha)$, then $\mathcal{G}$ is strongly regular with $k = (t+1)s$, $\lambda = t(\alpha-1)+s-1$, $\mu = (t+1)\alpha$ and $v = (s+1)(1+st/\alpha)$ (see Section 1.2.6(4)). The non-trivial eigenvalues of $\mathcal{G}$ are easy to compute using (i). We obtain $-(t+1)$ and $s-\alpha$. Their multiplicities can be computed by (ii.1) and (ii.2). Integral values are obtained only if

$$\text{(iii)} \qquad s+t+1-\alpha \text{ divides } (s-\alpha)(s+1)st/\alpha.$$

In particular, when Γ is a finite generalized quadrangle we obtain that $s+t$ divides $st(s^2-1)$.

Chapter 2

More Examples

2.1 Tessellations of surfaces

We first describe some 'small' rank 3 geometries. We say they are small not because they are 'globally' small (some of them in fact have infinitely many elements), but because they 'locally' look small: the residues of their elements are ordinary polygons or ordinary digons.

We shall use a few very elementary notions from topology in this section, which are presumably familiar to most readers and can be found in the very first pages of any textbook of algebraic topology.

2.1.1 Definition

A *tessellation of a surface* is a family $\mathcal{T}$ of ordinary polygons, called *faces* of $\mathcal{T}$, such that the following hold, where the edges and vertices of those polygons are taken as *edges* and *vertices* of $\mathcal{T}$:

(i) every edge belongs to precisely two faces;

(ii) if X, Y are distinct faces with $X \cap Y \neq \emptyset$, then $X \cap Y$ is either a single vertex or an edge;

(iii) if both vertices of an edge e are in a face X, then e is an edge of X;

(iv) for every vertex x of $\mathcal{T}$, the *star* $\mathcal{T}_x$ of x (that is the system of edges and faces of $\mathcal{T}$ containing x) is an ordinary polygon, where edges and faces play the role of vertices and edges, respectively;

(v) the vertex–edge system of $\mathcal{T}$ is a connected graph.

Our definition is motivated by the fact that every family $\mathcal{T}$ of polygons as above can in fact be drawn on some connected (possibly non-compact) topological surface without borders and with no singularities. However, this has only little to do with what we are going to say. It is anyway clear that a tessellation $\mathcal{T}$ as defined above is a rank 3 geometry. Vertices, edges and faces are the three types of elements. We denote them by 0, 1 and 2, respectively. The incidence relation is the natural one. The connectedness of the incidence graph is a trivial consequence of (v). Stars of vertices and faces are ordinary polygons. Residues of edges are ordinary digons. Every flag of rank 2 is in precisely 2 maximal flags.

2.1.2 Examples

We now list some well-known examples. The first five of them ((1), (2) and (3)) are tessellations of the sphere. Next, three tessellations of the Euclidean plane are described ((4) and (5)). We also give examples of tessellations of the topological projective plane (6), of the torus (7) and the cylinder (7).

(1) All faces and all stars are triangles. It is straightforward to check that this condition uniquely determines $\mathcal{T}$ as the tetrahedron: the reader may start making a picture of $\mathcal{T}$ and he will realize that drawing a tetrahedron is the only way to finish it, consistently with the above conditions.

The tetrahedron is self-dual. Indeed, for every element x, there is just one element at maximal distance from x in the incidence graph of $\mathcal{T}$, called the *antipodal* element of x. We denote it by $op(x)$. The function op mapping x onto $op(x)$ is an involutory automorphism of the incidence graph of $\mathcal{T}$, interchanging vertices with faces. We call it the *antipodal involution*. The following pictorial way to exhibit this self-duality is well known: represent faces by their barycentres and join two barycentres by an edge when they represent adjacent faces; then the four old vertices will be represented by the four new faces arising in this way, each old vertex being paired with the new face closest to it. Edges are still represented as edges, but they now get different positions. We will revisit this construction in the next section, from the point of view of chamber systems.

We recall that the tetrahedron is the smallest 3-dimensional (degenerate) projective geometry. It is well known that all finite projective geometries of dimension ≥ 3 are self-dual. The self-duality of the tetrahedron is a particular instance of this property.

(2) All faces are triangles and all stars are quadrangles. The reader may

check that this condition uniquely determines the octahedron.

An antipodal involution *op* can be defined in this case too, just as for the tetrahedron. However, *op* now maps faces onto faces and vertices onto vertices.

The dual of the octahedron is the cube: all faces are quadrangles and all stars are triangles. This duality is already evident from the fact that the definition of the cube is obtained just interchanging the words 'face' and 'star' in the definition of the octahedron; it can also be checked directly, drawing the incidence graphs of the octahedron and of the cube and noticing that the same graph is obtained in both cases. We can also exhibit it by a pictorial trick such as the one used above to elucidate the self-duality of the tetrahedron.

However, the way we choose to represent objects is of no consequence in our combinatorial perspective. Therefore, the octahedron and the cube are just the same geometry, represented in two different ways (by two different tessellations).

We recall that the octahedron is the smallest (weak) polar space of rank 3 (see Section 1.3.5).

(3) All faces are triangles, all stars are pentagons and $\mathcal{T}$ is drawn on a sphere. It is well known that these conditions uniquely determine the icosahedron. Its dual is the dodecahedron (all faces are pentagons, all stars are triangles and $\mathcal{T}$ is drawn on a sphere). Therefore, the icosahedron and the dodecahedron are the same geometry. The antipodal involution *op* can be defined as in the previous two cases. It maps faces onto faces and vertices onto vertices.

(4) The Euclidean plane can be tessellated by squares. Stars are quadrangles and this tessellation is self-dual. Its self-duality can be exhibited by the trick used for the tetrahedron. Indeed, we can represent squares by their centres and join centres representing adjacent squares. New squares arise in this way and we obtain a new tessellation of the plane in squares, which is actually the same tessellation as we started from, only shifted a bit. Each of the old vertices sits inside precisely one new square. Hence we can take that new square as representative of that old vertex. The self-duality of the tessellation is now evident.

(5) The Euclidean plane can also be tessellated by equilateral triangles. Stars are now hexagons. The geometry obtained in this way can be dually represented by the tessellation of the Euclidean plane in regular hexagons. The duality between these two tessellation can be made evident representing

faces by their centres and joining centres representing adjacent faces, as we have done in the previous cases.

(6) Let us start from the dodecahedron and take as new objects the orbits of the antipodal involution *op*, stating that two of these orbits are incident when they come from incident elements (vertices, edges or faces) of the dodecahedron. This amounts to identifying antipodal points of a sphere, if the dodecahedron is realized as usual as a regular polyhedron inscribed in a sphere. Thus, we obtain a tessellation of the (topological) projective plane. We call it *halved dodecahedron.* Faces and stars of the halved dodecahedron are pentagons and triangles, respectively, as in the dodecahedron, but we now have 10 vertices, 15 edges and 6 faces instead of 20, 30 and 12, respectively.

The vertex–edge system of the halved dodecahedron is the so-called *Petersen graph.* This graph contains 12 pentagons and there are precisely two ways to select 6 of them so as to obtain a representation for the faces of the halved dodecahedron. The reader may check this last claim, as an exercise.

Of course, the same can be done starting from the icosahedron. We just obtain the dual of the above (hence, the same geometry). We call it a *halved icosahedron.*

(7) The tessellations of the plane described in (4) and (5) also admit homorphic images, which can be drawn on a cylinder or a torus. Indeed, if $\mathcal{T}$ is one of those tessellations, there is a group G of translations of the plane leaving $\mathcal{T}$ invariant. Let G° be a non-trivial subgroup of G such that, for every vertex x of $\mathcal{T}$ and every $g \in G^\circ$, x and $g(x)$ have distance ≥ 5 in the incidence graph of the geometry $\mathcal{T}$. We obtain a homomorphic image of $\mathcal{T}$ identifying elements in the same orbit of G°. If G° has finite index in G, then a tessellation of the torus arises in this way. Otherwise, we obtain a tessellation of the cylinder.

The geometries we will describe now also arise from tessellations, but they are not themselves tessellations.

(8) The axioms (ii) and (iii) in the definition of tessellations have forced us to assume some restrictions when forming homomorphic images in (7). If we drop them, then more homomorphic images can be considered. They will not be tessellations, of course, but they are still geometries. For instance, we may only assume in (7) that x and $g(x)$ have distance ≥ 3 in the incidence graph of $\mathcal{T}$.

(9) We can also consider the homomorphic image of the octahedron (namely, of the cube) obtained identifying antipodal elements, as we have done in the dodecahedron (6). The geometry obtained in this way will be called the *halved octahedron* (also, *halved cube*). It is not a tessellation (both (ii) and (iii) are violated in it), but it can anyway be drawn on the real projective plane, since the octahedron is a topological model for the sphere and we have identified antipodal elements.

If we start from the cube, we can give the following model for this homomorphic image. We take the complete graph on four vertices as vertex–edge system and choose as faces the three quadrangular subgraphs in that graph. The incidence relation is the natural one. In particular, all vertices are incident to all faces.

(10) We mention one more geometry, which can be obtained by a tessellation of the plane, although it is not itself a tessellation.

Let us start from the tessellation $\mathcal{T}$ of the plane by equilateral triangles (5) and give the triangles two colours, say black and white, in such a way that adjacent triangles have opposite colours. The elements of the geometry Γ we want to form are the vertices and the triangles of $\mathcal{T}$, the latter ones being partitioned into two types, 'black' and 'white', according to their colour. Incidences between vertices and triangles are as in $\mathcal{T}$, whereas two triangles are said to be incident in Γ if they are adjacent. All residues of elements of Γ are triangles.

Homomorphic images of Γ can also be formed, as in (7) and (8).

The geometries described in (1)–(5) and (10) are instances of Coxeter complexes (to be defined in Chapter 11).

All previous examples are *regular*, in the following sense: residues of elements of the same type are of the same kind. Of course, non-regular examples might be considered, too: prisms with triangular bases, pyramids with non-triangular bases, etc.

It is well known that the tetrahedron, the octahedron (dually, the cube) and the icosahedron (dually, the dodecahedron) are the only regular tessellation of the sphere. They are said to be *spherical* for this reason.

2.1.3 Chamber systems of tessellations

Since every panel of a tessellation $\mathcal{T}$ is in precisely two chambers, a gallery of the geometry $\mathcal{T}$ is uniquely determined by its type and its first chamber. For instance, let γ be a gallery of type $(1,2,0)$ starting at a chamber $C = \{a_0, a_1, a_2\}$, where a_0, a_1, a_2 are respectively the vertex, the edge and the face of the chamber C. Then γ has four chambers (length 3), and its second, third and fourth chambers are $\{a_0, b_1, a_2\}$, $\{a_0, b_1, b_2\}$ and $\{b_0, b_1, b_2\}$, where a_1 and b_1 are the two edges of the face a_2 on the vertex a_0, the faces a_2 and b_2 are the two faces on the edge b_1 and a_0, b_0 are the two vertices of the edge b_1.

The flag complex $\mathcal{K}(\mathcal{T})$ of $\mathcal{T}$ can be represented by the barycentric subdivision of $\mathcal{T}$. Indeed, every element (face, edge or vertex) of $\mathcal{T}$ is a vertex in $\mathcal{K}(\mathcal{T})$ and the edges of $\mathcal{K}(\mathcal{T})$ are the flags of rank 2 of $\mathcal{T}$. The triangles of the barycentric subdivision of $\mathcal{T}$ are the chambers of $\mathcal{T}$, adjacent triangles representing adjacent chambers.

The dualities between the octahedron and the cube and between the icosahedron and the dodecahedron now become quite evident. The reader may make a picture of the barycentric subdivisions (flag complexes) of these polyedra and one glance will be sufficent to realize that just the same triangulations are obtained for the octahedron and the cube and for the icosahedron and the dodecahedron. The self-duality of the tetrahedron is clear as well: the antipodal involution is an automorphism of the triangulation arising from the barycentric subdivision of the tetrahedron.

Similarly, looking at barycentric subdivisions is sufficent to recognize the self-duality of the tessellation of the plane by squares and the duality between the tessellation of the plane by equilateral triangles and the tessellation of the plane by regular hexagons.

Exercise 2.1. We already know that the chamber systems of the tetrahedron and of the octahedron (of the cube) have diameters 6 and 9, respectively (Exercise 1.7). Prove that the chamber system of the icosahedron (of the dodecahedron) has diameter 15.

Exercise 2.2. Let $\mathcal{T}$ be the tetrahedron, the octahedron (the cube) or the icosahedron (the dodecahedron). Prove that two chambers of $\mathcal{T}$ have maximal distance if and only if they are represented by antipodal triangles in the barycentric subdivision of $\mathcal{T}$ (if and only if they consist of antipodal elements of $\mathcal{T}$).

Exercise 2.3. Prove that the flag complex of the geometry of black and white triangles (Section 2.1.2(10)) is the tessellation of the plane by equilateral triangles.

(Hint: since triangles are vertices in the flag complex, we can represent them by their centres and we can join centres of two adjacent triangles between them and with the two vertices of the common edge of those triangles...)

Exercise 2.4. Using the representation of the flag complex of a tessellation by its barycentric subdivision, show that for every tessellation $\mathcal{T}$ we can define a dual tessellation $\mathcal{T}^*$, and that both $\mathcal{T}$ and $\mathcal{T}^*$ are drawn on the same surface.

Exercise 2.5. Let Γ be a 4-dimensional simplex or a 4-dimensional cube. Check that the flag complexes of residues of elements of Γ are triangulations of the sphere.

Exercise 2.6. Prove that the tetrahedron, the octahedron, the cube, the icosahedron and the dodecahedron are the only regular tessellations of the sphere.

(Hint: the Euler characteristic of the sphere is 2.)

2.2 Removing something ...

2.2.1 ... from projective geometries

We know that affine geometries are obtained by deleting hyperplanes from non-degenerate projective geometries. This construction can also be described as follows. Given a hyperplane H of a non-degenerate projective geometry $\mathcal{P}$, we keep all proper subspaces X of $\mathcal{P}$ such that $X \cup H$ spans $\mathcal{P}$, and discard the rest. Dually, if we fix a point a of $\mathcal{P}$ and we delete a with its star, then we obtain a dual affine geometry. Indeed, the dual $\mathcal{P}^*$ of $\mathcal{P}$ is a projective geometry, the points of $\mathcal{P}$ being the hyperplanes of $\mathcal{P}^*$ (note that, if $\mathcal{P} = PG(n, K)$, then $\mathcal{P}^* = PG(n, K')$, with K' the dual of K).

These constructions can be generalized as follows. Fix a proper subspace U of $\mathcal{P}$ of dimension $u \geq 1$ and form a new geometry Γ taking as elements the proper subspaces X of $\mathcal{P}$ such that $X \cup U$ spans $\mathcal{P}$. The incidence relation is symmetrized inclusion, inherited from $\mathcal{P}$. The geometry Γ obtained in this way has rank $u + 1$ and is called an *attenuated space* (also, a $(u + 1)$-*net*). Thus, affine spaces are special cases of attenuated spaces (when U is a hyperplane). The proof that Γ is in fact a geometry is left as an exercise.

Since the incidence relation of Γ is defined by means of inclusion, we can speak of 'points', 'lines', 'planes', 'maximal subspaces' of Γ, as in matroids and polar spaces (the points of Γ are $(n - u - 1)$-dimensional subspaces of $\mathcal{P}$).

If $u = 1$, then Γ is a net. If $u \geq 2$, then the stars of the points of Γ are u-dimensional projective geometries, whereas the planes of Γ are nets. The maximal subspaces of Γ are the hyperplanes of $\mathcal{P}$ not containing U. If X is such a hyperplane and $\mathcal{X}$ is the projective geometry of subspaces of X, then the residue of X is the attenuated space of rank u obtained from $\mathcal{X}$, giving $U \cap X$ the role that U has in $\mathcal{P}$ in the definition of Γ (the proofs of these claims are left as exercises for the reader).

The dual of this construction can be described as follows. Choose a nonempty subspace V of $\mathcal{P}$ of dimension $v = n - u - 1 \leq n - 2$ and take all nonempty subspaces X of $\mathcal{P}$ such that $X \cap V = \emptyset$. Of course, this gives us the dual of an attenuated space of rank $u + 1$ (a dual affine space, when V is a point).

Affine analogues of dual attenuated spaces can be defined as follows. Choose V as above, let H be a hyperplane of $\mathcal{P}$ and take all proper nonempty subspaces X of $\mathcal{P}$ such that $X \cap V = \emptyset$ and $X \cup H$ spans $\mathcal{P}$ (namely, $X \not\subseteq H$). If Γ is the geometry obtained in this way and $v \leq n - 3$, then the stars of the points of Γ are the same as in the dual attenuated space defined by means of V, whereas maximal subspaces of Γ are affine geometries. We call Γ an *affine dual-attenuated space.*

We can do more, mixing the construction of attenuated spaces and dual attenuated spaces together, as follows. Choose two proper nonempty subspaces U and V of $\mathcal{P}$ of dimensions u and v respectively with $u - v \geq 1$ and take all proper nonempty subspaces X of $\mathcal{P}$ such that $X \cup U$ spans $\mathcal{P}$ and $X \cap V = \emptyset$. The geometry Γ obtained in this way has rank $u - v + 1$. If $u - v \geq 2$, then the residues of the points and of the maximal subspaces of Γ are dual attenuated spaces and attenuated spaces, respectively. In this case we call Γ an *attenuated–dual-attenuated space.*

In particular, if U is a hyperplane and V is a point, then this construction amounts to deleting U and the star of V. The geometry Γ now has rank n and, if $n \geq 3$, then the residues of the points and of the maximal subspaces of Γ are dual affine geometries and affine geometries, respectively. We call Γ a *bi-affine geometry.* More generally, affine dual-attenuated spaces are obtained when U is a hyperplane. Their duals (*attenuated dual-affine spaces*) are obtained by choosing a point for V.

We can also mix attenuated spaces and their duals together in another way. Choose a proper nonempty subspace U of $\mathcal{P}$ and take all proper nonempty subspaces X of $\mathcal{P}$ such that either $X \cap U = \emptyset$ or $X \cup U$ spans $\mathcal{P}$. This gives us a geometry Γ of rank n, which we call an *amalgam of attenuated spaces.* When U is a point or a hyperplane of $\mathcal{P}$, we just obtain a dual affine geometry or an affine geometry, respectively.

It is worth remarking that many interesting partial planes (in particular, some proper partial geometries) arise as point–line systems from the above geometries (see [51]).

2.2.2 Removing hyperplanes from polar spaces

Hyperplanes. Let $\mathcal{P}$ be a thick-lined polar space of rank $n > 2$. A *projective hyperplane* of $\mathcal{P}$ (a *hyperplane* of $\mathcal{P}$, for short) is a proper subset H of the set of points of $\mathcal{P}$ having the property that every line of $\mathcal{P}$ not contained in H meets H in precisely one point (it is easily seen that this forces H to be a nonempty proper subspace of the point–line system of $\mathcal{P}$, in the meaning of Section 1.6.2).

For instance, given a point p of $\mathcal{P}$, let $p^\perp$ be the set of points of $\mathcal{P}$ collinear with p in the natural point–line system of $\mathcal{P}$ (with the convention that $p \in p^\perp$, as in Section 1.2.1). Using the Buekenhout–Shult Property (Exercise 1.16) the reader can check that $p^\perp$ is a hyperplane of $\mathcal{P}$. It is called the hyperplane *tangent* to $\mathcal{P}$ at p. A hyperplane of $\mathcal{P}$ is said to be *secant* if it is not tangent to $\mathcal{P}$.

Let $\mathcal{P}$ be a classical polar space realized inside a vector space V, as in Section 1.3.4. Then the points of $\mathcal{P}$ contained in a given hyperplane of V form a hyperplane of $\mathcal{P}$, called a *hyperplane section* of $\mathcal{P}$. Actually, if $\mathcal{P}$ is classical, then all hyperplanes of $\mathcal{P}$ can be obtained as hyperplane sections in some vector space V as above [39]. For instance, if v is a non-zero vector of V representing the point p of $\mathcal{P}$, then the hyperplane tangent to $\mathcal{P}$ at p can be obtained by intersecting $\mathcal{P}$ with the hyperplane $v^\perp$ of V ($\perp$ now denotes the orthogonality relation with respect to the a form defining $\mathcal{P}$ in V, as in Appendix I of Chapter 1). When the characteristic of the division ring K over which $\mathcal{P}$ is defined is not 2, there is essentially only one way to realize $\mathcal{P}$ in a K-vector space V (see [222], Chapter 8). In that case all hyperplanes of $\mathcal{P}$ can be obtained as hyperplane sections in V. On the other hand, when $char(K) = 2$, it might happen that we can choose V in different ways. For instance, if $\mathcal{P} \cong \mathcal{Q}_{2n}(2) \cong \mathcal{S}_{2n-1}(2)$ (see Section 1.3.5), then $\mathcal{P}$ can be realized both in $V(2n+1, 2)$ (as $\mathcal{Q}_{2n}(2)$) and in $V(2n, 2)$ (as $\mathcal{S}_{2n}(2)$). In cases like this, there is only one choice of V where all hyperplanes of $\mathcal{P}$ can be obtained as hyperplane sections (compare [222], Theorem 8.6(ii) and [39]). For instance, all hyperplanes of $\mathcal{P} = \mathcal{Q}_{2n}(2)$ are hyperplane sections in $V(2n+1, 2)$, but only tangent hyperplanes can be obtained as hyperplane sections when $\mathcal{P}$ is realized as $\mathcal{S}_{2n-1}(2)$ in $V(2n, 2)$. Indeed, if X is a hyperplane of $V(2n, 2)$, then the alternating form defining $\mathcal{P}$ induces a degenerate form on X. The radical of that induced form is a point p of $\mathcal{P}$ and the hyperplane section defined by X is just $p^\perp$.

The reader wanting to know more on projective hyperplanes of polar spaces is referred to [39] and [25].

Affine polar spaces. Let H be a hyperplane of a thick-lined polar space $\mathcal{P}$ of rank $n > 2$. If we discard all subspaces of $\mathcal{P}$ contained in H, then we obtain a geometry of rank n. We denote it by $\mathcal{P} - H$ and we call it the *affine polar space* defined by H in $\mathcal{P}$.

We partition the elements of $\mathcal{P} - H$ into types according to the dimensions they have as subspaces of $\mathcal{P}$. According to the terminology we have chosen for polar spaces, elements of $\mathcal{P} - H$ of dimension 0, 1, 2 and $n-1$ will be called *points*, *lines*, *planes* and *maximal subspaces* respectively.

The residue of a maximal subspace X of $\mathcal{P} - H$ is an $(n-1)$-dimensional affine geometry. Indeed, when X is regarded as a maximal subspace of $\mathcal{P}$, $H \cap X$ is a hyperplane of the projective geometry X and the residue of X in $\mathcal{P} - H$ consists of all subspaces of X not contained in the hyperplane $H \cap X$ of the projective geometry X.

The residue of a point of $\mathcal{P} - H$ is evidently the same as in $\mathcal{P}$.

In order to prove that $\mathcal{P} - H$ is indeed a geometry, we only must show that the collinearity graph of $\mathcal{P} - H$ is connected. Then the conclusion will follow from the above description of the residues of points and maximal subspaces of $\mathcal{P} - H$.

Let a, b be points of $\mathcal{P} - H$ with $a^\perp \cap b^\perp = \emptyset$ in $\mathcal{P} - H$ (notation as in Section 1.6.2). This means that $a^\perp \cap b^\perp \subseteq H$ in $\mathcal{P}$. Let X be a maximal subspace of $\mathcal{P}$ on a. Since $a^\perp \cap b^\perp \subseteq H$, we have $b^\perp \cap X = X \cap H$ (this can be proved using Exercise 1.15 and recalling that $X \cap H$ is a hyperplane of the projective geometry X). By (v) of Section 1.3.5, we can take a line x of $\mathcal{P}$ on b not meeting $X \cap H$. Let d be a point on that line distinct from both b and $x \cap H$ (in order to prove that such a point exists, we can use Exercises 1.16 and 1.15, recalling that $\mathcal{P}$ is thick-lined by assumption). Since $d \neq b$ and $b^\perp \cap X = X \cap H$, we have $d^\perp \cap X \not\subseteq H$ (we again can use Exercise 1.11 to prove this). Let $c \in d^\perp \cap (X - H)$. Then c and d give us a path of length 3 from a to b in the collinearity graph of $\mathcal{P} - H$. Therefore this graph is connected of diameter ≤ 3.

We add a few more definitions. The hyperplane H dropped from the polar space $\mathcal{P}$ to form the affine polar space $\mathcal{P} - H$ is uniquely determined by $\mathcal{P} - H$, up to isomorphism [39]. We call it the *space at infinity* of $\mathcal{P} - H$ and, denoted $\mathcal{P} - H$ by Γ, we denote H by Γ^∞. We also say that the polar space $\Gamma \cup \Gamma^\infty$ $(= \mathcal{P})$ is the *projective completion* of Γ $(= \mathcal{P} - H)$. We say that the affine polar space Γ is of *tangent* (respectively, *secant*) *type* if Γ^∞ is a tangent (secant) hyperplane of the projective completion of Γ.

For more on affine polar spaces the reader is referred to [39] and [49].

2.3 Affine expansions

Let Γ be a geometry of rank $n \geq 2$ realized by a set S of points of $PG(m, K)$ $(m \geq n)$ together with certain lines of $PG(m, K)$ and (when $n \geq 3$) some higher dimensional subspaces of $PG(m,K)$, with the incidence relation inherited from $PG(m,K)$. Let the following hold in Γ:

($*$) the set S spans $PG(m,K)$ and, for every subspace X of $PG(m,K)$ representing an element of Γ, the set $S \cap X$ spans X.

Identifying $PG(m,K)$ with the geometry at infinity of $AG(m+1,K)$, we can form a new structure taking as elements the points of $AG(m+1,K)$ and those lines and subspaces of $AG(m+1,K)$ whose directions are elements of $\Gamma \subseteq PG(m,K)$. As incidence relation we take the one inherited from $AG(m+1,K)$. Let $Af(\Gamma)$ be the incidence structure defined in this way.

Using ($*$), we can prove that $Af(\Gamma)$ is residually connected, hence it is a geometry. We call it the *affine expansion* of Γ. It has rank $n+1$ and the residues of its points are isomorphic to Γ. The residues of its planes are nets (possibly, affine planes or grids).

We only show how the connectedness of $Af(\Gamma)$ can be proved, leaving the rest for the reader. Given distinct points a, b of $Af(\Gamma)$ (namely, points of $AG(m+1,K)$), we want to show that we can connect them by points and lines of $Af(\Gamma)$. let X be the line of $AG(m+1,K)$ through a and b and let x be the point at infinity of X. We can span $PG(m,K)$ in a sequence of steps, starting from S and taking points collinear with pairs of points already reached at some previous step. Let s be the minimal number of steps we need to reach x in the previous construction. We work by induction on s. If $s=0$, then $x \in S$, whence X is a line of $Af(\Gamma)$. Otherwise, x is collinear with a pair of points y, z that have already been reached in less than s steps. Let u be a plane of $AG(m+1,K)$ with the line of $PG(m,K)$ on x, y, z as the line at infinity. Let Y be the line of u on a with y as the point at infinity and let Z be the line of u on b with z as the point at infinity. These lines meet in a point c of u. By the inductive hypothesis, we can connect a to c and c to b inside $Af(\Gamma)$. We are done.

Needless to say, if $\Gamma = PG(m,K)$, then the affine expansion of Γ is just $AG(m+1,K)$.

If Γ is a polar space, then the affine expansion of Γ strongly resembles an affine polar space, but it is not an affine polar space in general. We will show in Chapter 8 that affine expansions of polar spaces are in fact homomorphic images of affine polar spaces.

If Γ is an affine dual-attenuated space obtained from a hyperplane–subspace pair (H,V) of $PG(n,K)$ as in Section 2.2.1 and if $V \subseteq H$, then Γ is the affine expansion of the dual attenuated space defined by V in the hyperplane H $(=PG(n-1,K))$. In particular, if V is a point of H, then Γ is (a bi-affine geometry and it is) the affine expansion of the dual of

$AG(n-1, K')$ (where K' is the dual of K).

Interesting examples of partial geometries arise as point-line systems of affine expansions of suitable finite linear spaces. We will now describe one of those examples, in fact a finite generalized quadrangle. However, proper partial geometries can also be obtained by some modification of the following construction (see [51]).

Let Γ be the circular space with $q+2$ points, $q = 2^r$ for some positive integer r. We can form a model of Γ by the $q+2$ points of a hyperoval O of $PG(2, q)$ (see [81]), taking the pairs of points of O as lines of Γ. The affine expansion of Γ has rank 3 and its planes are grids. Its point–line system is denoted by $T_2^*(O)$ and it is a finite generalized quadrangle, with lines of size q and $q+2$ points on every line. When $r = 1$, $T_2^*(O) = \mathcal{Q}_5^-(2)$ (see [166]), but when $r > 1$ the generalized quadrangle $T_2^*(O)$ is non-classical.

2.4 Enrichments

Let Γ be a geometry of rank 2 with set of points P, gonality $g \geq 2$ and diameter $d \geq 3$. We call the lines of Γ 'blocks'. We say that a clique X of the collinearity graph $(P, \perp)$ of Γ is contained in a block Y if all points of X are incident to Y. Let the following hold for some integer $t \geq 3$:

(i) every block of Γ has at least t points;

(ii) every t-clique of $(P, \perp)$ is contained in at most one block;

(iii) for every $i = 1, 2, ..., t-2$ and for every i-clique X of $(P, \perp)$, the system of $(t-1)$-cliques and blocks containing X is a geometry of rank 2 with respect to the symmetrized inclusion taken as incidence relation.

Then we can form a geometry $E(\Gamma)$ of rank t by inserting the i-cliques of Γ $(i = 2, 3, ..., t-1)$ as new elements between points and blocks of Γ, with the incidence relation defined as symmetrized inclusion (the reader may check that $E(\Gamma)$ is indeed a geometry). We call $E(\Gamma)$ the *enrichment* of Γ. Using the determinate article 'the' is not an abuse here. Indeed, every rank 2 geometry admits at most one enrichment (we leave the proof of this claim as an exercise).

Trivially, Γ is a truncation of its enrichment. The residues in $E(\Gamma)$ of the blocks of Γ are matroids. Their i-dimensional subspaces are the sets of size $i+1$ $(i = 0, 1, ..., t-2)$. When blocks have finitely many points, these matroids are just truncated simplices.

For instance, if $\mathcal{S}$ is a Steiner system $S(t, k, v)$ with $t > 2$ (see Section 1.2.6(5)), then we can define its enrichment $E(\mathcal{S})$, which is a t-dimensional matroid. The hyperplanes of the matroid $E(\mathcal{S})$ are truncated $(k-1)$-dimensional simplices. Stars of $(t-3)$-dimensional subspaces of $E(\mathcal{S})$ may have interesting structures. For instance, if $\mathcal{S}$ is the Steiner system for one of the Mathieu groups M_{22}, M_{23} or M_{24}, then those stars are isomorphic to $PG(2,4)$ (see [91]).

Mőbius, Laguerre and Minkowski planes (Section 1.2.6(2)) also admit enrichments. Their enrichments have rank 3. Residues of blocks now are complete graphs. Residues of points are nets. Actually, these latter residues are affine planes in enrichments of Mőbius planes. In the case of Laguerre planes, they are nets and dual nets at the same time.

Exercise 2.7. Let Γ admit an enrichment $E(\Gamma)$ and let t be the rank of $E(\Gamma)$. Prove that every i-clique of $(P, \perp)$ with $i < t$ is contained in at least two blocks and that distinct blocks of Γ have distinct sets of points.

Exercise 2.8. Let Γ and t be as in the previous exercise. Prove that, for every $(t-2)$-clique X of Γ, the star of X in $E(\Gamma)$ is a partial plane.

Exercise 2.9. Let Γ be a biplane [91]. Prove that Γ admits an enrichment of rank 3 and that $E(\Gamma)$ is the dual of the enrichment of the dual of Γ.

Exercise 2.10. (Double enrichments). Let Γ be a geometry of rank 2 satisfying the following:
(i) for every point x, the points collinear with x and the blocks on x form a geometry with respect to the incidence relation inherited from Γ and the dual of this geometry admits an enrichment of rank 2;

(ii) the analogue of (i) holds in the dual of Γ.

Prove that a geometry of rank 4 can be obtained from Γ inserting pairs of collinear points and pairs of distinct blocks with nonempty intersections as new elements between points and blocks of Γ.

Exercise 2.11. A set $\mathcal{P}$ of permutations of a finite nonempty set X is said to be *sharply t-transitive* for some positive integer t if, for every choice of elements x_1, x_2, ..., x_t and y_1, y_2, ..., y_t of X with $x_i \neq x_j$ and $y_i \neq y_j$ $(1 \leq i < j \leq t)$, there is just one mapping $f \in \mathcal{M}$ such that $f(x_i) = y_i$ for every $i = 1, 2, ..., t$.

Let X be a finite nonempty set and let $\mathcal{P}$ be a sharply t-transitive set of permutations of X, with $t \geq 3$. Permutations of X can be viewed as subsets of $X \times X$. If we take $X \times X$ as set of points and $\mathcal{P}$ as set of blocks, then we obtain a rank 2 geometry $\Gamma_{\mathcal{P}}$. Prove that $\Gamma_{\mathcal{P}}$ admits an enrichment of rank t.

Exercise 2.12. With the notation of the previous exercise, prove that, when $t = 3$, then $\Gamma_{\mathcal{P}}$ is a finite Minkowski plane, and prove that all finite Minkowski planes arise in this way.

Exercise 2.13. Let X, Y be finite nonempty sets. A set $\mathcal{M}$ of mappings from X to Y is said to be *sharply t-transitive* for some positive integer t if, for every choice of distinct elements x_1, x_2, ..., x_t of X and of (possibly non-distinct) elements y_1, y_2, ..., y_t of Y, there is just one mapping $f \in \mathcal{M}$ such that $f(x_i) = y_i$ for every $i = 1, 2, ..., t$.

Let $\mathcal{M}$ be a sharply t-transitive set of mappings from X to Y, with $t \geq 3$. Mappings from X to Y can be viewed as subsets of $X \times Y$. If we take $X \times Y$ as set of points and $\mathcal{M}$ as set of blocks, then we obtain a rank 2 geometry $\Gamma_{\mathcal{M}}$. Prove that $\Gamma_{\mathcal{M}}$ admits an enrichment of rank t.

(Remark. A set of mappings $\mathcal{M}$ as above is usually called an *optimal (m,t)-code*, where $m = |X|$.)

Exercise 2.14. With the notation of the previous exercise, prove that, when $t = 3$ and $|X| = |Y| + 1$, then $\Gamma_{\mathcal{M}}$ is a finite Laguerre plane, and prove that all finite Laguerre planes arise in this way.

Exercise 2.15. Prove that the enrichment of a classical Laguerre plane is the dual of the affine expansion of a dual circular space.

(Hint: a cone in $PG(3, K)$ is a hyperplane in the dual of $PG(3, K)$.)

Exercise 2.16. Given a plane H of $PG(3, q)$ (q even), a hyperoval O in H and a point p of $PG(3, q)$ not in H, let P be the set of points of $PG(3, q)$ distinct from p and belonging to lines joining p with points of O. Imitating the construction of classical Laguerre planes, we define a rank 2 geometry Γ taking P as set of points and the planes of $PG(3, q)$ not on p as blocks. Prove that Γ admits an enrichment $E(\Gamma)$ and that $E(\Gamma)$ is the dual of the affine expansion of a dual circular space.

(Remark: the above construction is an instance of the construction of Exercise 2.13, with $t = 3$, $X = O$ and $Y = GF(q)$.)

Exercise 2.17. Let Π be a partial plane realized in the star of a point p of $PG(3, q)$; that is, points and lines of Π are represented by some lines and planes of $PG(3, q)$ on p. We can define a rank 3 geometry $\Gamma(\Pi)$ as follows: as elements of type 0 we take the points of $PG(3, q)$ other than p and belonging to lines on p representing points of Π; as elements of type 1 we take the lines of $PG(3, q)$ not on p and contained in planes on p representing lines of Π; the planes of $PG(3, q)$ not on p are taken as elements of type 2. The incidence relation of $\Gamma(\Pi)$ is the natural one, inherited from $PG(3, q)$. Prove that the dual of Π admits an affine expansion and that $\Gamma(\Pi)$ is the dual of the affine expansion of the dual of Π.

Exercise 2.18. Let Γ be a geometry admitting an affine expansion $Af(\Gamma)$.

Prove that both $Af(\Gamma)$ and its dual admit an affine expansion.

Exercise 2.19. Let Γ be the bi-affine geometry obtained from $PG(m, K)$ dropping a hyperplane H and the star of a point $p \in H$. Prove that Γ is the affine expansion of a dual affine geometry.

Exercise 2.20. Let Γ be a geometry of rank n realized inside $PG(m, K)$ as in Section 2.3. Given a positive integer r, we can identify $PG(m, K)$ with an m-dimensional subspace S of $PG(m+r, K)$. Let $Af_r(\Gamma)$ consist of those subspaces X of $PG(m+r, K)$ such that $X \cup S$ spans $PG(m+r, K)$ and $X \cap S$ either represents an element of Γ or is empty. Prove that $Af_r(\Gamma)$ forms a geometry of rank $n+1$ with respect to the incidence relation inherited from $PG(m+r, K)$.

(Remark: We have $Af_1(\Gamma) = Af(\Gamma)$. Note that, when $\Gamma = PG(m, K)$, then $Af_r(\Gamma)$ is an attenuated space.)

Chapter 3

Diagrams and Orders

3.1 Diagrams

3.1.1 Definitions

According to the conventions stated in Section 1.2, we keep the symbols P and L as types for rank 2 geometries.

Let us fix a set I of types of size $|I| \geq 2$. Giving a *diagram* $\mathcal{D}$ over I is selecting a class $\mathcal{D}_{\{i,j\}}$ of rank 2 geometries for every pair of distinct types $i, j \in I$, establishing at the same time which of the types i, j is given the role of P. Thus, $\mathcal{D}$ is a labelled graph with set of vertices I where, for every choice of distinct types $i, j \in I$, a name of the class $\mathcal{D}_{\{i,j\}}$ is taken as label of the edge $\{i, j\}$ and, if the class $\mathcal{D}_{\{i,j\}}$ is not self-dual, then the correspondence between the types i, j and P, L must be implicit in the name we use to denote that class. We will always assume that the classes of rank 2 geometries used to form diagrams are closed under isomorphism.

More formally, a diagram over I is a system of pairs

$$\mathcal{D} = ((\mathcal{D}_{\{i,j\}}, \alpha_{\{i,j\}}) \mid i, j \in I, i \neq j)$$

where $\mathcal{D}_{\{i,j\}}$ is a class of rank 2 geometries closed under isomorphism and $\alpha_{\{i,j\}}$ is a bijection from $\{i, j\}$ to $\{P, L\}$.

The cardinality $n = |I|$ of I is called the *rank* of the diagram $\mathcal{D}$.

Let $\mathcal{D} = ((\mathcal{D}_{\{i,j\}}, \alpha_{\{i,j\}} \mid i, j \in I, i \neq j)$ be a diagram over I. We say that a geometry Γ over the set of types I *belongs* to the diagram $\mathcal{D}$ if, for every pair of distinct types $i, j \in I$, all residues of Γ of type $\{i, j\}$ are members of $\mathcal{D}_{\{i,j\}}$, the types i and j playing the roles that $\alpha_{\{i,j\}}$ gives them. If $\mathbf{C}$ is a class of geometries closed under isomorphism, then we say that $\mathbf{C}$ *belongs* to $\mathcal{D}$ if all members of $\mathbf{C}$ belong to $\mathcal{D}$.

We do not make any assumption on the classes $\mathcal{D}_{\{i,j\}}$ at this general level, allowing any choice for them, at least in principle. Thus, infinitely many diagrams could be given for the same geometry Γ, choosing as $\mathcal{D}_{\{i,j\}}$ any class of rank 2 geometries closed under isomorphism and containing all residues of Γ of type $\{i,j\}$. However, such a wild notion of diagrams would be quite useless in practise. Some restrictions are in fact implicitly assumed in 'real life' on those classes of rank 2 geometries that might be sensibly chosen in forming diagrams. On the other hand, we are not so brave as to dare drawing a sharp distinction between 'good' and 'bad' classes of rank 2 geometries. We only state the following very mild restriction: we will never consider classes of rank 2 geometries where generalized digons are gathered together with other geometries that are not generalized digons. According to this additional convention, we say that a geometry Γ over the set of types I *admits a diagram* if, for every 2-subset J of I, either all residues of Γ of type J are generalized digons or none of them is a generalized digon.

There are geometries which do not admit any diagram. The reader may find examples of this kind in [144] and in [5] (Examples 6 and 7).

We have assumed $|I|\geq 2$ in the definition of diagrams. However, allowing $|I|= 1$ is sometimes convenient. When $|I|= 1$ we state the convention that the only diagrams considered are either the class of all rank 1 geometries or the class of all rank 1 geometries of given size.

Remark. Geometries admitting a diagram were called *pure* in [144] and [19]. Geometries not admitting any diagram were called *mixed* in [144].

3.1.2 A lexicon for diagrams

According to the convention of representing a diagram as a labelled graph, we will now list the labels most frequently used for edges of diagrams. The letters i, j written under the nodes of the edge are the two types corresponding to those nodes.

The class of partial planes (over the set of types $\{i,j\}$) is represented as follows:

$$\overset{i}{\bullet}\overset{\pi}{\rule{5em}{0.4pt}}\overset{j}{\bullet}$$

The set of axioms defining partial planes (Section 1.2.1) is self-dual, that is we can interchange P and L in those axioms without changing anything (the dual of a partial plane is a partial plane). Therefore, we need not say which of the types i and j plays the role of P.

The class of linear spaces will be denoted as follows:

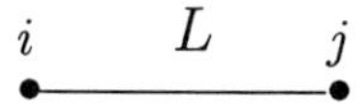

$$\overset{i}{\bullet}\!\overset{L}{\text{———————}}\!\overset{j}{\bullet}$$

where the types i and j correspond to points and lines, respectively. The class of affine planes is denoted writing Af instead of L on top of the edge:

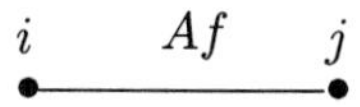

$$\overset{i}{\bullet}\!\overset{Af}{\text{———————}}\!\overset{j}{\bullet}$$

We use the symbol c instead of L to denote the class of circular spaces (Section 1.2.2):

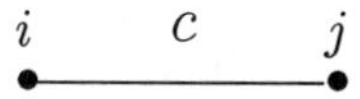

$$\overset{i}{\bullet}\!\overset{c}{\text{———————}}\!\overset{j}{\bullet}$$

Dual linear spaces, dual affine planes and dual circular spaces are denoted writing L^*, Af^* and c^* instead of L, Af, c, respectively:

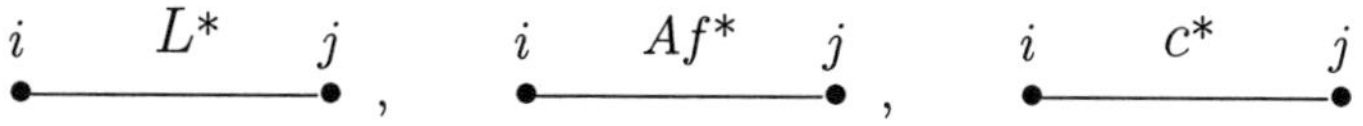

$$\overset{i}{\bullet}\!\overset{L*}{\text{———————}}\!\overset{j}{\bullet}\ ,\quad \overset{i}{\bullet}\!\overset{Af*}{\text{———————}}\!\overset{j}{\bullet}\ ,\quad \overset{i}{\bullet}\!\overset{c*}{\text{———————}}\!\overset{j}{\bullet}$$

The symbol N on top of the edge denotes the class of nets:

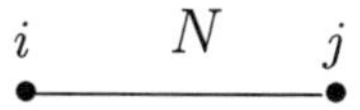

$$\overset{i}{\bullet}\!\overset{N}{\text{———————}}\!\overset{j}{\bullet}$$

The symbol N^* denotes the class of dual nets:

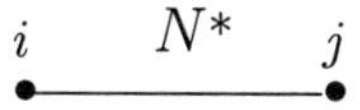

$$\overset{i}{\bullet}\!\overset{N*}{\text{———————}}\!\overset{j}{\bullet}$$

It is clear from the previous conventions that we read edges from left to right, when possible.

The next symbol

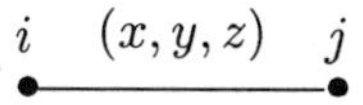

$$\overset{i}{\bullet}\!\overset{(x,y,z)}{\text{———————}}\!\overset{j}{\bullet}$$

is the name given to the class of rank 2 geometries of gonality $g = x$ and diameters $d_P = y$, $d_L = z$, where i and j correspond to P and L, respectively.

When $x = y = z = m$ (generalized m-gons) the following notation is normally used instead of the previous one:

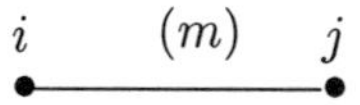

$$\overset{i}{\bullet}\!\overset{(m)}{\text{———————}}\!\overset{j}{\bullet}$$

Since the class of generalized m-gons is self-dual, specifying the correspondence between i, j and P, L is unnecessary here.

When $m = 4$, 3 or 2, the following notation is also used quite often instead of the above:

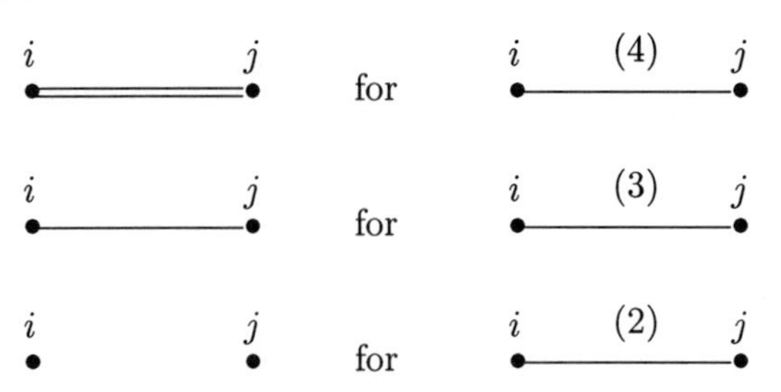

According to these conventions, if $\mathcal{D}_{\{i,j\}}$ is the class of generalized digons, we will indicate this by deleting the edge of the diagram that should join the types i and j. Henceforth, we will always represent generalized digons in this way, so that diagrams will resemble simple graphs, apart from the labels we put on top of their edges. Thus, we can speak of *connected* and *disconnected* diagrams, *trees*, *cycles*, *strings* etc., as we do for simple graphs.

Coxeter diagrams. A *Coxeter diagram* (also, a diagram *of Coxeter type*) is a diagram where, for every pair of distinct types i, j, the stroke joining i with j represents the class of generalized m_{ij}-gons for some $m_{ij} = 2, 3, 4, ..., \infty$. Namely, a Coxeter diagram is described by a symmetric matrix $M = (m_{ij})_{i,j \in I}$, where $m_{ii} = 1$ for every $i \in I$, by convention, and for $i, j \in I, i \neq j$ the entry m_{ij} of the matrix M means that all residues of type $\{i, j\}$ in a geometry represented by M are generalized m_{ij}-gons. Matrices of this kind are called *Coxeter matrices.*

3.1.3 Examples

We now give diagrams for some of the geometries of rank $n \geq 3$ that we have described in Chapters 1 and 2. Some of the diagrams we will draw have standard names. We write these names at the left side of the diagrams. We also compose a name for the remaining diagrams, according to certain tacit conventions which the reader will soon guess.

The proof that the geometries we consider indeed have the diagram we give is implicit in the arguments we have used in Chapters 1 and 2 to show that they are indeed geometries.

(1) **Projective geometries of rank n.** We may choose 0, 1, ..., $n - 1$ as types, denoting dimensions by them, as in Sections 1.3.1–1.3.3.

(A_n) 0 — 1 — 2 ..... $n-2$ — $n-1$

The name of this diagram is A_n. In particular, A_2 is a name for the class of (possibly degenerate) projective planes, whereas A_1 is a name for the class of all geometries of rank 1.

The diagram A_n is of Coxeter type.

(2) **Affine geometries of dimension n.** Types: 0, 1, ..., $n-1$, as in the case of projective geometries.

($Af.A_{n-1}$) 0 Af 1 — 2 ..... $n-2$ — $n-1$

We call this diagram $Af.A_{n-1}$. In particular, $Af.A_1$ means the same as Af.

(3) **Matroids of dimension** $n \geq 2$. We take 0, 1, ..., $n-1$ as types, as in the previous two cases.

(L_n) 0 L 1 L 2 ..... $n-2$ L $n-1$

We give this diagram the name L_n. In particular, L_2 is a name for the class of linear spaces; we will always write L for L_2, according to the notation of Section 3.1.2.

(4) **Polar spaces of rank n.** Types, 0, 1, ..., $n-1$, as above (see Section 1.3.5).

(C_n) 0 — 1 ..... $n-3$ — $n-2$ = $n-1$

The name of this diagram is C_n (some authors prefer to write B_n for C_n). In particular, C_2 denotes the class of generalized quadrangles.

The C_n diagram is of Coxeter type.

(5) **Buildings of type** D_n, $n \geq 4$ (see Section 1.3.6). As types we take 0, 1, ..., $n-3$, $+$, $-$, where 0, 1, ..., $n-3$ are dimensions and $+$, $-$ denote the two classes M^+ and M^- of maximal subspaces (see Section 1.3.6).

(D_n) 0 — 1 ..... $n-4$ — $n-3$ < $-$, $+$

D_n is the name of this diagram. It is a Coxeter diagram.

We have assumed $n \geq 4$. However, we could also allow $n = 3$, taking D_3 as a synonym of A_3, with types $+$, 0, $-$ instead of 0, 1, 2 respectively.

(6) **Tessellations**. We now give diagrams for the tessellations considered in Section 2.1.2. The types 0, 1, 2 denote vertices, edges and faces, respectively (that is, they are dimensions, as in the previous cases).

We have Coxeter diagrams in these cases. Indeed rank 2 residues are ordinary m-gons, where the gonality m depends only on the type of the residue (we will turn back to this later, Proposition 3.2).

(6.a) The tetrahedron belongs to the Coxeter diagram A_3. In fact, it is a degenerate 3-dimensional projective geometry.

(6.b) The octahedron is a weak polar space of rank 3. It belongs to the Coxeter diagram C_3. Of course, if we think of the cube, we only have to read the C_3 diagram from right to left instead of from left to right. The halved octahedron (see Section 2.1.2(9)) also belongs to C_3.

(6.c) The icosahedron belongs to the following Coxeter diagram, called H_3 (when we think of the dodecahedron, we should read this diagram from right to left). The halved icosahedron (Section 2.1.2(6)) belongs to the same diagram.

(H_3) 0 —— 1 —(5)— 2

(6.d) The tessellations of the Euclidean plane considered in (4) and (6) of Section 2.1.2 and their quotients (Section 2.1.2, (7) and (8)) belong the following Coxeter diagrams, called $\widetilde{B}_2$ and $\widetilde{G}_2$, respectively:

($\widetilde{B}_2$) 0 ═══ 1 ═══ 2 (tess. in squares)

($\widetilde{G}_2$) 0 —— 1 —(6)— 2 (tess. in triangles)

Of course, the tessellation in regular hexagons, being the dual of the tessellation in equilateral triangles, is also represented by the diagram $\widetilde{G}_2$, except that the diagram should now be read from right to left.

(6.e) The geometry of black and white triangles described in Section 2.1.2 (10) belongs to the following Coxeter diagram, called $\widetilde{A}_2$:

$(\widetilde{A}_2)$

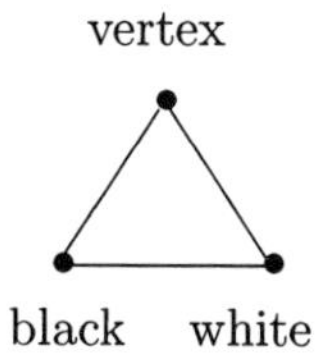

(7) **Affine polar spaces and affine expansions of polar spaces** (Sections 2.2.2 and 2.3). Both affine polar spaces and affine expansions of polar spaces have diagrams as follows:

$(Af.C_{n-1})$

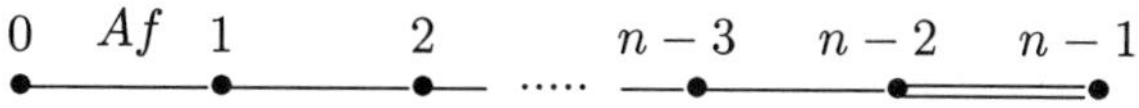

(8) **Diagrams for the geometries of Section 2.2.1**

(8.a) Attenuated spaces (m-nets) have the following diagram of rank $m = u + 1$ ($u = dim(U)$, U as in Section 2.2.1):

$(N.A_{m-1})$

The same diagram also describes dual attenuated spaces (we only must read it from right to left).

(8.b) An affine dual-attenuated space of rank $m \geq 3$ has the following diagram:

$(Af.A_{m-2}.N^*)$

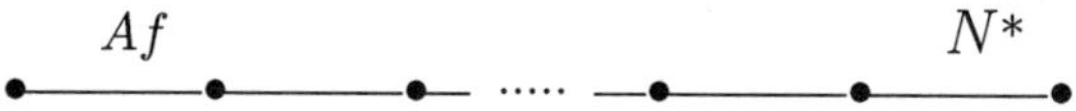

(8.c) Bi-affine geometries are special cases of the above. They can be represented by the following diagram:

$(Af.A_{n-2}.Af^*)$

(8.d) Attenuated–dual-attenuated spaces of rank $n \geq 3$ belong to the following diagram:

N N^*

$(N.A_{m-2}.N^*)$ ● — ● — ● — ..... — ● — ● — ●

(9) **Enrichments**

(9.a) Let Γ_{22}, Γ_{23}, Γ_{24} be the enrichments of the Steiner systems for M_{22}, M_{23} and M_{24}, respectively (see Section 2.4). The geometry Γ_{24} belongs to the following diagram:

c c c

$(c^3.A_2)$ ● — ● — ● — ● — ●

points pairs blocks

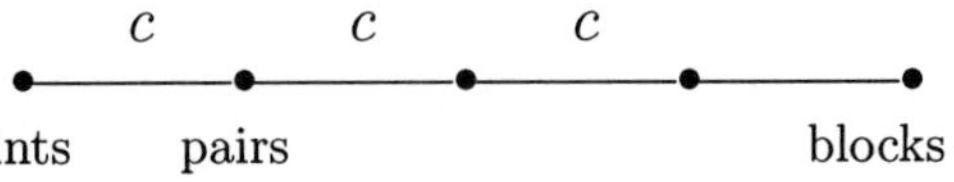

The geometry Γ_{23} is isomorphic to the residues of the points of Γ_{24}. Hence, it belongs to the following diagram:

c c

$(c^2.A_2)$ ● — ● — ● — ●

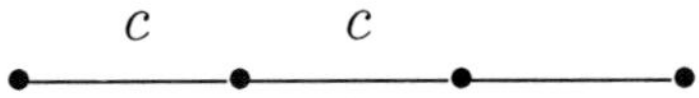

The geometry Γ_{22} is isomorphic to the residues of the point-pair flags of Γ_{24} (equivalently, to the residues of the points of Γ_{23}). Therefore, it belongs to the following diagram:

c

$(c.A_2)$ ● — ● — ●

Note that c^{n-1} (to be read as $c.c....c$, $n-1$ times) is a name for the following special case of L_n (of rank n):

c c c

(c^{n-1}) ● — ● — ● — ..... — ● — ●

We have preferred the notation c^{n-1} rather than c_n to avoid confusion with the Coxeter diagram C_n, and because c^{n-1} is the name currently used in the literature to designate the above diagram.

(9.b) The enrichments of Mőbius, Laguerre and Minkowski planes defined in Section 2.4 belong to diagrams of the following form:

c N_k

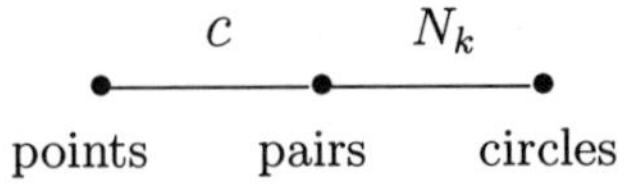

$(c.N_k)$ ● — ● — ●

points pairs circles

The symbol N_k here denotes the class of nets where, given a non-incident point–line pair (a,x), a is non-collinear with precisely k points of x. Mőbius, Laguerre and Minkowski planes correspond to $k = 0, 1$ and 2, respectively.

Indeed, N_0 is the class of affine planes, whereas N_1 is the class of nets which are also dual nets. Therefore, the diagram $c.N_0$ describing enrichments of Mőbius planes can also we written as follows:

$(c.Af)$ $\quad$ c $\quad$ Af

Exercise 3.1. Prove that the enrichments of the Steiner systems for M_{12} and M_{11} belong to the following diagrams:

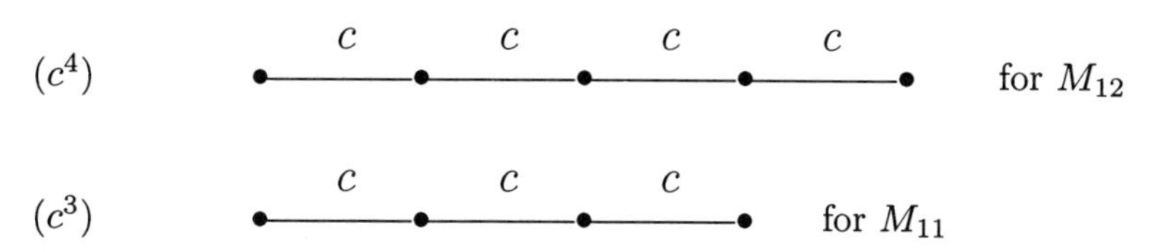

Exercise 3.2. Let Γ be an amalgam of attenuated spaces of rank $n \geq 3$ (Section 2.2.1). Prove that, if Γ is not an affine or dual affine geometry, then it belongs to a diagram of the following form:

Exercise 3.3. Let $\mathcal{P}$ be a classical polar space of rank n embedded in $PG(m, K)$ as in Section 1.3.4. Given a proper subspace U of $PG(m, K)$ of dimension $h \geq m-n+3$, let Γ be the intersection of $\mathcal{P}$ with the attenuated space defined by U in $PG(m, K)$. Prove that Γ is a geometry of rank $k = h + n - m$ belonging to the following diagram:

$(N.C_{k-1})$ $\quad$ N

Exercise 3.4. Check that an affine expansion of a circular space (Section 2.3) belongs to the following diagram:

$(C_2.c)$ $\quad$ c

Exercise 3.5. Let Γ be a biplane [91]. Define the enrichment $E(\Gamma)$ of Γ inserting pairs of points as new elements. Prove that $E(\Gamma)$ belongs to the following diagram:

$(c.c^*)$ $\quad$ c $\quad$ c^*

Exercise 3.6. Let Γ belong to a Coxeter diagram with $m_{ij} = m$ for all $i, j \in I$, $i \neq j$ (with m given). Prove that the flag complex $\mathcal{K}(\Gamma)$ belongs to a Coxeter diagram of the following form:

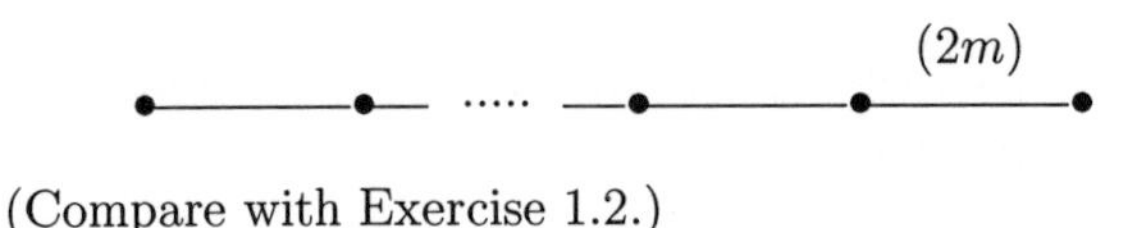

(Compare with Exercise 1.2.)

3.1.4 Residues and truncations

Let Γ be a geometry over the set of types I, belonging to a diagram $\mathcal{D}$. Let J be a proper nonempty subset of I and let $\mathcal{D}_J$ be the diagram induced by $\mathcal{D}$ on J. Specifically, J is the set of nodes of $\mathcal{D}_J$ and $\mathcal{D}_J$ is obtained from $\mathcal{D}$ keeping the edges with both nodes in J, with the labels they have in $\mathcal{D}$, and dropping the rest (in particular, when $|J| = 1$, $\mathcal{D}_J$ is the Coxeter diagram A_1 with only one node and no edges). It is clear that all residues of Γ of type J belong to $\mathcal{D}_J$.

Let us turn to truncations. Diagrams are not preserved under truncations, in general. Indeed, gonality may decrease and diameters often increase when we take truncations, as the following examples show.

(1) Let Γ be the halved octahedron (Section 2.1.2(9)) and take 0, 1, 2 as types, as in Section 3.1.3(6). We know that Γ belongs to the C_3 Coxeter diagram. The 2_--truncation $Tr_2^-(\Gamma)$ of Γ has gonality 2 whereas the residues of Γ of type $\{0, 1\}$ are triangles. $Tr_0^-(\Gamma)$ is the dual of the circular space on 4 points, whence it has gonality 3. On the other hand, the $\{1, 2\}$-residues of Γ are ordinary quadrangles. We also note that $Tr_1^-(\Gamma)$ is a generalized digon and $\{0, 2\}$-residues of Γ are ordinary digons: now the truncation and the 'corresponding' residues are of the same kind. However, this is an exceptional phenomenon, too. Indeed, when we truncate 'inner' nodes of string diagrams, diameters are generally higher in a truncation than in the corresponding residues. For instance, the $\{0, 1\}_+$-truncation of an n-dimensional matroid with $n \geq 3$ has (gonality 2 and) diameters $d_0 = 3$ and $d_{n-1} = 3$ or 4 (we have $d_{n-1} = 3$ when the matroid is a projective geometry).

(2) Let Γ be a building of type D_4, with types 0, 1, +, −, as in Section 3.1.3(5). Its truncation $Tr_1^-(\Gamma)$ (see Section 1.6.1(6)) belongs to this dia-

gram:

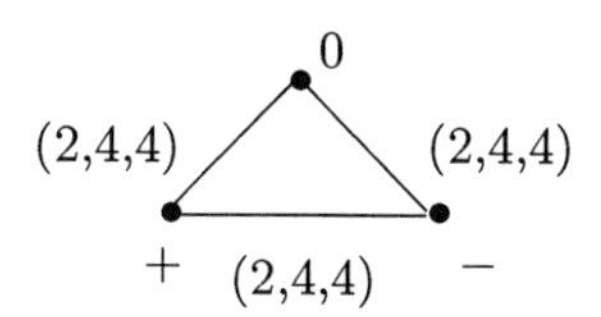

On the other hand, the $\{0, +, -\}$-residues of Γ belong to the completely disconnected Coxeter diagram of rank 3:

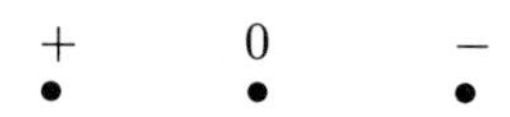

(3) The $\{k+1, k+2, ..., n-1\}_-$-truncation of an n-dimensional projective geometry ($2 \leq k < n$) belongs to the following diagram:

$(A_k.L)$ $\quad$ 0 — 1 ..... $k-2$ — $k-1$ L k

Therefore, the $\{k+1, k+2, ..., n-1\}_-$-truncation of an n-dimensional affine geometry ($3 \leq k < n$) belongs to the following diagram of rank $k+1$:

$(Af.A_{k-1}.L)$ $\quad$ Af L

(4) The $\{h, h+1, ..., k-1, k\}_+$-truncation of an n-dimensional projective geometry ($0 \leq h$, $k < n$ and $h+1 \leq k-1$) belongs to the following diagram $L^*.Af_{k-h-1}.L$, of rank $k-h+1$:

(5) Let $n \geq 4$ and $1 \leq k \leq n-3$. The $\{0, 1, ..., k-1\}_-$-truncation of a polar space of rank n belongs to the following diagram $L^*.C_{n-1-k}$ of rank $n-k$:

(6) The $\{0, 1, ..., k-1\}_-$-truncation of a D_n building ($1 \leq k \leq n-4$) belongs to the following diagram $L^*.D_{n-k-1}$ of rank $n-k$ (when $k = n-4$, we use the notation $L^*.D_3$ even if the symbol A_3 seems to be more appropriate than D_3; indeed the symbol $L^*.A_3$ has already been implicitly defined above, as the dual of $A_3.L$, representing a truncation of a projective geometry).

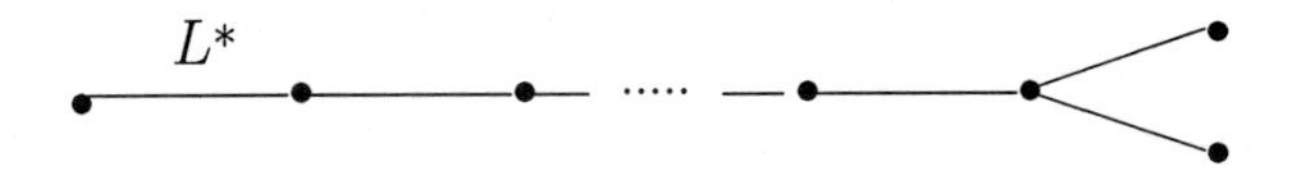

(7) The $\{n-3, +, -\}_+$-truncation of a D_n-building belongs to the following diagram of rank 3:

$(L.L^*)$ $\quad$ $+ \quad L \quad 0 \quad L^* \quad -$

(8) The $\{k, k+1, ..., n-1\}_-$-truncation of an n-dimensional matroid ($1 < k < n$) is a k-dimensional matroid, whence it belongs to the diagram L_k (see Section 3.1.3(3)).

3.2 Orders

3.2.1 Definitions and notation

Given a geometry Γ over the set of types I, a type $i \in I$ and a cardinal number $k > 0$, we say that Γ has *order* k at i if every panel of Γ of cotype i is contained in precisely $k+1$ chambers (of course, $k+1 = k$ if k is infinite).

We say that the geometry Γ *has order* $(k_i)_{i \in I}$ if it has order k_i at i, for every $i \in I$. In this case, we also say that Γ *admits order*, when we are not interested in the particular values k_i of its orders.

If Γ has order $(k_i)_{i \in I}$ and $k_i = k$ for some cardinal number k and for all $i \in I$, then we say that Γ has *uniform order* k, or that it has *order* k, for short.

A geometry Γ is said to be *thin* at the type i if it has order 1 at i. Γ is called *thin* if it has uniform order 1. A geometry Γ is *thick* at the type i if every panel of Γ of cotype i is contained in at least three chambers. Γ is said to be *thick* if it is thick at every type. The next lemma is evident:

Lemma 3.1 *The ordinary m-gon is the only thin geometry of rank 2 and gonality m.*

The following is a trivial consequence of the previous lemma:

Proposition 3.2 *Let $M = (m_{ij})_{i,j \in I}$ be a Coxeter matrix and let Γ be a thin geometry over the set of types I such that, for every choice of distinct types $i, j \in I$, all residues of Γ of type $\{i, j\}$ have gonality m_{ij}. Then Γ belongs to the Coxeter diagram represented by M.*

If every panel of Γ of cotype i is contained in at most k chambers for some given positive integer k (in particular, if Γ has finite order at i), then we say that Γ is *locally finite* at i. The geometry Γ is said to be *locally finite* if it is locally finite at every type (in particular, if it has finite orders).

It is clear that a locally finite geometry Γ of rank 2 is finite if it has diameter $d < \infty$ (see Section 1.2.3 for the definition of d). Indeed, if every element of Γ is contained in at most k chambers, then a straightforward computation shows that we have $|P \cup L| \leq 1 + k \cdot \sum_{i=0}^{d-1}(k-1)^i$.

Remark. Requiring that all panels of Γ of cotype i belong to a finite number of chambers is not sufficent to obtain the local finiteness of Γ at i, as the following example shows. Let Γ be an infinite tree with a vertex 0 of valency 2 such that every vertex of Γ at distance k from 0 has valency $k+2$. The graph Γ is a generalized ∞-gon with the words 'even' and 'odd' as types, a vertex of Γ being declared of even or odd type according to whether its distance from 0 is even or odd. It is obvious that Γ is not locally finite in our sense. Nevertheless, each of the elements of Γ is incident to a finite number of elements.

Let Γ belong to a diagram $\mathcal{D}$. If Γ has order k at the type i, then we write k under the node of $\mathcal{D}$ corresponding to i. For instance, the following pictures represent the class of projective planes with lines of size $s+1$, the class of affine planes with lines of size s and the class of grids, respectively:

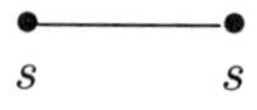

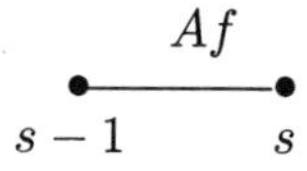

$$\underset{s}{\bullet}\!\!-\!\!\!-\!\!\!-\!\!\underset{s}{\bullet}\quad,\quad \underset{s-1}{\bullet}\!\!\overset{Af}{-\!\!\!-\!\!\!-}\!\!\underset{s}{\bullet}\quad \text{and} \quad \bullet\!=\!=\!=\!\underset{1}{\bullet}$$

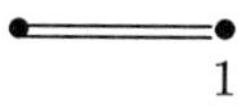

If Γ belongs to a string diagram and has orders, then we give those orders in a sequence, understanding that the ith order corresponds to the ith node of the diagram, counting the nodes of the diagram from left to right. For instance, if Γ has diagram and orders as follows, then we say that Γ has order (r, s, t):

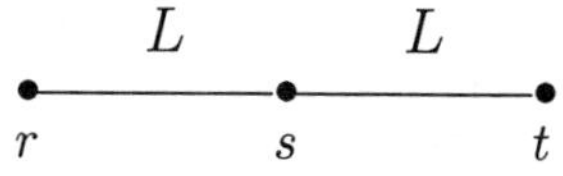

$$\underset{r}{\bullet}\!\!\overset{L}{-\!\!\!-\!\!\!-}\!\!\underset{s}{\bullet}\!\!\overset{L}{-\!\!\!-\!\!\!-}\!\!\underset{t}{\bullet}$$

In order to avoid confusion between orders and types when integers are chosen to denote types, we write types over the corresponding nodes of the diagram, as we have done till now. For instance, the following picture denotes the class of generalized quadrangles with set of types $\{0, 1\}$ admitting orders s and t at the types 0 and 1 respectively ($s+1$ points on every line and $t+1$ lines through every point, if the elements of type 0 and 1 are chosen as points and lines, respectively):

0 1

•══•

s t

Certain labels for edges of diagrams contain some information about orders. For instance, the label c for circular spaces contains the information that we have order 1 at the first node of the edge. In this cases, the orders implicit in the diagram may be omitted. For instance, we can write

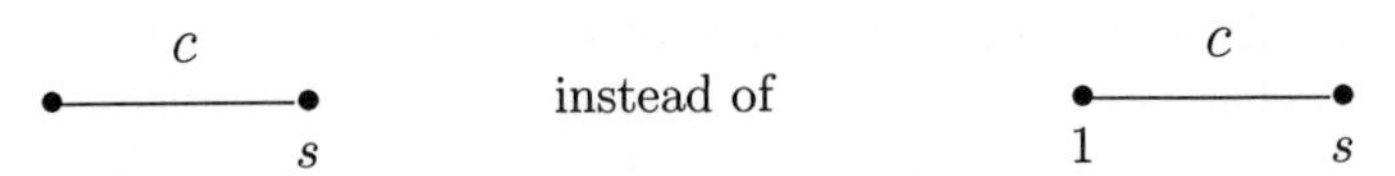

When diagrams are combined with orders, some combinations of orders and labels for edges are synonymous. For instance, the following are pairs of synonymous symbols:

c: •—• (1)	and	•—• (1, 1)	
c: •—• (2)	and	Af: •—• (1, 2)	
L: •—• (s, s)	and	•—• (s, s)	(with $s < \infty$)
L: •—• (s-1, s)	and	Af: •—• (s-1, s)	(with $s < \infty$)
L: •—• (1, s)	and	c: •—• (s)	(with $s < \infty$)
N: •—• (s, 1)	and	•══• (s, 1)	

The following strings are synonymous, too:

3.2.2 Examples of rank 2

All partial geometries admit finite orders, by definition. A t-(v, k, λ)-design Γ also admits orders. Indeed it is easily seen that every point of Γ is in precisely $h = \lambda[\binom{v-1}{t-1}/\binom{k-1}{t-1}]$ blocks. Hence Γ has order $(k-1, h-1)$. Non-thick nets are just grids; not all grids admit orders. All finite thick nets are partial geometries, hence they admit orders. It is not hard to prove that all infinite thick nets also admit orders.

Trivially, every generalized digon admits orders. It is also well known that a projective plane has uniform order if and only if it has an order at one of its two types and that every non-degenerate projective plane has a uniform order, say s ([90], Chapter 3; in Proposition 3.5 we will prove the same statement for generalized m-gons with m any odd integer > 1). The ordinary triangle (thin projective plane) is the only degenerate projective plane admitting an order.

Affine planes are obtained removing a line from a non-degenerate projective plane. Therefore every affine plane admits orders. In fact, if s is the order of the projective plane from which a given affine plane Γ has been obtained, then Γ has order $(s-1, s)$. Of course, we have $s - 1 = s$ if s is infinite.

Turning to generalized m-gons, we have the following propositions.

Proposition 3.3 *Every thick generalized m-gon with $m < \infty$ admits orders. Furthermore, every thick generalized m-gon with $m < \infty$ and m odd has uniform order.*

Proof. Let Γ be a thick generalized m-gon with $m < \infty$ and let a, b, a' be distinct elements of Γ with $a{*}b{*}a'$. Since Γ is thick, there is another element $a'' \neq a, a'$ in the residue Γ_b of b. Let $a_0, a_1, ..., a_m$ be a path of length m in the incidence graph of Γ, with $a_0 = a$, $a_1 = b$, $a_2 = a''$ and $a_{i-1} \neq a_{i+1}$ for $i = 1, 2, ..., m-1$. Such a path exists because Γ has gonality m. For the very same reason, a_m has distance m from a. Let a_{m+1} be an element in the residue Γ_{a_m} of a_m. Since Γ has both diameters equal to m, there is path $a_{m+1}, a_{m+2}, ..., a_{2m}$ with $a_{2m} = a_0$, and this path is unique, because Γ has gonality m. In particular, a_{2m-1} is the unique element of Γ_{a_0} at distance $m-2$ from a_{m+1}. For the same reason, every element $a_{2m-1} \in \Gamma_{a_0}$ in turn determines a unique element $a_{m+1} \in \Gamma_{a_m}$ at distance $m-2$ from a_{2m-1}. Therefore there is a bijection f from the residue of $a_0 = a$ to the residue of

a_m. Substituting a' for a in the role of a_0 and keeping the elements a_1, a_2, ..., a_m of the path we were considering, a bijection $f' : \Gamma_{a'} \longrightarrow \Gamma_{a_m}$ is also obtained. Therefore $f^{-1}f'$ is a bijection from $\Gamma_{a'}$ to Γ_a. The first claim of the proposition follows from this and the connectedness of Γ.

If m is odd, then a_m and a have distinct types. Hence the existence of the bijection f proves the second claim. □

Proposition 3.4 *A generalized m-gon with $m < \infty$ admits orders if it admits an order $s > 1$ at one type.*

Proof. Let Γ be a generalized m-gon ($m < \infty$) admitting order $s > 1$ at the type P (P and L are the two types of Γ, as usual). Given any $b \in L$ and any two elements $a, a' \in \Gamma_b$, we have $|\Gamma_a| = |\Gamma_{a'}|$. Indeed there is another element $a'' \in \Gamma_b$, because $s > 1$, and we can use the same argument as in the proof of Proposition 3.3. The conclusion follows by the connectedness of Γ. □

Proposition 3.5 *A generalized m-gon with $m < \infty$ and m odd has a uniform order if it admits an order at one type.*

Proof. Let Γ be a generalized m-gon ($m < \infty$ and odd) admitting order s at the type P. If $s > 1$, we can apply Proposition 3.4 and we have finished.

Let Γ be thin at P and assume that Γ is not thin, by way of conradiction. Then there is some $b \in P$ incident to at least three distinct elements $a, a', a'' \in L$. As in the proof of Proposition 3.3, we can choose a path $a_2, a_3, ..., a_m$ with $a_2 = a''$ and a_m at distance m from both a and a' and at distance $m - 2$ from a''. We obtain bijections $f : \Gamma_a \longrightarrow \Gamma_{a_m}$ and $f' : \Gamma_{a'} \longrightarrow \Gamma_{a_m}$ such that, for every $x \in \Gamma_a$ (for every $x' \in \Gamma_{a'}$), $f(x)$ (respectively, $f'(x')$) is the unique element of Γ_{a_m} at distance $m - 2$ from x (from x'), as in the proof of Proposition 3.3.

We have $a_m \in P$ because m is odd. Furthermore, Γ_{a_2} has precisely two elements, because Γ is thin at P and f, f' are bijections from Γ_a and $\Gamma_{a'}$ to Γ_{a_m}. The element b is one of the two elements of Γ_a and it is clear that $f(b) = a_{m-1}$. Therefore, if c is the other element of Γ_a and d is the other element of Γ_{a_m}, we have $f(c) = d$. Similarly, if c' is the element of $\Gamma_{a'}$ different from b, we have $f'(c') = d$. Hence, d has distance $m - 1$ from a and a'. However, this is impossible, because Γ has gonality m. This contradiction shows that Γ is thin. □

Exercise 3.7. Let Γ be a rank 2 geometry with uniform order s. Prove that its flag complex $\mathcal{K}(\Gamma)$ has order $(s, 1)$. In particular, a generalized $2m$-gon of order $(s, 1)$ arises in this way from every generalized m-gon with uniform order s (see Section 1.4.1).

Exercise 3.8. Prove by counterexamples that the statements of Propositions 3.3, 3.4 and 3.5 are false in the case of $m = \infty$.

3.2.3 Some higher rank examples

(1) It follows from Propositions 3.3 and 3.5 that every non-degenerate projective geometry has uniform order and that a projective geometry has uniform order if it admits an order at some type. In particular, simplices (namely, thin projective geometries) are the only degenerate projective geometries admitting orders.

All affine geometries admit orders. More precisely, if an affine geometry Γ is obtained from a projective geometry of order s, then Γ has order $(s-1, s, ..., s)$ (with $s-1 = s$ if $s = \infty$). We also say that the affine geometry Γ has *order* s, by abuse of notation. Evidently, Γ is thick if and only if $s > 2$. When $s = 2$, then Γ is thin at the type 0.

(2) Since a projective geometry has uniform order if it admits an order at some type, the same holds for any geometry belonging to a Coxeter diagram $M = (m_{ij})_{i,j \in I}$ with $m_{ij} = 2$ or 3 for all $i, j \in I$. The Coxeter diagram D_n is of this kind, for instance.

In particular, a D_n-building Γ over a field K has uniform order $s = |K|$.

(3) Every thick-lined polar space of rank ≥ 3 has order $(s, s, ..., s, t)$, with $s > 1$, by Propositions 3.3 and 3.4. For instance, $\mathcal{Q}^+_{2n-1}(K)$ has order $(s, s, ..., s, 1)$ with $s = |K|$.

If $\mathcal{P}$ is a weak polar space of rank n with all lines of size 2, then $\mathcal{P}$ admits orders at the first $n-1$ nodes of the C_n diagram. Every such polar space arises from some complete n-partite graph $\mathcal{G}$, as we have remarked in Section 1.3.5, and admits orders if and only if all classes of the n-partition of $\mathcal{G}$ have the same size, say $s+1$. In that case, $(1, 1, ..., 1, s)$ is the order of $\mathcal{P}$. When $s = 1$, we obtain the thin polar space of rank n (in particular, the octahedron, when $n = 3$).

(4) Let $\mathcal{D}$ be a diagram of the form $c^k.\mathcal{X}$, where $\mathcal{X}$ is some diagram of rank $n-k$ (see Sections 3.1.3, (9) and (10), and Exercises 3.1 and 3.5, for instance). Trivially, every geometry belonging to $\mathcal{D}$ is thin at the first k nodes of $\mathcal{D}$ (of course, we assume that $\mathcal{D}$ is drawn starting from the left with the series of c-edges). For instance, the enrichments of the Steiner systems for M_{22}, M_{23} and M_{24} have orders $(1, 4, 4)$, $(1, 1, 4, 4)$ and $(1, 1, 1, 4, 4)$ respectively.

(5) All geometries obtained from tessellations as in Section 2.1 are thin. Their quotients considered in Section 2.1.2 are also thin. Their flag complexes are also thin. More generally, the flag complex of a thin geometry is thin.

Exercise 3.9. Compute the orders for the geometries defined in 2.2.1, assuming that K is finite.

Exercise 3.10. Compute the orders for the enrichments of the Steiner systems for M_{11} and M_{12} defined in Section 2.4 and prove that these geometries also belong to the following diagrams:

$(c^3.Af)$ •—c—•—c—•—c—•—Af—• for M_{12}

$(c^2.Af)$ •—c—•—c—•—Af—• for M_{11}

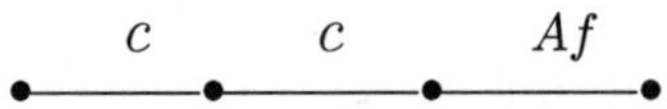

(Note that the affine plane of order 2 and the circular space on four points are the same geometry.)

Exercise 3.11. Let Γ be as in Exercise 3.6. Prove that $\mathcal{K}(\Gamma)$ is thin at all types except possibly the first one and that $\mathcal{K}(\Gamma)$ admits orders if and only if Γ has a uniform order.

Exercise 3.12. Assuming $K = GF(q)$, compute the orders of the truncation of the D_4-building over K considered in Section 3.1.4(2).

Exercise 3.13. Compute the orders of the truncations considered in Section 3.1.4(3)–(7), assuming that the geometries we truncate admit orders.

Exercise 3.14. Check that finite classical generalized quadrangles have orders as follows:

$\mathcal{S}_3(q)$ has uniform order q;
$\mathcal{Q}_4(q)$ is the dual of the above, hence it has order q;
$\mathcal{Q}_5^-(q)$ has order (q, q^2);
$\mathcal{H}_3(q^2)$ is the dual of the above, hence it has order (q^2, q);
$\mathcal{H}_4(q^2)$ has order (q^2, q^3).

Exercise 3.15. Check that finite classical polar spaces have orders as follows:

$(q, q, ..., q, q)$ for $\mathcal{S}_{2n-1}(q)$ and $\mathcal{Q}_{2n}(q)$,
$(q, q, ..., q, 1)$ for $\mathcal{Q}_{2n-1}^+(q)$,
$(q, q, ..., q, q^2)$ for $\mathcal{Q}_{2n+1}^-(q)$,
$(q^2, q^2, ..., q^2, q)$ for $\mathcal{H}_{2n-1}(q^2)$,
$(q^2, q^2, ..., q^2, q^3)$ for $\mathcal{H}_{2n}(q^2)$.

Exercise 3.16. Check that the generalized quadrangle $T_2^*(O)$ described in Section 2.3 has order $(q-1, q+1)$.

Exercise 3.17. Given a plane S of $PG(m,q)$, with $m \geq 4$ and q even, let O be a hyperoval of S. We can form a geometry Γ of rank $n = m-1$, as follows. For every $i = 0, 1, ..., m-3$, the i-dimensional subspaces of $PG(m,q)$ disjoint from S are taken as elements of type i. The elements of Γ of type $m-2$ are the $(m-2)$-dimensional subspaces X of $PG(m,q)$ such that $X \cap S$ is a point of O. The incidence relation is the natural one, inherited from $PG(m,q)$.

Prove that Γ has diagram and orders as follows:

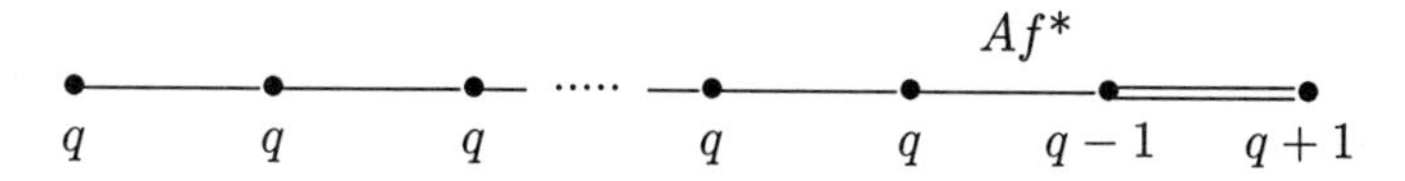

3.2.4 Restrictions on orders

Some restrictions on the orders of a finite linear space admitting orders and, more generally, of a partial geometry, have been discussed in Section 1.2.6(4) and Appendix II of Chapter 1.

Turning to non-degenerate finite projective planes, the order s of such a plane is a prime power in all known examples and it has been conjectured that no finite projective plane exists of order s with s not a prime power. However, no proof of this conjecture has been found till now. The most far reaching result in this direction is the celebrated Bruck–Ryser theorem (see [90]): if $s \equiv 1$ or 2 $(mod\ 4)$, then every odd prime p dividing s to an odd power satisfies $p \equiv 1$ $(mod\ 4)$. This immediately rules out the case of $s = 6$ (and infinitely many other cases, too). However, the next non-prime-power case that might arise, namely $s = 10$, safely passes through the Bruck–Ryser test. This case was ruled out in 1990, with the aid of a computer [115] (41 years after the paper by Bruck and Ryser appeared). $s = 12$ is the next case standing against us.

We have no problems of this kind for non-degenerate projective geometries of dimension $n > 2$. Indeed, they are all classical (Theorem 1.3). Hence, in the finite case, they have prime power order.

Concerning finite generalized m-gons, we have the following theorem:

Theorem 3.6 (Feit–Higman) *Let Γ be a generalized m-gon with $m < \infty$ and admitting finite orders s and t. Then one of the following holds:*

(i) we have $m = 2, 3, 4, 6$ or 8;
(ii) we have $s = 1$ or $t = 1$ and $m = 12$;

(iii) we have $s = t = 1$ (namely, Γ is an ordinary polygon).

The proof of the above theorem is a computation of eigenvalues of powers of adjacency matrices (see Appendix II of Chapter 1 for the definition of adjacency matrices). We do not give that proof here. The reader may find it in [66].

The next corollary is an easy consequence of Theorem 3.6.

Corollary 3.7 *Let Γ be a generalized m-gon ($m < \infty$) with finite uniform order $s > 1$. Then $m = 2$, 3, 4 or 6.*

Proof. We have $m = 2$, 3, 4, 6 or 8, by Theorem 3.6. However, the case of $m = 8$ is immediately ruled out. Indeed, if $m = 8$, then $\mathcal{K}(\Gamma)$ is a generalized 16-gon with orders $m < \infty$ and 1 (see Exercise 3.7) and this is impossible, by Theorem 3.6. □

We will now consider the case of $m = 4$ more closely, referring the reader to [106] (part A) for the case of $m = 6$ and 8.

Let Γ be a finite thick generalized quadrangle. Γ has finite orders s, t, by Proposition 3.3. Of course, $1 < s, t$, because we have assumed that Γ is thick. We have already remarked that $s + t$ divides $st(s + 1)(t + 1)$ (Appendix II of Chapter 1). Furthermore, the following inequalities hold ([166], Chapter 1):

(i) $s \leq t^2$ (dually, $t \leq s^2$);

(ii) if $s < t^2$, then $s \leq t^2 - t$ (and dually).

However, these conditions allow infinitely many possibilities for (s, t) for which no examples are known. We now list all pairs (s, t) for which examples are known. In what follows, q is a prime power. When only classical examples exist, we write 'classical only' beside the pair, also mentioning the classical examples with that order (see also Exercise 3.14). When both classical and non-classical examples are presently known, we mention the classical examples in parentheses. When only non-classical examples are known with those orders, we write 'non-classical'. The reader should consult [166] and [210] for non-classical examples. As the order in which types are given is in principle irrelevant, we do not mention dual cases separately.

(1) $s = t = q$, q odd. Classical only ($\mathcal{S}_3(q)$; dually, $\mathcal{Q}_4(q)$).

(2) $s = t = q$, q even. (Classical example: $\mathcal{S}_3(q)$, namely $\mathcal{Q}_4(q)$.)

(3) $s = q, t = q^2$. (Classical example: $\mathcal{Q}_5^-(q)$; dually, $\mathcal{H}_3(q^2)$.)

(4) $s = q^2, t = q^3$. Classical only ($\mathcal{H}_4(q^2)$).

(5) $s = q - 1, t = q + 1$, $q > 2$. Non-classical.

Exercise 3.18. Prove that no finite generalized quadrangles exist of order $(s, s+1)$, $s > 1$.

Exercise 3.19. Let p be a prime and let a, b positive integers with $p > 1$ and $a \leq b$. Prove that a generalized quadrangle of order (p^a, p^b) exists only if there are positive integers c, k such that $a = kc$ and $b = (k+1)c$.

Exercise 3.20. Prove that for every pair of positive integers s, t there is just one generalized ∞-gon of order (s, t).

3.3 From local to global

We can now come to the 'philosophy' of diagram geometry. Giving a diagram $\mathcal{D}$ for a class $\mathbf{C}$ of geometries and, possibly, attaching orders to it, is summarizing some 'local' information on $\mathbf{C}$. In some (with luck, many) cases that local information already implies some substantial 'global' information on $\mathbf{C}$. In the best case, $\mathcal{D}$ characterizes $\mathbf{C}$ (namely, $\mathbf{C}$ is precisely the class of all geometries belonging to $\mathcal{D}$). The problem we will deal with in this book is indeed the following one:

how to recover a great deal of global information on the members of $\mathbf{C}$ *from the local information embodied in* $\mathcal{D}$ *(and, possibly, in an assignment of orders for* $\mathcal{D}$*) ?*

Some results in this style will be given in Chapters 7, 12, 13, 14 and 15. However, before starting with this programme, we must first describe the structures of geometries belonging to disconnected diagrams. We will do this in the next chapter. This will allow us to restrict our investigation to connected diagrams and will provide a tool enabling us to analyse residues in geometries with connected diagrams. Then we will be ready to begin our job.

Chapter 4

Diagram Graphs

4.1 Definition

Given a geometry Γ over a set of types I, let $\mathcal{D}(\Gamma)$ be the graph defined on I as follows: two distinct types $i, j \in I$ are adjacent in $\mathcal{D}(\Gamma)$ when at least one of the residues of Γ of type $\{i, j\}$ is not a generalized digon. We call $\mathcal{D}(\Gamma)$ the *diagram graph* of Γ.

Let Γ belong to a diagram $\mathcal{D}$. According to the notation stated in Section 3.1.2, we represent generalized digons by drawing no edges in $\mathcal{D}$. Then $\mathcal{D}(\Gamma)$ is just the simple graph obtained from the labelled graph $\mathcal{D}$ by dropping all labels. For instance, in all examples of Section 3.1.3 except (5) and (6.e) the diagram graph is a string. The diagram graph of a building of type D_n looks like its Coxeter diagram (Section 3.1.3(5)). The diagram graph of the geometry of black and white triangles of Section 2.1.2(10) is a triangle, the same as its Coxeter diagram (Section 3.1.3(6.e)).

In order to avoid confusion with single strokes representing projective planes in Coxeter diagrams, we use empty circlets instead of dots to represent nodes of diagram graphs:

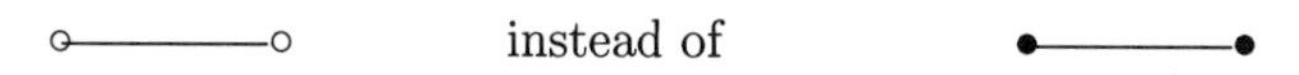

We keep the notation of Chapter 3 for types, when types are integers, writing them above the nodes of a diagram graph.

The geometry Γ admits a diagram if and only if $\mathcal{D}(\Gamma)$ can be viewed as a diagram for Γ, where only two classes of rank 2 geometries are chosen to label edges, namely the class of generalized digons and the class of all rank 2 geometries that are not generalized digons. However, there are 'bad' geometries that do not belong to any diagram, as we have remarked in

Section 3.1.1; their diagram graphs are not diagrams. Because of this, we have preferred the word 'diagram graph' to the word 'basic diagram' used in [19].

The following lemmas are quite easy to prove. We leave their proofs for the reader.

Lemma 4.1 *For every non-maximal flag F of Γ, the diagram graph $\mathcal{D}(\Gamma_F)$ of the residue Γ_F of F is a subgraph of the graph induced by $\mathcal{D}(\Gamma)$ on the cotype of F.*

Lemma 4.2 *The geometry Γ admits a diagram if and only if, for every non-maximal flag F, the diagram graph $\mathcal{D}(\Gamma_F)$ of Γ_F is the graph induced by $\mathcal{D}(\Gamma)$ on the cotype of F.*

Lemma 4.3 *For every proper subset J of the set I of types of Γ, the graph induced by $\mathcal{D}(\Gamma)$ on $I - J$ is a subgraph of the diagram graph $\mathcal{D}(Tr_J^-(\Gamma))$ of the truncation $Tr_J^-(\Gamma)$ of Γ (notation as in Section 1.6).*

Lemma 4.4 *Two distinct types $i, j \in I$ are joined in $\mathcal{D}(\Gamma)$ if and only if the i- and j-adjacency relations of $\mathcal{C}(\Gamma)$ do not commute (that is, if and only if $\Phi^i\Phi^j \neq \Phi^j\Phi^i$).*

4.2 Reducibility

Let I_1, I_2, ..., I_m $(m > 1)$ be pairwise disjoint finite nonempty sets and, for every $i = 1, 2, ..., m$, let $\Gamma_i = (V_i, *_i)$ be a geometry over the set of types I_i, with set of elements V_i and type function $t_i : V_i \longrightarrow I_i$. Assume $V_i \cap V_j = \emptyset$ for $i, j = 1, 2, ..., m$, $i \neq j$.

Let $*$ be the relation defined on $V = \bigcup_{i=1}^m V_i$ by the following clauses: $*$ induces $*_i$ on V_i for every $i = 1, 2, ..., m$, and we have $x * y$ for every choice of $x \in V_i$ and $y \in V_j$, with $i, j = 1, 2, ..., m$, $i \neq j$. Let us set $I = \bigcup_{i=1}^m I_i$ and let t be the mapping from V onto I inducing t_i on V_i for every $i = 1, 2, ..., m$.

It is easily seen that $\Gamma = (V, *)$ is a geometry, where we take I and t as set of types and type function, respectively. Γ is called the *direct sum* of the geometries Γ_1, Γ_2, ..., Γ_m, and we write $\Gamma = \oplus_{i=1}^m \Gamma_i$. We also say that Γ (or any geometry isomorphic to Γ) is *reducible* and that the summands Γ_1, Γ_2, ..., Γ_m form a *decomposition* of Γ.

We say that a geometry is *irreducible* if it is not reducible. Needless to say, all geometries of rank 1 are irreducible. Generalized digons are the only reducible geometries of rank 2.

If Γ is a direct sum of irreducible geometries $\Gamma_1, \Gamma_2, ..., \Gamma_m$, then these geometries are said to form an *irreducible decomposition* of Γ, or to be *irreducible components* of Γ. We extend this terminology to the case when Γ is irreducible; in this case we say that Γ is the *irreducible decomposition*, or the *irreducible component*, of itself.

The next lemma is quite trivial:

Lemma 4.5 *Let $\Gamma = \oplus_{i=1}^m \Gamma_i$, where $\Gamma_1, \Gamma_2, ..., \Gamma_m$ are geometries over the sets of types $I_1, I_2, ..., I_m$ respectively, as above.*

Then, for every $i = 1, 2, ..., m$ and for every flag F of Γ of cotype I_i, we have $\Gamma_i = Tr^+_{I_i}(\Gamma) = \Gamma_F$.

The diagram graph of Γ is the disjoint union of the diagram graphs of the summands $\Gamma_1, \Gamma_2, ..., \Gamma_m$.

The next theorem is also quite elementary; in spite of this, it is one of the most effective tools in diagram geometry. The earliest version of it appeared in [219].

Theorem 4.6 (Direct sum theorem) *Let Γ be a geometry over a set of types I and let $\{J, K\}$ be a partition of I into two classes. Then the following are equivalent:*

(i) we have $\Gamma = Tr^+_J(\Gamma) \oplus Tr^+_K(\Gamma)$;
(ii) the subsets J and K are not connected by any path in $\mathcal{D}(\Gamma)$;
(iii) we have $\Phi^J\Phi^K = \Phi^K\Phi^J$.

Proof. Trivially, (i) implies (ii); furthermore, (ii) implies (iii) by Lemma 4.4. Therefore, we only must prove that (iii) implies (i). Let (iii) hold. We must show that, given any two elements x, y of Γ with types in J and in K respectively, we have $x * y$. Let A and B be chambers containing x and y respectively. We have $\Phi^K\Phi^J = \Omega$ by (iii). Therefore, there is a chamber C with $A\Phi^K C$ and $C\Phi^J B$. Since x belongs to A and has type in J, we have $x \in C$. Similarly, $y \in C$. Since both x and y are in C, we have $x * y$. □

The following is a trivial consequence of the direct sum theorem:

Corollary 4.7 *A geometry is irreducible if and only if its diagram graph is connected.*

Corollary 4.8 *Let $I_1, I_2, ..., I_m$ be the connected components of $\mathcal{D}(\Gamma)$. Then $Tr^+_{I_1}(\Gamma), Tr^+_{I_2}(\Gamma), ..., Tr^+_{I_m}(\Gamma)$ form the unique irreducible decomposition of Γ.*

Proof. Use the direct sum theorem, then Lemma 4.5 to continue the decomposition process till the end and to prove the uniqueness of the irreducible decomposition of Γ. $\square$

Exercise 4.1. Given graphs $\mathcal{G}_i = (V_i, \sim_i)$, $i = 1, 2, ..., m$, the product $\mathcal{G} = \prod_{i=1}^m \mathcal{G}_i$ of the graphs $\mathcal{G}_i$ $(i = 1, 2, ..., m)$ is defined as follows: $\prod_{i=1}^m V_i$ is the set of vertices of $\mathcal{G}$ and two vertices $(x_1, x_2, ..., x_m)$, $(y_1, y_2, ..., y_m)$ of $\mathcal{G}$ are adjacent in $\mathcal{G}$ when for some $k = 1, 2, ..., m$ we have $x_i = y_i$ for all indices $i = 1, 2, ..., k-1, k+1, ..., m$ and $x_k \sim_k y_k$.

Let Γ be a geometry over a set of types I, let $\{I_1, I_2, ..., I_m\}$ be a partition of I and $\Gamma_i = Tr^+_{I_i}(\Gamma)$, for $i = 1, 2, ..., m$. Prove that we have $\Gamma = \oplus_{i=1}^m \Gamma_i$ if and only if $\mathcal{C}(\Gamma) = \prod_{i=1}^m \mathcal{C}(\Gamma_i)$.

Exercise 4.2. Let Γ be a geometry of rank 3 not admitting any diagram. Prove that one of the following holds:

(1) the diagram graph of Γ is a triangle;
(2) the diagram graph of Γ is a string and, if i, j are the two terminal nodes of the string, then, up to interchanging i and j if necessary, we can always find distinct elements x, y of type i incident to the same panel of Γ and such that all elements of type j incident to x are also incident to y.

Exercise 4.3 Let Γ be a geometry of rank ≥ 3, admitting a diagram. Assume that $\mathcal{D}(\Gamma)$ is a tree and let J be a set of terminal nodes of that tree. Prove that $Tr^-_J(\Gamma)$ admits a diagram and that its diagram graph is the tree obtained removing from $\mathcal{D}(\Gamma)$ the nodes in J.

4.3 String diagram graphs

In this section $\Gamma = (V, *)$ is a geometry of rank n and we assume that the diagram graph $\mathcal{D}(\Gamma)$ of Γ is a string:

(note that the diagram graphs of almost all geometries described in Chapters 1 and 2 are strings).

We take the nonnegative integers 0, 1, ..., $n-1$ as types, labelling the nodes of $\mathcal{D}(\Gamma)$ as follows:

0 1 2 $n-2$ $n-1$

An orientation of $\mathcal{D}(\Gamma)$ is implicit in the above convention, the nodes labelled by 0 and $n-1$ being naturally taken as the first and the last node, respectively. Needless to say that $\mathcal{D}(\Gamma)$ might be given the opposite orientation as well. However, changing the orientation of $\mathcal{D}(\Gamma)$ has no real influence on what we are going to say, except for the notation.

4.3.1 Posets

The set of types $I = \{0, 1, ..., n-1\}$ inherits a natural ordering from the ordered set of natural numbers. Hence the type function $t : V \longrightarrow I$ of Γ induces an ordering on every flag of Γ. Thus, if $F = \{x_1, x_2, ..., x_m\}$ is a flag of Γ with $t(x_1) < t(x_2) < ... < t(x_m)$, we identify F with the ordered m-tuple $(x_1, x_2, ..., x_m)$. According to this convention, when we say that an ordered pair (x, y) of elements of Γ *is a flag*, then we understand that $x * y$ and that $t(x) < t(y)$. We will write $x \leq_\Gamma y$ to mean that either the ordered pair (x, y) is a flag or $x = y$.

Theorem 4.9 *The relation $\leq_\Gamma$ is a partial order.*

Proof. We need only prove the transitivity of $\leq_\Gamma$. Let $x \leq_\Gamma y \leq_\Gamma z$, for distinct elements $x, y, z \in V$. By Lemma 4.1, $t(x)$ and $t(z)$ belong to distinct connected components of $\mathcal{D}(\Gamma_y)$. We have $x \leq_\Gamma z$, by the direct sum theorem. □

Therefore $\mathcal{P}_0(\Gamma) = (V, \leq_\Gamma)$ is a poset, in fact a graded poset of rank n with dimension function t. We call it the *poset of* Γ (relative to the choice of 0 as the initial node of $\mathcal{D}(\Gamma)$).

Given two distinct elements $x, y \in V$ with $x \leq_\Gamma y$, the open interval $]x, y[$ of $\mathcal{P}_0(\Gamma)$ is the residue of a flag of Γ of cotype $\{t(x)+1, t(x)+2, ..., t(y)-1\}$ containing x and y. Given an element $x \in V$, the open interval $]-, x[$ is the residue of a flag of type $\{t(x), t(x)+1, ..., n-1\}$ containing x. Similarly, the open interval $]x, +[$ is the residue of a flag of type $\{0, 1, ..., t(x)\}$ containing x.

When $t(x) > 0$, the open interval $]-, x[$ will be called the *lower residue* of x, also the *underlying structure* of x, or the *structure* of x for short. We have $]-, x[= \Gamma_x$ when $t(x) = n-1$. According to the notation used for many examples of Section 1.3, we write Γ_x^- for $]-, x[$. Trivially, $\Gamma_x^- = \Gamma_x$ if $t(x) = n-1$.

When $t(x) < n-1$, the open interval $]x, +[$ will be called the *upper residue* of x, also the *star* of x. We denote it by Γ_x^+. Trivially, we have $\Gamma_x^+ = \Gamma_x$ when $t(x) = 0$.

When $0 < t(x) < n-1$ we have $\Gamma_x = \Gamma_x^- \oplus \Gamma_x^+$, by the direct sum theorem.

We finish this section with a characterization of the posets arise from geometries with string diagram graphs.

Theorem 4.10 *Let $\mathcal{P} = (V, \leq)$ be a graded poset of rank n, with dimension function $t : V \longrightarrow \{0, 1, ..., n-1\}$.*

We have $\mathcal{P} = \mathcal{P}_0(\Gamma)$ for a (uniquely determined) geometry Γ of rank n with string diagram graph, if and only if all the following hold:

(i) every open interval of $\mathcal{P}$ of rank 1 has at least 2 elements;

(ii) every open interval of $\mathcal{P}$ of rank ≥ 2 is connected;

(iii) for every $i = 1, 2, ..., n-1$, there are elements $x, y \in V$ with types $t(x) = i-1$, $t(y) = i$ and such that $x \not< y$.

We leave the (easy) proof for the reader. We only remark that (ii) corresponds to the residual connectedness of geometries (Theorem 1.16), (i) corresponds to the clause (i) in the definition of geometries (Section 1.1; see also Lemma 1.8) and (iii) corresponds to the connectedness of the diagram graph, by the direct sum theorem. The reader might also compare the above with Exercise 1.18. Conditions (i) and (ii) of Theorem 4.10 respectively correspond to conditions (i) and (iii) of Exercise 1.18. Condition (ii) of Exercise 1.18 corresponds to the assumption that $\mathcal{P}$ is graded.

The uniqueness of the geometry $\Gamma = (V, *)$ satisfying $\mathcal{P}_0(\Gamma) = \mathcal{P}$ is also evident (provided that (i), (ii) and (iii) hold in the graded poset $\mathcal{P}$, of course). Indeed, the following is the only way to define the incidence relation $*$ of Γ: given any two elements $x, y \in V$, we have $x * y$ iff either $x \leq y$ or $y \leq x$ (the reader may notice that we have used this definition in almost all examples of Chapters 1 and 2).

We say that a graded poset $\mathcal{P}$ satisfying (i), (ii) and (iii) of Theorem 4.10 is *geometric*. If $\mathcal{P}$ is a geometric graded poset, then the geometry Γ such that $\mathcal{P}_0(\Gamma) = \mathcal{P}$ will be denoted by $\Gamma(\mathcal{P})$ and it will be called the *natural geometry* of $\mathcal{P}$.

We have distinguished between $\mathcal{P}$ and $\Gamma(\mathcal{P})$ (and hence between $\mathcal{P}_0(\Gamma)$ and Γ). However this distinction is seldom relevant.

Exercise 4.4. (Direct products). For $i = 1, 2, ..., m$, let $\mathcal{P}_i$ be a geometric poset of rank n_i and let $\overline{\mathcal{P}}_i$ be the graded poset of rank $n_i + 1$ obtained by adding to $\mathcal{P}_i$ one element Ω_i on top. Let $\overline{\mathcal{P}} = \prod_{i=1}^{m} \overline{\mathcal{P}}_i$ be the direct product of the posets $\overline{\mathcal{P}}_1$, $\overline{\mathcal{P}}_2$, ..., $\overline{\mathcal{P}}_m$. Let $\mathcal{P}$ be the poset obtained by dropping the maximal element $(\Omega_1, \Omega_2, ..., \Omega_m)$ from $\overline{\mathcal{P}}$.

Prove that $\mathcal{P}$ is a geometric graded poset of rank $n = \sum_{i=1}^{m} n_i$.

If $\Gamma_i = \Gamma(\mathcal{P}_i)$ for $i = 1, 2, ..., m$ and $\Gamma = \Gamma(\mathcal{P})$, then we say that Γ is the *direct product* of the geometries Γ_1, Γ_2, ..., Γ_m and we write $\Gamma = \prod_{i=1}^{m} \Gamma_i$.

4.3.2 Points, lines and planes

Let Γ be a geometry of rank $n \geq 2$, with a string diagram graph and set of types $I = \{0, 1, ..., n-1\}$ where 0 is chosen as the first node of the string $\mathcal{D}(\Gamma)$, according to the natural ordering of the set of integers $\{0, 1, ..., n-1\}$.

The elements of type 0 and those of type 1 are called *points* and *lines* of Γ, respectively. The $\{0,1\}_+$-truncation of Γ is the *natural point–line system* of Γ (Section 1.6.2). We will denote it by $\mathcal{L}_0(\Gamma)$ (the index 0 should remind us of the orientation we have chosen for $\mathcal{D}(\Gamma)$). The collinearity relation of $\mathcal{L}_0(\Gamma)$ is called the (*natural*) *collinearity relation* of Γ. We denote it by $\perp$, as in Sections 1.2.3 and 1.6.2. The collinearity graph of $\mathcal{L}_0(\Gamma)$ is called the (*natural*) *collinearity graph* of Γ.

A line of Γ is said to be *thick* (*thin*) if it is incident to at least 3 (precisely 2) points. We say that Γ is *thick-lined* if all lines of Γ are thick. We say that Γ is *thin-lined* if all lines of Γ are thin.

When $n \geq 3$, we call the elements of type 2 *planes* (sometimes called 'quads' in the literature; also 'blocks' or 'circles', in some contexts).

We call the elements of type $n-1$ *dual points* and those of type $n-2$ *dual lines*. They form the *dual point–line system* of Γ. We denote this by $\mathcal{L}_{n-1}(\Gamma)$. The collinearity relation and the collinearity graph of $\mathcal{L}_{n-1}(\Gamma)$ will be called the *dual collinearity relation* and the *dual collinearity graph* of Γ, respectively. We denote the dual collinearity relation by $\top$. If $n \geq 3$, then the elements of type $n-3$ will be called *dual planes*.

We say that Γ is *top-thin* (*top-thick*) if every dual line is incident to precisely 2 (at least 3) dual points.

We now mention some conditions on lines or planes, that hold in many geometries with string diagram graphs (for instance, they hold in almost all examples of Chapters 1 and 2).

Good system of lines. We say that Γ has a *good system of lines* if $\mathcal{L}_0(\Gamma)$ is a partial plane. We may also define the dual property. We say that Γ has a *good dual system of lines* if $\mathcal{L}_{n-1}(\Gamma)$ is a partial plane.

Good system of planes. We say that Γ has a *good system of planes* if it has a good system of lines and, given a line x and a plane u, if there are distinct points incident to both x and u, then $x * u$. It is clear how the dual property (to have a *good dual system of planes*) can be defined.

Triangular property. We say that Γ satisfies the *triangular property* if it has a good system of planes and every 3-clique of the collinearity graph of Γ is in (the residue of) some plane. Needless to say, we can also consider the dual of this property, the *dual triangular property.*

The reader may check that the triangular property holds in matroids of dimension $n \leq 3$, in polar spaces of rank $n \geq 3$ (see Exercise 1.16), in affine polar spaces (Section 2.2.2). It also holds in attenuated spaces of rank $n \geq 3$ (Section 2.2.1), but it does not hold in the remaining examples of Section 2.2.1. However, all those examples have a good system of planes (provided they have rank ≥ 3, of course). The triangular property holds in in affine expansions of linear spaces (Section 2.3) but it fails to hold in almost all affine expansions of rank 2 geometries and in many affine expansions of polar spaces. However, all affine expansions have good systems of planes. The triangular property also holds in tessellations of surfaces (Section 2.1.2).

The halved octahedron (Section 2.1.2(9)) does not have a good system of lines. Its dual (the halved cube) has a good system of lines, but it does not have a good system of planes.

Flatness. Let $n \geq 3$. We say that Γ is *flat* if all elements of Γ of type $< n-2$ are incident to all dual points of Γ (when $n = 3$, the above amounts to saying that all points of Γ are incident to all planes of Γ).

The reader may check that the halved octahedron is a flat geometry of rank 3. More examples of flat geometries will be given later (Section 6.4.2; Exercises 7.21 and 7.22; Exercise 8.56).

Let $n \geq 4$. We say that Γ is *almost flat* if all points of Γ are incident to all dual points of Γ. Examples of almost flat but non-flat geometries of rank $n \geq 4$ will be given in Exercises 8.48 and 8.56.

Exercise 4.5. Let Γ belong to the following diagram:

$$\bullet \overset{\pi}{\text{———}} \bullet \overset{L}{\text{———}} \bullet$$

(it is understood that the diagram is oriented from left to right). Prove that Γ has a good system of lines.

(Hint: if x, y are lines incident to a common point, say a, then we can take a plane u incident to both x and y in Γ_a, since Γ_a is a linear space. Then we pass to the partial plane Γ_u ...)

Exercise 4.6. Let Γ have the above diagram and a good system of planes. Prove that Γ satisfies the triangular property.

(Hint: given a 3-clique $\{a, b, c\}$ of the collinearity graph, let x, y be the lines through a and b and through a and c, respectively. We find a plane incident to both x and y in Γ_a ...)

Exercise 4.7. Let Γ have a good system of lines and diagram as follows (oriented from left to right, as usual):

$L \qquad X$

where X may denote any class of rank 2 geometries other than generalized digons. Prove that Γ has a good system of planes.

(Hint: given a line x and a plane u incident to two distinct points a, b of x, we can take the line y of u through x and y. Indeed Γ_u is a linear space. However, $x = y$ because Γ has a good system of lines.)

Exercise 4.8. Prove that flat geometries belonging to the following diagram does not have a good systems of lines:

$L \qquad \pi$

Exercise 4.9. Let Γ be a flat geometry of rank ≥ 4 and assume that the residues of the dual points of Γ have good systems of planes. Prove that Γ has a good system of planes.

Exercise 4.10. Let Γ be almost flat and let the following two conditions hold in Γ:

(i) residues of dual points are matroids;
(ii) for every line x, there are dual points not incident to x.

Prove that Γ does not have a good system of lines.

Exercise 4.11. Let Γ have rank $n \geq 4$ and assume that the star of every element of Γ of type $n-4$ is flat. Prove that Γ is flat.

4.3.3 Shadows

Let Γ and 0 be as in the previous section. Given an element x of Γ, the *shadow* of x is the set of points of Γ incident to x. We denote this set of points by $\sigma_0(x)$. The function σ_0, mapping x onto its shadow, will be called the *shadow operator* of Γ.

Trivially, the shadow operator is a homomorphism from the poset $\mathcal{P}_0(\Gamma)$ of Γ into the poset of all subsets of the set of points of Γ. The image of the set of elements of Γ under σ_0 is called the *shadow space* of Γ and it is denoted by $\sigma_0(\Gamma)$.

We often have $\sigma_0(\Gamma) \cong \mathcal{P}_0(\Gamma)$. This happens in many examples of Chapters 1 and 2 (actually, in many of them the shadow space *is* the geometry). However there are geometries Γ where $\sigma_0(\Gamma) \not\cong \mathcal{P}_0(\Gamma)$.

For instance, this happens in every flat or almost flat geometry. Indeed, if Γ is flat or almost flat, then $\sigma_0(\Gamma)$ is a proper homomorphic image of $\mathcal{P}_0(\Gamma)$, as all dual points have the same shadow.

Things may be even worse than in flat geometries. For instance, given a set S of size $n+1 \geq 3$, we can form a geometry $\Gamma(S)$ of rank 2 taking S as set of points and all subsets of S of size ≥ 2 as lines. We call $\Gamma(S)$ the *collapsed n-dimensional simplex*. The shadow space of $\Gamma(S)$ is the Boolean lattice of all subsets of S, deprived of its least element $\emptyset$, and is not even a homomorphic image of the poset $\Gamma(S)$. Many other counterexamples like this can easily be found.

Exercise 4.12. Prove that, if Γ has a good system of planes, then shadows of planes are subspaces of $\mathcal{L}_0(\Gamma)$ (in the meaning of Section 1.6.2).

Exercise 4.13. (Generalized Buekenhout–Shult axiom). Let Γ satisfy the triangular property and assume that the lower residues of the planes of Γ either are linear spaces (with the points of Γ as points) or have gonality $g \geq 4$. Prove that the following holds in Γ (compare Exercise 1.16):

for every point–line pair (a, x) with $a^{\perp} \cap \sigma_0(x) \neq \emptyset$, either $\sigma_0(x) \subseteq a^{\perp}$ or $|a^{\perp} \cap \sigma_0(x)| = 1$.

Exercise 4.14. Let Γ have a good system of lines and diagram as in Exercise 4.7. Assume now that the label X in that diagram denotes some class of partial planes. Prove that we have $\sigma_0(\Gamma) \cong \mathcal{P}_0(\Gamma)$.

(Hint: let u, v be planes with $\sigma_0(u) \subseteq \sigma_0(v)$; we can take a point a of u and two lines x, y of u on a; since the points of u are points of v, both x and y are lines of v (Exercise 4.7). In Γ_a we see that $u = v$.)

Exercise 4.15. Let Γ be the halved octahedron, with types 0, 1, 2 as follows

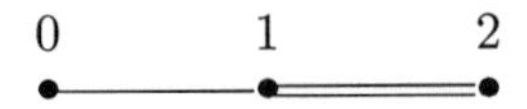

Prove that $\sigma_0(\Gamma)$ is the Boolean lattice on three atoms, deprived of its least element $\emptyset$, whereas $\sigma_2(\Gamma)$ is the affine plane of order 2, enriched with the set of four points of the plane as the greatest element.

Exercise 4.16. Let Γ be a geometry of rank 3 with string diagram graph but not admitting any diagram. Prove that $\sigma_0(\Gamma) \not\cong \mathcal{P}_0(\Gamma)$ for at least one of the two possible choices of the initial node 0.

(Hint: see Exercise 4.2.)

Exercise 4.17. Prove that a geometry Γ of rank $n \geq 3$ admits a diagram if and only if all residues of Γ of rank 3 admit diagrams.

4.4 Revisiting the definition of geometries

In the next chapters we will often need the assumption that the geometries we consider admit diagrams. Furthermore, this condition is often a trivial consequence of other hypotheses (such as the flag-transitivity, Chapter 9). Therefore, we state the following convention for the rest of this book:

only geometries admitting diagrams will be considered from now on.

In some sense, we are thus modifying the definition of geometries: a geometry is a graph as defined in Section 1.1, admitting a diagram in the meaning of Chapter 3.

As we only consider geometries admitting diagrams, a diagram graph is now just a diagram formed using the two most general classes of rank 2 geometries allowed for diagrams: the class of generalized digons and the class of all the rest.

14.1 Revisiting the definition of geometries

Chapter 5

Grassmann Geometries

In this chapter we show how to construct geometries with string diagram graphs starting from geometries with arbitrary diagram graphs. This will allow us to extend to geometries with non-string diagram graphs the concepts defined for geometric graded posets in Sections 4.3.2 and 4.3.3. However, in order to explain that construction we need to state some lemmas on topological properties of sets of vertices of graphs. We will do this in the next section.

5.1 Separation and dimension in a graph

In this section $\mathcal{D}$ will always be a given finite graph and I will be the vertex set of $\mathcal{D}$.

5.1.1 Separation and reduction

Given $A, B, C \subseteq I$, we say that A *separates* B from C, and we write $B \mid_A C$, if every path of $\mathcal{D}$ from B to C must cross A somewhere. It is understood that, if B and C are not connected by any path in $\mathcal{D}$ (in particular if one of them is empty), then every subset of I separates B from C.

The next lemma is quite easy. We leave its proof for the reader.

Lemma 5.1 *Both of the following hold for any $A, B, C, D \subseteq I$:*

(i) if $B \mid_A C$ and $B \mid_C D$, then $B \mid_A D$;
(ii) if $B \mid_A C$ and $D \subseteq C$, then $B \mid_A D$.

Given $J \subseteq I$, we say that a subset A of I is *J-reduced* if none of the proper subsets of A separates A from J. Trivially, all subsets of J are J-reduced.

In particular, $\emptyset$ and J are J-reduced. A vertex $x \in I$ is J-reduced if and only if the connected component of x in $\mathcal{D}$ meets J.

Lemma 5.2 *For every $J \subseteq I$, every subset of a J-reduced set is J-reduced.*

Proof. Let $B \subseteq A \subseteq I$ and let C be a proper subset of B with $B \mid_C J$. Then $(A - B) \cup C$ is a proper subset of A separating A from J, by Lemma 5.1(ii). Hence A is not J-reduced if B is not J-reduced. □

Lemma 5.3 *Let $A, B, J \subseteq I$, let A, B separate each other from J and let A be J-reduced. Then $A \subseteq B$.*

Proof. Let $A \not\subseteq B$, if possible. Let $a \in A - B$. Since A is J-reduced, $A - a$ does not separate A from J. Hence there is a path $x_0, x_1, ..., x_m$ with $x_0 \in J$, $x_m = a$ and $x_i \notin A$, for $i = 0, 1, ..., m - 1$. However, this path must cross B at some point $x_k \in B$, as $A \mid_B J$. We have $k < m$ (hence $x_k \in B - A$), since $x_m = a \notin B$. On the other hand, we also have $B \mid_A J$. Therefore the path $x_0, x_1, ..., x_k$ must cross A somewhere. This is impossible, because x_m is the only vertex of this path that belongs to A. □

Lemma 5.4 *Given $A, J \subseteq I$, there is just one J-reduced subset of A separating A from J. Furthermore, this subset of A is contained in all subsets of A separating A from J.*

Proof. We first prove the existence claim. If A is J-reduced, then there is nothing to prove. Otherwise, there is a proper subset $A_1 \subseteq A$ separating A from J. If A_1 is J-reduced, then we are done. Otherwise, we can take a proper subset A_2 of A_1 separating A_1 from J. By Lemma 5.1(i), A_2 also separates A from J. If A_2 is J-reduced then we are done. Otherwise, we repeat the above argument substituting A_2 for A_1. Since I is finite, we cannot continue this process up to infinity. However, the process stops only when we reach a J-reduced subset of A separating A from J. The existence of such a subset is proved. Let us call it B.

By Lemma 5.1(ii), we have $X \mid_B J$ for every $X \subseteq A$ separating A from J. Therefore, B is contained in every subset of A separating A from J, by Lemma 5.3. The uniqueness of B follows from this. □

The J-reduced subset of A separating A from J is called the *J-reduction* of A. We denote it by $Red_J(A)$.

Trivially, we have $Red_J(A) = \emptyset$ if and only if A and J are not connected by any path in $\mathcal{D}$. If $x \in I$, then we have $Red_J(x) = x$ if the connected component of x meets J; otherwise, we have $Red_J(x) = \emptyset$.

5.1.2 The posets $\mathcal{P}_J(\mathcal{D})$ and $\mathcal{P}_J^+(\mathcal{D})$

Given a nonempty subset J of I, we write $A \leq_J B$ for $B \mid_A J$. By Lemma 5.1(i), the relation $\leq_J$ is transitive. Since $\leq_J$ is also reflexive, it is a preorder.

Let $P_J(\mathcal{D})$ be the family of all J-reduced subsets of I. By Lemma 5.3, the preorder $\leq_J$ induces a partial order on $P_J(\mathcal{D})$. We denote the poset $(P_J(\mathcal{D}), \leq_J)$ by $\mathcal{P}_J(\mathcal{D})$ and we call it the *J-poset* of $\mathcal{D}$.

If we only consider nonempty J-reduced subsets of I, then we use the symbols $P_J^+(\mathcal{D})$ and $\mathcal{P}_J^+(\mathcal{D})$ for $P_J(\mathcal{D})$ and $\mathcal{P}_J(\mathcal{D})$ respectively. We call $\mathcal{P}_J^+(\mathcal{D})$ the *J^+-poset* of $\mathcal{D}$.

5.1.3 Interiors and closures

The *frontier* of a subset $A \subseteq I$ is the set of vertices in $I - A$ adjacent in $\mathcal{D}$ to some vertex of A. We denote it by $fr(A)$. In particular, if $x \in I$, then $fr(x)$ is the adjacency of x. Trivially, we have $fr(A) = \emptyset$ if and only if either $A = \emptyset$ or A is a join of connected components of $\mathcal{D}$.

Given $A, J \subseteq I$, let $Int_J(A)$ be the set of those vertices in $I - A$ that can be connected to $J - A$ by paths not meeting A. In the graph induced by $\mathcal{D}$ on $I - A$, the set $Int_J(A)$ can be described as the join of the connected components meeting $J - A$. We call $Int_J(A)$ the *J-interior* of A. Trivially, we have $Int_J(A) = \emptyset$ if and only if $A \supseteq J$.

The complement $I - Int_J(A)$ of $Int_J(A)$ will be called the *J-closure* of A. We denote it by $Cl_J(A)$. Trivially, $A \subseteq Cl_J(A)$. We say that A is *J-closed* if $Cl_J(A) = A$.

Lemma 5.5 *We have* $(A \cap J) \cup fr(Int_J(A)) = Red_J(A) = Red_J(Cl_J(A))$.

Proof. Trivially, $A \cap J = Cl_J(A) \cap J$ and $Int_J(A) = Int_J(Cl_J(A))$. Hence proving the first equality of the lemma is enough to obtain the rest.

A path from A to J either reaches $A \cap J$ or enters $Int_J(A)$, thus crossing $fr(Int_J(A))$ somewhere. Therefore $(A \cap J) \cup fr(Int_J(A)) \leq_J A$. Furthermore, we can reach $J - A$ from any point of $fr(Int_J(A))$ entering $Int_J(A)$ at the first step, then moving inside $Int_J(A)$.

Hence $(A \cap J) \cup fr(Int_J(A))$ is also J-reduced. Since it is a subset of A, it is the J-reduction of A. □

Lemma 5.6 *Given any two subsets $A, B \subseteq I$, we have $A \leq_J B$ if and only if $Cl_J(A) \supseteq Cl_J(B)$.*

Proof. Let $A \leq_J B$ and let $x \in Int_J(A)$. Then there is a path α from x to some vertex $y \in J - A$, not crossing A. Let $x \notin Int_J(B)$, if possible.

Then every path from x to J must cross B somewhere (possibly, just ending at some vertex of $B \cap J$). Hence α meets B at some vertex z. However, $A \leq_J B$. Hence the subpath of α from z to y also crosses A somewhere. We have a contradiction. Therefore $x \in Int_J(A)$. Thus, we have proved that $Int_J(A) \subseteq Int_J(B)$. That is, $Cl_J(A) \supseteq Cl_J(B)$.

Conversely, let $Cl_J(A) \supseteq Cl_J(B)$. Then $Cl_J(A) \leq_J Cl_J(B)$. By Lemma 5.5 and by the transitivity of $\leq_J$ we obtain that $A \leq B$. □

Lemma 5.7 *For every $A \subseteq I$, we have $Cl_J(A) = Cl_J(Red_J(A))$.*

Proof. The sets A and $Red_J(A)$ separate each other from J. The conclusion follows from Lemma 5.6. □

Lemma 5.8 *Let $B \subseteq I$ be J-closed and $a \in A = Red_J(B)$. Then $B - a$ is J-closed and we have $Red_J(B - a) = (A - a) \cup fr(Int_J(A) \cup a)$.*

Proof. Let us set $C = (A - a) \cup fr(Int_J(A) \cup a)$. Since $a \in A$ and A is J-reduced, either $a \in A \cap J$ or there is a path going from a to $J - A$ through $Int_J(A)$. Therefore, $Int_J(C) = Int_J(A) \cup a$. Hence $Cl_J(C) = Cl_J(A) - b$. Since $Cl_J(A) = B$ by Lemma 5.7, we have $B - a = Cl_J(C)$. Hence $B - a$ is J-closed. The set $(A - a) - J$ is contained in the following set:

$$(fr(Int_J(A)) \cup fr(a)) - (fr(Int_J(A)) \cup a),$$

which is equal to $fr(Int_J(A) \cup a)$. We have already proved that the latter is equal to $fr(Int_J(C))$. Therefore, $C = (C \cap J) \cup fr(Int_J(C))$. By Lemma 5.5, we have $C = Red_J(Cl_J(C)) = B - a$. □

Lemma 5.9 *Let B be J-closed with $Int_J(B) \neq \emptyset$. Then there is some element $a \in Int_J(B)$ such that $B \cup a$ is J-closed.*

Proof. Given elements $x, y \in Int_J(B)$, let us write $x \leq_{J-B} y$ to mean that x separates y from $J - B$ in the graph induced by $\mathcal{D}$ on $Int_J(B)$. The relation $\leq_{J-B}$ is a partial order relation. Indeed the vertices of $Int_J(B)$ are $(J - B)$-reduced. Since I is finite, there are elements of $Int_J(B)$ maximal for $\leq_{J-B}$. Let a be one of them. We must prove that $B \cup a$ is J-closed, namely that $I - (B \cup a) = Int_J(B \cup a)$. Trivially $Int_J(B \cup a) \subseteq I - (B \cup a)$. Let us prove the reverse inclusion.

Let $x \in I - (B \cup a)$. Then $x \in I - B$ and, since B is J-closed, we also have $x \in Int_J(B)$. Furthermore $x \neq a$ and, by the maximality of a with respect to $\leq_{J-B}$, there is a path in $Int_J(B)$ from x to $J - B$ avoiding a. Therefore $x \in Int_J(B \cup a)$. Hence $I - (B \cup a) \subseteq Int_J(B \cup a)$. □

5.1.4 Dimensions

Let $J \neq \emptyset$ and let n be the number of vertices joined to J by some path in $\mathcal{D}$. In particular, if J meets all connected components of $\mathcal{D}$, then $n = |I|$.

For every $A \subseteq I$ we set $d_J(A) = |Int_J(A)|$. The number $d_J(A)$ will be called the *J-dimension* of A. Trivially, we have $d_J(A) = n$ if and only if $A = \emptyset$ and $d_J(A) = 0$ if and only if $A \supseteq J$.

By Lemmas 5.6, 5.8 and 5.9, every maximal chain of the J-poset $\mathcal{P}_J(\mathcal{D})$ of $\mathcal{D}$ has $n+1$ elements. Therefore $\mathcal{P}_J(\mathcal{D})$ is a graded poset of rank $n+1$. The function d_J mapping every J-reduced set onto its J-dimension is the dimension function of the graded poset $\mathcal{P}_J(\mathcal{D})$.

If we drop the empty set, passing to the J^+-poset $\mathcal{P}_J^+(\mathcal{D})$, then we obtain a graded poset of rank n.

Lemma 5.10 *The elements of $\mathcal{P}_J^+(\mathcal{D})$ can be constructed recursively as follows:*

(i) the set J is the only element of $\mathcal{P}_J^+(\mathcal{D})$ of J-dimension 0;
(ii) for $i = 1, 2, ..., n-1$, the elements of $\mathcal{P}_J^+(\mathcal{D})$ of J-dimension i are those of the form $(A - a) \cup fr(Int_A \cup a)$, where $d_J(A) = i - 1$ and $a \in A$.

Proof. The initial clause (i) is trivial. The inductive clause (ii) is a consequence of Lemmas 5.8 and 5.9. □

Lemma 5.11 *Let d be a positive integer $< n-1$ and let A, B, C be distinct elements of $\mathcal{P}_J^+(\mathcal{D})$ with $d_J(B) = d_J(C) = d_J(A) + 1 = d$ and $A \leq_J B, C$. Then there is a unique element D of $\mathcal{P}_J^+(\mathcal{D})$ such that $B, C \leq_J D$ and $d_J(D) = d$ and we have $D = (A - \{b, c\}) \cup fr(Int_J \cup \{b, c\})$.*

Proof. By Lemma 5.10, we have $B = (A - b) \cup fr(Int_J(A \cup b))$ and $C = (A - c) \cup fr(Int_J(A) \cup c)$ for suitable elements $b, c \in A$. As $B \neq C$, we have $b \neq c$. Hence $b \in A - c \subseteq C$, $c \in A - b \subseteq B$, $b \cup Int_J(A) = Int_J(B)$ and $c \cup Int_J(A) = Int_J(C)$.

Therefore, $D = (B - c) \cup fr(Int_J(B) \cup c) = (C - b) \cup fr(Int_J(C) \cup b) = (A - \{b, c\}) \cup fr(Int_J \cup \{b, c\})$ has the properties required in the lemma. Let us prove that D is the unique element with those properties. If $d_J(D') = d + 1$ and $B, C \leq_J D'$, then $Int_J(D') = Int_J(B) \cup d = Int_J(A) \cup \{b, d\}$ and $Int_J(D') = Int_J(C) \cup e = Int_J(A) \cup \{c, e\}$ for suitable vertices d, e, by Lemma 5.10. Hence $d = c$ and $e = b$. Therefore $D' = D$. □

5.1.5 Intervals

Given $A \subseteq I$ with $0 < d_J(A) < n$, we denote by $\mathcal{D}_{A,J}^+$ and $\mathcal{D}_{A,J}^-$ the graphs induced by $\mathcal{D}$ on $Cl_J(A)$ and $Int_J(A)$ respectively.

Lemma 5.12 *Let $A, B \subseteq I$ with $A \leq_J B$, $0 < d_J(B)$ and $d_J(A) < n$. Then the following hold:*

(i) if $A \subseteq Int_J(B) \cup B$ and $d_J(B) < n$, then $Int_J(A)$ is the $(J-B)$-interior of $A - B$ in $\mathcal{D}^-_{B,J}$;
(ii) if $A \subseteq Int_J(B) \cup (B \cap J) \cup (Red_J(B) - Cl_J(A - B))$ and $d_J(B) < n$, then $A - B$ is $(J - B)$-reduced in $\mathcal{D}^-_{B,J}$ if and only if A is J-reduced in $\mathcal{D}$;
(iii) if $0 < d_J(A)$, then $B \subseteq Cl_J(A)$ and $Cl_J(B)$ is the $Red_J(A)$-closure of B in $\mathcal{D}^+_{A,J}$;
(iv) if $0 < d_J(A)$, then B ($\subseteq Cl_J(A)$) is $Red_J(A)$-reduced in $\mathcal{D}^+_{A,J}$ if and only if it is J-reduced in $\mathcal{D}$.

Proof. Since $A \leq_J B$, we have $A \cap J \subseteq B \cap J$. Furthermore, $Int_J(A) \subseteq Int_J(B)$ and $Cl_J(A) \supseteq Cl_J(B)$, by Lemma 5.6.

The proof of (i) is easy.

Let us turn to (ii). By (i), the set $A - B$ is $(J - B)$-reduced iff we can reach $J - A$ ($\subseteq J - B$) starting from any element of $A - B$ and going along paths through $Int_J(A)$. If $x \in A - B$ ($\subseteq Red_J(B) - (Cl_J(A - B) \cup B)$), then we can reach the set $J - B$ starting from x and going along a path with no vertices in $A - B$ and no vertices in $A \cap B$, because $x \notin Cl_J(A - B)$ and $x \in Red_J(B) - B$. This proves that $A - B$ is $(J - B)$-reduced if and only if A is J-reduced. (ii) is proved.

As $A \leq_J B$, we also have $Red_J(A) \leq_J Cl_J(B)$. The statements (iii) and (iv) easily follow from this remark. □

Given a J-reduced subset $A \subseteq I$ with $\emptyset \neq A \neq J$, the A^+-poset of $\mathcal{D}^+_{A,J}$ and the $(J \cap Int_J(A))^+$-poset of $\mathcal{D}^-_{A,J}$ will be denoted by $\mathcal{P}^+_{J,A}$ and $\mathcal{P}^-_{J,A}$, respectively. $[A, +[$ and $]-, A[$ are respectively the closed upper interval and the open lower interval defined by A in $\mathcal{P}^+_J(\mathcal{D})$.

Lemma 5.13 *We have $[A, +[= \mathcal{P}^+_{J,A}$ and $]-, A[\cong \mathcal{P}^-_{J,A}$.*

Proof. By Lemma 5.12(iv), the posets $[A, +[$ and $\mathcal{P}^+_{J,A}$ have precisely the same elements. By Lemma 5.6, the separation relation can be replaced by the inclusion relation between closures. Closures of elements above A are the same in $\mathcal{D}$ and $\mathcal{D}^+_{A,J}$, by Lemma 5.12(iii). Hence the posets $[A, +[$ and $\mathcal{P}^+_{J,A}$ are the same.

For every $X \in]-, A[$, let us set $f(X) = X - A$. By Lemmas 5.5 and 5.6, we have $X \subseteq Int_J(A) \cup A$, for every $X \in]-, A[$. Therefore, we can apply Lemma 5.12(ii), obtaining that $f(X) \in \mathcal{P}^-_{J,A}$ for every $X \in]-, A[$. Conversely, for every $Y \in \mathcal{P}^-_{J,A}$, we set $g(Y) = Y \cup (A \cap J) \cup (A - Cl_J(Y))$. We have $g(Y) \leq_J A$. Furthermore, $g(Y)$ is J-reduced by Lemma 5.12(ii).

Trivially, $g = f^{-1}$. Therefore f is a bijection from $]-, A[$ to $\mathcal{P}^{-}_{J,A}$. We still must prove that f is an isomorphism of posets, that is that f and g preserve separation relations. This follows from Lemma 5.12(i), using Lemma 5.6 to interpret separations as inclusions between interiors. □

Corollary 5.14 *Let $A \leq_J B \leq_J C$, with A J-reduced. Then $A \mid_B C$.*

Proof. We may assume $\emptyset \neq A \neq J$, otherwise there is nothing to prove. We have $A \leq_J Red_J(B) \leq_J Red_J(C)$, since $\leq_J$ is preserved under taking J-reductions. By Lemma 5.12(iv), $Red_J(B)$ and $Red_J(C)$ are A-reduced sets in $\mathcal{D}^{+}_{A,J}$. By Lemma 5.13 applied in $[A, +[$, we have that $Red_J(B)$ separates $Red_J(C)$ from A in $\mathcal{D}^{+}_{A,J}$. Hence B separates the A-closure of $Red_J(C)$ from A in $\mathcal{D}^{+}_{A,J}$. However, by Lemma 5.12(iii) the A-closure of $Red_J(C)$ in $\mathcal{D}^{+}_{A,J}$ is $Cl_J(C)$ and it contains C. Hence B separates C from A in $\mathcal{D}^{+}_{A,J}$. Furthermore, a path of $\mathcal{D}$ starting from a vertex of $\mathcal{D}^{+}_{A,J}$ either is entirely contained in $Cl_J(A)$ or crosses A as soon as it gets out from $Cl_J(A)$. Therefore B also separates C from A in $\mathcal{D}$. □

Exercise 5.1. Prove the following claims:

(a) The poset $\mathcal{P}_I(\mathcal{D})$ is the dual Boolean lattice of all subsets of I.
(b) Let $\mathcal{D}$ be a string and give it an orientation choosing one of its two terminal nodes as the initial node 0. Then $\mathcal{P}^{+}_0(\mathcal{D}) = \mathcal{D}$.
(c) Let $\mathcal{S}$ be a string with initial node 0 and terminal node v ($v = 0$ if $\mathcal{S}$ consist of a single node) and let $\mathcal{G}$ be any graph. Form a new graph $\mathcal{D}$ joining v to all vertices of $\mathcal{G}$. Then $\mathcal{P}_0(\mathcal{D})$ consists of the string $\mathcal{S}$ followed by the dual Boolean lattice of all sets of vertices of $\mathcal{G}$.

Exercise 5.2 Let $\mathcal{D}$ be a disconnected graph with connected components $\mathcal{D}_1$, $\mathcal{D}_2$, ..., $\mathcal{D}_m$. For every $i = 1, 2, ..., m$, let J_i be a subset of the set of vertices of $\mathcal{D}_i$. Let us set $J = \bigcup_{i=1}^{m} J_i$ and add a new element as greatest element to each of the posets $\mathcal{P}_J(\mathcal{D})$ and $\mathcal{P}_{J_i}(\mathcal{D}_i)$ ($i = 1, 2, ..., m$). Let $\overline{\mathcal{P}}_J(\mathcal{D})$ and $\overline{\mathcal{P}}_{J_i}(\mathcal{D}_i)$ be the posets obtained in this way.

Prove that we have $\overline{\mathcal{P}}_J(\mathcal{D}) = \prod_{i=1}^{m} \overline{\mathcal{P}}_{J_i}(\mathcal{D}_i)$

5.2 Grassmann geometries

We can now explain how to transform geometries with non-string diagram graphs into geometries with string diagram graphs.

In this section $\Gamma = (V, *)$ is a geometry of rank n over a set of types I, with type function t and diagram graph $\mathcal{D} = \mathcal{D}(\Gamma)$, and J is a subset of I meeting all connected components of the graph $\mathcal{D}$. According to the convention stated in Section 4.4, the geometry Γ admits a diagram.

5.2.1 The Grassmann poset

Let $F_J(\Gamma)$ be the set of flags F of Γ such that $t(F) \in \mathcal{P}_J^+(\mathcal{D})$ (see Section 5.1.2; note that the flags in $F_J(\Gamma)$ are nonempty, because $\emptyset \notin \mathcal{P}_J^+(\mathcal{D})$).

Let $\leq_J$ be the relation defined on $F_J(\Gamma)$ as follows: given $F, G \in F_J(\Gamma)$, we have $F \leq_J G$ if $F * G$ in Γ and $t(F) \leq_J t(G)$ in $\mathcal{P}_J^+(\mathcal{D})$.

Lemma 5.15 *The relation $\leq_J$ is a partial order.*

Proof. Trivially, $\leq_J$ is reflexive. Let us prove that it is transitive. Let $F \leq_J G \leq_J H$ (with $F, G, H \in F_J(\Gamma)$). Then $t(F) \leq_J t(G) \leq_J t(H)$ (whence $t(F) \leq_J t(G)$) and $F * G * H$. By Corollary 5.14, $t(G)$ separates $t(H)$ from $t(F)$. Hence $t(F) - t(G)$ and $t(H) - t(G)$ are not connected in the diagram graph $\mathcal{D}(\Gamma_G)$, by Lemma 4.2. By the direct sum theorem, we have $(F - G) * (H - G)$. Hence $F * H$. Therefore $F \leq_J H$.

Let us prove that $\leq_J$ is anti-symmetric. Let $F \leq_J G \leq_J F$. Then $F * G$ and $t(F) = t(G)$. Therefore $F = G$. □

We call $(F_J(\Gamma), \leq_J)$ the *J-Grassmann poset* of Γ, and we denote it by $Gr_J(\Gamma)$. It is clear that the type function t induces an epimorphism from $Gr_J(\Gamma)$ to $\mathcal{P}_J^+(\mathcal{D})$.

Lemma 5.16 *For every chain φ of $Gr_J(\Gamma)$ and every maximal chain ξ of $\mathcal{P}_J^+(\mathcal{D})$ containing $t(\varphi)$, there is a maximal chain γ of $Gr_J(\Gamma)$ containing φ and such that $t(\gamma) = \xi$.*

Proof. Let $\varphi = (F_1, F_2, ..., F_m)$ be a chain in $Gr_J(\Gamma)$. Since the flags $F_1, F_2, ..., F_m$ are pairwise incident, $\bigcup_{i=1}^m F_i$ is a flag, hence it is contained in some chamber C of Γ. Let $\xi = (J = X_0, X_1, ..., X_{n-1})$ be a maximal chain of $\mathcal{P}_J^+(\mathcal{D})$ containing $t(\varphi)$ and let us set $G_i = C \cap t^{-1}(X_i)$, for $i = 1, 2, ..., n-1$. Then $\gamma = (G_0, G_1, ..., G_{n-1})$ is a chain of $Gr_J(\Gamma)$ containing φ and we have $t(\gamma) = \xi$.

Trivially, if $F, G \in F_J(\Gamma)$ are such that $F \leq_J G$ and $t(F) = t(G)$, then $F = G$. Therefore, every maximal chain of $Gr_J(\Gamma)$ meets each of the fibres of t in precisely one element. It follows from this that the chain γ constructed above is maximal. □

Corollary 5.17 *The Grassmann poset $Gr_J(\Gamma)$ is graded of rank n. Its dimension function is the composition $t_J = d_J t$ of the epimorphism t from $Gr_J(\Gamma)$ to $\mathcal{P}_J^+(\mathcal{D})$ with the dimension function d_J of $\mathcal{P}_J^+(\mathcal{D})$.*

Trivial, by the previous lemma.

5.2.2 Interlude: a connectedness property

Given two disjoint nonempty subsets $X, Y \subseteq I$, let $\Gamma^{X,Y}$ be the bipartite graph having the flags of Γ of type X and Y as vertices, two such flags being adjacent in $\Gamma^{X,Y}$ precisely when they are incident in Γ. The next lemma generalizes Theorem 1.18 (strong connectedness).

Lemma 5.18 *The graph $\Gamma^{X,Y}$ is connected.*

Proof. We work by induction on the rank n of Γ. When $n = 2$, the statement is trivial.

Let $n > 2$. Then there are nonempty sets of types not containing any of X or Y. Let us call them (X, Y)-*free* sets of types. Trivially, every nonempty subset of an (X, Y)-free set of types is (X, Y)-free. In particular, (singletons of) elements of a (X, Y)-free set of types are (X, Y)-free. We call them (X, Y)-free types.

Let F, G be vertices of $\Gamma^{X,Y}$. We must prove that there is a path of $\Gamma^{X,Y}$ going from F to G. We can assume that $t(F) = X$ and $t(G) = Y$. Indeed, if $t(F) = t(G) = X$, then we can take a chamber $C \supseteq G$ and, considering the flag $C \cap t^{-1}(Y)$, we are led back to the above.

We must consider two cases.

Case 1. There are at least two (X, Y)-free types, say h and k. Let a, b be elements of type h and k respectively, incident to F and G respectively. By Theorem 1.18, there is a path $a = a_1 * b_1 * ... * a_m * b_m = b$ in Γ with $t(a_i) = h$ and $t(b_i) = k$ for $i = 1, 2, ..., m$. For every $i = 1, 2, ..., m$, let C_i be a chamber containing a_i and b_i. Let us set $G_i = C_i \cap t^{-1}(X)$. For every $i = 1, 2, ..., m - 1$, let D_i be a chamber containing b_i and a_{i+1}. Let us set $F_i = D_i \cap t^{-1}(Y)$ $(i = 1, 2, ..., m - 1)$, $F_0 = F$ and $F_m = G$. By the inductive hypothesis in the residue Γ_{a_i} of a_i, we can connect F_{i-1} to G_i by a path of flags of type X and Y containing a_i. Similarly, by the inductive hypothesis in Γ_{b_i}, we can connect G_i to F_i by a path of flags of type Y and X containing b_i. Taking all these paths together we obtain a path of $\Gamma^{X,Y}$ from F to G.

Case 2. There is only one (X, Y)-free type. Then $n = 3$ and $|X| = |Y| = 1$. The conclusion now follows from Theorem 1.18. □

5.2.3 The Grassmann geometry

Lemma 5.19 *Let $n = 2$. Then either Γ is not a generalized digon and $Gr_J(\Gamma)$ is the poset of Γ, or $Gr_J(\Gamma)$ is the dual of the flag complex of Γ.*

The proof is easy. We leave it for the reader.

Lemma 5.20 *Let $F \in F_J(\Gamma)$, with $t(F) \neq J$. The lower open interval $]-,F[$ of F in $Gr_J(\Gamma)$ is isomorphic to the J-Grassmann poset of Γ_G, where G is any flag of Γ of type $Cl_J(t(F))$ and containing F.*

Proof. Let G be as in the statement of the lemma. If $H \leq_J F$, then $t(F)$ separates $t(H)$ from $t(G)$, by Corollary 5.14 and because $t(F) = Red_J(Cl_J(t(F))) \leq_J Cl_J(t(F))$. Therefore $(H-F) * (G-F)$, by Lemma 4.2 and the direct sum theorem. Hence $H - F \in \Gamma_G$. Furthermore, we have $\mathcal{D}(\Gamma_G) = \mathcal{D}^-_{t(G),J}$, by Lemma 4.2 and because Γ admits a diagram.

Therefore, $H - F \in F_J(\Gamma_G)$ by Lemma 5.12(ii). Exploiting the equality $\mathcal{D}(\Gamma_G) = \mathcal{D}^-_{t(G),J}$ and mimicking the proof of Lemma 5.1 we see that, if $H' \in F_J(\Gamma)$, then $H = H' \cup t^{-1}((t(F) \cap J) \cup (t(F) - Cl_J)) \in]-,F[$.

As $H - F = H'$, we have a bijection between $]-,F[$ and $Gr_J(\Gamma_G)$. This bijection is an isomorphism, by Lemma 5.13 and because it preserves the incidence relation between flags. □

Theorem 5.21 *The Grassmann poset $Gr_J(\Gamma)$ is geometric and its natural geometry admits a diagram.*

Proof. We work by induction on n. When $n = 1$ there is nothing to prove. When $n = 2$, the conclusion follows from Lemma 5.19, recalling that the flag complex of Γ is a geometry (Section 1.5.1; also, Exercise 1.1).

Let $n > 2$. We know that $Gr_J(\Gamma)$ is graded, by Corollary 5.17. We must prove that (i), (ii) and (iii) of Theorem 4.10 hold in it.

Condition (i) of Theorem 4.10 holds for all intervals of the form $]F,G[$ with $t_J(F) + 1 = t_J(G)$ or $]-,G[$ with $t_J(G) = 1$. Indeed, in these cases we can use Lemma 5.20 to apply the inductive hypothesis to the interval $]-,G[$. As for intervals of the form $]F,+[$ with $t_J(F) = n-2$, if there are at least 2 maximal elements X, Y of $\mathcal{P}^+_J(\mathcal{D})$ containing $t(F)$, then we can choose flags of type X and Y in $]F,+[$ by Lemma 5.16, thus establishing (i) in this case. If there is just one maximal element X above $t(F)$ in $\mathcal{P}^+(\mathcal{D})$ and $t(F) \not\supseteq X$, then there are at least two flags of type X incident to F, by Lemma 1.8. Thus (i) is proved in this case, too. We are left with the case where X is the unique maximal element above $t(F)$ and $X \subseteq t(F)$. However, this is impossible, by Lemma 5.2 and because $X \neq \emptyset$.

Let us turn to (iii) of Theorem 4.10. Suppose on the contrary that, for some $i = 1, 2, ..., n-1$, we have $F \leq_J G$ for any two flags F, G with $t_J(F) + 1 = t(G) = i$. Let X be a set of types with $d_J(X) = i - 1$ and let $Y = (X - x) \cup fr(Int_J \cup x)$ for some $x \in X$. Then $d_J(Y) = i$, by Lemma 5.10. By the previous assumption, all flags of type X are incident to all flags of type Y. By the direct sum theorem, X and Y are not connected in $\mathcal{D}(Tr^+_{X \cup Y}(\Gamma))$. Hence X and Y are not connected in $\mathcal{D}$, by Lemma 4.3.

However, this is impossible, since $Y = (X - x) \cup fr(Int_J(X) \cup x)$ with $x \in X$. Therefore (iii) of Theorem 4.10 holds.

Using Lemmas 5.19 and 5.20 and the inductive hypothesis we can also prove (ii) of Theorem 4.10 in all cases except when the intervals we consider are either $]-,+[$ (that is $F_J(\Gamma)$) or $]F,+[$ for some flag F with $t_J(F) < n-2$.

Therefore, in order to finish the proof that $Gr_J(\Gamma)$ is geometric, we must prove that $]-,+[$ is connected and that $]F,+[$ is connected for every F with $t_J(F) < n-2$. Let us consider $]-,+[$ first.

Let F, G be distinct atoms of $Gr_J(\Gamma)$. Let us first assume $J \neq I$. Since J meets all connected components of $\mathcal{D}$, there is a type $h \in fr(J)$. By Lemma 5.18, there are flags G_0, $G_1, ..., G_m$ of type J and elements $a_1, a_2, ..., a_m$ of type h such that $G_0 = F$, $G_m = G$ and $G_{i-1} * a_i * G_i$ for every $i = 1, 2, ..., m$. Trivially, $G_{i-1} \leq_J a_i$ and $G_i \leq_J a_i$ for all $i = 1, 2, ..., m$. Therefore $G_0, a_1, G_1, ..., a_m, G_m$ give us a connection in $Gr_J(\Gamma)$ from F to G, as we wanted.

Let now assume $J = I$. Then the atoms F and G of $Gr_J(\Gamma)$ are chambers of Γ. By Theorem 1.9, there is a gallery from F to G. The chambers and the panels of this gallery give us a path from F to G in $Gr_J(\Gamma)$.

Since $Gr_J(\Gamma)$ is graded, every element of $Gr_J(\Gamma)$ lies above some atoms of $Gr_J(\Gamma)$. Thus, the connectedness of $]-,+[$ is proved.

Let us turn to $]F,+[$, with $t_J(F) = i < n-2$. Let G, H be elements of $]F,+[$ with $t_J(G) = t_J(H) = i+1$ (they exist by Lemma 5.16). If $t(G) = t(H)$, then an argument similar to that used in the case of $]-,+[$ gives us a path of $]F,+[$ from G to H (we leave the details for the reader; needless to say, we now must work inside Γ_F).

Let $t(G) \neq t(H)$. By Lemma 5.11, there is an element $X \in \mathcal{P}_J^+(\mathcal{D})$ with $d_J(X) = i+2$ and $t(G), t(H) \leq_J X$. By Lemma 5.16, there is a flag K of type X with $G \leq_J K$. Again by Lemma 5.16, there is flag H' of type $t(H)$ with $F \leq_J H' \leq_J K$. The flags H' and H are as in the previous case. We have already seen in that case that flags as H' and H can be connected in $]F,+[$. Since K connects G to H' in $]F,+[$, the connectedness of $]F,+[$ is proved.

We still must prove that the natural geometry of $Gr_J(\Gamma)$ admits a diagram. By Lemma 4.30 and the inductive hypothesis, $]-,F[$ admits a diagram for every $F \in Gr_J(\Gamma)$ with $t_J(F) = n_J - 1$. We finish the proof showing that $]F,+[$ is not a generalized digon for every $F \in Gr_J(\Gamma)$ with $t_J(F) = n-3$.

Let $t_J(F) = n-3$, that is $|Cl_J(t(F))| = 3$. Let i, j, h be the three types in $Cl_J(t(F))$. The following are easy to check:

(1) if $t(F) = Cl_J(t(F))$, then $]F,+[$ is a triangle;
(2) if $t(F) = \{i,j\}$, then $]F,+[$ is a degenerate projective plane;

(3) if $t(F) = i$ and $i \mid_j h$, then j and h are joined in $\mathcal{D}(\Gamma)$ and $]F, +[$ is isomorphic to a residue of Γ of type $\{j, h\}$;
(4) if $t(F) = i$ but none of j, h separates the other one from i, then $]F, +[$ is isomorphic to the dual flag complex of a residue of Γ of type $\{j, h\}$.

All the possibilities that can occur for $t(F)$ are mentioned above and none of them gives us a generalized digon. This is quite clear in cases (1) and (2). Generalized digons do not arise in (3) because j and h are joined in $\mathcal{D}(\Gamma)$ and Γ admits a diagram. They cannot arise in (4) because flag complexes of rank 2 geometries have gonality ≥ 4 (see Section 1.5.1). □

The natural geometry of the poset $Gr_J(\Gamma)$ is called the *J-Grassmann geometry* of Γ (also, the *Grassmann geometry* of Γ with respect to J). We denote it by the same symbol $Gr_J(\Gamma)$ used for its poset.

The integers 0, 1, ..., $n-1$ and the dimension function t_J of the poset $Gr_J(\Gamma)$ are the set of types and the type function of the geometry $Gr_J(\Gamma)$.

We have assumed that J meets all connected components of $\mathcal{D}(\Gamma)$ to make the notation easier. However, we can also drop this assumption rewriting, the previous theory for the general case. Thus, we can define the *J-Grassmann geometry* $Gr_J(\Gamma)$ of Γ for any nonempty set of types J. However, there is really no need for such a generalization. Indeed, let $[J]$ be the join of the connected components of $\mathcal{D}(\Gamma)$ meeting J. Then, by Lemma 4.5 and Corollary 4.8, we have $Gr_J(\Gamma) = Gr_J(Tr^+_{[J]}(\Gamma))$. By the direct sum theorem, the diagram graph of $Tr^+_{[J]}(\Gamma)$ is the graph induced by $\mathcal{D}(\Gamma)$ on $[J]$. Thus, we can always reduce ourselves to the case where J meets all connected components of the diagram graph.

Remark. Other expressions are often used in the literature instead of 'Grassmann geometry'; for instance: 'shadow geometry' [185], [162]; also 'canonical linearization' [146]. We prefer 'Grassmann geometry' for a number of reasons. This is the name traditionally given to certain instances of the construction we have described. The expression 'canonical linearization' is clumsy. The expression 'shadow geometry' might suggest the wrong idea that Grassmann geometries and shadows spaces (Sections 4.3.3 and 5.2.4) always coincide. This actually happens quite often, but not always (counterexamples have been given in Section 4.3.3).

Exercise 5.3. Let A be a J-closed set of types. Prove that $\mathcal{D}(Tr^+_A(\Gamma))$ and $\mathcal{D}(\Gamma)$ induce the same graph on $A - Red_J(A)$.

Exercise 5.4. Using the above, prove the following analogue of Lemma 5.20 for open upper intervals:

Let $F \in F_J(\Gamma)$ with $t_J(F) < n-1$ and let $A = Cl_J(t(F))$. Then the open upper interval $]F, +[$ of F in $Gr_J(\Gamma)$ is equal to the open upper interval of F in $Gr_{t(F)}(Tr^+_A(\Gamma))$.

Exercise 5.5. Rewrite the proof of Theorem 5.21 exploiting the statement of Exercise 5.4.

Exercise 5.6. Let $C_J(\mathcal{D})$ be the set of maximal chains of $\mathcal{P}^+_J(\mathcal{D})$. We can define a coloured graph $\mathcal{C}_J(\mathcal{D})$ on $C_J(\mathcal{D})$ with colours 0, 1, ..., $n-1$ declaring that two maximal chains $X, Y \in C_J(\mathcal{D})$ are j-adjacent in $\mathcal{C}_J(\mathcal{D})$ when X and Y only differ at the element of dimension j ($j = 0, 1, ..., n-1$).

We define a coloured graph $\mathcal{C}_J(\Gamma)$ with set of colours $\{0, 1, ..., n-1\}$, as follows: the vertices of $\mathcal{C}_J(\Gamma)$ are the pairs (C, X) where C is a chamber of Γ and $X \in C_J(\mathcal{D})$ and we state that two pairs (C, X) and (D, Y) are j-adjacent in $\mathcal{C}_J(\Gamma)$ ($j = 0, 1, ..., n-1$) if either $C = D$ and X, Y are j-adjacent in $\mathcal{C}_J(\mathcal{D})$ or C and D are adjacent in $\mathcal{C}(\Gamma)$, $d_J(C - C \cap D) = j$ and $X = Y$.

Prove that $\mathcal{C}_J(\Gamma)$ is isomorphic to the chamber system of $Gr_J(\Gamma)$.

5.2.4 Points, lines and shadows

Using Grassmann geometries, we can now extend to any geometry the notions and the notation stated in Sections 4.3.2 and 4.3.3 for geometries with string diagram graphs.

Points and lines. Given a geometry Γ and a nonempty set J of types such that $Gr_J(\Gamma)$ has rank ≥ 2, the natural point–line system $\mathcal{L}_0(Gr_J(\Gamma))$ of $Gr_J(\Gamma)$ will be called the *J-Grassmann point–line system* of Γ (also, the *point–line system* of Γ with respect to J). We denote it by $\mathcal{L}_J(\Gamma)$. Points and lines of $\mathcal{L}_J(\Gamma)$ will be called *J-points* and *J-lines*, respectively.

Most of the structures traditionally called 'Grassmannians' (see [206] for instance) are actually J-Grassmann point-line systems, with J consisting of a single type.

In many important cases we can recover a geometry Γ from some of its Grassmann point-line systems, or even characterize those rank 2 geometries that are J-Grassmann point-line systems of geometries in given classes, with a given choice of J. For instance, characterizations of this kind with $|J| = 1$ are known for all thick buildings of Lie type (to be defined in Chapter 13) for almost all choices of the type J (see [206], [42], [21], [38], [78], [33]; also [36] and [37]).

Shadows. If F is a nonempty flag of Γ of J-reduced type, then the shadow of F in $Gr_J(\Gamma)$ is called the *J-shadow* of F in Γ and it is denoted by $\sigma_J(F)$. That is, $\sigma_J(F)$ is the set of flags of Γ of type J incident to F.

The shadow space of $Gr_J(\Gamma)$ is called the *J-shadow space* of Γ (also, the *shadow space* of Γ with respect to J) and it is denoted by $\sigma_J(\Gamma)$. The shadow operator of $Gr_J(\Gamma)$ is called the *J-shadow operator* of Γ.

5.3 Examples

We will not give detailed proofs for everything we will say in this section. When one of our claims will be neither quite evident nor backed by any reference, this will be because its proof is not difficult and it is left for the reader to find it.

5.3.1 Strings

Let $\mathcal{D}(\Gamma)$ be a string and let 0 be one of the two extremal nodes of $\mathcal{D}(\Gamma)$. Then $Gr_0(\Gamma) = \Gamma$. Indeed the nonempty 0-reduced sets of types are precisely the types (compare with Exercise 5.1(b); as usual, we write 0 for $\{0\}$, for short). The 0-shadow space as defined in Section 5.2.4 is just the shadow space $\sigma_0(\Gamma)$ of Section 4.3.3. We call it the *natural* shadow space of Γ, as we did for the point–line system $\mathcal{L}_0(\Gamma)$ in Section 4.3.2.

However, we can also form Grassmann geometries with respect to inner nodes of $\mathcal{D}(\Gamma)$. Needless to say, Grassmann geometries obtained in this way are quite different from Γ in general (however, exceptions exist for this 'rule': see Exercise 5.12). For instance, let 1 be the central node of the Coxeter diagram A_3

(A_3) 0 •——• 1 ——• 2

and let Γ belong to this diagram. Then $Gr_1(\Gamma)$ belongs to the Coxeter diagram C_3

(C_3) •——•══•

In particular, if Γ is the tetrahedron, then $Gr_1(\Gamma)$ is the octahedron. If $\Gamma = PG(3, K)$ for some field K, then $Gr_1(\Gamma) = \mathcal{Q}_5^+(K)$; more generally, if Γ is a (possibly degenerate) projective geometry, then $Gr_1(\Gamma)$ is a (possibly weak) polar space (see also Section 1.3.6).

5.3.2 Direct products

Let $\Gamma = \oplus_{i=1}^{m} \Gamma_i$ be the irreducible decomposition of Γ (Section 4.2) and, for every $i = 1, 2, ..., m$, let J_i be a nonempty set of types of Γ_i. Let us set $J = \bigcup_{i=1}^{m} J_i$.

Then $Gr_J(\Gamma)$ is the direct product of $Gr_{J_1}(\Gamma)$, $Gr_{J_2}(\Gamma)$, ..., $Gr_{J_m}(\Gamma)$, as defined in Exercise 4.4:

$$(1) \qquad \prod_{i=1}^{m} Gr_{J_i}(\Gamma) \;=\; Gr_J(\Gamma)$$

In particular, if Γ_1, Γ_2, ..., Γ_m have string diagram graphs and, for every $i = 1, 2, ..., m$, we have given $\mathcal{D}(\Gamma_i)$ an orientation, thus identifying Γ_i with one of its two posets, then we have

$$(2) \qquad \prod_{i=1}^{m} \Gamma_i \;=\; Gr_J(\oplus_{i=1}^{m} \Gamma_i)$$

where J is the set of the initial nodes of the (oriented) diagram graphs $\mathcal{D}(\Gamma_1)$, $\mathcal{D}_2(\Gamma_2)$, ..., $\mathcal{D}(\Gamma_m)$. This gives us another way to prove the claim of Exercise 4.4. We might also take (2) as a definition of direct products of geometries with (oriented) diagram graphs instead of the definition stated in Exercise 4.4.

An Associative property. Let $\{X_1, X_2, ..., X_k\}$ be a partition of the set of indices $\{1, 2, ..., m\}$. For $j = 1, 2, ..., k$, let us set $K_j = \bigcup_{i \in X_j} J_i$ and let 0_j be the initial node of the diagram graph of $Gr_{X_j}(\oplus_{i \in X_j} \Gamma_i)$. Let us set $K = \{0_1, 0_2, ..., 0_m\}$ and $J = \bigcup_{i=1}^{m} J_i$. The following is not hard to prove:

$$(3) \qquad Gr_K(\oplus_{i=1}^{k} Gr_{K_j}(\oplus_{i \in X_j}(\Gamma))) \;\cong\; Gr_J(\oplus_{i=1}^{m}(\Gamma_i))$$

An associative property for direct products of geometries with string diagram graphs is contained in the above isomorphism. Indeed, if Γ_1, Γ_2, ..., Γ_m have string diagram graphs, then (3) can be rewritten as follows:

$$(4) \qquad \prod_{i=1}^{k} (\prod_{j \in X_i} \Gamma_j) \;\cong\; \prod_{i=1}^{m} \Gamma_i$$

Direct products of polar spaces. A polar space is usually viewed as a poset, as in Chapter 1. Its dual poset is called a *dual polar space.* This change of terminology only means that we have reversed the orientation of the C_n diagram

(C_n) 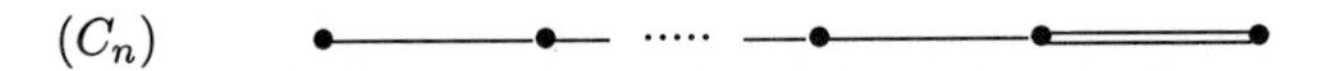

reading from right to left instead of from left to right. We now come to products of polar spaces. Let $\mathcal{P}_1$, $\mathcal{P}_1$, ..., $\mathcal{P}_m$ be polar spaces and let $\mathcal{P}'_1$, $\mathcal{P}'_2$, ..., $\mathcal{P}'_m$ be their duals. It is not hard to check that the product $\Gamma = \prod_{i=1}^m \mathcal{P}'_i$ of $\mathcal{P}'_1$, $\mathcal{P}'_2$, ..., $\mathcal{P}'_m$ (defined as in (2)) is a dual polar space. That is, the dual of the geometric poset Γ is a polar space. We call it the *direct product* of the polar spaces $\mathcal{P}_1$, $\mathcal{P}_2$, ..., $\mathcal{P}_m$, by abuse.

Geometries of rank 1 can also be taken as factors, instead of dual polar spaces. For instance, the direct product of n geometries of rank 1 is a thin-lined polar space of rank n. Conversely, every thin-lined polar space of rank n is a direct product of n geometries of rank 1. This easily follows from the characterization of thin-lined polar spaces of rank n as system of cliques of complete n-partite graphs (see Section 1.3.5).

All products of polar spaces are weak polar spaces (Section 1.3.5). Conversely, as we said in Section 1.3.5, all weak polar spaces arise as products of thick-lined polar spaces and/or rank 1 geometries (see [29] for a proof of this claim).

Remark. Products of polar spaces have been defined in [29] in a way seemingly different from the above, mimicking the construction of direct products of projective geometries (Section 1.3.3). However, the definition of [29] and ours are equivalent.

Products of projective geometries. We have defined products of projective geometries in Section 1.3.3. They are not direct products of geometries in the meaning we have stated. However, they are related to our products, as we will see in a few lines.

Let $\mathcal{P}_1$, $\mathcal{P}_2$, ..., $\mathcal{P}_m$ be geometric posets of ranks $n_1, n_2, ..., n_m$ respectively and let $\widehat{\mathcal{P}}_i$ be the poset obtained by adding a least element $\mho_i$ and a greatest element Ω_i to $\mathcal{P}_i$, for $i = 1, 2, ..., m$. Let us form the direct product $\widehat{\mathcal{P}} = \Pi_{i=1}^m \widehat{\mathcal{P}}_i$ of these posets and let $\overline{\mathcal{P}}$ be the poset obtained from $\widehat{\mathcal{P}}$ by dropping its least and greatest elements, namely $(\mho_1, \mho_2, ..., \mho_m)$ and $(\Omega_1, \Omega_2, ..., \Omega_m)$. Then $\overline{\mathcal{P}}$ is geometric of rank $m - 1 + \sum_{i=1}^m n_i$.

We say that the natural geometry of $\overline{\mathcal{P}}$ is the *extended direct product* of the natural geometries $\Gamma(\mathcal{P}_1)$, $\Gamma(\mathcal{P}_2)$, ..., $\Gamma(\mathcal{P}_m)$ of $\mathcal{P}_1$, $\mathcal{P}_2$, ..., $\mathcal{P}_m$ with respect to the orientations induced on the diagram graphs of $\Gamma(\mathcal{P}_1)$, $\Gamma(\mathcal{P}_2)$, ..., $\Gamma(\mathcal{P}_m)$ by the posets $\mathcal{P}_1$, $\mathcal{P}_2$, ..., $\mathcal{P}_m$.

We can also extend the above definition allowing empty factors. Thus, if $\mathcal{P}_i = \emptyset$, then $\widehat{\mathcal{P}}_i = \{\mho_i, \Omega_i\}$ and $\mathcal{P}_i$ contributes something to the product

$\widehat{\mathcal{P}}$ even if $\mathcal{P}_i = \emptyset$. The poset $\overline{\mathcal{P}}$ is still geometric even if some or all factors are empty, except in the trivial case of $m = 1$ with $\mathcal{P}_1 = \emptyset$.

The usual product of projective geometries (Section 1.3.3) is in fact an extended direct product in the meaning stated above. Factors consisting of single points in the usual definition correspond to empty posets in our definition. For instance, an n-dimensional simplex is the extended direct product of $n-1$ empty posets. A degenerate projective plane is the extended direct product of an empty poset and a line. A line with just two points is the extended direct product of two empty posets. A degenerate 3-dimensional projective space is the extended direct product either of two lines or of an empty poset and a (possibly degenerate) projective plane.

5.3.3 The case of $J = I$

Let I be the set of types of Γ. Then $Gr_I(\Gamma)$ is the dual of the flag complex $\mathcal{K}(\Gamma)$ of Γ (compare with Exercise 5.1(a)). This gives us another way to prove the claim of Exercise 1.1.

In particular, if Γ is a direct sum of geometries of rank 1, then the dual of $Gr_I(\Gamma)$ is a thin-lined polar space. Conversely, every thin-lined polar space can be obtained in this way (Exercise 1.2).

If furthermore all irreducible components of Γ have just two elements, then $Gr_I(\Gamma)$ is the n-dimensional cube, where n is the rank of Γ (Section 1.3.5).

More generally, if Γ belongs to a Coxeter diagram where all edges are labelled by m, then $Gr_I(\Gamma)$ belongs to the following Coxeter diagram (Exercise 3.6):

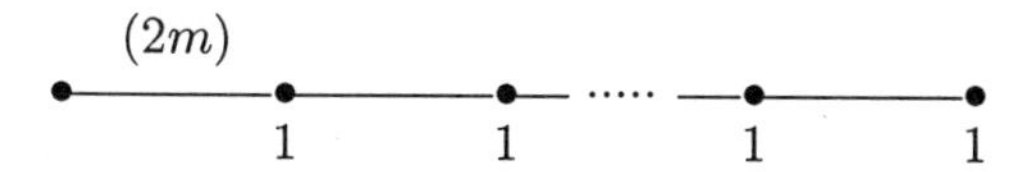

where 1, 1, ..., 1 are orders. For instance, if Γ is the geometry of black and white triangles of 2.1.2(10), then $Gr_I(\Gamma)$ is the tessellation of the plane in regular hexagons (Exercise 2.3). In this case we pass from the Coxeter diagram $\widetilde{A}_2$ to the Coxeter diagram $\widetilde{G}_2$ (see Section 3.1.3, (6.e) and (6.d)).

5.3.4 Buildings of type D_n

The hyperbolic quadric $\mathcal{Q}^+_{2n-1}(K)$ is the 0-Grassmann geometry of the building of type D_n over K (see Section 1.3.6) with respect to the 'initial' node 0 of the D_n diagram:

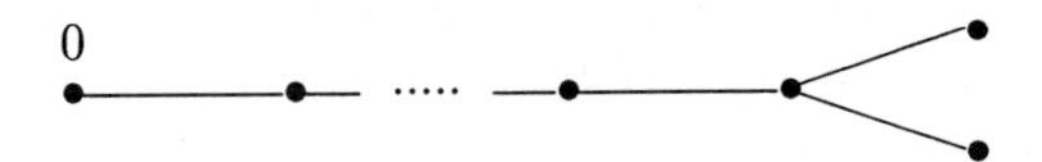

We also know that $\mathcal{Q}_5^+(K)$ is the Grassmann geometry of $PG(3, K)$ with respect to the central node of the A_3 diagram (Section 5.3.1).

Finally, $\mathcal{Q}_3^+(K)$, being a grid, is the dual flag complex of a generalized digon (Section 5.3.3).

We recall that hyperbolic quadrics (of rank $n \geq 3$, if we want to exclude grids) are the only classical top-thin polar spaces. By Theorems 1.5 and 1.6, a non-classical top-thin polar space must be a grid, or weak, or infinite of rank 3. The construction of buildings of type D_n can be repeated for these non-classical polar spaces, too. Indeed the main step of that construction was finding a suitable partition of the set of maximal subspaces into two disjoint classes, called M^+ and M^- in Section 1.3.6. This can be done in general, as we will see in a few lines.

Let $\mathcal{P}$ be a top-thin polar space of rank n.

As in Section 1.3.6, given maximal subspaces X, Y of $\mathcal{P}$, we write $X\Theta Y$ to mean that $n - 1 - dim(X \cap Y)$ is even.

Lemma 5.22 *The relation Θ is an equivalence relation with just two equivalence classes.*

Proof. In Section 1.3.6 we encouraged the reader to find an algebraic proof of this claim. We now give a geometric proof of it. We neglect some details in this proof, referring to Exercise 1.15 for them.

The statement of the lemma is a trivial consequence of the following:

(∗) given any three maximal subspaces X, Y, Z, an odd number of pairs (X, Y), (Y, Z), (Z, X) are in Θ.

Thus, we need only prove (∗). We work by induction on n. If $n = 2$, then $\mathcal{P}$ is a grid and (∗) holds. Let $n \geq 3$.

Case 1. Two of X, Y, Z have nonempty intersection, say $X \cap Y \neq \emptyset$. Let $a \in X \cap Y$. If $a \in Z$, then the statement of (∗) follows from the inductive hypothesis in the residue $\mathcal{P}_a$ of the point a.

Let $a \notin Z$ and let Z_a be the maximal subspace on a such that $Z_a \cap Z = a^\perp \cap Z$ (see Exercise 1.15). Let Θ_a be the analogue of Θ in $\mathcal{P}_a$. We have $X\Theta Y$ if and only if $X\Theta_a Y$. On the other hand, $X \cap Z_a$ is the subspace generated by $a \cup (X \cap Z)$ in X. Hence X corresponds to Z_a in Θ_a if and only if X does not correspond to Z in Θ. Similarly for Y, Z_a and Z. Therefore, the number of pairs (X, Y), (Y, Z), (Z, X) belonging to Θ is odd or even according to whether the number of pairs (X,Y), (X,Z_a), (Y,Z_a) in Θ_a is odd or even. However, the latter number is odd, by the inductive hypothesis in $\mathcal{P}_a$. Therefore, an odd number of pairs (X,Y), (Y,Z), (Z,X) are in Θ, as claimed in $(*)$.

Case 2. The maximal subspaces X, Y, Z are pairwise disjoint. Let Θ contains an even number of pairs (X,Y), (Y,Z), (Z,X), if possible. Then none of these pairs is in Θ and n is odd, since X, Y, Z are pairwise disjoint. Given a point $a \in X$, let Y_a and Z_a be the maximal subspaces such that $Y \cap Y_a = a^\perp \cap Y$ and $Z \cap Z_a = a^\perp \cap Z$ (see Exercise 1.15). We have $X \cap Y_a = X \cap Z_a = Y_a \cap Z_a = a$, because X, Y, Z are pairwise disjoint. Since n is odd, (X,Y_a), (X,Z_a) and (Y_a,Z_a) are in Θ. On the other hand, neither (Y_a,Y) nor (Z_a,Z) are in Θ, because $Y_a \cap Y$ and $Z_a \cap Z$ are hyperplanes in the projective geometries $\mathcal{P}_Y$ and $\mathcal{P}_Z$, respectively. Since $Z \cap Z_a \neq \emptyset$, we have $Y \Theta Z_a$ by the conclusion obtained in Case 1, applied to Z, Z_a and Y. However, this gives us $Y_a \Theta Z_a \Theta Y$ and $(Y_a,Y) \notin \Theta$, which contradicts the conclusion of Case 1 (indeed $Y_a \cap Y \neq \emptyset$). A contradiction is reached: $(*)$ is proved. □

A geometry $\Gamma(\mathcal{P})$ can now be defined as in Section 1.3.6, by dropping the $(n-2)$-dimensional subspaces of $\mathcal{P}$ and partitioning the set of maximal subspaces of $\mathcal{P}$ into the two equivalence classes of Θ. The incidence relation of $\Gamma(\mathcal{P})$ is the one inherited from $\mathcal{P}$, with the additional clause that two maximal subspaces of $\mathcal{P}$ are incident in $\Gamma(\mathcal{P})$ when they intersect in an $(n-2)$-dimensional subspace of $\mathcal{P}$.

Proposition 5.23 *The structure $\Gamma(\mathcal{P})$ is a geometry of rank n.*
If $n \geq 4$, then $\Gamma(\mathcal{P})$ belongs to the Coxeter diagram D_n

and $\mathcal{P}$ is the Grassmann geometry of $\Gamma(\mathcal{P})$ with respect to the initial node of this diagram.

If $n = 3$, then $\Gamma(\mathcal{P})$ is a (possibly degenerate) 3-dimensional projective geometry and $\mathcal{P}$ is the Grassmann geometry of $\Gamma(\mathcal{P})$ with respect to the central node of the A_3 Coxeter diagram.

If $n = 2$, then $\Gamma(\mathcal{P})$ is a generalized digon and $\mathcal{P}$ is the dual flag complex of $\Gamma(\mathcal{P})$.

Sketch of the Proof. The proof is just a rephrasing of arguments used in Section 1.3.6. Thus, we only give the main lines of the proof, leaving the details for the reader.

The first step is to prove the connectedness of the incidence graph of $\Gamma(\mathcal{P})$. This can now be done more quickly than in Section 1.3.6, exploiting Theorem 1.18. Then we can prove that, given a maximal subspace X of $\mathcal{P}$, the residues of X in $\Gamma(\mathcal{P})$ and in $\mathcal{P}$ are isomorphic. Finally, we remark that, given a point x of $\mathcal{P}$, if $n \geq 3$ then $\Gamma(\mathcal{P}_x)$ is the residue of x in $\Gamma(\mathcal{P})$. This is enough to obtain the conclusion, except for some additional work when $n = 3$ to check that $\Gamma(\mathcal{P})$ is in fact a projective geometry. □

Corollary 5.24 *Top-thin polar spaces of rank* 3 *are precisely the Grassmann geometries of (possibly degenerate)* 3*-dimensional projective geometries, with respect to the central node of the* A_3 *diagram.*

This is trivial, by the previous theorem and Section 5.3.1.

The geometry $\Gamma(\mathcal{P})$ will be called the *unfolding* of $\mathcal{P}$. Unfoldings of top-thin polar spaces of rank $n \geq 4$ are called *buildings of type* D_n.

We will prove in Chapter 7 that all geometries belonging to Coxeter diagrams of type D_n and A_3 are in fact buildings of type D_n and 3-dimensional projective geometries, respectively. For the moment, we only remark that, if Γ is a geometry belonging to D_n or to A_3, and 0 is the initial node of the D_n Coxeter diagram (when $n \geq 5$) or any of the three extremal nodes of D_4 (when $n = 4$)

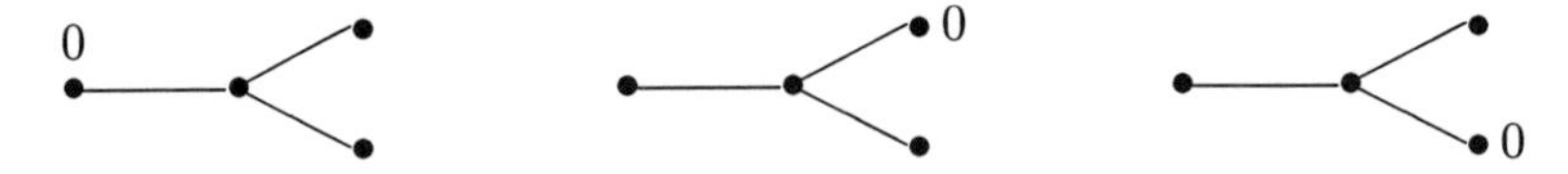

or the central node of A_3 (when $n = 3$), then $Gr_0(\Gamma)$ has diagram C_n.

Remark. The unfolding of a top-thin polar space $\mathcal{P}$ is called the *oriflamme complex* of $\mathcal{P}$ by Tits in [222] (Chapter 7).

We warn the reader that we have not yet given any good motivation to use the word 'building' in this context, since we have not said anything about buildings in general.

From D_4 to F_4. Let Γ be a building of type D_4 and let 0 be the central node of the D_4 diagram:

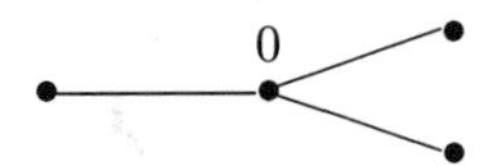

It is easy to check that $Gr_0(\Gamma)$ belongs to the following Coxeter diagram, denoted by F_4:

(F_4) 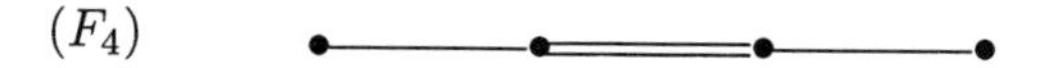

and that $Gr_0(\Gamma)$ is thin at the third and the fourth node of the diagram (in the meaning of Section 3.2.1). We call Γ the *unfolding* of $Gr_0(\Gamma)$. The geometry $Gr_0(\Gamma)$ is called the *metasymplectic space* associated with the D_4 building Γ.

5.3.5 Exercises

The next exercises are grouped as follows. Exercises 5.7–5.16 consider rank 3 geometries. Exercises 5.17–5.20 deal with products and Exercises 5.21–5.27 deal with flag complexes. Exercises 5.28–5.32 are related to C_n, D_n and F_4. A class of diagrams, including strings, forks (such as D_n) and complete diagrams (such as $\widetilde{A}_2$) is considered in Exercises 5.33–5.35. Exercises 5.36–5.39 deal with the Coxeter diagrams A_n, C_n, D_n and E_n (the last diagram will be defined in Exercise 5.36).

Exercise 5.7. Let Γ belong to the following diagram:

$(X^*.X)$ 0 X^* 1 X 2

where X is some class of rank 2 geometries other than generalized digons and X^* is the dual class of X.

Prove that $Gr_1(\Gamma)$ is top-thin (Section 4.3.2) and that it belongs to the following diagram:

$(X.C_2)$ 0 X 1 2

Exercise 5.8. Given Γ as in the previous exercise, prove that the dual point–line system of $Gr_1(\Gamma)$ is a partial plane of gonality 4 and that the dual of $Gr_1(\Gamma)$ also has a good system of planes.

Exercise 5.9. Let Γ be as in Exercise 5.7 and let X be a class of partial planes. Prove that $Gr_1(\Gamma)$ has a good system of lines.

Exercise 5.10. Let Γ be as in Exercise 5.7 and let X be the class L of linear spaces. Prove that $Gr_1(\Gamma)$ has a good system of planes and that $\sigma_1(\Gamma) \cong Gr_1(\Gamma)$.

(Hint: see Exercises 4.7 and 4.14.)

Exercise 5.11. Keeping the hypotheses of Exercise 5.7 on Γ, prove that $Gr_1(Gr_1(\Gamma)) \cong Gr_{0,2}(\Gamma)$.

Exercise 5.12. Let Γ be the tessellation of the Euclidean plane in squares (Section 2.1.2(4)). We recall that this tessellation belongs to the Coxeter diagram $\widetilde{B}_2$:

$(\widetilde{B}_2)$ 0 ═══ 1 ═══ 2

Prove that $Gr_1(\Gamma) \cong \Gamma$.

(Hint: represent the faces of the tessellation Γ by their centres, then join the centre of every face with the four vertices of that face...)

Exercise 5.13. Let Γ_0 be the tetrahedron and define $\Gamma_{n+1} = Gr_1(\Gamma_n)$ (for $n = 0, 1, 2, ...$). Thus, Γ_1 is the octahedron. Prove that, when $n \geq 2$, Γ_n is a tessellation of the sphere into eight triangles and $4 \cdot 3^{n-1} - 6$ squares, with $8 \cdot 3^{n-2}$ vertices. Stars of vertices are squares.

Exercise 5.14. Let Γ be the geometry of black and white triangles of Section 2.1.2(10) and let 0 be any type of Γ. Prove that $Gr_0(\Gamma)$ is the tessellation of the Euclidean plane into equilateral triangles (hence it is the flag complex of Γ; see Exercise 2.3).

(Hint: let 0 correspond to a type of triangles, to the white ones say; represent white triangles by their centres and join these centres forming a new tessellation into equilateral triangles...)

Exercise 5.15. Let Γ be the amalgam of attenuated spaces defined by a line L of $PG(3, K)$ (see Section 2.2.1), where K is a field. The geometry Γ belongs to the following diagram (Exercise 3.2):

$(Af^*.Af)$ 0 —Af^*— 1 —Af— 2

Note that L is a point in $\mathcal{Q}_5^+(K)$. Prove that $Gr_1(\Gamma)$ is the affine polar space obtained by removing the tangent hyperplane $L^\perp$ from $\mathcal{Q}_5^+(K)$.

Exercise 5.16. Starting from the symplectic variety $\mathcal{S}_3(K)$ embedded in $PG(3, K)$, form a new geometry Γ with set of types $\{0, 1, 2\}$, as follows.

The elements of Γ of type 0 and 2 are the points and the planes of $PG(3, K)$ respectively. As elements of type 1 we take the lines of $PG(3, K)$ not belonging to $\mathcal{S}_3(K)$. The incidence relation is the natural one, inherited from $PG(3, K)$, except that we also require $\pi \neq p^\perp$ for a point–plane flag (p, π) of $PG(3, K)$ to be an incident pair in Γ.

The geometry Γ defined in this way belongs to the diagram $Af^*.Af$ of

Exercise 5.15. Prove that $Gr_1(\Gamma)$ is an affine polar space, obtained from the Klein quadric $\mathcal{Q}_5^+(K)$ by removing a secant hyperplane.

(Hint: consider a secant hyperplane section of $\mathcal{Q}_5^+(K)$ and check that, if $\mathcal{Q}_5^+(K)$ is viewed as the Grassmann geometry of lines of $PG(3,K)$, then that section appears as a symplectic variety in $PG(3,K)$.)

Exercise 5.17. Describe the chamber system of a direct product of geometries with string diagram graphs.

(Hint: use Exercises 4.1, 5.2 and 5.6.)

Exercise 5.18. Let $\mathcal{G}$ be a graph having strings $\mathcal{S}_1$, $\mathcal{S}_2$, ..., $\mathcal{S}_m$ as connected components. For every $i = 1, 2, ..., m$, let 0_i be one of the two terminal nodes of $\mathcal{S}_i$, or the unique node of $\mathcal{S}_i$, if $\mathcal{S}_i$ has just one node. Let us set $J = \{0_1, 0_2, ..., 0_m\}$ and form an overgraph $\mathcal{D}$ of $\mathcal{G}$ by joining some pairs of vertices in J (possibly all pairs or none of them).

Let Γ be a geometry having $\mathcal{D}$ as diagram graph and let F be a flag of Γ of type J. For every $i = 1, 2, ..., m$, let a_i be the element of F of type 0_i and let $\overline{\Gamma}_i$ be the $(\mathcal{S}_i - 0_i)_+$-truncation of the residue of a_i in Γ (where it is understood that $\overline{\Gamma}_i = \emptyset$ if $\mathcal{S}_i = 0_i$).

Prove that the residue of the point F of $Gr_J(\Gamma)$ is the extended direct product of $\overline{\Gamma}_1$, $\overline{\Gamma}_2$, ..., $\overline{\Gamma}_m$.

(Hint: use Exercise 5.4.)

Exercise 5.19. Given $\mathcal{S}_1$, $\mathcal{S}_2$, ..., $\mathcal{S}_m$ and 0_1, 0_2, ..., 0_m as in the previous exercise, let Γ_i be a geometry with diagram graph $\mathcal{S}_i$ and let a_i be an element of Γ_i of type 0_i $(i = 1, 2, ..., m)$. Prove that the extended direct product of the residues Γ_{a_1}, Γ_{a_2}, ..., Γ_{a_m} is the residue of the point $F = \{a_1, a_2, ..., a_m\}$ of the direct product $\Pi_{i=1}^m \Gamma_i$.

Exercise 5.20. Prove that the direct product of an ordinary m-gon $(3 \leq m < \infty)$ and a thin geometry of rank 1 is a pair of m-gonal pyramids glued at theirs bases (in particular, when $m = 4$ we obtain the octahedron).

(An *m-gonal pyramid* is a polyhedron with $m + 1$ faces where one face, called the *basis* of the pyramid, is an ordinary m-gon and all other faces are triangles.)

Exercise 5.21. Let Γ belong to a diagram $\mathcal{D}$ where all edges have the same label, say X. Prove that $Gr_I(\Gamma)$ belongs to the following diagram

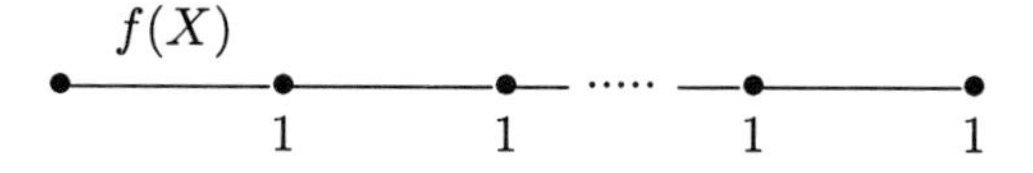

where $f(X)$ denotes the class of dual flag complexes of the geometries belonging to the class X.

Exercise 5.22. Let Γ have rank $n \geq 3$. Prove that the flag complex of Γ satisfies the triangular property and that $\sigma_I(\Gamma) \cong Gr_I(\Gamma)$ ($= \mathcal{K}_C(\Gamma)$; see Lemma 1.22).

Exercise 5.23. Let Γ be a geometry of rank $n \geq 2$ with string diagram graph. Prove that Γ is the flag complex of a geometry of rank n if and only if all the following hold:

(i) the collinearity graph of Γ is n-partite;
(ii) the shadow space of Γ is the poset of nonempty cliques of the collinearity graph of Γ;
(iii) the shadow operator of Γ is an isomorphism.

Exercise 5.24. Prove that every generalized $2m$-gon with thin lines and $m \geq 2$ is the flag complex of a generalized m-gon.

(Hint: show that the collinearity graph is bipartite.)

Exercise 5.25. Prove that the dual flag complex of the tetrahedron is the so-called truncated octahedron, obtained cutting an octahedron by 6 planes, one for every vertex of the octahedron (14 faces: 8 hexagons and 6 squares).

Exercise 5.26. Let 0, 1 be the first two nodes of the C_3 Coxeter diagram:

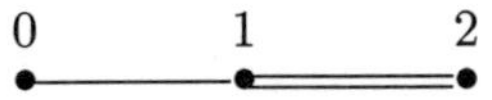

0 1 2
•———•═══•

and let Γ be the octahedron. Prove that $Gr_{\{0,1\}}(\Gamma)$ is the truncated octahedron (see Exercise 5.25). Hence it is isomorphic to the dual flag complex of the tetrahedron.

Exercise 5.27. Let Γ be a geometry with string diagram graph of rank 3 and let i be the central node of $\mathcal{D}(\Gamma)$. Let 0, 1 be the first two types of $Gr_i(\Gamma)$. Prove that $Gr_{\{0,1\}}(Gr_i(\Gamma))$ is isomorphic to the dual flag complex of Γ.

Exercise 5.28. Let Γ belong to the Coxeter diagram C_4:

0 1 2 3
(C_4) •———•———•═══•

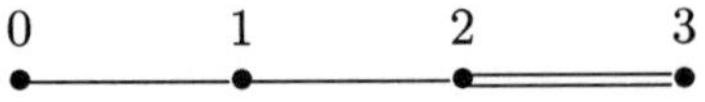

Check that $Gr_1(\Gamma)$ belongs to the Coxeter diagram F_4:

0 1 2 3
(F_4) •———•═══•———•

Exercise 5.29. Check that the unfolding of a top-thin polar space $\mathcal{P}$ is thick (is thin) if and only if $\mathcal{P}$ is thick-lined (is thin-lined).

Exercise 5.30. Let Γ be a thick building of type D_n. Without using Theorem 1.5, prove that the rank 2 residues of Γ other than generalized digons are isomorphic to $PG(2, K)$, for some field K.

(Hint: given a Desarguesian projective plane Π over a pair of types $\{i, j\}$, let K_i^j be the division ring coordinatizing Π with the type i representing points. With the notation of Section 3.1.3(5) and looking at residues of type A_3, we see that $K_{n-4}^{n-3} = K_{n-3}^{+} = K_{n-3}^{-} = K_{+}^{n-3}$. Hence $K_{+}^{n-3} = K_{n-3}^{+}$ ...)

Exercise 5.31. Let $\mathcal{P}$ be a thick-lined top-thin polar space of rank $n \geq 4$. Without using Theorem 1.5, prove that the planes of $\mathcal{P}$ are isomorphic to $PG(2, K)$, for some field K.

(Hint: unfold $\mathcal{P}$ and use Exercise 5.30.)

Exercise 5.32. Let Γ be a building of type D_4, let 0 be the central node of the D_4 diagram and let i be any of the three extremal nodes of that diagram. Let 1 be the second node of the C_4 diagram, as in Exercise 5.28. Prove that $Gr_0(\Gamma) \cong Gr_1(Gr_i(\Gamma))$.

Exercise 5.33. Let $\mathcal{D}$ be a graph with $n = m + k$ vertices, consisting of a string $\mathcal{S}$ with $m \geq 1$ nodes, labelled by 0, 1, ..., $m - 1$ from left to right as usual, and a set K of $k \geq 2$ further vertices forming a complete graph or a trivial graph, all joined to the last node $m - 1$ of $\mathcal{S}$ but not joined to any other node of $\mathcal{S}$. For instance, when $m = 2$ and $k = 3$ the graph $\mathcal{D}$ is one of the following:

Given a geometry Γ with $\mathcal{D}(\Gamma) = \mathcal{D}$, let us form its Grassmann geometry $Gr_0(\Gamma)$ with respect to the initial node 0 of $\mathcal{S}$. Let x be an element of Γ having as type the terminal node $m - 1$ of $\mathcal{S}$.

Prove that the upper residue of x in $Gr_0(\Gamma)$ is the dual flag complex of the K_+-truncation of the residue of x in Γ.

(Remark. The two connected diagram graphs of rank 3 (the string with three nodes and the triangle) are instances of the above, with $m = 1$ and $k = 2$; the underlying diagram graph of D_n is another example, with $m = n - 2$ and $k = 2$; in D_4 we can also choose $m = 1$ and $k = 3$.)

Exercise 5.34. Let Γ be a geometry of rank $n = m+k$, with string diagram graph and types 0, 1, ..., $n - 1$, labelling the nodes of $\mathcal{D}(\Gamma)$ from left to right as usual. Prove that Γ is the 0-Grassmann geometry of a geometry with a diagram graph as in Exercise 5.33 if and only if both the following hold (compare with Exercise 5.8 and 5.23):

(i) the dual collinearity graph of Γ is k-partite;
(ii) the dual shadow operator σ_{n-1} of Γ, restricted to the set of elements of type $n-k$, $n-k+1$, ..., $n-1$, is a bijection between this set of elements and the set of nonempty cliques of the dual collinearity graph and, given elements x, y of type $t(x) < n-k$ and $t(y) \geq n-k$, we have $x * y$ if $\sigma_{n-1}(y) \subseteq \sigma_{n-1}(x)$.

Exercise 5.35. Let $\mathcal{S}$ be a string diagram of rank m (possibly $m = 1$) with nodes labelled by 0, 1, ..., $m-1$ from left to right. Let $\mathcal{G}$ be another diagram, of rank $n-m$, and let $\mathcal{D}$ be a diagram consisting of $\mathcal{S}$ and $\mathcal{G}$, with some edges from nodes of $\mathcal{G}$ to the node $m-1$ but no edges from nodes of $\mathcal{G}$ to any node of $\mathcal{S}$ other than $m-1$. Let K be the set of nodes in $\mathcal{G}$ and let $J = fr(m-1) \cap K$, where $fr(m-1)$ is the neighbourhood of $m-1$ in the underlying graph of $\mathcal{D}$.

Let Γ be a geometry belonging to $\mathcal{D}$ and let a be an element of Γ of type $m-1$. Prove that the upper residue of a in $Gr_0(\Gamma)$ is the J-Grassmann geometry of the K_+-truncation of Γ_a.

(Hint: use Exercise 5.4.)

Exercise 5.36. Let us consider the following Coxeter diagram of rank $n \geq 4$, denoted by E_n:

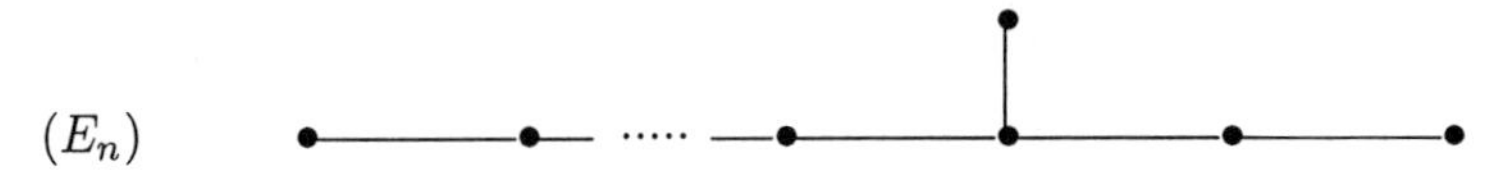

(note that $E_4 = A_4$ and $E_5 = D_5$; the symbol $\widetilde{E}_8$ is currently used instead of E_9).

Let Γ be a geometry with string diagram graph of rank $n \geq 4$ and types labelled by 0, 1, ..., $n-1$ as usual. Let the following hold in Γ:

(i) for every element u of Γ of type $n-1$, the residue Γ_u of u is either a non-degenerate $(n-1)$-dimensional projective geometry or a thick-lined top-thin polar space of rank $n-1$;
(ii) for every element v of Γ of type $n-3$, the upper residue Γ_v^+ of v is a degenerate projective plane with one thick line.

Prove that Γ is the O-Grassmann geometry of a thick geometry belonging to E_n, where O is the initial node of the diagram E_n:

Exercise 5.37. Let Γ be an n-dimensional projective geometry, with $n \geq 3$, and let i be any node of the A_n diagram. Prove that the triangular property

holds in $Gr_i(\Gamma)$ and that the planes of $Gr_i(\Gamma)$ are projective planes.

Exercise 5.38. Let Γ be a polar space of rank $n \geq 3$ and let i be any node of the C_n diagram of Γ. Prove that the Triangular Property holds in $Gr_i(\Gamma)$. Prove furthermore that the planes of $Gr_i(\Gamma)$ are projective planes in all cases except when i is one of the last two nodes of the C_n diagram of Γ. When i is the node before the last one, then some planes of $Gr_i(\Gamma)$ are generalized quadrangles (trivially, all planes are generalized quadrangles when i is the last node).

Exercise 5.39. Let Γ be a building of type D_n ($n \geq 4$) and let i be any node of the D_n diagram. Prove that the triangular property holds in $Gr_i(\Gamma)$ and that the planes of $Gr_i(\Gamma)$ are projective planes.

5.4 Recovering Γ from $Gr_J(\Gamma)$

We say that a diagram (or a diagram graph) $\mathcal{D}$ *admits recovery from Grassmann geometries* with respect to a set J of types if, given any two geometries Γ_1, Γ_2 in $\mathcal{D}$ with isomorphic J-Grassmann posets, every isomorphism between the posets $Gr_J(\Gamma)$ and $Gr_J(\Gamma)$ is induced by a unique isomorphism between (the incidence graphs of) Γ_1 and Γ_2.

Some diagrams exist that do not admit recovery from Grassmann geometries with respect to some set of types (see the hint we will give for Exercise 5.41). However, many diagrams admit recovery with respect to some set of types. Trivially, this is the case when $\mathcal{D}$ is a string and J is the initial node of $\mathcal{D}$. Also, every diagram over a set of types I admits recovery from Grassmann geometries with respect to I (this is an easy consequence of Theorem 1.25). Let us examine some examples less trivial than these.

If (the underlying diagram graph of) $\mathcal{D}$ is as in Exercise 5.33 and 0 is the first node of the initial string $\mathcal{S}$ of $\mathcal{D}$, then $\mathcal{D}$ admits recovery from Grassmann geometries with respect to 0. Actually, in this case we can also characterize 0-Grassmann geometries of geometries in $\mathcal{D}$ (Exercise 5.34) and that characterization shows us the way to recover Γ from $Gr_0(\Gamma)$. To see this, we may only remark that the k-partition of the dual collinearity graph of $Gr_0(\Gamma)$ is unique. This follows from the fact that the dual collinearity graph of $Gr_0(\Gamma)$ is nothing but the incidence graph of the K_+-truncation of Γ (notation as in Exercise 5.33) and this latter graph admits just one k-partition by Theorem 1.25. Hence the K_+-truncation of Γ is uniquely determined by $Gr_0(\Gamma)$. On the other hand, Γ and $Gr_0(\Gamma)$ have the same $\{0, 1, ..., m-1\}_+$-truncations and the incidence relation in $Gr_0(\Gamma)$ between elements of type $i \leq m-1$ and dual points is the same as the incidence relation in Γ between elements of type $i = 0, 1, ..., m-1$ and elements of type $j \in K$. Therefore Γ can be recovered from $Gr_0(\Gamma)$.

In particular, if $\mathcal{D} = D_n$ and 0 is the initial node of D_n when $n \geq 5$ or any node when $n = 4$, or if $\mathcal{D} = A_3$ and 0 is the central node, then $\mathcal{D}$ admits recovery with respect to 0.

We have obtained the above as a consequence of Exercise 5.34. However, it is not difficult to prove in general that A_n and D_n admit recovery with respect to every type (we leave this as an exercise for the reader).

As we have remarked in Section 5.2.4, characterizations of i-Grassmann point–line systems of thick buildings of Lie type are known for almost every choice of the type i. Those characterizations entail positive answers to the recovering problem, provided that the thickness is assumed. The thickness hypothesis is essential here: if we drop it, then counterexamples may arise. For instance, there are non-isomorphic weak polar spaces of rank 4 that have the same Grassmann geometry with respect to the second node of the C_4 diagram (see the hint we will give for Exercise 5.41).

Exercise 5.40. (Partially Grassmann geometries). Let $\mathcal{D}$ be a connected diagram graph and 0 a node of $\mathcal{D}$. Let $\mathcal{D} - 0$ be disconnected and let $\mathcal{D}_1$, $\mathcal{D}_2$, ..., $\mathcal{D}_m$ be the connected components of $\mathcal{D} - 0$. Given a proper nonempty subset X of $\{1, 2, ..., m\}$, let $J = 0 \cup (\bigcup_{i \in X} \mathcal{D}_i)$.

Given a geometry Γ belonging to $\mathcal{D}$, a new geometry $Gr_{0,X}(\Gamma)$ can be formed as follows.

The elements of $Gr_{0,X}(\Gamma)$ are the elements of Tr_J^- and the flags of Γ representing elements of $Gr_0(Tr_J^+(\Gamma))$. As incidence relation we take the relation inherited from the incidence relation of $Gr_0(Tr_J^+(\Gamma))$ and the incidence relation of Γ restricted to elements of Tr_J^- or to element-flag pairs (x, F) of Γ with $x \in Tr_J^-(\Gamma)$ and $F \in Gr_0(Tr_J^+(\Gamma))$. Check that the structure $Gr_{0,X}(\Gamma)$ defined in this way is in fact a geometry. Prove that $Gr_0(Gr_{0,X}(\Gamma)) \cong Gr_0(\Gamma)$.

(Hint: use Exercise 5.4 and (3) of Section 5.3.2.)

Exercise 5.41. Find two geometries Γ_1, Γ_2 belonging to the same diagram and such that $Gr_0(\Gamma_1) \cong Gr_0(\Gamma_2)$ for some type 0 but $\Gamma_1 \not\cong \Gamma_2$.

(Hint: use Exercise 5.40. Note that, for any two proper nonempty subsets $X, Y \subseteq \{1, 2, ..., m\}$, we have $Gr_0(Gr_{0,X}(\Gamma)) \cong Gr_0(Gr_{0,Y}(\Gamma))$. In order to find an example as required, we must choose X and Y in such a way that $Gr_{0,X}(\Gamma) \not\cong Gr_{0,Y}(\Gamma)$ but $Gr_{0,X}(\Gamma)$ and $Gr_{0,Y}(\Gamma)$ belong to the same diagram. This goal can be reached as follows: choose as Γ the unfolding of a weak top-thin polar space $\mathcal{P}$ of rank 4, where $\mathcal{P}$ is obtained as direct product of a rank 1 geometry and a thick-lined top-thin polar space of rank 3; then choose as X and Y two distinct pairs of extremal nodes of the D_4 diagram.)

Exercise 5.42. Take $X = \{2, 3, ..., m\}$ in Exercise 5.40 and assume that $\mathcal{D}_1 \cup 0$ is a string. Let t be the extremal node of this string other than 0. Prove that $Gr_{0,X}(\Gamma) \cong Gr_t(\Gamma)$.

(Compare with Exercise 5.35).

Exercise 5.43. Let $\mathcal{D}$ be a tree and let us attach orders > 1 to the nodes of $\mathcal{D}$, in order to include a thickness information in it. Prove that $\mathcal{D}$, equipped with those orders, admits recovery with respect to every type.

(Hint: let Γ belong to $\mathcal{D}$ and let 0 be a node of $\mathcal{D}$. The dual points of $Gr_0(\Gamma)$ are the elements of Γ whose type is a terminal node of $\mathcal{D}$, where $\mathcal{D}$ is viewed as a rooted tree with root 0. The dual lines of $Gr_0(\Gamma)$ are of two kinds: thick or thin. The thick dual lines are elements of Γ whereas the thin dual lines of $Gr_0(\Gamma)$ are the edges of the restriction of the incidence graph of Γ to the set of elements whose type is a terminal node of $\mathcal{D}$. In this way, a piece of Γ can be recovered from $Gr_0(\Gamma)$. We can now pass to residues of dual points of $Gr_0(\Gamma)$ ...)

Exercise 5.44. Let $\mathcal{D}$ be a cycle and let us attach orders > 1 to the nodes of $\mathcal{D}$. Prove that $\mathcal{D}$, equipped with those orders, admits recovery with respect to every type.

Exercise 5.45. Prove that the diagram E_n (Exercise 5.36) admits recovery with respect to every type.

Chapter 6

Intersection Properties

In this chapter Γ is a geometry of rank n over a set of types I with type function t, and J is a nonempty subset of I. According to the convention stated in Section 4.4, the geometry Γ admits a diagram.

We recall that, if J does not meet all connected components of the diagram graph $\mathcal{D}(\Gamma)$ of Γ, then the J-Grassmann geometry $Gr_J(\Gamma)$ of Γ is the J-Grassmann geometry of the $[J]_+$-truncation of Γ, where $[J]$ is the join of the connected components of $\mathcal{D}(\Gamma)$ meeting J (see Section 5.2.3). We denote by n_J the number of elements of $[J]$. That is, n_J is the rank of $Gr_J(\Gamma)$. Hence $n_J = n$ if and only if J meets all connected components of $\mathcal{D}(\Gamma)$.

We keep the notation of Section 5.1.1 for the separation relation in $\mathcal{D}(\Gamma)$. The symbol $\leq_J$ denotes the order relation in the poset $Gr_J(\Gamma)$, as in Section 5.2.1.

6.1 Shadows

The empty flag $\emptyset$ has J-reduced type but it is not an element of $Gr_J(\Gamma)$. However, we can also define its shadow $\sigma_J(\emptyset)$ as the set of all flags of Γ of type J. Shadows of flags of non-J-reduced type are defined by means of shadows of flags of J-reduced type, as follows. We first define the *J-reduction* $Red_J(G)$ of a flag G as the subflag of G of type $Red_J(t(G))$. A flag G is said to be J-reduced if $G = Red_J(G)$, that is if either $G \in Gr_J(\Gamma)$ or $G = \emptyset$. Then, given any flag G, we take $\sigma_J(Red_J(G))$ as the *J-shadow* of G, denoting it by $\sigma_J(G)$.

Proposition 6.1 *The J-shadow of a flag G is the set of flags of type J incident to G.*

Proof. Trivially, all flags of type J incident to G are also incident to $Red_J(G)$, whence they are in $\sigma_J(G)$.

Conversely, let $H \in \sigma_J(G)$. Then $H * Red_J(G)$. Since $t(Red_J(G)) = Red_J(t(G))$, we have $(H - H \cap Red_J(G)) * (G - Red_J(G))$ by Lemma 4.1 and the direct sum theorem in the residue of $Red_J(G)$. Hence $H * G$. □

The previous proposition has the following trivial consequences:

Corollary 6.2 *Let F, G be flags with $F \subseteq G$. Then $\sigma_J(F) \supseteq \sigma_J(G)$.*

Corollary 6.3 *Let F_1, F_2, ..., F_m be pairwise incident flags. Then*

$$\sigma_J(\bigcup_{i=1}^{m} F_i) = \bigcap_{i=1}^{m} \sigma_J(F_i).$$

Corollary 6.4 *For every nonempty flag F, we have $\sigma_J(F) = \bigcap_{x \in F} \sigma_J(x)$.*

Given a flag F with $J \not\subseteq t(F)$ and a flag $G * F$, the $(J - t(F))$-shadow of $G - F$ in the residue Γ_F of F will be denoted by $\sigma^F_{J-t(F)}(G - F)$. We also set $\sigma^F_J(G) = \{H \cup (F \cap t^{-1}(J)) \mid H \in \sigma^F_{J-t(F)}(G - F)\}$.

The next corollary is also a trivial consequence of Proposition 6.1:

Corollary 6.5 *Given F and G as above, we have $\sigma^F_J(G) = \sigma_J(F \cup G) = \sigma_J(F) \cap \sigma_J(G)$.*

Lemma 6.6 (Separation lemma) *If F and G are incident flags with $J \mid_{t(F)} t(G)$ (notation as in Section 5.1), then $\sigma_J(F) \subseteq \sigma_J(G)$.*

Proof. Let $Red_J(G) \neq \emptyset$. Then $Red_J(F) \neq \emptyset$ and $Red_J(F) \leq_J Red_J(G)$ in the geometric poset $Gr_J(\Gamma)$. Hence $\sigma_J(Red_J(F)) \subseteq \sigma_J(Red_J(G))$. That is, $\sigma_J(F) \subseteq \sigma_J(G)$.

Let $Red_J(G) = \emptyset$. Then the J-shadow of G is the set of all flags of Γ of type J and the inclusion $\sigma_J(F) \subseteq \sigma_J(G)$ is trivial. □

The shadow space $\sigma_J(\Gamma)$ contains all J-shadows of flags of Γ, except possibly the set $\sigma_J(\emptyset)$ of all flags of type J. However, examples exist where $\sigma_J(\emptyset) \in \sigma_J(\Gamma)$. For instance, the natural shadow spaces of the halved octahedron and of the collapsed n-dimensional simplex (Section 4.3.3) are examples of this kind.

It is sometimes convenient to add $\sigma_J(\emptyset)$ to $\sigma_J(\Gamma)$. We set $\widehat{\sigma}_J(\Gamma) = \sigma_J(\Gamma) \cup \{\sigma_J(\emptyset)\}$. Needless to say, $\sigma_J(\emptyset)$ is the greatest element of the poset $\widehat{\sigma}_J(\Gamma)$ and $\widehat{\sigma}_J(\Gamma) = \sigma_J(\Gamma)$ when $\sigma_J(\emptyset) \in \sigma_J(\Gamma)$.

Consistent with this, $\widehat{Gr}_J(\Gamma)$ will be the poset obtained adding the empty flag $\emptyset$ to the poset $Gr_J(\Gamma)$ as the greatest element.

The homomorphism $\sigma_J : Gr_J(\Gamma) \longrightarrow \sigma_J(\Gamma)$ can naturally be extended to a homomorphism $\widehat{\sigma}_J : \widehat{Gr}_J(\Gamma) \longrightarrow \widehat{\sigma}_J(\Gamma)$.

Lemma 6.7 *The shadow operator $\sigma_J : Gr_J(\Gamma) \longrightarrow \sigma_J(\Gamma)$ is an isomorphism if and only if $\widehat{\sigma}_J : \widehat{Gr}_J(\Gamma) \longrightarrow \widehat{\sigma}_J(\Gamma)$ is an isomorphism.*

Proof. Trivially, if $\widehat{\sigma}_J$ is an isomorphism then σ_J is an isomorphism. Thus, we need only prove that, if σ_J is an isomorphism, then $\widehat{\sigma}_J$ is an isomorphism.

Suppose on the contrary that σ_J is an isomorphism but $\widehat{\sigma}_J$ is not. Then $\sigma_J(F) = \sigma_J(\emptyset)$ for some $F \in Gr_J(\Gamma)$. Since σ_J is an isomorphism, F is the greatest element of the geometric poset $Gr_J(\Gamma)$. However, the existence of a greatest element in $Gr_J(\Gamma)$ forces a panel of cotype $n_J - 1$ to be contained in precisely one chamber, contradicting Lemma 1.8 (or (i) of Theorem 4.10, equivalently). A contradiction has been achieved. □

In this chapter we are mostly interested in the case when σ_J is an isomorphism. In view of Lemma 6.7, in that case we may interchange $\widehat{\sigma}_J$ with σ_J at our convenience.

The next lemma is a partial converse of the separation lemma.

Lemma 6.8 *The J-shadow operator σ_J is an isomorphism if and only if the following holds:*

(**Separation property**) *for every J-reduced flag F and every flag G, if $\sigma_J(F) \subseteq \sigma_J(G)$, then $F * G$ and $J \mid_{t(F)} t(G)$.*

Proof. Trivially, σ_J is an isomorphism if and only if the statement of the separation property holds whenever G is J-reduced. This remark immediately gives us the 'if' part of the lemma and shows that to prove the 'only if' part we need only prove that, if $F \leq_J Red_J(G)$, then $J \mid_{t(F)} t(G)$ and $F * G$.

Let $F \leq_J Red_J(G)$. Then $J \mid_{t(F)} t(G)$, $t(G) \leq Cl_J(Red_J(t(G)))$ and $t(F) - Red_J(t(G)) \subseteq Int_J(Red_J(t(G)))$ by Lemmas 5.1(i), 5.7, 5.5 and 5.6. Therefore, $t(F) \mid_{Red_J} (t(G))t(G)$. Hence $F * G$, by the direct sum theorem applied to the residue of $Red_J(G)$. □

Corollary 6.9 *Let σ_J be an isomorphism. Then for every flag F the J-reduction of F is the smallest subflag $G \subseteq F$ such that $\sigma_J(G) = \sigma_J(F)$.*

Proof. This follows from Lemmas 6.8 and 5.3.

6.2 The property $\mathbf{IP}_J$

We say that the *intersection property* holds in Γ with respect to the set of types J, or that it holds for $\sigma_J(\Gamma)$ (or that $\mathbf{IP}_J$ holds in Γ, for short) if both of the following hold:

(**Isomorphism property**) The shadow operator σ_J is an isomorphism;

(**Weak intersection property**) The poset $\sigma_J(\Gamma) \cup \{\emptyset\}$ is a semilattice with respect to $\cap$.

If the isomorphism property holds, then σ_J induces a grading on $\sigma_J(\Gamma)$, consistent with the grading of the poset $Gr_J(\Gamma)$. Therefore, if the intersection property $\mathbf{IP}_J$ holds, then $\sigma_J(\Gamma)\cup\{\emptyset\}$ is a (graded) complete semilattice and $\widehat{\sigma}_J(\Gamma)\cup\{\emptyset\}$ is a (graded) atomic complete lattice. The atoms of $\widehat{\sigma}_J(\Gamma)\cup\{\emptyset\}$ are the flags of Γ of type J.

The next proposition is trivial.

Proposition 6.10 *The intersection property* $\mathbf{IP}_J$ *holds in* Γ *if and only if the intersection property holds for the natural shadow space of* $Gr_J(\Gamma)$.

In the next two sections we prove that the Intersection Property $\mathbf{IP}_J$ is preserved under the taking of suitable residues and truncations.

6.2.1 Residues

In this section F is a flag with $J \not\subseteq t(F)$. The next lemma is an easy consequence of Lemma 4.2. We leave its proof for the reader.

Lemma 6.11 *Let* A, B *be subsets of* $I - t(F)$. *We have* $J - t(F) \mid_A B$ *in* $\mathcal{D}(\Gamma_F)$ *if and only if* $J \mid_A B$ *in* $\mathcal{D}(\Gamma)$.

Theorem 6.12 *Let* $\mathbf{IP}_J$ *hold in* Γ. *Then* $\mathbf{IP}_{J-t(F)}$ *holds in* Γ_F.

Proof. Let us prove that the separation property of Lemma 6.8 holds in Γ_F. By Lemma 6.8, this will be enough to establish the isomorphism property for Γ_F.

Let G and H be flags of Γ_F with $\sigma^F_{J-t(F)}(G) \subseteq \sigma^F_{J-t(F)}(H)$ and let G be $J - t(F)$-reduced. By Lemma 6.11, we have $G \subseteq Red_J(G \cup F)$. By Corollary 6.5, we have $\sigma_J(Red_J(G \cup F)) \subseteq \sigma_J(H \cup F)$.

Therefore $Red_J(G \cup F) * (H \cup F)$ and $t(Red_J(G \cup F))$ separates J from $t(H \cup F)$ in $\mathcal{D}(\Gamma)$, by the separation property in Γ (Lemma 6.8). Since

$G \subseteq Red_J(G \cup F)$, we have $G * H$ and $J - t(G) \mid_{t(G)} t(H)$ in $\mathcal{D}(\Gamma_F)$ (by Lemma 6.11). The separation property is proved for Γ_F.

Let us prove that the weak intersection property holds in Γ_F. Let G, H be flags in Γ_F with $\sigma^F_{J-t(F)}(G) \cap \sigma^F_{J-t(F)}(H) \neq \emptyset$. By Corollary 6.5, we have $\sigma_J(G \cup F) \cap \sigma_J(H \cup F) \neq \emptyset$. By the weak intersection property in Γ, there is a flag K such that $\sigma_J(K) = \sigma_J(G \cup F) \cap \sigma_J(H \cup F)$. By Corollary 6.2, we have $\sigma_J(K) \subseteq \sigma_J(F)$. We may assume that K is J-reduced. Hence, by the separation property, K is incident to both $G \cup F$ and $H \cup F$. Therefore, $K * F$ and, since $\sigma_J(K) \subseteq \sigma_J(F)$, we have $\sigma_J(K \cup F) = \sigma_J(K)$, again by Corollary 6.2. Finally, by Corollary 6.5 we obtain $\sigma^F_{J-t(F)}(K - F) = \sigma^F_{J-t(F)}(G) \cap \sigma^F_{J-t(F)}(H)$. The weak intersection property is proved for Γ_F.
□

6.2.2 Truncations

In this section K is a J-closed subset of I, not meeting J.

Lemma 6.13 *Let A, B be subsets of $I - K$. We have $J \mid_A B$ in $\mathcal{D}(Tr^-_K(\Gamma))$ if and only if $J \mid_A B$ in $\mathcal{D}(\Gamma)$.*

Proof. By Lemma 4.3, the graph induced by $\mathcal{D}(\Gamma)$ on $I - K$ is a subgraph of $\mathcal{D}(Tr^-_K(\Gamma))$. The 'only if' part follows from this. Let us prove the 'if' part.

Let $J \mid_A B$ in $\mathcal{D}(\Gamma)$ and let $\gamma = (i_1, i_2, ..., i_m)$ be a path from B to J in $\mathcal{D}(Tr^-_K(\Gamma))$ avoiding A, if possible. Since $J \mid_A B$ in $\mathcal{D}(\Gamma)$, γ is not a path of $\mathcal{D}(\Gamma)$. Hence there are consecutive nodes i_k, i_{k+1} of γ that are not joined in $\mathcal{D}(\Gamma)$. By the direct sum theorem applied to residues of flags of Γ of cotype $K \cup \{i, j\}$, there is a path γ_k in $\mathcal{D}(\Gamma)$ from i_k to i_{k+1} entirely contained in K but for its first and last nodes i_k and i_{k+1}. Trivially, γ_k avoids A. We can substitute the edge $\{i_k, i_{k+1}\}$ of γ with γ_k and, doing this for all edges of γ that are not edges of $\mathcal{D}(\Gamma)$, we obtain a path of $\mathcal{D}(\Gamma)$ from $i_1 \in B$ to $i_m \in J$, avoiding A. This contradicts the hypothesis that $J \mid_A B$ in $\mathcal{D}(\Gamma)$. Therefore, no path γ as above exists. That is, $J \mid_A B$ in $\mathcal{D}(Tr^-_K(\Gamma))$. □

Theorem 6.14 *Let $\mathbf{IP}_J$ hold in Γ and let $Tr^-_K(\Gamma)$ admit a diagram. Then $\mathbf{IP}_J$ holds in $Tr^-_K(\Gamma)$.*

Proof. Let us prove that the separation property of Lemma 6.8 holds in $Tr^-_K(\Gamma)$. Let F, G be flags of $Tr^-_K(\Gamma)$, let F be J-reduced in $Tr^-_K(\Gamma)$ and $\sigma_J(F) \subseteq \sigma_J(G)$ (note that the J-shadows of F and G are the same in $Tr^-_K(\Gamma)$ as in Γ, by Proposition 6.1). The flag F is J-reduced in Γ by Lemma 6.13. Hence we have $F * G$ and $J \mid_{t(F)} t(G)$ in $\mathcal{D}(\Gamma)$ by the separation property in Γ. By Lemma 6.13, we also have $J \mid_{t(F)} t(G)$ in $\mathcal{D}(Tr_K^-(\Gamma))$. The separation property is proved for $Tr_K^-(\Gamma)$.

Let us prove that the weak intersection property holds in $Tr_K^-(\Gamma)$. Let F, G be flags of $Tr_K^-(\Gamma)$ with $\sigma_J(F) \cap \sigma_J(G) \neq \emptyset$. By the weak intersection property in Γ, there is a flag H of Γ such that $\sigma_J(H) = \sigma_J(F) \cap \sigma_J(G)$. We need only prove that H is in $Tr_K^-(\Gamma)$, namely that $t(H) \subseteq I - K$. We may assume that H is J-reduced. Hence $t(H)$ separates both $t(F)$ and $t(G)$ from J. Therefore H is incident to both F and G by the separation property. On the other hand, $t(F)$ and $t(G)$ are subsets of $Int_J(K) = I - K$. Hence, if $H' = H - t^{-1}(K)$, the set $t(H')$ separates both $t(F)$ and $t(G)$ from J. Therefore $\sigma_J(H') \subseteq \sigma_J(F) \cap \sigma_J(F) = \sigma_J(H)$ by the separation lemma. On the other hand, we have $\sigma_J(H') \supseteq \sigma_J(H)$ by Proposition 6.1. Hence $\sigma_J(H') = \sigma_J(H)$ and we have $H' = H$ by Corollary 6.9. Therefore $t(H) \subseteq I - K$. $\square$

Corollary 6.15 *Let $\mathcal{D}(\Gamma)$ be a string with nodes labelled by $0, 1, ..., n-1$ from left to right and let $K = \{m, m+1, ..., n-1\}$, with $m > 1$. Let $\mathbf{IP}_0$ hold in Γ. Then $\mathbf{IP}_0$ holds in $Tr_K^-(\Gamma)$.*

This is trivial, by Theorem 6.14.

Problem. We had to assume that $Tr_K^-(\Gamma)$ admits a diagram in Theorem 6.14, because this property has been assumed since the beginning in our theory of Grassmann geometries and intersection properties. We will make a similar assumption in Propositions 6.28, 6.29 and 6.31. Of course, we could make that assumption extending the definitions of Grassmann posets and intersection properties so as to allow geometries not admitting diagrams, but the theory we would get in this way would be too weak. We should do better to strive to prove that the property of admitting diagrams is preserved when taking truncations, possibly modulo some general conditions (yet to be discovered). Exercise 4.16 might give us some hints for this.

It is worth remarking that, if $\mathcal{D}(\Gamma)$ is a string and K is a connected piece of $\mathcal{D}(\Gamma)$ containing just one of the two extremal nodes of $\mathcal{D}(\Gamma)$, then $Tr_K^-(\Gamma)$ admits a diagram. This is a straightforward consequence of the direct sum theorem (Theorem 4.6). We will also prove later (Lemma 6.36) that, if $\mathcal{D}(\Gamma)$ is a string and the intersection property holds in the natural shadow space of Γ, then all truncations of Γ admit diagrams. Something can also be said when $\mathcal{D}(\Gamma)$ is a tree (see Exercise 6.42).

6.3 Examples

6.3.1 The rank 2 case

In this Section Γ is a geometry of rank 2, with set of types $\{P, L\}$, as in Section 1.2.

Proposition 6.16 *The following are equivalent:*

(i) the intersection property holds in Γ with respect to one of the two types P, L of Γ;
(ii) the intersection property holds in Γ with respect to each of the two types of Γ;
(iii) the geometry Γ is either a partial plane or a generalized digon.

Proof. The reader may check that (iii) implies (ii). Trivially, (ii) implies (i). Let us prove that (i) implies (iii).

Let $\mathbf{IP}_P$ hold and let Γ have gonality $g = 2$. Then there are distinct lines x, y such that $|\sigma_P(x) \cap \sigma_P(y)| \geq 2$. By the weak intersection property (Section 6.1), there is a flag F such that $\sigma_P(F) = \sigma_P(x) \cap \sigma_P(y)$. Since there are at least two points in $\sigma_P(x) \cap \sigma_P(y)$, the flag F does not contain any point. That is, F is either a line or empty. We may also assume that F is P-reduced. Therefore F is incident to both x and y by the separation property. However, no line can be incident to both x and y, since $x \neq y$ and incidence between lines means equality. Hence $F = \emptyset$ and x and y are incident to all points. Therefore $\emptyset = Red_P(x) = Red_P(y)$ by Corollary 6.9. This forces $Red_P(L) = \emptyset$, that is, Γ is a generalized digon. □

Corollary 6.17 *Let $\mathbf{IP}_J$ hold in Γ and let $n_J \geq 2$. Then the J-Grassmann point-line system $\mathcal{L}_J(\Gamma)$ of Γ is a partial plane.*

This is an easy consequence of Proposition 6.10, Corollary 6.15 and Proposition 6.16.

6.3.2 Reducible geometries and direct products

Let $\mathcal{D}(\Gamma)$ be disconnected with connected components I_1, I_2, ..., I_m. Then $\Gamma = \oplus_{i=1}^m \Gamma_i$, where we write Γ_i for $Tr^+_{I_i}(\Gamma)$, for short.

Let I_1, I_2, ..., I_k be the connected components of $\mathcal{D}(\Gamma)$ met by J. We write J_i for $J \cap I_i$ for short, for $i = 1, 2, ..., k$.

We have $Gr_J(\Gamma) = \prod_{i=1}^k Gr_{J_i}(\Gamma_i)$ (see Section 5.3.2(1)). By the definition of direct products of geometric posets (Exercise 4.4), this is equivalent to saying that the poset $\widehat{Gr}_J(\Gamma)$ (see Section 6.1) is the direct product of the posets $\widehat{Gr}_{J_1}(\Gamma_1)$, $\widehat{Gr}_{J_2}(\Gamma_2)$, ..., $\widehat{Gr}_{J_k}(\Gamma_k)$.

Given a flag F of Γ, let us set $F_i = F \cap t^{-1}(I_i)$, for $i = 1, 2, ..., k$ and let σ_i be the J_i-shadow operator in Γ_i. By the direct sum theorem, we have $\sigma_J(F) = \prod_{i=1}^{k} \sigma_i(F_i$. Therefore, $\widehat{\sigma}_J(\Gamma) = \prod_{i=1}^{k} \widehat{\sigma}_i(\Gamma)$.

The next proposition is now evident:

Proposition 6.18 *The property* $\mathbf{IP}_J$ *holds in* Γ *if and only if the property* $\mathbf{IP}_{J_i}$ *holds in* Γ_i *for every* $i = 1, 2, ..., k$.

In particular:

Corollary 6.19 *The intersection property holds for the natural shadow space of a direct product of geometric graded posets if and only if it holds for the natural shadow space of every factor of that product.*

6.3.3 The case of $J = I$

Proposition 6.20 *The intersection property always holds with respect to the full set of types of a geometry.*

Proof. The I-Grassmann geometry $Gr_I(\Gamma)$ is the dual flag complex of Γ and the I-shadow space is the dual of the complex $\mathcal{K}_{\mathcal{C}}(\Gamma)$ defined in Section 1.6.2. The conclusion follows from Lemma 1.22. □

6.3.4 Examples from Chapters 1, 2 and 5

We will only consider examples of rank $n \geq 3$ (the rank 2 case is covered by Proposition 6.16).

(1) **String diagrams.** The intersection property holds for the natural shadow spaces of all examples with string diagram graphs and rank $n \geq 3$ given in Chapters 1 and 2, except for the following ones:

the halved octahedron (and its dual, the halved cube); see Section 4.3.3;

the 'small' homomorphic images of tessellation of the Euclidean plane considered in Section 2.1.2(8);

amalgams of attenuated spaces (Section 2.2.1);

many affine expansions of rank 2 geometries (Section 2.3); in particular, the intersection property does not hold in the affine expansions of the circular spaces, considered in Section 2.3.

However, there are many affine expansions where the intersection property holds. For instance, it holds in affine expansions of polar spaces.

The job of checking the above claim in every case is left for the reader. Some cases are quite easy: matroids (in particular, projective and affine geometries), polar spaces, tessellations of surfaces, truncations meeting the hypotheses of Corollary 6.15. The enrichment construction (Section 2.4) can now be seen as a trick to recover a geometry satisfying the intersection property inside a rank 2 geometry without that property.

(2) **Buildings of type D_n.** Let Γ be a building of type D_n (see Section 5.3.4) and let 0 be the initial node of the D_n diagram. The 0-Grassmann geometry $Gr_0(\Gamma)$ is a polar space, whence the intersection property holds in the natural shadow space of $Gr_0(\Gamma)$. By Proposition 6.10, the property $\mathbf{IP}_0$ holds in Γ.

Let $n-3$ be the central node of the $\mathbf{D}_n$ diagram, as in Section 3.1.3(5). The reader can check that $\mathbf{IP}_{n-3}$ holds in Γ. Therefore, the intersection property holds for the natural shadow space of $Gr_{n-3}(\Gamma)$. In particular, it holds for natural shadow spaces of metasymplectic spaces associated with D_4 buildings.

The reader can also check that the intersection property holds in a D_4 building with respect to each of the types + and −

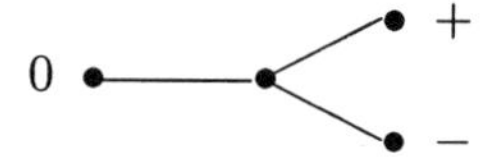

Therefore, the intersection property holds in a D_4 building with respect to every type.

(3) **The $\widetilde{A}_2$ geometry of Section 2.1.2(10).** Let Γ be the geometry of black and white triangles of Section 2.1.2(10). For every type i, the i-Grassmann geometry of Γ is the tessellation of the Euclidean plane in equilateral triangles (Exercise 5.14). Since the intersection property holds for tessellations, $\mathbf{IP}_i$ holds in Γ for every type i, by Proposition 6.20.

We shall say more on the previous examples in Section 6.6.

6.3.5 Exercises

Exercise 6.1. Use Proposition 6.20 to prove that the intersection property holds for the tessellation of the Euclidean plane in regular hexagons (even if a direct proof of this claim is quite easy).

Exercise 6.2. Let Γ be an amalgam of attenuated spaces, of rank 3. Check that the intersection property holds in Γ with respect to the central node of the diagram.

(Hint: use Exercise 5.15, Proposition 6.20 and the fact that the intersection property holds for natural shadow spaces of affine polar spaces; see also Exercise 6.7.)

(Remark: however, the intersection property does not hold in Γ for any of the two extremal nodes of the diagram.)

Exercise 6.3. Let Γ be the geometry constructed in Exercise 5.14. Check that the intersection property holds in Γ with respect to the central node of the diagram.

(Hint: as in Exercise 6.2.)

Exercise 6.4. Let Γ be as above. Check that the intersection property does not hold in Γ, with respect to any of the two extremal nodes of the diagram.

Exercise 6.5. Let Γ belong to the following diagram

$(L.\pi)$ — diagram: •—L—•—π—•

Prove that the intersection property holds for the natural shadow space of Γ if and only if Γ has a good system of lines.

(Hint. Let Γ have a good system of lines. Then Γ also has a good system of planes (Exercise 4.7). If u, v are planes with a common point a, for every other common point b of u and v, lines through a and b exist in Γ_u and Γ_v. Since Γ has a good system of lines, these lines coincide ...)

Exercise 6.6. Let Γ belong to the following diagram:

(L_3) — diagram: •—L—•—L—•

Check that the intersection property holds in the natural shadow space of Γ.

(Hint: combine Exercises 4.5 and 6.5.)

Exercise 6.7. Let Γ belong to the following diagram:

$(L^*.L)$ — diagram: •(0)—L^*—•(1)—L—•(2)

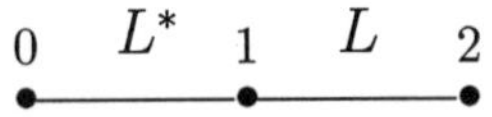

Check that $\mathbf{IP}_1$ holds in Γ.

(Hint: Γ has both a good system of lines and a good dual system of lines (Exercise 4.5); then $Gr_1(\Gamma)$ has a good system of lines; now use Exercise

6.5.)

Exercise 6.8. Let Γ be a geometry of rank $n \geq 3$ with string diagram graph. Prove that Γ has a good system of planes if the intersection property holds for the natural shadow space of Γ (see also Section 6.7.2). Show by counterexamples that the converse does not hold.

(Hint: amalgams of attenuated spaces, many affine expansions ...).

Exercise 6.9. Show by counterexamples that the analogue of Corollary 6.15 may fail to hold if K is disconnected in $\mathcal{D}(\Gamma)$.

(Hint: see Section 1.6.1(4).)

Exercise 6.10. Find a geometry Γ where the property $\mathbf{IP}_J$ does not hold for some set of types J but it holds in the K_+-truncation of Γ, for some K properly containing J.

(Hint: consider the halved cube ...)

Exercise 6.11. Let Γ be a geometry of rank $n = m + k$, with string diagram graph and types 0, 1, ..., $n-1$, labelling the nodes of $\mathcal{D}(\Gamma)$ from left to right as usual. Let $\mathbf{IP}_{n-1}$ hold in Γ.

Prove that Γ is the 0-Grassmann geometry of a geometry with a diagram graph as in Exercise 5.33 if and only both of the following hold:

(i) the dual collinearity graph is k-partite;
(ii) the $n-1$-shadows of elements of type $n-k$, $n-k+1$, ..., $n-1$ are precisely the nonempty cliques of the dual collinearity graph.

Exercise 6.12. Let Γ be a geometry of rank n, with string diagram graph and types 0, 1, ..., $n-1$, labelling the nodes of $\mathcal{D}(\Gamma)$ from left to right. Let $\mathbf{IP}_{n-1}$ hold in Γ. Prove that the following are equivalent:

(i) the geometry Γ is top-thin and the dual collinearity graph of Γ is bipartite;
(ii) we have $\Gamma = Gr_0(\Gamma')$ for some geometry Γ' with one of the following diagram graphs:

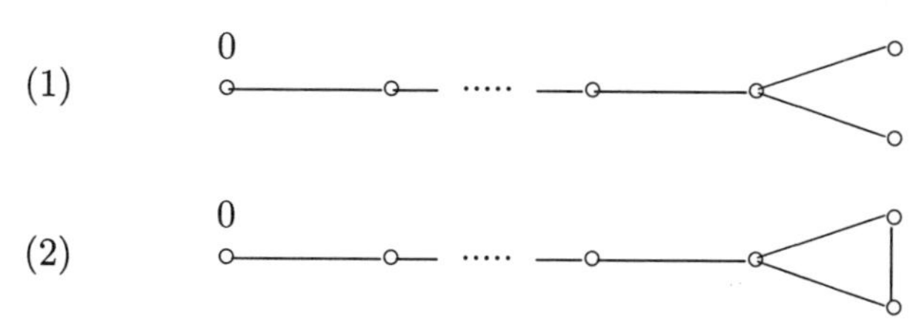

Exercise 6.13. Let Γ be the geometry of black and white triangles of Section 2.1.2(10). Describe the smallest homomorphic image of Γ that satisfies $\mathbf{IP}_i$ with respect to every type i.

(Hint: that homomorphic image of Γ has 7 elements of each type and its rank 2 truncations are isomorphic to $PG(2,2)$; thus, they realize a model of $PG(2,2)$ on a torus; see [145].)

6.4 Counterexamples

In this section we consider examples satisfying only one of the two properties forming the intersection property, namely the isomorphism property and the weak intersection property (Section 6.1).

6.4.1 Some rank 2 examples

There many rank 2 geometries satisfying the isomorphism property but not the weak intersection property, with respect to each of the two types P and L (notation as in Sections 1.2 and 6.3.1). For instance, t-(v, k, λ)-designs with $t > 2$ or $\lambda > 1$ (see Section 1.2.6(5)), Mőbius, Laguerre and Minkowski planes (Section 1.2.6(2)), point–hyperplane systems of n-dimensional matroids ($n > 2$), are examples of this kind.

Examples also exist satisfying the weak intersection property but not the isomorphism property. For instance, the collapsed n-dimensional simplex (Section 4.3.3) satisfies the weak intersection property but not the isomorphism property with respect to P (however, it satisfies the isomorphism property but not the weak intersection property with respect to L). The point–line system of the halved octahedron satisfies the weak intersection property but not the isomorphism property, with respect to each of the two types P and L.

6.4.2 Flat C_3 geometries

In this section we consider flat geometries (Section 4.3.2) belonging to the Coxeter diagram C_3 (flat C_3 *geometries*, for short)

$$(C_3) \qquad \overset{0}{\bullet}\!\!-\!\!-\!\!-\!\!\overset{1}{\bullet}\!\!=\!\!=\!\!=\!\!\overset{2}{\bullet}$$

The halved octahedron is an example of flat C_3 geometry. We now describe another example.

The $Alt(7)$**-geometry.** We start with a set S of seven elements. We take S as set of points and the 3-subsets of S as lines of the geometry Γ we want to define, with the containment relation as incidence relation.

The automorphism group $L_3(2)$ of $PG(2,2)$ has index 30 in the symmetric group S_7. Therefore, we can form 30 copies of $PG(2,2)$, taking S as set of points of each of them. However, $L_3(2)$ is also contained in the alternating group A_7. Therefore A_7 has 2 orbits of size 15 on that set of 30 copies of $PG(2,2)$.

We take one of those two orbits as set of planes for Γ. The incidence relation between these planes and the points and lines previously chosen for Γ is the natural one: all points are incident to all planes and a line is incident to a plane if it is one of the seven lines of that projective plane.

It is straightforward to check that, given a point a, a 3-subset x of S (line of Γ) containing a and a plane u, if x is not a line of u, then there is just one line y of u on a such that $x \cap y = a$ and just two projective planes of order 2 can be drawn on S having x and y as lines. Furthermore, these two planes are permuted if we permute a suitable pair of points. Hence those planes are in distinct orbits of A_7. Therefore, only one of them is in the orbit we have selected for Γ.

It follows from the above that residues of points of Γ are generalized quadrangles of uniform order 2, whence they are isomorphic to $\mathcal{S}_3(2)$, since $\mathcal{S}_3(2)$ is the unique generalized quadrangle of order 2, up to isomorphisms (see [166], Chapter 6). Residues of planes of Γ are isomorphic to $PG(2,2)$. Hence Γ is a C_3 geometry of uniform order 2. Trivially, Γ is flat. The geometry Γ constructed above will be called the *Alt*(7)-*geometry* (we prefer this name to the more popular name 'A_7-geometry', in order to avoid confusions with the Coxeter diagram A_7). The *Alt*(7)-geometry has been discovered by Neumaier [136].

Properties of flat C_3 geometries. Let Γ be a flat C_3 geometry. Trivially, the isomorphism property does not hold in Γ with respect to any of the types 0 or 2.

However, $\mathcal{L}_2(\Gamma)$ is a partial plane (see Exercise 4.5) and $\sigma_2(\Gamma)$ is simply the partial plane $\mathcal{L}_2(\Gamma)$ plus the set $\sigma_2(\emptyset)$ of all planes. It is clear from this that the weak intersection property holds in Γ with respect to the type 2.

On the other hand, the weak intersection property may or may not hold in Γ with respect to 0. For instance, it holds in the halved octahedron but it does not hold in the *Alt*(7)-geometry.

The point–line system $\mathcal{L}_0(\Gamma)$ has gonality 2 (see Exercise 4.8). Hence $\mathbf{IP}_1$ does not hold in Γ, by Theorem 6.14. On the other hand, the isomorphism property holds in Γ with respect to the type 1. Indeed, given any two distinct elements x, y of type 0 or 2, we never have $\sigma_1(x) \subseteq \sigma_1(y)$. When x and y have the same type $i \in \{0,2\}$, we can choose an element z of the other type $j \in \{0,2\}$ and remark that the 1-shadow of x is not contained in the 1-shadow of y in the partial plane Γ_z. If x and y have distinct types, we see in the partial plane Γ_x that $\sigma_1(x) \not\subseteq \sigma_1(y)$.

Since $\mathbf{IP}_1$ does not hold in Γ but the isomorphism property holds in Γ with respect to the type 1, the weak intersection property does not hold in Γ with respect to the type 1.

6.4.3 Examples from Chapters 1 and 2

The isomorphism property (but not the weak intersection property) holds for the natural shadow spaces of amalgams of attenuated spaces of rank $n > 3$ and for natural shadow spaces of many affine expansions of rank 2 geometries.

The same can be said for truncations of matroids, of polar spaces, of attenuated spaces, of dual attenuated spaces, of attenuated–dual-attenuated spaces, of affine polar spaces, even if the set of deleted types consists only of inner nodes of the diagram. It is left for the reader to check the above claims.

Certain truncations of D_4 buildings are considered in Section 1.6.1(6). Their rank 2 residues have gonality 2 but they are not generalized digons. Hence the intersection property fails to hold in them. By Theorem 6.12, the intersection property does not hold in those truncated D_4 buildings with respect to any type or any pair of types (but it holds with respect to the full set of types, by Proposition 6.20). However, it is not difficult to check the isomorphism property holds in them, with respect to every type.

Exercise 6.14. Let $\mathcal{R}$ be a commutative ring with unit 1 and let M be a left module over $\mathcal{R}$, satisfying the following properties:

(i) for every $x \in M$, we have $1x = x$;
(ii) we have $ax = bx$ only if $a = b$, for all $x \in M - 0$ and all $a, b \in \mathcal{R}$.

Take M as set of points and the cosets of the submodules $\mathcal{R}x$ ($x \in M - 0$) as lines. Let Γ be the rank 2 geometry obtained in this way.

Prove that the intersection property holds in Γ if and only if the isomorphism property holds in Γ with respect to the type P, and that this happens if and only if $\mathcal{R}$ is a division ring. Prove also that, if $\mathcal{R}$ is a principal ideal domain, then Γ satisfies the weak intersection property with respect to P.

Find an example where Γ does not satisfy the weak intersection property with respect to P.

Exercise 6.15. Let $\mathbf{R}$ be the set of real numbers and take $\mathbf{R}$ as the set of points and the half-lines of $\mathbf{R}$ as lines. Let Γ be the rank 2 geometry obtained in this way. Prove that neither the isomorphism property nor the weak intersection property holds in Γ, with respect to either of the two types P and L.

Exercise 6.16. Let Γ be an affine expansion of a linear space $\mathcal{L}$ realized inside $PG(2,K)$. Check that the isomorphism property holds for the natural shadow space of Γ. Prove also that the weak intersection property holds for the natural shadow space of Γ if and only if $\mathcal{L} = PG(2,K)$ (that is, if and only if $\Gamma = AG(3,K)$).

6.5 The properties $\mathbf{IP}'_J$ and $\mathbf{IP}''_J$

In the proofs of Theorems 6.12 and 6.14 we have implicitly shown that the following property is a consequence of $\mathbf{IP}_J$:

($\mathbf{IP}'_J$) Given any two flags F and G, if $\sigma_J(F) \cap \sigma_J(G) \neq \emptyset$, then there is a flag H incident to both F and G and such that $\sigma_J(F) \cap \sigma_J(G) = \sigma_J(H)$.

We may also consider the following generalization of $\mathbf{IP}'_J$:

($\mathbf{IP}''_J$) Given any family $(F_k)_{k \in K}$ of flags, if $\bigcap_{k \in K} \sigma_J(F_k) \neq \emptyset$, then there is a flag G incident to all flags F_k ($k \in K$) and such that $\sigma_J(G) = \bigcap_{k \in K} \sigma_J(F_k)$.

6.5.1 The properties $\mathbf{IP}_J$, $\mathbf{IP}'_J$ and $\mathbf{IP}''_J$ are equivalent

Trivially, $\mathbf{IP}''_J$ implies $\mathbf{IP}'_J$ and $\mathbf{IP}'_J$ implies the weak intersection property with respect to J.

Lemma 6.21 *The intersection property* $\mathbf{IP}_J$ *implies* $\mathbf{IP}''_J$.

Proof. Let $\mathbf{IP}_J$ hold in Γ and let $(F_k)_{k \in K}$ be a family of flags as in the hypotheses of $\mathbf{IP}''_J$. By $\mathbf{IP}_J$, the poset $\sigma_J(\Gamma) \cup \{\emptyset\}$ is a complete semilattice. Hence there is a flag G such that $\sigma_J(G) = \bigcap_{k \in K} \sigma_J(F_k)$. We can assume that G is J-reduced. Therefore G is incident to all flags F_k ($k \in K$), by the separation property of Lemma 6.8. □

Lemma 6.22 *Let* $\mathbf{IP}'_J$ *hold and let* F *be a flag with* $J \not\subseteq t(F)$. *Then* $\mathbf{IP}'_{J-t(F)}$ *holds in* Γ_F.

Proof. Let X, Y be flags of Γ_F with $\sigma^F_{J-t(F)}(X) \cap \sigma^F_{J-t(F)}(Y) \neq \emptyset$. By Corollary 6.5 and $\mathbf{IP}'_J$, there is a flag H incident to both $X \cup F$ and $Y \cup F$ and such that $\sigma_J(H) = \sigma_J(X \cup F) \cap \sigma_J(Y \cup F)$. By Corollary 6.2, we can assume that $H \supseteq F$. Trivially, $H - F \in \Gamma_F$ and $H - F$ is incident to both X and Y. By Corollary 6.5, we have $\sigma^F_{J-t(F)}(X) \cap \sigma^F_{J-t(F)}(Y) = \sigma^F_{J-t(F)}(H - F)$. □

The next lemma is a partial converse of the direct sum theorem.

Lemma 6.23 *Let* $\mathbf{IP}'_J$ *hold and let x be an element of Γ incident to all flags of type J. Then the connected component of $\mathcal{D}(\Gamma)$ containing $t(x)$ does not meet J.*

Proof. Since $\mathbf{IP}'_J$ is preserved under taking residues (Lemma 6.22), we can work by induction on the rank n of Γ.

If $n = 1$ there is nothing to prove. Let $n > 1$ and let x be as in the hypotheses of the lemma. We have $t(x) \not\in J$, otherwise all flags of type J would contain x, which is impossible by Lemma 1.8 applied to flags of cotype $t(x)$.

Let y be another element of type $t(x)$, distinct from x. By $\mathbf{IP}'_J$, there is a flag H incident to both x and y and such that $\sigma_J(H) = \sigma_J(x) \cap \sigma_J(y)$ $(= \sigma_J(y))$. We have $t(x) \not\in t(H)$, since $y \neq x$. We have two cases to examine.

Case 1. $H \neq \emptyset$. Since $t(x) \not\in t(H)$ and $\sigma_J(H) = \sigma_J(y)$, we have $J \cap t(H) = \emptyset$, otherwise all flags of type J incident to y would contain some common elements, which is impossible by Lemma 1.8 in Γ_y. We have $\sigma^H_J(y) = \sigma^H_J(\emptyset)$. Therefore, by the inductive hypothesis in Γ_H, the connected component of $\mathcal{D}(\Gamma_H)$ containing $t(x)$ $(= t(y))$ does not meet J. We also have $\sigma^y_J(h) = \sigma^y_J(\emptyset)$ for all $h \in H$. Hence, by the inductive hypothesis in Γ_y, there are no paths in $\mathcal{D}(\Gamma_y)$ from $t(H)$ to J. By Lemma 4.3, $t(H)$ and $t(x)$ separate each other from J in $\mathcal{D}(\Gamma)$. Since $t(H) \cap J = \emptyset$ and $t(x) \not\in t(H) \cup J$, the above happens only if none of the connected components of $\mathcal{D}(\Gamma)$ meets both $t(H) \cup t(x)$ and J. In particular, the connected component of $\mathcal{D}(\Gamma)$ containing $t(x)$ does not meet J.

Case 2. For every y as above, $H = \emptyset$ is the only possible choice for H. Therefore, all elements of type $t(x)$ are incident to all flags of type J and $\emptyset$ is the only flag incident to all flags of type J and to two distinct elements of type $t(x)$, one of which is x. However, we can now replace x with any other element y of type $t(x)$, repeating the above argument for y instead of x. Thus, we are led to the case where $\emptyset$ is the only flag incident to at least two elements of type $t(x)$ and to all flags of type J.

If $|J| = n - 1$, then $t(x)$ is not joined in $\mathcal{D}(\Gamma)$ with any other type. In this case we have finished.

Let $|J| < n-1$ and let us choose a type $i \notin J \cup t(x)$. Let z be an element of type i. Since z is incident to at least two elements of type $t(x)$, we have $\sigma_J(z) \neq \sigma_J(\emptyset)$, because $\emptyset$ is now the only flag incident to all flags of type J and with at least two elements of type $t(x)$.

By $\mathbf{IP}'_J$, for every element y of type $t(x)$ there is a flag K incident to both z and y and such that $\sigma_J(K) = \sigma_J(z) \cap \sigma_J(y)$ $(= \sigma_J(z))$. If y can be chosen not incident to z, then $z \notin K$ and we can repeat the argument of Case 1 replacing K and z for H and y. Thus, the connected component of $\mathcal{D}(\Gamma)$ containing i does not meet J. However, this contradicts the inequality $\sigma_J(z) \neq \emptyset$, by the direct sum theorem. Therefore z is incident to all elements y of type $t(x)$. Since this holds for every type $i \notin J \cup t(x)$ and for every element x of type i and since all elements of type $t(x)$ are incident to all flags of type J, the type $t(x)$ is not joined with any other type in $\mathcal{D}(\Gamma)$. □

Lemma 6.24 *Let* $\mathbf{IP}'_J$ *hold and let* F, G *be flags such that* $F * G$ *and* $\sigma_J(F) \subseteq \sigma_J(G)$. *Then* $J \mid_{t(F)} t(G)$.

Proof. The statement is trivial if $J \subseteq t(F)$. Let $J \not\subseteq t(F)$. The property $\mathbf{IP}'_{J-t(F)}$ holds in Γ_F by Lemma 6.22. Applying Lemma 6.23 in Γ_F with respect to all elements $x \in G - F$ and going back to $\mathcal{D}(\Gamma)$ by Lemma 4.3, we obtain that $J \mid_{t(F)} t(G)$. □

Corollary 6.25 *Let* $\mathbf{IP}'_J$ *hold and let* F, G *be* J*-reduced flags such that* $F * G$ *and* $\sigma_J(F) = \sigma_J(G)$. *Then* $F = G$.

Proof. By Lemma 6.24, $t(F)$ and $t(G)$ separate each other from J in $\mathcal{D}(\Gamma)$. Since they are both J-reduced, we have $t(F) = t(G)$. Hence $F = G$, as $F * G$. □.

Lemma 6.26 *Let* $\mathbf{IP}'_J$ *hold. Then the separation property of Lemma 6.8 holds with respect to* J.

Proof. Let F, G be as in the hypotheses of the separation property. By $\mathbf{IP}'_J$ there is a flag H incident to both F and G and such that $\sigma_J(H) = \sigma_J(F) \cap \sigma_J(G)$ $(= \sigma_J(F))$. We can assume that H is J-reduced. Hence $H = F$ by Corollary 6.26. Therefore $F * G$. We have $J \mid_{t(F)} t(G)$ by Lemma 6.24. The separation property is proved. □

By Lemmas 6.21, 6.26 and 6.8, and recalling that $\mathbf{IP}''_J$ implies $\mathbf{IP}'_J$ and that $\mathbf{IP}'_J$ implies the weak isomorphism property, we have the following:

Theorem 6.27 *The properties* $\mathbf{IP}_J$, $\mathbf{IP}'_J$ *and* $\mathbf{IP}''_J$ *are pairwise equivalent.*

6.5.2 Some consequences of Theorem 6.27

The results of this section depend on the equivalence between $\mathbf{IP}_J$ and $\mathbf{IP'}_J$, stated in Theorem 6.27. Most of these results will be used in Section 6.7.1, for the proof of Theorem 6.38.

Proposition 6.28 *Let I_1, I_2, ..., I_m be the connected components of the graph induced by $\mathcal{D}(\Gamma)$ on $I - J$ and let $Tr^+_{I_i \cup J}(\Gamma)$ admit a diagram, for $i = 1, 2, ..., m$. Then the property $\mathbf{IP}_J$ holds in Γ if and only if it holds in $Tr^+_{I_i \cup J}(\Gamma)$ for every $i = 1, 2, ..., m$.*

Proof. The 'only if' part follows from Theorem 6.14. Let us prove the 'if' part.

Let $\mathbf{IP}_J$ hold in $Tr^+_{I_i \cup J}$, for all $i = 1, 2, ..., m$ and let F, G be flags of Γ with $\sigma_J(F) \cap \sigma_J(G) \neq \emptyset$.

Let us set $F_i = F \cap t^{-1}(I_i \cup J)$ and $G_i = G \cap t^{-1}(I_i \cup J)$. By Corollary 6.3, we have $\sigma_J(F) = \bigcap_{i=1}^m \sigma_J(F_i)$ and $\sigma_J(G) = \bigcap_{i=1}^m \sigma_J(G_i)$. Therefore $\sigma_J(F) \cap \sigma_J(G) = \bigcap_{i=1}^m (\sigma_J(F_i) \cap \sigma_J(G_i))$. Hence $\sigma_J(F_i) \cap \sigma_J(G_i) \neq \emptyset$ for $i = 1, 2, ..., m$.

By $\mathbf{IP'}_J$ in $Tr^+_{I_i \cup J}(\Gamma)$, there is a flag H_i of $Tr^+_{I_i \cup J}(\Gamma)$ incident to both F_i and G_i and such that $\sigma_J(H_i) = \sigma_J(F_i) \cap \sigma_J(G_i)$. By the direct sum theorem applied to the residue of a flag $K \in \sigma_J(F) \cap \sigma_J(G)$ ($\subseteq \sigma_J(F_i) \cap \sigma_J(G_i) = \sigma_J(H_i)$), the flag H_i is incident to F_j, G_j, H_j, for all $j = 1, 2, ..., m$. Hence $H = \bigcup_{i=1}^m H_i$ is a flag and it is incident to F (since $F = \bigcup_{i=1}^m F_i$) and to G (since $G = \bigcup_{i=1}^m G_i$). Furthermore, $\sigma_J(H) = \bigcap_{i=1}^m \sigma_J(H_i)$ by Corollary 6.3. Therefore $\sigma_J(H) = \sigma_J(F) \cap \sigma_J(G)$. □

Proposition 6.29 *Let J, K be nonempty sets of types such that every type $i \notin J \cup K$ separates J from K in $\mathcal{D}(\Gamma)$. Assume that $Tr^-_K(\Gamma)$ and $Tr^-_J(\Gamma)$ admit diagrams and that $\mathbf{IP}_J$ and $\mathbf{IP}_K$ hold in $Tr^-_K(\Gamma)$ and $Tr^-_J(\Gamma)$, respectively. Then $\mathbf{IP}_{J \cup K}$ holds in Γ.*

Proof. Let F, G be flags such that $\sigma_{J \cup K}(F) \cap \sigma_{J \cup K}(G) \neq \emptyset$.

If $t(F) \subseteq J \cup K$, then $F \subseteq X$ for every $X \in \sigma_{J \cup K}(F) \cap \sigma_{J \cup K}(G)$. Hence $F * G$ and $\sigma_{J \cup K}(F \cup G) = \sigma_{J \cup K}(F) \cap \sigma_{J \cup K}(G)$. Thus, the flag $H = F \cup G$ has the property required in the statement of $\mathbf{IP}'_{J \cup K}$. The same conclusion is achieved if $t(G) \subseteq J \cup K$.

Let us assume that neither $t(F)$ nor $t(G)$ are contained in $J \cup K$. Then both $t(F) - (J \cup K)$ and $t(G) - (J \cup K)$ separate J from K, by our hypotheses on J and K. Let us set $F' = F - t^{-1}(K)$, $G' = G - t^{-1}(K)$, $F'' = F - t^{-1}(J)$ and $G'' = G - t^{-1}(J)$.

By $\mathbf{IP}_J$ and $\mathbf{IP}_K$ in $Tr^-_K(\Gamma)$ and $Tr^-_J(\Gamma)$, there are flags H_J and H_K of $Tr^-_K(\Gamma)$ and $Tr^-_J(\Gamma)$ respectively, incident to F' and G' and to F'' and

G'' respectively and such that $\sigma_J(H_J) = \sigma_J(F') \cap \sigma_J(G')$ and $\sigma_K(H_K) = \sigma_K(F'') \cap \sigma_K(G'')$. We can assume that H_J is J-reduced in $Tr_K^-(\Gamma)$ and that H_K is K-reduced in $Tr_J^-(\Gamma)$.

By the separation property and by Lemma 6.13, $t(H_J)$ separates J from $t(F') \cup t(G')$ and $t(H_K)$ separates K from $t(F'') \cup t(G'')$ in $\mathcal{D}(\Gamma)$. Furthermore, $\sigma_J(F') = \sigma_J(F)$, $\sigma_J(G') = \sigma_J(G)$, $\sigma_K(F'') = \sigma_K(F)$ and $\sigma_K(G'') = \sigma_K(G)$, because $t(F) - (J \cup K)$ and $t(G) - (J \cup K)$ separate J from K. Using Corollary 5.14 and the assumption that $J \mid_i K$ for every $i \in I - (J \cup K)$, we can easily prove that $t(F) \cup t(G)$ separates $t(H_K)$ from J, $t(H_J)$ from K and $t(H_J)$ from $t(H_K)$, as it separates J from K. Therefore, H_J and H_K are incident, H_J is incident to $F'' \cup G''$ and H_K is incident to $F' \cup G'$, by the direct sum theorem. Hence $H = H_J \cup H_K$ is a flag incident to both F and G.

The set of types $t(H_J) \cup t(H_K)$ separates $J \cup K$ from $t(F) \cup t(G)$, since $t(H_J)$ separates J from $t(F) \cup t(G)$, $t(H_K)$ separates K from $t(F) \cup t(G)$ and $J \mid_i K$ for all $i \in I - (J \cup K)$. Hence, $\sigma_{J \cup K}(H) \subseteq \sigma_{J \cup K}(F) \cap \sigma_{J \cup K}(G)$ by the separation lemma.

Furthermore, all flags in $\sigma_J(H_J) = \sigma_J(F) \cap \sigma_J(G)$ are incident to H_K and to all flags in $\sigma_K(H_K) = \sigma_K(F) \cap \sigma_K(G)$, and these in turn are incident to H_J and to all flags in $\sigma_J(H_J)$, by the direct sum theorem. It is now quite evident that $\sigma_{J \cup K}(H) = \sigma_{J \cup K}(F) \cap \sigma_{J \cup K}(G)$. □

Proposition 6.30 *Let J, K be nonempty sets of types such that every type $i \notin J \cup K$ separates J from K in $\mathcal{D}(\Gamma)$. Let $\mathbf{IP}_J$ and $\mathbf{IP}_K$ hold in Γ. Then $\mathbf{IP}_{J \cup K}$ holds.*

The proof is similar to that of Proposition 6.30. We leave it for the reader.

Note that Proposition 6.29 is not a trivial consequence of Proposition 6.30. Indeed, in order to use Proposition 6.30 for a proof of Proposition 6.29, we should first prove that the hypotheses of Proposition 6.29 imply that $Tr_K^-(\Gamma)$ and $Tr_J^-(\Gamma)$ admit diagrams. Actually, this can be done, but it needs some work.

Proposition 6.31 *Let K be a join of connected components of the graph induced by $\mathcal{D}(\Gamma)$ on $I - J$. Let $Tr_K^-(\Gamma)$ admit a diagram and let $\mathbf{IP}_J$ hold in it. Then $\mathbf{IP}_{J \cup K}$ holds in Γ.*

Proof. Let F, G be flags of Γ such that $\sigma_{J \cup K}(F) \cap \sigma_{J \cup K}(G) \neq \emptyset$. Let $A = (F \cup G) \cap t^{-1}(J \cup K)$. Trivially, A is a flag and it is contained in all flags of $\sigma_{J \cup K}(F) \cap \sigma_{J \cup K}(G)$. Hence $F * A * G$.

Let us set $F' = F - t^{-1}(K)$ and $G' = G - t^{-1}(K)$. By $\mathbf{IP}_J$ in $Tr_K^-(\Gamma)$ there is a flag H' of $Tr_K^-(\Gamma)$ incident to both F' and G' and with $\sigma_J(H') = \sigma_J(F') \cap \sigma_J(G')$. Let $X \in \sigma_{J \cup K}(F) \cap \sigma_{J \cup K}(G)$ and let $X' = X \cap t^{-1}(J)$.

Trivially, X' is incident to A, F' and G'. Hence $X' * H'$. Since J separates K from $I - (J \cup K)$, we obtain $H' * A$ by the direct sum theorem applied to the residue of X'. Therefore $F * H' * G$.

Let us set $H = (H' - t^{-1}(K)) \cup A$. We have $F * H * G$, because both H' and A are incident with both F and G.

It is implicit in the above that, given a flag X of type $J \cup K$ incident to both F and G, the flag $X \cap t^{-1}(J)$ is incident to H. On the other hand, J separates K from $I - (J \cup K)$. Hence $X \cap t^{-1}(K)$ is incident to $H - A$, by the direct sum theorem applied to the residue of $X \cap t^{-1}(J)$. Trivially, $X * A$. Hence $X * H$. Therefore $\sigma_{J \cup K}(H) \supseteq \sigma_{J \cup K}(F) \cap \sigma_{J \cup K}(G)$.

We can also assume that H' is J-reduced in $Tr_{\overline{K}}^{-}(\Gamma)$. Hence $t(H')$ separates $t(F')$ and $t(G')$ from J, by the separation property in $Tr_{\overline{K}}^{-}(\Gamma)$ and by Lemma 6.13. Therefore $t(H)$ separates $t(F)$ and $t(G)$ from $J \cup K$, since $H \supseteq A$ and J separates $I - (J \cup K)$ from K. Hence $\sigma_{J \cup K}(H) \subseteq \sigma_{J \cup K}(F) \cap \sigma_{J \cup K}(G)$ by the separation lemma.

Thus, $\sigma_{J \cup K}(H) = \sigma_{J \cup K}(F) \cap \sigma_{J \cup K}(G)$. □

Proposition 6.32 *Let K be a join of connected components of the graph induced by $\mathcal{D}(\Gamma)$ on $I - J$ and let $\mathbf{IP}_J$ hold in Γ. Then $\mathbf{IP}_{J \cup K}$ holds.*

The proof is similar to that of Proposition 6.31. We leave it for the reader. Note that Proposition 6.32 is not a trivial consequence of Proposition 6.31.

Proposition 6.33 *Let $\mathbf{IP}_J$ hold, let K be a nonempty set of types and let F, G be J-reduced flags such that $t(F) \cup t(G)$ separates J from K in $\mathcal{D}(\Gamma)$ and $\sigma_K(F) \cap \sigma_K(G) \neq \emptyset$. Then there is a flag H incident to both F and G and such that $\sigma_K(H) = \sigma_K(F) \cap \sigma_K(G)$.*

Proof. We have $\sigma_J(F) \cap \sigma_J(G) \subseteq \sigma_J(X)$ for every flag $X \in \sigma_K(F) \cap \sigma_K(G)$, by the separation lemma. Therefore, we have

$$\bigcap(\sigma_J(X) \mid X \in \sigma_J(F) \cap \sigma_J(G)) \supseteq \sigma_J(F) \cup \sigma_J(G) \neq \emptyset$$

By $\mathbf{IP}''_J$, there is a flag H incident to all flags $X \in \sigma_K(F) \cap \sigma_K(G)$ and such that

$$\sigma_J(H) = \bigcap(\sigma_J(X) \mid X \in \sigma_J(F) \cap \sigma_J(G)) \supseteq \sigma_J(F) \cup \sigma_J(G)$$

Since both F and G are J-reduced, they are incident to H and both $t(F)$ and $t(G)$ separate $t(H)$ from J, by the separation property. Furthermore, we can assume that $t(H)$ is J-reduced. Hence, $J \mid_{t(H)} K$ by the separation property (we recall that the flags $X \in \sigma_K(F) \cap \sigma_K(G)$ have type $t(X) = K$). By Corollary 5.14, $t(H)$ separates both $t(F)$ and $t(G)$ from K. Hence

$\sigma_K(H) \subseteq \sigma_K(F) \cap \sigma_K(G)$, by the separation lemma. On the other hand, H has been chosen incident to all flags $X \in \sigma_K(F) \cap \sigma_K(G)$. Therefore $\sigma_K(H) = \sigma_K(F) \cap \sigma_K(G)$. □

The following modification of $\mathbf{IP}'_{IP}$ allows us to save some work when we want to check if $\mathbf{IP}_J$ holds:

($\mathbf{IP}_J^{Red}$) given nonempty J-reduced flags F and G, if $\sigma_J(F) \cap \sigma_J(G) \neq \emptyset$, then there is a nonempty J-reduced flag H such that $H \leq_J F, G$ in $Gr_J(\Gamma)$ and $\sigma_J(H) = \sigma_J(F) \cap \sigma_J(G)$.

Proposition 6.34 *The properties $\mathbf{IP}_J$ and $\mathbf{IP}_J^{Red}$ are equivalent.*

Proof. Let $\mathbf{IP}_J$ hold. Let F and G be as in the hypotheses of $\mathbf{IP}_J^{Red}$. Then $\sigma_J(\emptyset) \neq \sigma_J(F), \sigma_J(G)$ by Corollary 6.9 and there is a flag H incident to both F and G and such that $\sigma_J(H) = \sigma_J(F) \cap \sigma_J(G)$, by $\mathbf{IP}_J$. We have $H \neq \emptyset$, because $\sigma_J(F), \sigma_J(G) \neq \emptyset$. Furthermore, we can assume that H is J-reduced. Hence $H \leq_J F, G$ by the separation property.

Conversely, let $\mathbf{IP}_J^{Red}$ hold. Let 0 be the initial node of $\mathcal{D}(Gr_J(\Gamma))$. Trivially, $\mathbf{IP}_J^{Red}$ in Γ is equivalent to $\mathbf{IP}_0^{Red}$ in $Gr_J(\Gamma)$. Hence $\mathbf{IP}_0^{Red}$ holds in $Gr_J(\Gamma)$. By Proposition 6.10, $\mathbf{IP}_J$ in Γ is equivalent to $\mathbf{IP}_0$ in $Gr_J(\Gamma)$.

To prove that $\mathbf{IP}_0$ holds in $Gr_J(\Gamma)$, we work by induction on the rank n_J of $Gr_J(\Gamma)$. Trivially, $\mathbf{IP}_0$ holds in $Gr_J(\Gamma)$ when $n_J = 1$. Hence $\mathbf{IP}_J$ holds in Γ in this case. When $n_J = 2$, it is easily seen that $\mathbf{IP}_0^{Red}$ forces $Gr_J(\Gamma)$ to be a partial plane. Hence $\mathbf{IP}_0$ holds in $Gr_J(\Gamma)$ by Proposition 6.16. Thus, $\mathbf{IP}_J$ holds in Γ in this case, too.

Let $n_J > 2$. We firstly prove that $\emptyset$ is the unique J-reduced flag of Γ incident to all flags of type J. Let $\sigma_J(F) = \sigma_J(\emptyset)$ for some nonempty J-reduced flag F, if possible. If F' is any nonempty J-reduced flag of Γ with $F \leq_J F'$ in $Gr_J(\Gamma)$, then $\sigma_J(F) \subseteq \sigma_J(F')$ by the separation lemma. Therefore we can assume that $t_J(F) = n_J - 1$. Given $G \in Gr_J(\Gamma)$, there is $H_G \in Gr_J(\Gamma)$ such that $H_G \leq_J G, F$ in $Gr_J(\Gamma)$ and $\sigma_J(H_G) = \sigma_J(G)$ in Γ. It is easy to check that $\mathbf{IP}_0^{Red}$ is preserved when taking residues of elements of $Gr_J(\Gamma)$ of type $n_J - 1$. Hence $\mathbf{IP}_0$ holds in these residues by the inductive hypothesis. Let G be any element of $Gr_J(\Gamma)$ with $t_J(G) = n_J - 1$ and $G \neq F$. Trivially, $t_J(H_G) < n_J - 1$. By the separation property in the residue $Gr_J(\Gamma)_G$ of G, we have that $t_J(H_G)$ and 0 belong to distinct connected components of $\mathcal{D}(Gr_J(\Gamma)_G)$, since $\sigma_J(H_G) = \sigma_J(G)$. However, this implies that $\mathcal{D}(\Gamma)$ is disconnected, by Lemma 4.3. The contradiction is achieved. Therefore, we have $\sigma_J(F) \neq \sigma_J(\emptyset)$ for every nonempty J-reduced flag F of Γ. We can now prove that $\mathbf{IP}_J$ holds in Γ.

Let F, G be flags of Γ as in the hypotheses of $\mathbf{IP}'_J$. If $\sigma_J(F) = \sigma_J(\emptyset)$, then none of the connected components of $\mathcal{D}(\Gamma)$ meets both J and $t(F)$, according to our hypotheses. Therefore $Red_J(G) * F$ and the statement of $\mathbf{IP}'_J$ holds with $H = Red_J(G)$. The same if $\sigma_J(G) = \sigma_J(\emptyset)$.

Let $\sigma_J(\emptyset) \neq \sigma_J(F), \sigma_J(G)$. There is a nonempty J-reduced flag $H \leq_J Red_J(F), Red_J(G)$ such that $\sigma_J(H) = \sigma_J(Red_J(F)) \cap \sigma_J(Red_J(G))$, by $\mathbf{IP}^{Red}_J$. By the direct sum theorem applied to the residues of $Red_J(F)$ and $Red_J(G)$, the flag H is also incident to F and G. We are done. □

6.5.3 Exercises

Exercises 6.17–6.23 examine properties equivalent to or weaker than $\mathbf{IP}_J$. Exercises 6.24–6.33 consider the diagrams $L^*.L$ and D_n.

Exercise 6.17. Prove that $\mathbf{IP}_J$ is equivalent to the following:

($\mathbf{IP}^*_J$) for every flag F and every element x, if $\sigma_J(F) \cap \sigma_J(x) \neq \emptyset$, then there is a flag H incident to both F and x and such that $\sigma_J(H) = \sigma_J(F) \cap \sigma_J(x)$; the same holds in residues of flags G with $J \not\subseteq t(G)$, with respect to $J - t(G)$.

(Remark. The assumption that $\mathbf{IP}^*_i$ holds for every type i is the Intersection Property of [18] and [19].)

Exercise 6.18. Prove that the following properties are together equivalent to $\mathbf{IP}_J$:

($\mathbf{IP}^\circ_J$) given any two elements x and y, if $\sigma_J(x) \cap \sigma_J(y) \neq \emptyset$, then there is a flag H incident to both x and y and such that $\sigma_J(H) = \sigma_J(x) \cap \sigma_J(y)$; the same holds in residues of flags G with $J \not\subseteq t(G)$, with respect to $J - t(G)$.

($\mathbf{RP}_J$) (**Reduction property**) For every flag F, the set of subflags of F with the same J-shadow as F has a least element.

(Remark. $\mathbf{IP}^\circ_J$ is the intersection property considered by Tits in [227]; it is a specialization of $\mathbf{IP}^*_J$. The property $\mathbf{RP}_J$ is an analogue of the statement of Corollary 6.9.)

Exercise 6.19. Check that the reduction property $\mathbf{RP}_i$ holds in every rank 2 geometry, with respect to any of the two types $i = P, L$ (hence $\mathbf{IP}_i$ and $\mathbf{IP}^\circ_i$ are equivalent in rank 2 geometries, by Exercise 6.18).

Exercise 6.20. Prove that $\mathbf{IP}^*_i$ holds in a geometry Γ for every type $i \in I$ if and only if $\mathbf{IP}^\circ_i$ holds in Γ for every $i \in I$.

Exercise 6.21. Prove that a flag F is J-reduced if $\sigma_J(G) \neq \sigma_J(F)$ for every $G \subseteq F$. Show by counterexamples that the converse may fail to hold.

(Hint: consider flat geometries.)

Exercise 6.22. The next special case of the isomorphism property has been considered in Corollary 6.25:

(**Weak isomorphism property**) Given any two J-reduced flags F and G, if $\sigma_J(F) = \sigma_J(G)$, then $F = G$.

Examples have been given in Section 6.4.1 showing that the weak isomorphism property does not imply the isomorphism property. Prove that the weak isomorphism property implies the reduction property $\mathbf{RP}_J$. Show by counterexamples that the reduction property does not imply the weak isomorphism property.

(Hint: consider flat geometries.)

Exercise 6.23. Prove that the isomorphism property implies the conclusion of Lemma 6.23:

if x is an element of Γ incident to all flags of type J, then the connected component of $t(x)$ in $\mathcal{D}(\Gamma)$ does not meet J; the same holds in the residue of every flag F such that $J \not\subseteq t(F)$, with respect to $J - t(F)$.

Exercise 6.24. Use Proposition 6.28 to prove that the intersection property holds with respect to the central node of the D_4 diagram in a D_4 building (recall that $\mathbf{IP}_i$ holds for every extremal node i of D_4; see Section 6.3.4(2); also Exercises 6.31 and 6.33).

Exercise 6.25. Use Proposition 6.28 and Exercise 4.5 to prove the statement of Exercise 6.7.

Exercise 6.26. Let Γ be as in Exercise 6.7. Use Proposition 6.31 and Exercise 4.5 to prove that the intersection property holds in Γ with respect to $J = \{0, 1\}$ and $J = \{1, 2\}$.

Exercise 6.27. Let Γ be a rank 3 geometry with string diagram graph and having both a good system of lines and a good dual system of lines (Section 4.3.2). Use Proposition 6.28 to prove that the intersection property holds in Γ with respect to the central node of the diagram (compare with Exercise 6.25).

Exercise 6.28. Let Γ be a building of type D_n

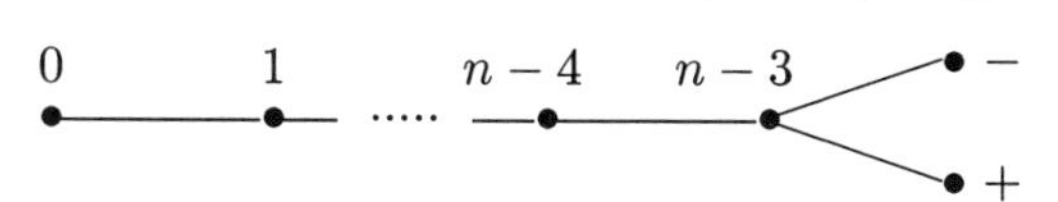

Using Proposition 6.33 and the information that $\mathbf{IP}_0$ holds in Γ, prove that $\mathbf{IP}_J$ holds in Γ, with $J = \{+, -\}$ or $J = \{k, k+1, ..., n-3, +, -\}$,

$0 \leq k \leq n-3$.

(Suggestion: exploit Proposition 6.34 to save work).

Exercise 6.29. Let Γ be as in Exercise 6.28. Use that exercise and Proposition 6.33 to prove that $\mathbf{IP}_J$ holds in Γ, with $J = \{0, 1, ..., h\}$, $0 \leq h \leq n-3$.

Exercise 6.30. Let Γ be as in Exercise 6.28 and $J = \{0, 1, ..., h, +, -\}$ with $0 \leq h \leq n-3$ or $J = \{0, 1, ..., h, k, k+1, ..., n-3, +, -\}$ with $0 \leq h < k \leq n-3$. Using Exercises 6.28 and 6.29 and Proposition 6.30, prove that $\mathbf{IP}_J$ holds in Γ.

Exercise 6.31. Let Γ be as in Exercise 6.28. Using that exercise, the information that $\mathbf{IP}_0$ holds in Γ, Theorem 6.14, Proposition 6.33 and Proposition 6.28, prove that $\mathbf{IP}_{n-3}$ holds in Γ.

Exercise 6.32. Let Γ be as in Exercise 6.28. Using Exercise 6.31, prove that both $Gr_+(\Gamma)$ and $Gr_-(\Gamma)$ have good systems of lines.

Exercise 6.33. Use Exercises 6.28–6.32 to prove that $\mathbf{IP}_J$ holds in a D_4 building, for every nonempty set of types J.

6.6 The property IP

We say that the (*strong*) *intersection property* holds in Γ (that **IP** *holds* in Γ, for short) if $\mathbf{IP}_J$ holds in Γ for every nonempty set of types J.

When Γ has rank 2, then either of $\mathbf{IP}_P$ or $\mathbf{IP}_L$ is sufficient to obtain **IP**, by Propositions 6.16 and 6.20 (P and L are the two types of Γ, as in Section 6.3.1). We will see in the next section (Theorem 6.38) that a similar result can be stated for any geometry with a string diagram graph: if the diagram graph $\mathcal{D}(\Gamma)$ of Γ is a string with initial node 0 and if $\mathbf{IP}_0$ holds in Γ, then **IP** also holds in Γ (needless to say, the same can be stated for the terminal node of the string $\mathcal{D}(\Gamma)$). Therefore, **IP** holds in almost all examples considered in Chapters 1 and 2, as almost all of them have string diagram graphs and satisfy $\mathbf{IP}_0$ (see Section 6.3.4(1)). For instance, **IP** holds in matroids (in particular, in projective or affine geometries) and in polar spaces.

On the other hand, if $\mathcal{D}(\Gamma)$ is a string but i is now a node of the string $\mathcal{D}(\Gamma)$ other than the initial or the terminal one, then $\mathbf{IP}_i$ is not sufficient to obtain **IP**, in general (see Exercises 6.2, 6.3, 6.4, 6.7, 6.26; also Proposition 6.20).

We have not yet defined buildings in general. We will do this in Chapter 13. However, many of the geometries we have described in Chapters 1, 2 and 5 are in fact buildings: projective geometries, polar spaces, buildings of type D_n, metasymplectic spaces of buildings of type D_4 (Section 5.3.4), the icosahedron (and the dodecahedron) the tessellations of the euclidean plane of Section 2.1.2, (1)–(5), the geometry of black and white triangles of Section 2.1.2(10).

It is proved in [222] (Chapter 12) that the strong intersection property **IP** holds in all buildings (see also Proposition 13.4).

Problem. Is it true that **IP** holds in a geometry Γ whenever $\mathbf{IP}_i$ holds in Γ for every type i ?

Remark. Actually, the intersection property most frequently considered in the literature is the assumption that $\mathbf{IP}_i$ holds for every type i, rather than **IP**.

6.7 String diagrams

In Sections 6.7.1 and 6.7.2 we assume that Γ is a geometry with string diagram graph $\mathcal{D}(\Gamma)$ and set of types $I = \{0, 1, ..., n-1\}$ with 0 chosen as the initial node of $\mathcal{D}(\Gamma)$:

$$\overset{0}{\circ}\!\!-\!\!-\!\!\overset{1}{\circ}\!\!-\!\!-\!\!\overset{2}{\circ}\!\!-\; \cdots\cdots \;-\!\!\overset{n-2}{\circ}\!\!-\!\!-\!\!\overset{n-1}{\circ}$$

Non-string cases will be considered in Section 6.7.3.

6.7.1 From $\mathbf{IP}_0$ to IP

As I can be viewed as a poset, we can use for it the notation stated for posets in Section 0.3. Thus, given $h, k \in I$ with $h \leq k$, the interval $[h, k]$ is the subset $\{h, h+1, ..., k\}$ of I. We define $[h, k]$ even if $h > k$, stating that $[h, k] = \emptyset$ in this case.

Given a nonempty subset J of I, the connected components of J and of $I - J$ in the string $\mathcal{D}(\Gamma)$ are intervals. We denote the number of connected components of J (of $I - J$) by the symbol $d^+(J)$ (by $d^-(J)$, respectively). Since $J \neq \emptyset$, we have $d^+(J) \geq 1$. Trivially, $d^-(J) = 0$ if and only if $J = I$. The set J is an interval if and only if $d^+(J) = 1$. The set J is the complement of an interval if and only if $d^-(J) = 1$. We have $d^+(J) = d^-(J)$ if and only if only one of 0 or $n-1$ belongs to J. We have $d^+(J) = d^-(J)+1$ if and only if both 0 and $n-1$ are in J. We have $d^-(J) = d^+(J)+1$ if and only if none of 0 or $n-1$ is in J.

Lemma 6.35 *Let* $\mathbf{IP}_0$ *hold in* Γ. *Then* $\mathbf{IP}_J$ *holds in* Γ *for every nonempty set of types* J *with* $d^-(J) \leq 1$.

Proof. The property $\mathbf{IP}_I$ holds in Γ (Proposition 6.20). By Proposition 6.33 with $J = 0$ we obtain that $\mathbf{IP}^{Red}_{[k,n-1]}$ holds in Γ for every $k \leq n-1$. Hence $\mathbf{IP}_{[k,n-1]}$ holds for every $k \leq n-1$, by Proposition 6.34. In particular, $\mathbf{IP}_{n-1}$ holds.

Thus, we can substitute $n-1$ for J in Proposition 6.33 and, again using Proposition 6.34, we obtain that $\mathbf{IP}_{[0,h]}$ holds in Γ for every $h \leq n-1$. By Proposition 6.30 and by the above, $\mathbf{IP}_{I-[h,k]}$ holds in Γ for every choice of $h, k \in I$ with $h \leq k$ and $k - h < n$. □

Lemma 6.36 *Let $\mathbf{IP}_0$ hold in Γ. Then, for every nonempty subset J of I, the truncation $Tr^+_J(\Gamma)$ admits a diagram and its diagram graph is the string defined by the order relation induced on J by the natural ordering of the set of integers $I = \{0, 1, ..., n-1\}$.*

Proof. Let $[h, k]$ be a connected component of $I - J$ such that $0 < h$. Then $\mathbf{IP}_{[0,h-1]}$ holds in Γ, by Lemma 6.35. Hence, given a flag F of type $[0, h-2] \cup [k+2, n-1]$, the property $\mathbf{IP}_h$ holds in Γ_F, by Theorem 6.12. By Lemma 4.3 and Corollary 6.9 in Γ_F, the truncation $Tr^-_{[h,k]}(\Gamma_F)$ is not a generalized digon. By the direct sum theorem, $Tr^-_{[h,k]}(\Gamma_F)$ is the residue in $Tr^+_J(\Gamma)$ of $F \cap t^{-1}(J)$. Hence this residue is not a generalized digon.

If $0 \notin J$, let $[0, k]$ be the connected component of $I - J$ containing 0. Then $k < n-1$ because $J \neq \emptyset$. Since $\mathbf{IP}_{n-1}$ holds in Γ by Lemma 6.35, we can repeat the previous argument replacing $n-1$ for 0 and we obtain that, for every flag F of type $[k+2, n-1]$, the residue in $Tr^+_J(\Gamma)$ of $F \cap t^{-1}(J)$ is not a generalized digon.

Therefore, for every connected component $[h, k]$ of $I - J$ and for every flag F' of $Tr^+_J(\Gamma)$ of type $J - \{h-1, k+1\}$, the residue of F' in $Tr^+_J(\Gamma)$ is not a generalized digon.

By the direct sum theorem, if $i, i+1 \in J$, then all residues in $Tr^+_J(\Gamma)$ of type $\{i, i+1\}$ are also residues in Γ of type $\{i, i+1\}$, whence they are not generalized digons.

Finally, let i, j, t be distinct types in J with $i < t < j$. By the direct sum theorem, every residue of $Tr^+_J(\Gamma)$ of type $\{i, j\}$ is a generalized digon. The lemma is proved. □

Lemma 6.37 *Let $\mathbf{IP}_0$ hold in Γ. Then $\mathbf{IP}_J$ holds for every nonempty set of types J with $d^+(J) = 1$.*

Proof. We may assume that $d^-(J) = 2$, otherwise we can apply Lemma 4.32 and we have finished. Hence $J = [h, k]$ with $0 < h \leq k < n-1$.

Let us set $L = [0, h-1]$ and $R = [k+1, n-1]$. By Lemma 6.36, the R_--truncation $Tr^-_R(\Gamma)$ of Γ admits a diagram. Hence we can apply Theorem

6.14 to $Tr^-_R(\Gamma)$, obtaining that $\mathbf{IP}_0$ holds in it. Similarly, $\mathbf{IP}_{n-1}$ holds in $Tr^-_L(\Gamma)$ (we recall that $\mathbf{IP}_{n-1}$ holds in Γ by Lemma 6.35). Since the diagram graphs of $Tr^-_L(\Gamma)$ and $Tr^-_R(\Gamma)$ are the strings induced on $[0, k]$ and $[h, n-1]$ respectively (Lemma 6.36), we can apply Lemma 6.35 obtaining that $\mathbf{IP}_J$ holds in both of these truncations. Finally, $\mathbf{IP}_J$ holds in Γ, by Proposition 6.28. □

Theorem 6.38 *Let* $\mathbf{IP}_0$ *hold in* Γ. *Then* $\mathbf{IP}$ *holds in* Γ.

Proof. The property $\mathbf{IP}_J$ holds when $d^+(J) + d^-(J) \leq 3$, by Lemmas 6.35 and 6.37. We go on by induction on $d^+(J)+d^-(J)$. Let $d^+(J)+d^-(J) > 3$. Given a connected component $X = [h, k]$ of $I - J$ we set $L^X = J \cap [0, h-1]$ and $R^X = J \cap [k+1, n-1]$. We also set

$$\Gamma^X = Tr^+_{X \cup J}(\Gamma), \quad \Gamma^{R,X} = Tr^-_{L^X}(\Gamma^X), \quad \Gamma^{L,X} = Tr^-_{R^X}(\Gamma^X).$$

By Lemma 6.36, the above truncations admit diagrams and their diagram graphs are the strings induced on $X \cup J$, $X \cup R^X$ and $X \cup L^X$, respectively. We have three cases to examine.

Case 1. Both 0 and $n-1$ belong to J. By the inductive hypothesis, $\mathbf{IP}_{L^X}$ and $\mathbf{IP}_{R^X}$ hold in Γ for every connected component X of $I - J$. These properties also hold in $\Gamma^{L,X}$ and $\Gamma^{R,X}$ respectively, by Theorem 6.14. Hence $\mathbf{IP}_J$ holds in Γ^X, by Proposition 6.29. Therefore $\mathbf{IP}_J$ holds in Γ by Proposition 6.28.

Case 2. We have $0 \in J$ but $n-1 \notin J$ (the case of $n-1 \in J$ but $0 \notin J$ is quite similar to this, since $\mathbf{IP}_{n-1}$ holds in Γ by Lemma 6.35). We can work just as in Case 1, except when X is the connected component of $I - J$ containing $n-1$. In this case, the proof that $\mathbf{IP}_J$ holds in Γ^X must be modified as follows. Let Y be the connected component of J containing the greatest element of J. By Lemma 6.37, the property $\mathbf{IP}_Y$ holds in Γ. By Lemma 6.36, the $(X \cup Y)_+$-truncation of Γ admits a diagram. Hence $\mathbf{IP}_J$ holds in it by Theorem 6.14. By Proposition 6.31, $\mathbf{IP}_J$ holds in Γ^X.

Case 3. Neither 0 nor $n-1$ are in J. We can work as above, using the same argument as in Case 2 to settle the case where X contains 0 or $n-1$.
□

6.7.2 The property $\mathbf{IP}_0^{Red}$

The nonempty 0-reduced flags of Γ are just the elements of Γ because 0 is the initial node of the string $\mathcal{D}(\Gamma)$. Hence the property $\mathbf{IP}_0^{Red}$ of Section 6.5.2 can be stated as follows, where the elements of type 0 are called points, as in Section 4.3.2, and $\leq_\Gamma$ is the partial order in the poset of Γ, as in Section 4.3.1:

($\mathbf{IP}_0^{Red}$) Given any two elements x, y of Γ, if there are distinct points incident to both x and y, then there is an element $z \leq_\Gamma x, y$ such that $\sigma_0(z) = \sigma_0(x) \cap \sigma_0(y)$.

By Proposition 6.34 and Theorem 6.38, we have the following:

Corollary 6.39 *Let* $\mathbf{IP}_0^{Red}$ *hold in* Γ. *Then* **IP** *holds.*

We have not considered the case of $|\sigma_0(x) \cap \sigma_0(y)| = 1$ in the above version of $\mathbf{IP}_0^{Red}$, because this case is trivial: if there is just one point incident to both x and y, then we can (and must) always choose that point for z to obtain the conclusion of $\mathbf{IP}_0^{Red}$.

Since we assume $|\sigma_0(x) \cap \sigma(y)_0| > 1$, an element z as in the statement of $\mathbf{IP}_0^{Red}$ must have type ≥ 1. In particular, if x is a line, then the condition $z \leq_\Gamma x, y$ forces $z = x$. Hence $x \leq_\Gamma y$.

The following special cases of the statement of $\mathbf{IP}_0^{Red}$ are sometimes sufficent to obtain $\mathbf{IP}_0^{Red}$ (hence **IP**, by Corollary 6.39):

($\mathbf{IP}_0^1$) Given a line x and an element y, if y is incident to at least two points of x, then $x \leq_\Gamma y$.

($\mathbf{IP}_0^{1,2}$) Given a line x and a plane y, if y is incident to at least two points of x, then $x \leq_\Gamma y$.

($\mathbf{IP}_0^{1,1}$) Given any two distinct points, there is at most one line incident to both of them.

The property $\mathbf{IP}_0^{1,1}$ is simply the property of having a good system of lines (see Section 4.3.2). The properties $\mathbf{IP}_0^{1,1}$ and $\mathbf{IP}_0^{1,2}$ together are the property of having a good system of planes (see Section 4.3.2).

As we said above, these properties are sometimes sufficent to obtain **IP** (see Exercise 6.5, for instance). Furthermore, the property **IP** is sometimes a consequence of the diagram (see Exercise 6.6, for instance). We will return to this in the next chapter.

6.7.3 A few remarks on non-string cases

A systematic investigation of non-string cases in the spirit of Theorem 6.38 is not easy. We only remark that, if $\mathbf{IP}_J$ holds in Γ, then **IP** holds in $Gr_J(\Gamma)$, by Proposition 6.10 and Theorem 6.38. In some cases this gives us some information on intersection properties in Γ even if $\mathcal{D}(\Gamma)$ is not a string, or it is a string but J is not one of its two extremal nodes. Examples of what can be done in this way are given in the next exercises.

The amount of information contained in $\mathbf{IP}_J$ depends on J, in general. For instance, this is information is null when $J = I$ (Proposition 6.20). On the other hand, extremal nodes of string diagrams contain a maximal amount of information (Theorem 6.38). Exercises 6.7 and 6.26 give other examples with null information (compare with Exercises 6.2–6.4).

Problem. Examining examples where $|J|= 1$ and $\mathcal{D}(\Gamma)$ is a tree, we may get the impression that the higher the degree of a node i is, the less information is contained in $\mathbf{IP}_i$. It would be nice to be able to say something less vague about this.

Exercise 6.34. Prove that the hypotheses of Proposition 6.30 force $Tr_{\overline{K}}^-(\Gamma)$ and $Tr_{\overline{J}}^-(\Gamma)$ to admit diagrams.

(Hint: J and K are joined by a string in $\mathcal{D}(\Gamma)$; the arguments used for Lemma 6.36 work in this case, too.)

Exercise 6.35. Let $\mathcal{D}(\Gamma)$ consist of a string $\mathcal{S}$ with extremal nodes 0 and t and of a connected graph $\mathcal{G}$ disjoint from $\mathcal{S}$, with no edges joining vertices of $\mathcal{G}$ to nodes of $\mathcal{S}$ other than t, but with some edges joining t to $\mathcal{G}$. Let $\mathbf{IP}_0$ hold in Γ. Prove that $\mathbf{IP}_J$ holds in Γ for every nonempty set J of nodes of $\mathcal{S}$.

(Hint: apply Theorem 6.38 in $Gr_0(\Gamma)$ and note that $Gr_J(Gr_0(\Gamma)) \cong Gr_J(\Gamma)$ for every nonempty set J of nodes of $\mathcal{S}$.)

Exercise 6.36. Let Γ, $\mathcal{G}$, $\mathcal{S}$ and 0 be as in the previous exercise and let G be the set of vertices of $\mathcal{G}$. Let $\mathbf{IP}_G$ hold in Γ. Prove that $\mathbf{IP}_0$ holds in Γ.

(Hint: $\mathbf{IP}$ holds in $Gr_G(\Gamma)$ by Proposition 6.10 and Theorem 6.38; the type 0 of Γ corresponds to the terminal node of the diagram of $Gr_G(\Gamma)$ ...)

Exercise 6.37. Let $\mathcal{D}(\Gamma)$ be connected and $i \in I$. Prove that $\mathbf{IP}_{I-i}$ implies $\mathbf{IP}_i$.

Exercise 6.38. Let Γ, $\mathcal{S}$, $\mathcal{K}$, 0 be as in Exercise 5.33 and let K be the set of vertices of $\mathcal{K}$. Prove that $\mathbf{IP}_0$ holds in Γ if and only if $\mathbf{IP}_K$ holds in it.

(Hint: use Exercise 6.36 for the 'if' part; as for the 'only if' part, remark that K corresponds to one type in $Gr_0(\Gamma)$ ...)

(Remark: the statement of Exercise 6.28 is a special case of the above.)

Exercise 6.39. Prove that $\mathbf{IP}$ holds in the geometry of black and white triangles of Section 2.1.2(10).

(Hint: we have remarked in Section 6.3.4 that $\mathbf{IP}_i$ holds in this geometry, for every type i; use Exercise 6.38.)

Exercise 6.40. Let $\mathcal{D}(\Gamma)$ be a tree with just one node 0 of degree > 2. Assume that the Intersection Property holds in Γ with respect to every terminal node of the tree $\mathcal{D}(\Gamma)$. Prove that $\mathbf{IP}_i$ holds in Γ for every type i.

(Hint: use Exercise 6.35.)

(Remark: the Coxeter diagrams D_n and E_n are trees as above.)

Exercise 6.41. Prove that **IP** holds in every D_n building (Section 5.3.4).

(Remark: the case of $n = 4$ has been settled in Exercise 6.33; some work in the general case has been done in Exercises 6.28–6.32.)

Exercise 6.42. Let $\mathcal{D}(\Gamma)$ be a tree and let P be a simple path in it. Assume that the intersection property holds in Γ with respect to the first node of P. Prove that $Tr_P^+(\Gamma)$ admits a diagram and satisfies the strong intersection property **IP**.

(Hint: use Theorem 6.12 to prove that $Tr_P^+(\Gamma)$ admits a diagram and has a string diagram graph; then use Theorems 6.14 and 6.38.)

Chapter 7

Recovering Geometries from Diagrams

We now start the job we promised to do at the end of Chapter 3: given a diagram, we try to describe the geometries belonging to that diagram, either classifying them, or at least characterizing them by other axioms of 'global' nature which are presumably more familiar to the reader. The tools we can use at this stage are very few and seemingly weak: we only have some results on disconnected diagrams (in particular, the direct sum theorem), the concept of geometric poset for string diagrams, the construction of Grassmann geometries and a number of results on intersection properties. However, we can already obtain many characterizations, in spite of the poverty of our equipment.

We will not give detailed proofs for all results stated in this chapter. We will only choose a few theorems to discuss in detail, just to give examples of how we can work with the tools we have. For other results we will only give references and sometimes a brief hint of the proof.

We will avoid theorems involving group-theoretic hypotheses or whose proof needs concepts that we have not yet explained (such as covers, for instance). We keep results of that kind for subsequent chapters. Anyway, we will not mention everything that could be mentioned (we do not intend to take a census of all geometries in this book).

We state one more convention in addition to those of Section 3.1.1, to make some statements more concise: given a diagram $\mathcal{D}$ and a class $\mathbf{C}$ of geometries, we say that $\mathcal{D}$ *characterizes* $\mathbf{C}$ if $\mathbf{C}$ is the class of all geometries belonging to $\mathcal{D}$.

7.1 Matroids

In this section we consider the diagram L_n, of rank n:

$$(L_n) \qquad \bullet \overset{L}{\text{———}} \bullet \overset{L}{\text{———}} \bullet\!\text{—} \cdots\cdots \text{—}\bullet \overset{L}{\text{———}} \bullet$$

The types are the nonnegative integers 0, 1, ..., $n-1$, labelling the nodes of the diagram from left to right as usual, according to the 'natural' orientation of the diagram. Thus, type 0 represents points, as in Section 4.3. Concerning points, lines, planes, lower and upper residues, collinearity relations and shadows, we use the notation stated in Section 4.3.

Most of what we will say in this section is taken from [18].

7.1.1 Lemmas

Henceforth Γ is a geometry belonging to L_n. In the following lemmas we do in the general case what the reader was asked to do in Exercise 6.6 for the rank 3 case.

Lemma 7.1 *The collinearity graph of Γ is a complete graph.*

Proof. The proof is by induction on n. If $n = 2$, there is nothing to prove. Let $n > 2$.

Given points a, b, c with $a \perp b \perp c$, let x, y be lines through a and b and through b and c, respectively. In the star Γ_b of b we find a plane u incident to both x and y, by the inductive hypothesis. The plane u is also incident to a and b, since Γ can be viewed as a poset (Section 4.3.1). The underlying structure Γ_u^- of u is a linear space. In Γ_u^- we see that $a \perp b$.

Therefore, every path of length 2 in the collinearity graph can be shortened to a path of length 1. The conclusion follows by the connectedness of the collinearity graph. $\square$

Lemma 7.2 *The natural point–line system of Γ is a linear space.*

Proof. By Lemma 7.1, we need only prove that Γ has a good system of lines. If $n = 2$, there is nothing to prove. Let $n > 2$.

Let a, b be distinct points of Γ and x, y lines incident to both a and b. In the star of a we find a plane u incident to both x and y. In Γ_u^- we see that $x = y$. $\square$

Lemma 7.3 *The property $\mathbf{IP}_0^1$ of Section 6.7.2 holds in Γ.*

Proof. Let a, b be distinct points, x a line incident to both a and b and z an element of type $t(z) > 1$ incident to both a and b. By Lemma 7.1 applied to Γ_z^-, there is a line y incident to z and to both a and b. We have $y = x$ by Lemma 7.2. Hence $x * z$. □

Lemma 7.4 *The Strong Intersection Property* **IP** *holds in* Γ.

Proof. By Corollary 6.39, we need only prove that $\mathbf{IP}_0^{Red}$ holds in Γ.

Let x, y be elements of Γ such that $|\sigma_0(x) \cap \sigma_0(y)| > 1$. We must prove that there is an element $z \leq x, y$ such that $\sigma_0(z) = \sigma_0(x) \cap \sigma_0(y)$.

We work by induction on n. If $n = 2$, there is nothing to prove. Let $n > 2$. Trivially, none of x, y is a point, as $|\sigma_0(x) \cap \sigma_0(y)| > 1$. Let us fix a point $a \in \sigma_0(x) \cap \sigma_0(y)$. For every element $u \in \Gamma_a$, we have the following:

$$(1) \qquad \sigma_0(u) = \bigcup(\sigma_0(w) \mid w \in \sigma_1^a(u))$$

(the symbol σ_1^a denotes the 1-shadow operator in Γ_a, according to the notation of Section 6.1). The inclusion $\supseteq$ in (1) is trivial. Let us prove the inclusion $\subseteq$. Given $b \in \sigma_0(u) - a$, let v be the line through a and b (this line is uniquely determined, by Lemma 7.2). We have $v * u$, by Lemma 7.3. This proves the inclusion $\supseteq$ in (1). By almost the same argument we can also prove that

$$(2) \qquad \sigma_0(x) \cap \sigma_0(y) = \bigcup(\sigma_0(w) \mid w \in \sigma_1^a(x) \cap \sigma_1^a(y))$$

By (2) we have $\sigma_1^a(x) \cap \sigma_1^a(y) \neq \emptyset$. By the inductive hypothesis in Γ_a, there is an element $z \in \Gamma_a$ such that $z \leq x, y$ and $\sigma_1^a(z) = \sigma_1^a(x) \cap \sigma_1^a(y)$. By this and by (1) and (2) we have $\sigma_0(z) = \sigma_0(x) \cap \sigma_0(y)$. □

Lemma 7.5 *Given an element x of type $i < n - 1$ and a point $a \notin \sigma_0(x)$, there is just one element y of type $i + 1$ incident to both a and x.*

Proof. The existence of such an element y can be proved by induction on n. When $n = 2$, there is nothing to prove. Let $n > 2$. If $i = 0$, then x is a point and y can be chosen to be a line through a and x (Lemma 7.1). Let $i > 0$.

Given a point $b \in \sigma_0(x)$, let z be a line through a and b (Lemma 7.1). By the inductive hypothesis in Γ_b, there is an element y of type $i + 1$ incident to both z and x. Trivially, y is also incident to a.

Let us prove the uniqueness of y. Let y' be another element of type $i+1$ incident to both a and x, if possible. The strong intersection property holds in Γ, by Lemma 7.4. By Theorem 6.14 (with $J = 0$) and Theorem 6.38, **IP** holds in the truncation $Tr^+_{\{0,1,\ldots,i+1\}}(\Gamma)$. Applying $\mathbf{IP}_0^1$ of Section 6.7.2 to y, y', a, x in the dual of $Tr^+_{\{0,1,\ldots,i+1\}}(\Gamma)$, we obtain $a * x$, contradicting our hypotheses on a and x. Therefore y is the unique element of type $i + 1$ incident to both a and x. □

7.1.2 The main theorem for L_n

Theorem 7.6 *The diagram L_n characterizes the class of n-dimensional matroids.*

Proof. We have remarked in Chapter 3 that all n-dimensional matroids belong to L_n. Conversely, let Γ belong to L_n.

By Lemma 7.4, we can identify Γ with its natural shadow space $\sigma_0(\Gamma)$ (Isomorphism Property, Section 6.2). We need only prove that $\widehat{\sigma}_0(\Gamma) \cup \{\emptyset\}$ (notation as in Section 6.1) can be viewed as the family of closed sets of an n-dimensional matroid with the same points as Γ. We recall that $\sigma_0(\emptyset)$ is the set of points of Γ.

By Lemma 7.4, $\widehat{\sigma}_0(\Gamma) \cup \{\emptyset\}$ is a complete atomic lattice with the set-theoretic intersection $\cap$ as meet operation and $\sigma_0(\emptyset)$ as set of atoms. Let $\langle ... \rangle$ be the usual closure operator of $\widehat{\sigma}_0(\Gamma) \cup \{\emptyset\}$:

$$\langle X \rangle = \bigcap (Y \mid Y \in \widehat{\sigma}_0(\Gamma) \cup \{\emptyset\}, Y \subseteq X) \qquad (X \subseteq \sigma_0(\emptyset))$$

Trivially, we have $\langle X \rangle = X$ if and only if $X \in \widehat{\sigma}_0(\Gamma)$. It is easily seen that $\langle ... \rangle$ satisfies (i)–(iii) of Section 1.3.2.

Since the diagram of Γ is a string, every element X of $\sigma_0(\Gamma)$ is the 0-shadow of a suitable element x of Γ. That element x is uniquely determined by X, since the posets Γ and $\sigma_0(\Gamma)$ are isomorphic. We denote x by $\sigma_0^{-1}(X)$ and we write $d(X) = t(\sigma_0^{-1}(X))$, where t is the type function of Γ. We also set $d(\emptyset) = -1$ and $d(\sigma_0(\emptyset)) = n$.

Let $X \in \widehat{\sigma}_0(\Gamma) \cup \{\emptyset\}$ and $a \in \sigma_0(\emptyset) - X$. Using Lemma 7.5 and the isomorphism $\widehat{Gr}_0(\Gamma) \cong \widehat{\sigma}_0(\Gamma)$, it is easy to prove that $d(\langle X \cup a \rangle) = d(X)+1$. It is now an easy exercise to check that (iv) and (v) of Section 1.3.2 also hold (hence $\langle ... \rangle$ defines a matroid) and that the matroid defined by $\langle ... \rangle$ on $\sigma_0(\emptyset)$ has dimension n. □

7.1.3 Projective and affine geometries

We now consider the following important special cases of L_n:

(A_n) •———•———•— ····· —•———•

Af

$(Af.A_{n-1})$ •———•———•— ····· —•———•

Corollary 7.7 *The diagram A_n characterizes (possibly degenerate) projective geometries of dimension n.*

Proof. Let Γ belong to A_n. Then Γ is an n-dimensional matroid, by Theorem 7.6. However, since A_n can also be viewed as a special case of the dual of L_n, Γ is also a dual matroid (the hyperplanes of Γ are the points of the dual of Γ). According to the definition of projective geometry given in Section 1.3.3, we must prove that, if X is a proper subspace of the matroid Γ and x is a point not in X, then every line on x contained in $\langle x \cup X \rangle$ meets X in a point. If X is a hyperplane (namely $t(X) = n-1$), then we obtain the conclusion we want, applying Lemma 7.5 to the dual of Γ. Otherwise, we consider the subspace $Y = \langle x \cup X \rangle$. It has dimension $t(Y) = t(X) + 1$ (see also Lemma 7.5), hence X is a hyperplane of the matroid Γ_Y^-. Hence we can work as above replacing Γ by Γ_Y^-. □

Corollary 7.8 *The n-dimensional simplex is the only thin geometry belonging to the Coxeter diagram A_n.*

(Trivial, by Corollary 7.7.)

Corollary 7.9 *The diagram $Af.A_{n-1}$ characterizes n-dimensional affine geometries.*

Proof. In Section 1.3.3 we have defined affine geometries as matroids of dimension at least 2 where planes are affine planes and stars of points are projective geometries. By Theorem 7.6 and Corollary 7.7, these conditions hold in every geometry belonging to $Af.A_{n-1}$. □

7.1.4 Truncations of projective and affine geometries

We will now consider the following special cases of L_n, assuming $n \geq 3$:

$(A_{n-1}.L)$ •——•——•— ..... —•——•—L—•

$(Af.A_{n-2}.L)$ •—Af—•——•— ..... —•——•—L—•

When $n = 3$ we write $Af.L$ for $Af.A_1.L$, for short.

Infinite dimensional matroids. In order to characterize geometries belonging to these diagrams, we must modify the definition of matroids a bit, so as to allow infinite dimensional matroids. For this purpose, it is sufficent to restrict (iv) of Section 1.3.2 to subsets $X \subseteq S$ such that $\langle X \rangle \neq S$, keeping the rest of (i)–(v) of Section 1.3.2 unchanged. Thus, if $\mathcal{M}$ is a matroid in this more general sense, all proper subspaces (namely, proper closed subsets) of $\mathcal{M}$ with at least two elements are finite dimensional matroids as in

Section 1.3.2, whereas it may be that infinitely many generators are needed for the improper subspace S. If this happens, then we say that $\mathcal{M}$ has *infinite dimension* $d = \infty$. Otherwise, $\mathcal{M}$ is a finite dimensional matroid as in Section 1.3.2.

If $\mathcal{M}$ is a matroid of dimension d (possibly $d = \infty$) and $n \leq d$ is a positive integer, then we define the *upper n-truncation* of $\mathcal{M}$ as the n-dimensional matroid consisting of all subspaces of $\mathcal{M}$ of dimension $i < n$. Needless to say, if $d < \infty$, then these truncations are truncations in the meaning of Section 1.7. We have allowed $n = d$ in this definition. Trivially, if $n = d$ then ($d < \infty$ and) $\mathcal{M}$ is its own upper n-truncation.

Points, *lines*, *planes* and, more generally, *i-dimensional subspaces* ($i < \infty$) of an infinite dimensional matroid $\mathcal{M}$ are defined just as in finite dimensional matroids. The points and the lines of $\mathcal{M}$ form a linear space.

Infinite dimensional projective and affine geometries. Infinite dimensional projective and affine geometries are defined just as in the finite dimensional case (Section 1.3.3), except for being infinite dimensional matroids. As in the case of finite dimension $d \geq 3$ (Theorems 1.3 and 1.4), a non-degenerate infinite dimensional projective geometry is the system of finitely generated linear subspaces of an infinite dimensional vector space and an infinite dimensional affine geometry is the system of cosets of finitely generated linear subspaces of an infinite dimensional vector space. (Actually, Theorem 1.3 is stated in [234] for possibly infinite dimensional projective geometries; the proof we have sketched for Theorem 1.4 works in the infinite dimensional case as well.)

The $n-1$-dimensional subspaces of a matroid $\mathcal{M}$ of dimension $n < \infty$ have been called hyperplanes in Section 1.3.2, but this definition does not make any sense in the infinite dimensional case. However, the following remark suggests to us a way to define hyperplanes of infinite dimensional projective geometries: it easily follows from the definition of (finite dimensional) projective geometries given in Section 1.3.3 that a hyperplane of a projective geometry $\mathcal{P}$ of finite dimension $n \geq 2$ is a proper subspace of the point–line system $\mathcal{L}(\mathcal{P})$ of $\mathcal{P}$ meeting every line of $\mathcal{P}$ not conained in it (see Section 1.6.2 for the definition of subspaces of partial planes). Thus, given an infinite dimensional projective geometry $\mathcal{P}$, a proper subspace H of the point–line system $\mathcal{L}(\mathcal{P})$ of $\mathcal{P}$ will be called a *projective hyperplane* (a *hyperplane*, for short) if every line of $\mathcal{P}$ not contained in H meets H (when $\mathcal{P}$ is the geometry of linear subspaces of a vector space V, this means that H is just a hyperplane of V). Note that the hyperplanes of an infinite dimensional projective geometry $\mathcal{P}$ are never subspaces (namely, closed subsets) of the matroid $\mathcal{P}$. They are only subspaces of the linear space $\mathcal{L}(\mathcal{P})$. Actually, they are the maximal proper subspaces of $\mathcal{L}(\mathcal{P})$ (the proof of this claim is not very difficult; we leave it for the reader).

As in the finite dimensional case, by deleting a hyperplane from a thick-lined infinite dimensional projective geometry we obtain an infinite dimensional affine geometry and every infinite dimensional affine geometry can be obtained in this way.

All planes of an infinite dimensional projective or affine geometry have the same order, called the *order* of the geometry, just as in the finite dimensional case.

Remarks. We warn the reader that different definitions of infinite dimensional projective and affine geometries are often used in the literature (see [69], for instance).

We also warn that, according to our definition of geometries (Section 1.1), all geometries have finite rank. Thus, infinite dimensional matroids are not geometries. In particular, infinite dimensional projective or affine geometries are not really geometries.

Truncations of projective geometries. It is quite easy to check that, given a projective geometry $\mathcal{P}$ of dimension $d \geq 2$ (possibly $d = \infty$), the point–line system $\mathcal{L}(\mathcal{P})$ of $\mathcal{P}$ satisfies the following property:

(**P**) (**Pasch axiom**) if a, b, c are pairwise distinct points and d, e are distinct points on the lines through a and b and through a and c, respectively, then the line through b and c meets the line through d and e.

Using this property, it is not difficult to prove that the proper subspaces (i.e. proper closed subsets) of the matroid $\mathcal{P}$ are precisely the finitely generated proper subspaces of the linear space $\mathcal{L}(\mathcal{P})$ and that all subspaces of $\mathcal{L}(\mathcal{P})$ are finitely generated if $d < \infty$. On the other hand, if $d = \infty$, then there are infinitely many subspaces of $\mathcal{L}(\mathcal{P})$ that are not closed subsets of $\mathcal{P}$.

Actually, one can prove more than the above. Given a linear space $\mathcal{L}$ satisfying **P**, the finitely generated proper subspaces of $\mathcal{L}$ are the proper subspaces of a projective geometry $\mathcal{P}(\mathcal{L})$ having $\mathcal{L}$ as point–line system.

We leave the proofs of the previous claims for the reader.

Proposition 7.10 *Let $\mathcal{M}$ be an n-dimensional matroid with $3 \leq n < \infty$ and let $\mathcal{M}$ induce projective geometries on its hyperplanes. Then $\mathcal{M}$ is an upper n-truncation of a (possibly infinite dimensional) projective geometry.*

Sketch of the Proof. Since the planes of $\mathcal{M}$ are projective planes, the Pasch axiom holds in the point–line system $\mathcal{L}(\mathcal{M})$ of $\mathcal{M}$. Therefore, the finitely generated subspaces of $\mathcal{L}(\mathcal{M})$ form a projective geometry, which we call $\mathcal{P}(\mathcal{M})$. The proper subspaces of $\mathcal{M}$ are the subspaces of $\mathcal{L}(\mathcal{M})$ generated by at most n points. Therefore $\mathcal{M}$ is the upper n-truncation of $\mathcal{P}(\mathcal{M})$. □

By this, by Theorem 7.6 and by Corollary 7.7 we immediately obtain the following:

Corollary 7.11 *The diagram $A_{n-1}.L$ $(n \geq 3)$ characterizes the class of upper n-truncations of projective geometries of dimension $d \geq n$ (possibly $d = \infty$).*

Truncations of affine geometries. Let $\mathcal{A}$ be an affine geometry of dimension $d \geq 2$ (possibly, $d = \infty$). Given two lines X, Y of $\mathcal{A}$, we write $X \parallel Y$ and we say that X and Y are *parallel*, if either X and Y are coplanar and do not intersect or $X = Y$. As in the finite dimensional case, $\parallel$ is an equivalence relation and satisfies the following properties [117]:

(**A1**) for every line X and every point a, there is precisely one line Y on a such that $Y \parallel X$;

(**A2**) if $X \parallel Y$ and a, b, c, d are pairwise distinct points with $a \in X$, $b, c \in Y$ and d on the line joining a with b, then the line through c and d meets X;

(**A3**) if X, Y, X', Y' are pairwise distinct lines such that $X \parallel X'$, $Y \parallel Y'$, $X \cap Y \neq \emptyset$, $X' \cap Y \neq \emptyset$ and $X \cap Y' \neq \emptyset$, then $X' \cap Y' \neq \emptyset$.

Using these properties, it is not difficult to prove that a finitely generated subspace X of the point–line system $\mathcal{L}(\mathcal{A})$ is a subspace of $\mathcal{A}$ if it satisfies the following property:

(**A4**) if $x \in X$, if Y is a line contained in X and Z is the line on x such that $Z \parallel Y$, then $Z \subseteq X$.

Conversely, every proper subspace (i.e. proper closed subset) of the matroid $\mathcal{A}$ is a finitely generated subspace of the linear space $\mathcal{L}(\mathcal{A})$ satisfying **A4**. It is easily seen that all subspaces of $\mathcal{L}(\mathcal{A})$ are finitely generated if $d < \infty$. Therefore, if $d < \infty$, then the subspaces of the matroid $\mathcal{A}$ are just the subspaces of $\mathcal{L}(\mathcal{A})$ satisfying **A4**.

It is not difficult to prove that **A1**–**A3** force all lines of $\mathcal{A}$ to have the same number of points. It is also worth remarking that, if the lines of $\mathcal{A}$ have more than two points, then we can obtain **A3** as a formal consequence of **A2** and all subspaces of $\mathcal{L}(\mathcal{A})$ satisfy **A4** (see [117]). Therefore, **A4** says nothing in this case.

Conversely, let $\mathcal{L}$ be a linear space equipped with an equivalence relation $\|$ between lines satisfying **A1**–**A3**. Then the finitely generated proper subspaces of $\mathcal{L}$ satisfying **A4** are the proper subspaces of an affine geometry $\mathcal{P}(\mathcal{L})$ having $\mathcal{L}$ as point–line system and $\|$ as parallelism relation [117].

Using the above, we obtain the following:

Proposition 7.12 *Let $\mathcal{M}$ be an n-dimensional matroid with $4 \leq n < \infty$ and let $\mathcal{M}$ induce affine geometries on its hyperplanes. Then $\mathcal{M}$ is an upper n-truncation of a (possibly infinite dimensional) affine geometry.*

Sketch of the Proof. Write $X \| Y$ to mean that X and Y are coplanar non-intersecting lines of $\mathcal{M}$ or $X = Y$. The relation $\|$ defined in this way is trivially reflexive and symmetric. The key point of the proof is proving that $\|$ is also transitive. Once this has been done, checking that **A2** and **A3** hold is quite easy. The rest is as in (the sketch of) the proof of Proposition 7.10.

Thus we only spend a few words on the transitivity of $\|$. Let X, Y, Z be pairwise distinct lines with $X \| Y \| Z$. By the definition of $\|$, these three lines are contained in a common subspace W of $\mathcal{M}$ of dimension $i \leq 3$. However, W is an affine plane or a 3-dimensional affine geometry, because $\mathcal{M}$ has dimension $n \geq 4$ and all hyperplanes of the matroid $\mathcal{M}$ are $n-1$-dimensional affine geometries. Therefore, we have $X \| Z$, by well-known properties of affine planes and 3-dimensional affine geometries (compare with the sketch of the proof of Theorem 1.4). □

By this, by Theorem 7.6 and Corollary 7.9, we immediately obtain the following:

Corollary 7.13 *The diagram $Af.A_{n-2}.L$ $(n \geq 4)$ characterizes upper n-truncations of affine geometries of dimension $d \geq n$ (possibly $d = \infty$).*

The hypothesis $n \geq 4$ plays an essential role in the proof we have sketched for Proposition 7.12. When $n = 3$, the same conclusion as above can be obtained, provided we assume that lines have enough points (see the next proposition).

Since we are going to get involved with the number of points on a line, we remark that, if the planes of a matroid $\mathcal{M}$ of dimension ≥ 2 are affine planes, then they have the same order. The proof of this claim is quite trivial. If X, Y are lines of $\mathcal{M}$, then we can take a line Z meeting both X and Y, then planes u and v containing X and Z and Y and Z, respectively. In the affine planes u and v we see that $|X| = |Z| = |Y|$.

Theorem 7.14 (Buekenhout [16]) *Let $\mathcal{M}$ be a 3-dimensional matroid and let the planes of $\mathcal{M}$ be affine planes of order $s \geq 4$. Then $\mathcal{M}$ is an upper 3-truncation of a (possibly infinite dimensional) affine geometry.*

We omit the proof of this statement. Note that the statement of this theorem is false when $s = 2$ or 3 (see [76]; also Section 7.1.5).

By theorems 7.6 and 7.14 we obtain the next corollary:

Corollary 7.15 *The diagram $Af.L$ together with the assumption that the planes have order $s \geq 4$ characterizes upper 3-truncations of (possibly infinite dimensional) affine geometries of order $s \geq 4$.*

Exercise 7.1. Let $\mathcal{M}$ be a 3-dimensional matroid satisfying both the following:

(i) the planes of $\mathcal{M}$ are affine planes of order $s \geq 3$;
(ii) the Pasch Axiom **P** holds in stars of points of $\mathcal{M}$.

Prove that $\mathcal{M}$ is an upper truncation of an affine geometry.

(Hint: showing that the relation $\parallel$ is transitive is the main step of the proof. Let $L \parallel M \parallel N \not\parallel L$, if possible, and let a be a point of L. Let X, Y, Z be the planes spanned by L and M, by M and N and by N and a, respectively. If X and Y meet in a line, then a contradiction can be reached as in the proof of Theorem 1.4. In order to prove that X and Z meet in a line, we can take two intersecting lines R, R' on Y both intersecting each of M and N. Since Y has order $s \geq 3$, we can do this in such a way that $R \cap R' \notin N \cup M$. Projecting this configuration of lines from a and applying the Pasch axiom in the star of a we force X and Z to meet in a line.)

7.1.5 The diagram $Af.L$ with $s = 2$ or 3

If we drop the assumption that all planes of $\mathcal{M}$ have order $s \geq 4$ from the hypotheses of Theorem 7.14, then counterexamples arise. For instance, the enrichment of the Mőbius plane arising from $\mathcal{Q}_3^-(3)$ has diagram and orders as follows:

$$\underset{1}{\bullet}\overset{Af}{\rule{2cm}{0.4pt}}\underset{2}{\bullet}\overset{L}{\rule{2cm}{0.4pt}}\underset{3}{\bullet} \quad = \quad \bullet\overset{c}{\rule{2cm}{0.4pt}}\underset{2}{\bullet}\overset{Af}{\rule{2cm}{0.4pt}}\underset{3}{\bullet}$$

This has evidently nothing to do with truncations of affine geometries. (We remark that this Mőbius plane is the unique geometry belonging to the above diagram; this follows from Theorem C of [208].)

More generally, it is easily seen that the following diagram with orders 1, 2, t (where $2 \leq t < \infty$) characterizes enrichments of Steiner systems $S(3,4,v)$ (with $v = 2t+4$):

$$\overset{Af}{\underset{1 \quad\quad 2}{\bullet\!\!-\!\!-\!\!\bullet}}\overset{L}{\underset{\quad\quad t}{-\!\!-\!\!\bullet}} \quad = \quad \overset{c}{\underset{\quad\quad 2}{\bullet\!\!-\!\!-\!\!\bullet}}\overset{L}{\underset{\quad\quad t}{-\!\!-\!\!\bullet}}$$

The Steiner system $S(3,4,v)$ arises from $AG(n,2)$ only if $v = 2^n$. However, a number of examples exist where v is not a power of 2 (see [76], [71]); for instance, we have $v = 10$ in the Mőbius plane mentioned above.

Turning to the case of $s = 3$, the diagram $Af.L$ with orders 2, 3, t

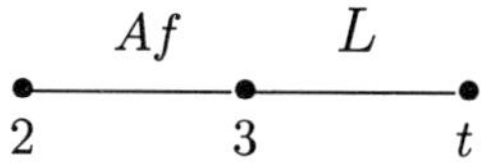

$$\overset{Af}{\underset{2 \quad\quad 3}{\bullet\!\!-\!\!-\!\!\bullet}}\overset{L}{\underset{\quad\quad t}{-\!\!-\!\!\bullet}}$$

characterizes Steiner systems $S(2,3,v)$ where any three points not in the same block generate a subsystem isomorphic to $AG(2,3)$ (points, blocks and $AG(2,3)$-subsystems are now taken as points, lines and planes, respectively).

These Steiner systems have been systematically investigated by M. Hall [76]. Thus, we will call them *Hall–Steiner systems.* The information we give on them is taken from [76] (also [61], Section 2.4, 8–19).

The number v of points of a Hall–Steiner system must be a power of 3 and, given any four points of a Hall–Steiner system, not in the same plane, they generate either an $S(2,3,27)$-subsystem or an $S(2,3,81)$ subsystem. Point–line–plane systems of n-dimensional affine geometries with $n \geq 3$ are characterized by the following property: any four points not in the same plane generate an $S(2,3,27)$-subsystem (hence $AG(3,3)$ is the only Hall–Steiner system with $v = 27$; of course, this could also be obtained by Proposition 7.8).

However, examples exist where four points generate an $S(2,3,81)$. By the above, they are not truncations of affine geometries. Thus, if $X \parallel Y$ means that either X and Y are coplanar non-intersecting lines or $X = Y$, then the relation $\parallel$ is not transitive in those examples; equivalently, the Pasch axiom fails to hold in the star of some (possibly, every) point (see Exercise 7.1).

Exercise 7.2. A *ternary loop* is an algebraic structure with one ternary operation f satisfying the following identities:

$$f(x,x,y) = y$$
$$f(x,y,z) = f(y,x,z) = f(x,z,y)$$

$$f(x, y, f(x, y, z)) = z$$

Let $\mathcal{L}$ be a ternary loop with at least four elements. Its subloops have 1, 2, 4 or at least 8 elements. Take the subloops with 1, 2 and 4 elements as points, lines and blocks respectively and define the incidence relation in the natural way, by symmetrized inclusion. Check that the structure defined in this way is a 3-dimensional matroid with diagram and orders as follows:

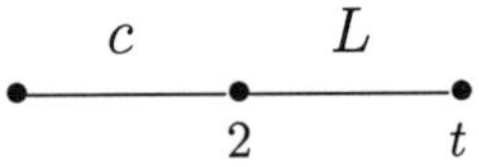

c L
•——•——•
2 t

where $2t + 4$ is the number of elements of $\mathcal{L}$ (t might be infinite, of course).

Prove that, conversely, every geometry with diagram and orders as above arises from some ternary loop by the previous construction.

Exercise 7.3. Check that the diagram $c.Af$ with finite orders 1, $s - 1$, s

($c.Af$) c Af
•——•——•
$s-1$ s

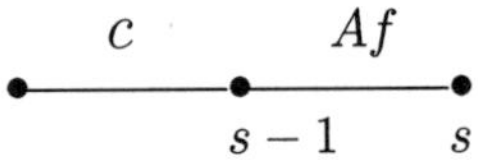

characterizes enrichments of inversive planes of order s (that is, Steiner systems $S(3, s + 1, s^2 + 1)$; see Section 1.2.6(5)).

Exercise 7.4. Check that the following diagram of rank n, with finite orders 1, 1, ..., 1, s, t

($c^{n-1}.L$) c L
•——•— ..... —•——•——•——•
t

characterizes enrichments of Steiner systems $S(n, s + n - 1, st + s + n - 1)$.

7.2 Locally finite matroids

In this section Γ is an n-dimensional matroid with finite orders s_0, s_1, ..., s_{n-1} and $n \geq 2$, as follows:

(L_n) L L L
•——•——•— ..... —•——•
s_0 s_1 s_2 s_{n-2} s_{n-1}

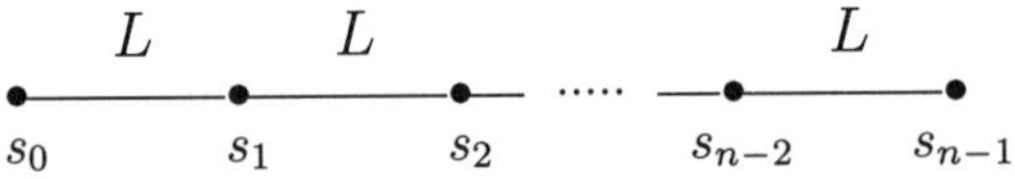

where $s_0 \leq s_1 \leq ... \leq s_{n-1}$ by a property of linear spaces with finite orders (Section 1.2.6(4)). If $s_0 = s_1 = ... = s_{n-1}$, then we have the Coxeter diagram A_n. In this case Γ is a projective geometry, by Corollary 7.7. If

$s_0 + 1 = s_1 = s_2 = ... = s_{n-1}$, then we have the diagram $Af.A_{n-1}$. In this case Γ is an affine geometry (Corollary 7.9).

7.2.1 Computing numbers of elements

Given Γ, s_0, s_1, ..., s_{n-1} as above, let N_m be the number of m-dimensional subspaces of Γ, $m = 0, 1, ..., n-1$.

Lemma 7.16 *The following holds for every $m = 0, 1, ..., n-2$:*

$$N_m = \prod_{k=0}^{m} \frac{\sum_{h=k}^{n} \prod_{i=k}^{h-1} s_i}{\sum_{h=k}^{m} \prod_{i=k}^{h-1} s_i}$$

(where $\prod_{i=k}^{k-1} s_i = 1$, by convention).

Proof. We first prove the following:

$$(1) \quad N_0 = \sum_{h=0}^{n} \prod_{i=0}^{h-1} s_i$$

We have already remarked in Section 1.2.6(4) that (1) holds when $n = 2$. We continue by induction on n.

Let $n > 2$. For every point a of Γ, the star Γ_a of a is an $n-1$-dimensional matroid with orders s_1, s_2, ..., s_{n-1}. If N'_0 is the number of lines of Γ on a, then we have

$$(1.i) \quad N'_0 = \sum_{h=0}^{n-1} \prod_{i=1}^{h} s_i$$

by the inductive hypothesis in Γ_a. We also have:

$$(1.ii) \quad N_0(N_0 - 1) = N_1(s_0 + 1)s_0$$

because every pair of distinct points of Γ determines a unique line, whereas every line of Γ has $s_0 + 1$ points. We can now count point-line flags in two ways: first choosing a point, next a line on it; or choosing a line first, next a point on it. We obtain the following:

$$(1.iii) \quad N_0 N'_0 = N_1(s_0 + 1).$$

By (1.ii) and (1.iii) we have $N_0 = 1 + s_0 N'_0$. Substituting (1.i) in this last relation we obtain (1).

We now prove that

$$(2) \qquad N_{n-1} \;=\; \prod_{k=0}^{n-1} \frac{\sum_{h=k}^{n} \prod_{i=k}^{h-1} s_i}{\sum_{h=k}^{n-1} \prod_{i=k}^{h-1} s_i}$$

We again work by induction on n. Let $n = 2$. Counting point–line flags in two ways as for (1.*iii*), we obtain $N_0(s_1 + 1) = N_1(s_0 + 1)$. Substituting (1) in this relation we obtain (2).

Let $n > 2$ and let N'_{n-1} be the number of hyperplanes on a point a of Γ. By the inductive hypothesis in Γ_a we have:

$$(2.i) \qquad N'_{n-1} \;=\; \prod_{k=0}^{n-2} \frac{\sum_{h=k}^{n-1} \prod_{i=k}^{h-1} s_{i+1}}{\sum_{h=k}^{n-2} \prod_{i=k}^{h-1} s_{i+1}}$$

Let N''_0 be the number of points in the residue of a hyperplane. By (1) we have:

$$(2.ii) \qquad N''_0 \;=\; \sum_{h=0}^{n-1} \prod_{i=0}^{h-1} s_i$$

Counting point–hyperplane flags in two ways, we obtain that $N_0 N'_{n-1} = N_{n-1} N''_0$. Substituting (2.*i*) and (2.*ii*) in this relation we obtain (2).

We can now compute N_m for $m = 1, 2, ..., n-2$. Let $Tr_m(\Gamma)$ be the truncation of Γ consisting of all elements of dimension $i \leq m$. Trivially, the first $m-1$ orders of $Tr_m(\Gamma)$ are s_0, s_1, ..., s_{m-2}, as in Γ. Applying (1) to stars of $m-1$-dimensional subspaces of Γ we see that the following is the last order of $Tr_m(\Gamma)$:

$$\sum_{h=1}^{n-m} \prod_{i=0}^{h-1} s_{i+m}.$$

The m-dimensional subspaces of Γ are the hyperplanes of $Tr_m(\Gamma)$. Thus, we can compute N_m applying (2) to $Tr_m(\Gamma)$. □

Corollary 7.17 *The geometry* Γ *is finite.*

Trivial, by the previous lemma.

The reader may check that the well-known relations

$$N_0 \;=\; N_{n-1} \;=\; \sum_{i=o}^{n} q^i$$

for the number N_0 of points and the number N_{n-1} of hyperplanes of $PG(n,q)$ can be obtained as a special case of Lemma 7.16.

Needless to say, many divisibility conditions are implicit in Lemma 7.16. Indeed N_m is a positive integer. For instance, when $n = 2$ we obtain the divisibility condition for linear spaces with finite orders mentioned in Section 1.2.6(4):

$$s_0 + 1 \mid s_1(s_1 + 1)$$

(here and henceforth we write | for 'divides'.)

7.2.2 The diagram $L.A_{n-1}$

In this section Γ is a geometry belonging to the following special case of L_n:

$(L.A_{n-1})$

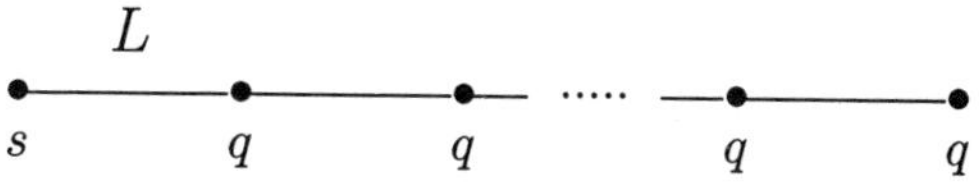

where s, q are finite orders and $n \geq 3$. Namely, Γ is an n-dimensional matroid where planes are finite linear spaces of order (s,q) (hence $s \leq q$) and stars of points are $n-1$-dimensional projective geometries (Corollary 7.7).

If $q = 1$, then $s = q = 1$, as $s \leq q$. We have the thin case of A_n and Γ is the n-dimensional simplex, by Corollary 7.8.

We assume $q > 1$ for the rest of this section.

We do not assume that q is a prime power. However, q is in fact a prime power if $n \geq 4$, by Theorem 1.3 applied to stars of points.

We will state two theorems, earlier obtained by Hughes [85] for the special case of $s = 1$ and later generalized by Doyen and Hubaut [64] for any $s \geq 1$.

Theorem 7.18 *Let Γ be as above with $n \geq 4$. Then Γ is either the projective geometry $PG(n,q)$ or the affine geometry $AG(n,q)$.*

Proof. If $s = q$, then $\Gamma = PG(n,q)$ by Corollary 7.7 and Theorem 1.3. If $s+1 = q$, then $\Gamma = AG(n,q)$, by Corollary 7.9 and Theorem 1.4.

Let $s+1 < q$. We have assumed $n \geq 4$. Therefore we can use Lemma 7.16 to compute the number of 3-dimensional subspaces in a 4-dimensional subspace of Γ (when $n > 4$) or the number of 3-dimensional subspaces of Γ (when $n = 4$). This number is a positive integer. Hence, by easy manipulations we obtain the following divisibility condition:

(1) $\quad (q^2+q+1)s+1 \mid q^3(q^3+q^2+q+1)$

Since $n \geq 4$, the order q is a prime power. That is, we have $q = p^m$ for some positive integer m and some prime p. Let p^k be the maximal power of p dividing $(q^2+q+1)s+1$. If $k \geq m$, then $q \mid p^k$, hence $q \mid s+1$ and $q \leq s+1$, contradicting the assumption that $s+1 < q$. Therefore $k < m$ and the divisibility condition (1) can be refined as follows:

$$(2) \quad (q^2+q+1)s+1 \mid q(q^3+q^2+q+1)$$

Multiplying $q(q^3+q^2+q+1)$ by s in (2) we obtain that

$$(3) \quad (q^2+q+1)s+1 \mid q(q-s)$$

On the other hand, $q-s > 0$, because we have assumed $s+1 < q$. Hence (3) gives us $(q^2+q+1)s+1 < q(q-s)$. However, this is evidently a contradiction. □

The information given us by Lemma 7.16 when $n = 3$ is not strong enough to achieve any satisfactory characterization. However, more divisibility conditions can be obtained considering the dual collinearity graph of Γ (see Section 4.3.2). It is worth remarking that Γ has a good dual system of lines, as it is a matroid.

Before to come to arithmetic properties of the dual collinearity graph of Γ, we state an elementary lemma which can be proved with no use of the hypothesis that Γ admits finite orders. Because of this, we state it for any $n \geq 3$, even if we will only exploit its special case with $n = 3$.

Lemma 7.19 *Two distinct hyperplanes of Γ either have no points in common or intersect in an $n-2$-dimensional subspace.*

Proof. Let u, v be distinct hyperplanes of Γ and let a be a point in $u \cap v$. The star Γ_a of a is an $n-1$-dimensional projective geometry. Therefore, there is an $n-2$-dimensional subspace of Γ in Γ_a incident to both u and v. □

The reader is referred to Appendix II of Chapter 1 for the definition of strongly regular graphs. The symbols k, λ and μ have the meaning stated there.

Lemma 7.20 *Let $n = 3$ and $s < q$. Then the dual collinearity graph of Γ is strongly regular with k, λ, μ as follows:*

$$k = q[(q+1)^2 - q(q+1)/(s+1)]$$

$$\lambda = q^2 - 1 + (sk+q)/(s+1)$$

$$\mu = [k((q+1)s+1)]/[q(s+1)]$$

Proof. If the dual collinearity graph of Γ is a complete graph, then the dual point–line system of Γ is a linear space with orders q and t where

$$t = q^2 + 2q - q(q+1)/(s+1)$$

(every plane of Γ has $(q+1)^2 - q(q+1)/(s+1)$ lines, by Lemma 7.16). Hence Γ has $1 + s[(q+1)^2 - q(q+1)/(s+1)]$ planes. On the other hand, Γ has $[(q^2+q+1)s+1](q^2+q+1)/[(q+1)s+1]$ planes, by Lemma 7.16. Hence we have

$$1 + s[(q+1)^2 - \frac{q(q+1)}{s+1}] = \frac{[(q^2+q+1)s+1](q^2+q+1)}{(q+1)s+1}$$

This implies $s = q$, contradicting the assumption $s < q$.

Therefore, there are non-collinear planes in Γ, that is planes not intersecting in a line. Let u, v be two non-collinear planes. We have $u \cap v = \emptyset$ by Lemma 7.19. Hence, given a point $a \in u$ and a line $x \subseteq v$, the point a and the line x span a plane, call it w. The plane w is collinear with v, as $x = w \cap v$. It is also collinear with u by Lemma 7.19, because $a \in w \cap u$. Therefore, u and v have distance 2 in the dual collinearity graph of Γ and this graph has diameter 2.

Every plane u of Γ has $(q+1)^2 - q(q+1)/(s+1)$ lines and every line in u is incident to q planes other than u. Therefore

$$k = q[(q+1)^2 - \frac{q(q+1)}{s+1}]$$

Let us compute λ. Let u, v be distinct planes of Γ intersecting in a line $x = u \cap v$. Let λ_1 be the number of planes on x distinct from both u and v. Let λ_2 be the number of planes intersecting x in one point (hence collinear with both u and v, by Lemma 7.19). Let λ_3 be the number of planes collinear with both u and v but not intersecting x in any point. We have $\lambda = \lambda_1 + \lambda_2 + \lambda_3$ and

$$(1) \quad \lambda_1 = q - 1$$

We also have

$$(2) \quad \lambda_2 = (s+1)q^2$$

because x has $s+1$ points, for every point $a \in x$ there are q^2 planes on a not on x and a plane not on x meets x in at most one point.

We compute λ_3 as follows. Let $\mathcal{A}$ be the set of point–plane flags (a, w) of Γ where w is a plane collinear with both u and v but not on x, whereas a is a point in $u \cap w - x$. Let us set $\alpha = |\mathcal{A}|$.

In order to compute α we first choose a point $a \in u - x$ and a line y of v other than x. There are sq choices for a and $q^2 + 2q - q(q+1)/(s+1) = (k/q) - 1$ choices for y. We have $a \notin v$, otherwise $a \in u \cap v = x$, whereas we have chosen a in $u - x$. Hence $a \notin y$. Let w be the plane of Γ spanned by $a \cup y$. Trivially, $w \neq u, v$ and $x \not\subseteq w$. The planes u and w are collinear by Lemma 7.19, since $a \in u \cap w$. Therefore $(a, w) \in \mathcal{A}$. Hence we have

$$(3) \quad \alpha = sq[(k/q) - 1] = s(k - q)$$

However, we can also count α in another way. If w is one of the λ_3 planes collinear with both u and v but not meeting x in any point, then $(a, w) \in \mathcal{A}$ for every point $a \in w \cap u$. Hence there are $(s+1)\lambda_3$ point–plane flags $(a, w) \in \mathcal{A}$ with $w \cap x = \emptyset$. On the other hand, if w is a plane meeting x in one point, we have $(a, w) \in \mathcal{A}$ for every point $a \in w \cap u - x$. Hence there are $s\lambda_2 = s(s+1)q^2$ point–plane flags $(a, w) \in \mathcal{A}$ with w meeting x in one point. Therefore

$$(4) \quad \alpha = s(s+1)q^2 + (s+1)\lambda_3$$

By (3) and (4) we obtain the following:

$$(5) \quad \lambda_3 = \frac{s(k-q)}{s+1} - sq^2$$

Finally, by (1), (2) and (5):

$$\lambda = q^2 - 1 + \frac{sk + q}{s+1}$$

We now compute μ. Let u, v be non-collinear planes of Γ and let $\mathcal{B}$ be the family of point–plane flags (a, w) where w is a plane collinear with both u and v and a is a point in $u \cap w$. Let us set $\beta = |\mathcal{B}|$. We can compute β in two ways.

We can first choose a point $a \in u$ and a line y of v. As $u \cap v = \emptyset$, we have $a \notin y$; hence a and y span a plane, call it w. The plane w is collinear with v. It is also collinear with u, by Lemma 7.19, since $a \in w \cap u$. Hence $(a, w) \in \mathcal{B}$. There are $(q+1)s + 1$ points in u and $(q+1)^2 - q(q+1)/(s+1) = k/q$ lines in v. Therefore:

$$(6) \quad \beta = ((q+1)s + 1)k/q$$

On the other hand, if a plane w is chosen collinear with both u and v, there are $s+1$ choices for a point $a \in u \cap w$. Hence

$$(7) \quad \beta = (s+1)\mu$$

By (6) and (7) we obtain the following:

$$\mu = \frac{k[(q+1)s+1]}{q(s+1)}$$

The lemma is proved. □

Theorem 7.21 *Let $n = 3$. Then one of the following holds:*

(i) we have $s = q$ and $\Gamma = PG(3, q)$;
(ii) we have $s + 1 = q$ and $\Gamma = AG(3, q)$;
(iii) we have $q = (s + 1)^2$;
(iv) we have $q = (s + 1)^3 + (s + 1)$ (q is not a prime power in this case).

Proof. If $q = s$ or $q = s + 1$, then $\Gamma = PG(3, q)$ or $AG(3, q)$ respectively, as in the proof of Theorem 7.18.

Let $s + 1 < q$. We know that

$$(1) \quad s+1 \mid q(q+1)$$

Using Lemma 7.16 to compute the number of planes of Γ, we also obtain the following divisibility condition, by some easy manipulations:

$$(2) \quad (q+1)s+1 \mid q^2(q^2+q+1)$$

The numbers λ and μ computed in Lemma 7.20 are positive integers. Hence we obtain the following:

$$(3) \quad (s+1)^2 \mid q^2(q-s)$$

$$(4) \quad (s+1)^2 \mid q(q+1)(2(s+1)+t(2s+1))$$

Assembling (4) and (1) we also get the next divisibility condition:

$$(5) \quad (s+1)^2 \mid q^2(q+1)$$

The conditions (5) and (3) give us the following:

$$(6) \quad s+1 \mid q^2$$

and (6) and (2) imply the next condition:

$$(7) \quad s+1 \mid q$$

Let us set $t = q/(s + 1)$. Substituting $(s + 1)t$ for q in (2), by easy manipulations we obtain the following:

$$(8) \quad ts+1 \mid (s+1)+(t-1)^2$$

Therefore $ts + 1 \leq (s + 1) + (t - 1)^2$. Hence either $t = 1$ or $t \geq (s + 1)$. If $t = 1$, then $q = s + 1$, contradicting the hypothesis $s + 1 < q$. Therefore

$$(9) \quad t \geq s+1$$

that is $q \geq (s+1)^2$. Multiplying $(s+1)+(t-1)^2$ by s in (8) we see that

$$(10) \quad ts+1 \mid s^2+2s+2-t$$

If $s^2+2s+2 > t$, then $ts+1 \leq s^2+2s+2-t$ by (10), hence $t \leq s+1$. Therefore $t = s+1$, by (9). That is

$$q = (s+1)^2$$

We have (iii) of the theorem.

Let $t \geq s^2+2s+2$, that is $q \geq (s+1)^3+(s+1)$. Multiplying $q^2(q^2+q+1)$ by s^3 in (2), it is easy to see that

$$(11) \quad (q+1)s+1 \mid q-(s+1)^3-(s+1)$$

We have assumed that $q \geq (s+1)^3+(s+1)$. On the other hand, the inequality $(q+1)s+1 < q-(s+1)^3-(s+1)$ is evidently impossible. Hence (11) implies that

$$q = (s+1)^3+(s+1)$$

We have case (iv) of the theorem. □

Remark. The enrichment of the Steiner system $S(3,6,22)$ for M_{22} is the only example known for (iii) of Theorem 7.21. Let us denote it by $S(M_{22})$. We have $s=1$ in $S(M_{22})$; hence $q=4$ in it and the diagram is as follows:

$(c.A_2)$

Every 3-dimensional matroid of order $(1,4,4)$ as above is the enrichment of a Steiner system $S(3,6,22)$ (Exercise 7.4) and it is well known that the Steiner system for M_{22} is the unique Steiner system of type $S(3,6,22)$ [91]. Hence $S(M_{22})$ is the unique example for (iii) of Theorem 7.21 with $s=1$.

No examples are known for (iv) of Theorem 7.21. It is likely that this case is impossible, since q cannot be a prime power here and no finite projective plane is known with order that is not a prime power. Actually, when $s=1$ we see that $q=10$. However, there are no projective planes of order 10 (see [115]). Therefore $s>1$ in (iv) of Theorem 7.21.

We also notice that $q \equiv 2$ $(mod\ 4)$ in (iv) of Theorem 7.21 when $s \not\equiv 3$ $(mod\ 4)$. The Bruck–Ryser test (see Section 3.2.4 or [90]) can be applied to this case. If $s \equiv 2$ $(mod\ 4)$ or $s \equiv 5$ $(mod\ 8)$, then $q \equiv 6$ $(mod\ 8)$, hence there is a prime p dividing q to an odd exponent and such that $p \equiv 3$

(*mod* 4). However, this is forbidden by the Bruck–Ryser test. Therefore, we have the following restrictions on s in (iv) of Theorem 7.21:

$s \equiv 0, 1, 3, 4$ or 7 $(mod\ 8)$ (and $s > 1$).

However, we are still far from proving that (iv) is impossible at all. We warn the reader that no improvement of Theorem 7.21 can be obtained computing the eigenvalues of the dual collinearity graph of Γ and their multiplicities (see Appendix II of Chapter 1 for how to compute them). We obtain integral values in both (iii) and (iv).

7.2.3 The diagram $L_k.A_m$

In this section Γ has diagram and finite orders as follows, where $m \geq 2$, $k \geq 2$ and $n = k + m - 1 \geq 3$ is the rank of Γ:

$$(L_k.A_m) \quad \overset{L}{\underset{s_0}{\bullet}}\!\!-\!\!\!-\!\!\!-\!\!\underset{s_1}{\bullet} \cdots\cdots \underset{s_{k-2}}{\bullet}\overset{L}{-\!\!\!-\!\!\!-}\underset{q}{\bullet}-\!\!\!-\!\!\!-\underset{q}{\bullet} \cdots\cdots \underset{q}{\bullet}-\!\!\!-\!\!\!-\underset{q}{\bullet}$$

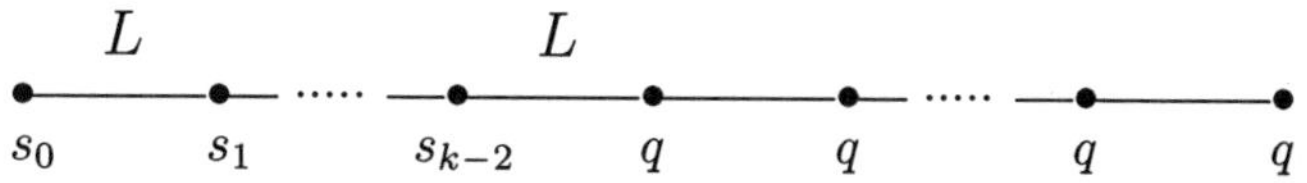

When $k = 2$ we have the diagram $L.A_{n-1}$ of the previous section. It is also worth noticing that $L_{k-1}.A_{m+1}$ (A_n when $k = 2$) is the special case of $L_k.A_m$ (of $L.A_{n-1}$, respectively) with $s_{k-2} = q$. Therefore the diagram $L_{n-1}.A_2$ includes A_n and $L_k.A_m$ as special cases, for $k = 2, ..., n-1$.

Lemma 7.22 *Let $k \geq 3$ and $s_{k-2} < q$. Then one of the following holds:*

(i) we have $m = k - 1 = 2$, $s_0 = s_1 = 1$ and $q = 4$;
(ii) we have $m = k - 2 = 2$, $s_0 = s_1 = s_2 = 1$ and $q = 4$;
(iii) we have $m = k - 1 = 2$, $s_0 = 1$, $s_1 = 3$ and $q = 68$.

Proof. We have $q = s_{k-2}+1, (s_{k-2}+1)^2$ or $(s_{k-2})^3+(s_{k-2}+1)$, by Theorems 7.18 and 7.21 and because we have assumed $q > s_{k-2}$. Furthermore, we have $q = s_{k-2} + 1$ if $m \geq 3$, by Theorem 7.18.

Let us consider the case of $k - 1 = m = 2$ first (the one of minimal rank). We can use Lemma 7.16 to compute numbers of subspaces of Γ of given dimensions and numbers of subspaces of given dimension in a given subspace of Γ. By the divisibility conditions obtained in this way we can see that the case of $q = s_1 + 1$ is impossible, that we have $s_0 = 1$ in any case, that $s_1 = 1$ if $q = (s_1 + 1)^2$ (hence $q = 4$) and that $s_1 = 3$ if $q = (s_1 + 1)^3 + (s_1 + 1)$ (hence $q = 68$). The details of these computations are left as exercises for the reader.

Since the case with $k - 1 = m = 2$ appears as a residue in any higher rank case, we have $m = 2$ in any case (hence $k = n - 1$) and either $s_0 =$

$s_1 = ... = s_{n-3} = 1$ and $q = 4$ or $s_0 = s_1 = ... = s_{n-4} = 1$, $s_{n-3} = 3$ and $q = 68$.

Again using Lemma 7.16, we can see that the case of $k = 4$ and $q = 68$ and the case of $k = 5$ and $q = 4$ are impossible. The lemma is proved. □.

Remark. The special case of Lemma 7.22 with $s_0 = s_1 = ... = s_{k-2} = 1$ is due to Hughes [85].

In (i) and (ii) of Lemma 7.22 the diagrams are as follows:

$(c^2.A_2)$

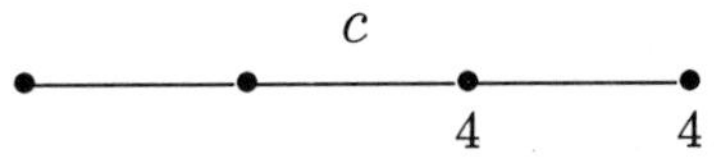

$(c^3.A_2)$

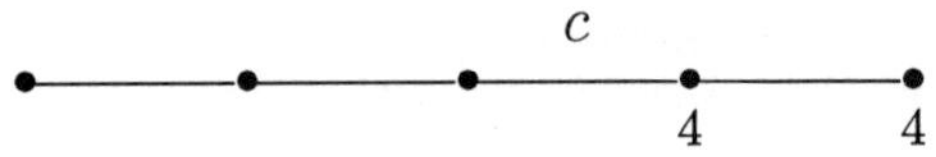

The enrichments of the Steiner systems $S(4, 7, 23)$ and $S(5, 8, 24)$ for the Mathieu groups M_{23} and M_{24} are in fact geometries (of rank 4 and 5, respectively) belonging to the above diagrams. Let us denote them by $S(M_{23})$ and $S(M_{24})$. These geometries are the only examples for (i) and (ii) of Lemma 7.22. Indeed, a matroid Γ of dimension 4 or 5 and order $(1, 1, 4, 4)$ or $(1, 1, 1, 4, 4)$ is the enrichment of a Steiner system $S(4, 7, 23)$ or $S(5, 8, 24)$, respectively (Exercise 7.4); it is well known [91] that the Steiner systems for M_{23} and M_{24} are the only Steiner systems of type $S(4, 7, 23)$ and $S(5, 8, 24)$. Hence $\Gamma = S(M_{23})$ or $S(M_{24})$, according to whether Γ has rank 4 or 5.

Needless to say, no example is known for (iii) of the above lemma.

By theorems 7.18 and 7.21, by Lemma 7.22 and by the remarks following Theorem 7.21 and Lemma 7.22, we have

Theorem 7.23 *Let Γ have diagram $L_{n-1}.A_2$:*

$(L_{n-1}.A_2)$

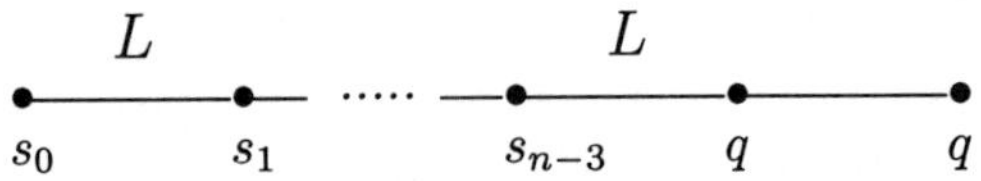

Then Γ is one of the following:

(i) an n-dimensional projective geometry (possibly, a simplex);
(ii) an n-dimensional affine geometry;
(iii) one of the geometries $S(M_{22})$, $S(M_{23})$ or $S(M_{24})$, enrichments of Steiner systems for the Mathieu groups M_{22}, M_{23} and M_{24} respectively;

(iv) an (unknown) 3-dimensional matroid with orders s, q, q *with* $s > 1$ *and* $q = (s+1)^2$*;*
(v) an (unknown) 3-dimensional matroid with orders s, q, q *with* $s > 1$, $s \equiv 0, 1, 3, 4$ *or* 7 *(mod* 8*) and* $q = (s+1)^3 + (s+1)$*;*
(vi) an (unknown) 4-dimensional matroid with orders 1, 3, 68, 68.

Since $L_{n-1}.A_2$ includes $L_k.A_m$ as a special case for $k = 2, 3, ..., n-1$, the above also gives us a classification for $L_k.A_m$ with $m \geq 2$.

7.2.4 The diagram $L_{n-1}.Af$

We now consider the following diagram of rank $n \geq 3$, with finite orders s_0, s_1, ..., s_{n-3}, $q-1$, q:

$(L_{n-1}.Af)$

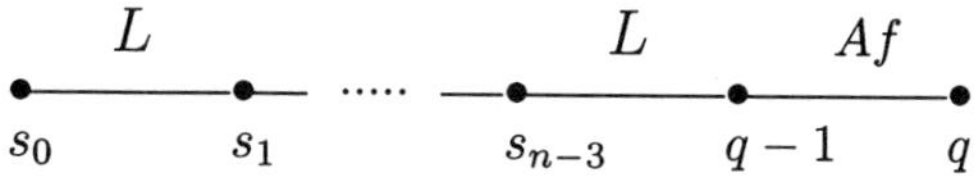

We first mention some examples. When $n = 3$ and $s_0 = 1$, then we have the diagram $c.Af$

$(c.Af)$

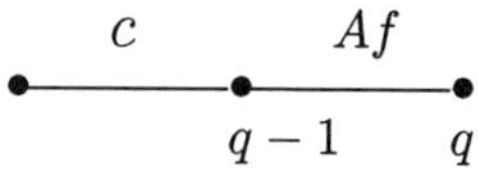

which characterizes enrichments of inversive planes (Exercise 7.3).

Steiner systems of type $S(4,5,11)$ and $S(5,6,12)$ are well known, related to the Mathieu groups M_{11} and M_{12} respectively. Let $S(M_{11})$ and $S(M_{12})$ denote their enrichments. The matroids $S(M_{11})$ and $S(M_{12})$ respectively belong to the following two special cases of $L_{n-1}.Af$:

$(c^2.Af)$

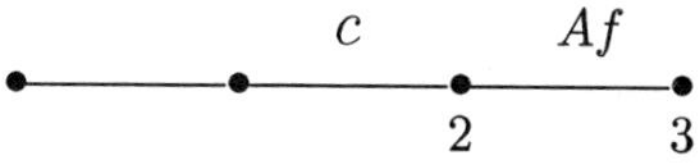

$(c^3.Af)$

Furthermore, $S(M_{11})$ and $S(M_{12})$ are the only geometries with diagrams and orders as above, by Exercise 7.4 and because the Steiner systems for M_{11} and M_{12} are the only Steiner systems of type $S(4,5,11)$ and $S(5,6,12)$ respectively [91].

When $q = 2$, $s_0 = s_1 = ... = s_{n-3} = q - 1 = 1$ and we have just a truncated $n+1$-dimensional simplex, by Corollary 7.13. The n-dimensional faces are the elements dropped in this truncation.

Theorem 7.24 *Let* Γ *belong to* $L_{n-1}.Af$, *with* $n \geq 3$ *and finite orders* s_0, s_1, ..., s_{n-3}, $q-1$, q. *Then* Γ *is one of the following:*

(i) a truncated $n+1$*-dimensional simplex;*
(ii) the enrichment of an inversive plane;
(iii) one of the two matroids $S(M_{11})$ *and* $S(M_{12})$*;*
(iv) an (unknown) matroid with $n = 4$, $s_0 = s_1 = 1$ *and* $q = 13$.

Sketch of the Proof. Using the divisibility conditions implicit in Lemma 7.16 we can prove that $s_0 = 1$ when $n = 3$. Hence the diagram $L_{n-1}.Af$ is in fact $c^{n-2}.Af$:

($c^{n-1}.Af$)

When $n = 3$, Γ is the enrichment of an inversive plane (in particular, a truncated 4-dimensional simplex if $q = 2$). Let $n \geq 4$. The following are easily obtained by Lemma 7.16:

if $n = 4$, then $q = 2, 3, 4, 8, 10, 13, 18, 28$ or 58;
if $n = 5$, then $q = 2, 3$ or 8;
if $n \geq 6$, then $q = 2$.

When $q = 2$ we have a truncated $n+1$-dimensional simplex.

Let $q > 2$. Hence $n = 4$ or 5. The stars of the $n-4$-dimensional elements of Γ (points or lines, according to whether $n = 4$ or 5) are enrichments of inversive planes. The cases of $q = 10, 18, 28$ or 58 can be shown to be impossible using this information and a theorem of Dembowski ([61], 6.2.14), stating that the order of a finite inversive plane is either odd or a power of 2. The cases of $q = 4$ or 8 are also impossible ([103], Section 5, Remark 2). Thus, the following are in fact the only possibilities:

$q = 3$ and $n = 4$ or 5;
$q = 13$ and $n = 4$.

When $q = 3$ we obtain $S(M_{11})$ and $S(M_{12})$. No examples are known for $q = 13$. □

Problem. Prove that the unknown matroids mentioned in Theorem 7.23, (iv)–(vi) and in Theorem 7.24(iv) do not exist.

We remark that, if Γ is a matroid as in (iv) of Theorem 7.24, then the affine planes obtained as stars of lines of Γ are non-Desarguesian. Indeed, Penttila [167] and Thas [211] have independently proved that, if Γ is as above, then the inversive planes arising from stars of points of Γ are non-classical. Therefore, stars of lines of Γ are non-Desarguesian affine planes, by Theorem C of [208].

Thus, the existence of an example for Theorem 7.24(iv) would imply the existence of an (unknown) non-Desarguesian affine plane of order 13.

7.3 One lemma on IP in $L_{n-1}.\pi$

We consider the following diagram of rank $n \geq 3$ in this section:

$$(L_{n-1}.\pi) \qquad \bullet \overset{L}{\text{———}} \bullet\text{—} \cdots\cdots \text{—}\bullet \overset{L}{\text{———}} \bullet \overset{\pi}{\text{———}} \bullet$$

where the label π denotes the class of partial planes, as in Chapter 3. As usual, we take the nonnegative integers 0, 1, ..., $n-1$ as types, labelling the nodes of the diagram by them from left to right.

We will consider many special cases of this diagram. Thus, it will be useful to have stated some general results on it.

Let Γ be a geometry belonging to $L_{n-1}.\pi$. We say that Γ has *good residual systems of lines* if it has a good system of lines (Section 4.3.2) and, for every $i = 0, 1, ..., n-4$ and every element x of Γ of type i, the star of x has a good system of lines. When $n = 3$, this property is just the same as having a good system of lines.

Lemma 7.25 *The strong intersection property* **IP** *holds in* Γ *if and only if* Γ *has good residual systems of lines.*

Proof. If **IP** holds in Γ, then it holds in all residues of Γ (Theorem 6.12), hence Γ has good residual systems of lines.

Conversely, let Γ have good residual systems of lines. By Corollary 6.39, in order to show that **IP** holds in Γ we only need to prove that $\mathbf{IP}_0^{Red}$ holds in Γ.

We first prove that the property $\mathbf{IP}_0^1$ of Section 6.7.2 holds (compare with Lemma 7.3). Let x and y be a line and an element of Γ and a, b distinct points in $\sigma_0(x) \cap \sigma_0(y)$. If y is a line, then $x = y$ because Γ has a good system of lines. Otherwise, there is a line x' through a and b in the lower residue of y, because the lower residue of y is a matroid by Theorem 7.6. Again, $x = x'$ because Γ has a good system of lines. That is, $x * y$. $\mathbf{IP}_0^1$ is proved.

We can now prove $\mathbf{IP}_0^{Red}$ by induction on n. The case of $n = 3$ is easy; we leave it for the reader (see also Exercise 6.5). When $n > 3$, we can use the same argument as in the proof of Lemma 7.4, but with some minor changes. This exercise is left for the reader. □

Corollary 7.26 *The strong intersection property* **IP** *holds in* Γ *if and only if* Γ *has a good system of lines and* **IP** *holds in stars of points of* Γ.

This is a trivial consequence of Lemma 7.25.

Two special cases of $L_{n-1}.\pi$ are considered in the next corollaries.

Corollary 7.27 *Let* Γ *belong to the following diagram:*

$(L.A_{n-2}.L^*)$

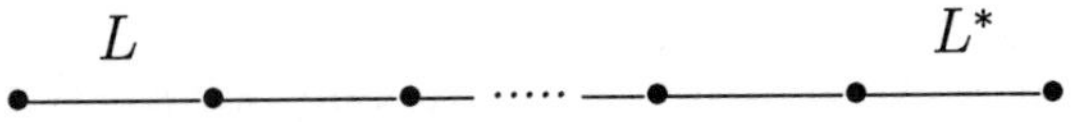

Then **IP** *holds in* Γ *if and only if* Γ *has a good system of lines.*

Proof. Trivial, by Corollary 7.26 and by Lemma 7.4, applied to stars of points of Γ. □

Corollary 7.28 *Let* Γ *belong to the following diagram:*

$(L.C_{n-1})$

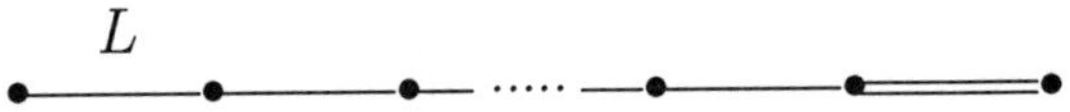

If Γ *has a good system of lines and stars of points of* Γ *are polar spaces, then* **IP** *holds in* Γ.

Proof. Easy, by Corollary 7.26 and because **IP** holds in polar spaces (Section 6.3.4 and Theorem 6.38).

7.4 The diagram C_n

We consider the Coxeter diagram C_n in this section:

(C_n)

Geometries belonging to C_n will be called *C_n geometries*, for short.

As usual, we take the nonnegative integers 0, 1, ..., $n-1$ as types, labelling the n nodes of the diagram C_n by them from left to right. As in Section 4.3.2, we call the elements of type $n-1$, $n-2$ and $n-3$ (if $n > 2$) dual points, dual lines and dual planes, respectively, and $\top$ denotes the dual collinearity relation.

7.4.1 Elementary properties of C_n geometries

Throughout this section Γ is a C_n geometry.

Lemma 7.29 *The geometry* Γ *has a good dual system of lines.*

Proof. If $n = 2$ there is nothing to prove. Let $n > 2$ and let us take the $\{n-3, n-2, n-1\}_+$-truncation of Γ. Residues of dual points of Γ are projective geometries, by Corollary 7.7. Hence this truncation belongs to the following diagram:

$(C_2.L)$ $\quad$ $n-1$ •══• $n-2$ —L— • $n-3$

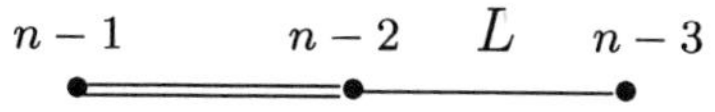

($n-1$, $n-2$, $n-3$ are the types saved in the truncation). The conclusion follows from the statement of Exercise 4.5. □

The previous lemma allows us to use the definite article 'the' when speaking of the dual line incident to two dually collinear dual points.

Lemma 7.30 *Given a dual point u and a point $a \notin \sigma_0(u)$, there is a dual point v such that $a * v \top u$.*

Proof. The statement is trivial if $n = 2$. Let $n > 2$. Let v be a dual point incident to a. By the connectedness of the dual collinearity graph of Γ, there is a path $v = v_0 \top v_1 \top \ldots \top v_m = u$ in that graph from v to u. We may choose v and this path so that m is minimal.

If $m = 1$, then $v \top u$ and we are done. Let $m > 1$. The point a is not incident to any of v_1, v_2, ..., v_{m-1}, by the minimality of m. Let x_1, x_2 be the dual lines incident to v_0 and v_1 and to v_1 and v_2, respectively. The residue Γ_{v_1} of v_1 is a projective geometry, by Corollary 7.7, and the dual lines x_0, x_1 of Γ are hyperplanes in this projective geometry. Hence a dual plane z of Γ incident to both x_0 and x_1 can be found in Γ_{v_1}. Trivially, $v * z * v_2$. In the projective geometry Γ_v we find a hyperplane y (dual line of Γ) incident to both a and z. The star Γ_z^+ of z is a generalized quadrangle. In Γ_z^+ we find a dual point w and a dual line x of Γ such that $y * w * x * v_2$. Hence $a * w \top v_2$ (we have $w \neq v_2$ because $a \notin \sigma_0(v_2)$). Substituting w for v and dropping v_1, we obtain a shorter path $w \top v_2 \top \ldots \top v_m = u$, contradicting the minimality of m.

Therefore $m = 1$ and $v \top u$. □

Corollary 7.31 *The collinearity graph of Γ has diameter $d \leq 2$.*

Proof. The statement is trivial if $n = 2$. Let $n > 2$. Given any two points a, b of Γ, we can choose a dual point $u * b$. If $a * u$, then we see that $a \perp b$ in Γ_u, because Γ_u is a projective geometry (Corollary 7.7). Otherwise, there are a dual point v and a dual line x such that $a * v * x * u$, by Lemma 7.30. Let c be any point incident to x. In the projective geometries Γ_v and Γ_u we see that $a \perp c$ and $c \perp b$. □

Corollary 7.32 *The geometry Γ is finite if it is locally finite.*

Sketch of the Proof. Let Γ be locally finite. Then an upper bound can be computed for the number of points of Γ using Lemma 7.30. The same can be done for the number of elements of type $i+1$ in stars of elements of type $i = 0, 1, 2, \ldots$. The finiteness of Γ is now evident. □

Lemma 7.30 shows that a weak version of axiom (iv) for polar spaces (Section 1.3.5) holds in any C_n geometry. The uniqueness claim is what Lemma 7.30 lacks of that axiom. In fact, axiom (iv) for polar spaces just says that the statement of Lemma 7.30 holds, that moreover the dual point v in that statement is unique, and that the same holds in stars of elements. We will show in the next section (Corollary 7.34) that, if we assume the strong intersection property, then we obtain that uniqueness claim as a by-product.

Exercise 7.5. Prove that, if Γ has a good system of lines and, for every type $i = 2, 3, \ldots, n-2$, distinct elements of type i have distinct 0-shadows, then **IP** holds in Γ.

(Hint: prove that, under the above hypotheses, Γ has good residual systems of lines; notice that lower residues of elements of Γ are projective geometries, by Corollary 7.7.)

Exercise 7.6 Let x be an element of Γ of type $i \leq n-2$ and let u be a dual point of Γ. Prove that there are a dual point v and an element y of type $n-i-2$ such that $x * v * y * u$.

Exercise 7.7 Let i, j be types with $j > i$ and let x, y be elements of Γ of type i and j respectively. Prove that there are elements u and v of type j and $j-i-1$ respectively, such that $x * u * v * y$.

(Remark: when $i = 0$ and $j = 1$, the above statement is a weaker version of the Buekenhout–Shult property (see Exercise 1.16).)

Exercise 7.8. Let Γ admit finite orders $q, q, \ldots, q, t$ and, for every $m = 0, 1, \ldots, n-1$, let N_m be the number of elements of Γ of type m.

Prove that

$$N_m \;\leq\; \prod_{k=0}^{m} \frac{(q^{n-k-1}t - 1)\sum_{i=0}^{n-k-1} q^i}{\sum_{i=0}^{k} q^i}$$

(Hint: use Lemma 7.30 to compute the upper bound for N_{n-1}. Then work by induction. Or start with an inductive argument, using Corollary 7.31 and Exercise 7.7 (with $i = 0$ and $j = 1$) to compute upper bounds for numbers of points.)

Exercise 7.9. Let Γ be a classical polar space defined over a finite field or a weak polar space admitting orders 1, 1, ..., 1, t. Prove that equality holds in the relations of Exercise 7.8.

Exercise 7.10. Prove that a flat C_n geometry with $n \geq 4$ has a good system of planes.

(Hint: use Exercise 4.9.)

Exercise 7.11. Let Γ be a flat C_n geometry with $n \geq 4$ and admitting finite orders q, q, ..., q, t. Prove that $q = 1$.

(Remark: flat C_n geometries with $n \geq 4$ and orders 1, 1, ..., 1, t actually exist; examples of this kind are described in [164]. This fact and Exercise 7.10 show that the property of having a good system of lines is not sufficent to obtain **IP** in C_n geometries when $n \geq 4$.)

Exercise 7.12. Let Γ be a flat C_3 geometry with finite orders q, q, t. Prove that $q \leq t$.

(Hint: since the dual point–line system of Γ is a partial plane, Γ has at least $(q^2+q+1)t+1$ planes; on the other hand, Γ has precisely $(qt+1)(t+1)$ planes, because it is flat and by Section 1.2.6(4))

Exercise 7.13. Prove that the following are equivalent in a C_3 geometry Γ:

(i) the geometry Γ does not have a good system of lines;
(ii) the dual collinearity graph of Γ contains a triangle formed by three distinct planes not on the same line.

(Hint. Proof that (i) implies (ii). Given distinct lines x, y and distinct points a, b with $a * x * b * y * a$, take a plane u on x; in Γ_a we can take a plane u_a and a line x_a such that $y * u_a * x_a * u$; similarly, in Γ_b ...

Proof that (ii) implies (i). Given distinct planes u, v, w and distinct lines x, y, z with $u * x * v * y * w * z * u$, take the intersection point a of x and z in Γ_u and the intersection point b of y and z in Γ_w; in Γ_v there is a line joining a with b ...)

7.4.2 Lemmas on C_n geometries satisfying IP

In this section Γ is a C_n geometry satisfying the strong intersection property **IP**.

We will prove in the next section that a C_n geometry is a polar space if and only if **IP** holds in it. That characterization of polar spaces will be obtained as a trivial consequence of the lemmas of this section.

Lemma 7.33 *Given a point a and a dual point u, we have $a * u$ if and only* *if there are distinct dual points v, w such that $v * a$, $w * a$, $v \top u$ and $w \top u$. The same statement holds in stars of elements of* Γ *of type* $i \leq n-3$.

Proof. When $n = 2$ the statement is a trivial consequence of the fact that generalized quadrangles have gonality 4. Let $n \geq 3$.

Let a be a point and u, v_1, v_2 be pairwise distinct dual points such that $a * v_i \top u$ $(i = 1, 2)$. Let x_i be the dual line incident to both v_i and u $(i = 1, 2)$. If $x_1 = x_2$, then $a * x_1 (= x_2)$ by the dual of $\mathbf{IP}_0^1$ (Section 6.7.2); hence $a * u$.

Let $x_1 \neq x_2$. Since x_1, x_2 are hyperplanes in the projective geometry Γ_u, a dual plane z of Γ incident to both x_1 and x_2 is uniquely determined inside Γ_u. Trivially, $z * v_i$ $(i = 1, 2)$. If $a * z$, then $a * u$.

Let $a \notin \sigma_0(z)$. The point a and the element z span a hyperplane y_i of the projective geometry Γ_{v_i} $(i = 1, 2)$. If $y_1 \neq y_2$, then we have $a * z$ by the dual of $\mathbf{IP}_0^1$ applied to the $\{0, 1, ..., n-2\}_+$-truncation of Γ (the property **IP** still holds in this truncation, by Theorem 6.14). However, we have assumed that $a \notin \sigma_0(z)$. Therefore $y_1 = y_2$. Now, passing to the star of z, we see that u, v_1, v_2 and x_1, x_2, y_1 $(= y_2)$ form a proper triangle in that star. However, this cannot be, as the star of z is a generalized quadrangle. Therefore $a * z$. Hence $a * u$.

Since **IP** is preserved under taking residues (Theorem 6.12), the same can be proved for stars of elements of type $i \leq n-4$ (stars of elements of type $n-3$ are generalized quadrangles, hence there is nothing to prove for them). □

Corollary 7.34 *Given a dual point u and a point $a \notin \sigma_0(u)$, there is just one dual point v such that $a * v \top u$. The same statement holds in stars of elements of* Γ *of type* $i \leq n-3$.

This is a trivial consequence of Lemmas 7.30 and 7.33.

If a and u are as in the hypotheses of the previous corollary, if v is the (unique) dual point such that $a * v \top u$ and x is the (unique) dual line incident to both u and v, then x will be called the *projection* of a onto u and it will be denoted by $pr_u(a)$.

The next lemma describes $\sigma_0(pr_u(a))$ (compare Exercise 1.15):

Lemma 7.35 *Let u be a dual point and let a be a point not incident to u. Then we have $a^{\perp} \cap \sigma_0(u) = \sigma_0(pr_u(a))$. The same statement (with $i+1$ instead of 0) holds in stars of elements of* Γ *of type* $i \leq n-3$.

Proof. If $n = 2$, then the statement is trivial. Let $n \geq 3$.

Let a, u be as above and let v be the unique dual point such that $a * v \top u$ (Corollary 7.34). In the projective geometry Γ_v we see that $\sigma_0(pr_u(a)) \subseteq a^{\perp} \cap \sigma_0(u)$. We must prove the converse inclusion.

Let $b \in a^{\perp} \cap \sigma_0(u)$ and let z be the line joining a with b. We have $z \notin \sigma_1(u)$ because $a \notin \sigma_0(u)$. By Lemma 7.30 in Γ_b, there are a dual point w and a dual line y in Γ_b such that $z * w * y * u$. We have $a * w$, as $a * z$. Therefore, $w = v$ by Lemma 7.33. Hence $y = pr_u(a)$, by Lemma 7.29. Therefore $b * pr_u(a)$.

Since the lemmas we have used also hold in stars of elements of Γ of type $i \leq n-3$, the same can be proved in those stars. □

Remark. The statement of Lemma 7.33 trivially holds in every flat C_3 geometry (Section 6.4.2). Actually, it says nothing in that case (the reader may notice that the statements of Lemma 7.30, Corollary 7.34 and Lemma 7.35 are empty in flat C_3 geometries.)

Thus, the statement of Lemma 7.33 does not imply **IP**. However, it can be proved that, if $n = 3$ and the statement of Lemma 7.33 holds in Γ, then either **IP** holds in Γ or Γ is flat ([150], Proposition 3). Unfortunately, things are not so easy in higher rank cases (see [168], Section 3).

Lemma 7.36 *Let $n \geq 3$. Then the Triangular Property (Section 4.3.3) holds in Γ and in stars of elements of Γ of type $i \leq n-4$.*

Proof. The geometry Γ has a good system of lines and a good system of planes because **IP** holds in it. Let a, b, c be distinct pairwise collinear points. We must prove that there is a plane incident to all of them.

Let x, y, z be the lines through a and b, through b and c and through c and a, respectively. As Γ has a good system of lines and the points a, b, c are pairwise distinct, either $x = y = z$ or the lines x, y, z are pairwise distinct. In the first case the statement is trivial. Therefore, we assume that x, y, z are pairwise distinct.

Let u be any dual point incident to x. Then $u * a, b$. Let $u * c$. If $n = 3$, then u is a plane incident to all of a, b, c, as we were looking for. Otherwise, that plane can be found inside the projective geometry Γ_u.

Let $c \notin \sigma_0(u)$. Then u is not incident to any of y or z. Applying Lemma 7.30 in Γ_a and Γ_b, we find dual points v_a and v_b respectively, with $y * v_b \top u$ and $z * v_a \top u$. As $c \notin \sigma_0(u)$, we have $v_a = v_b$, by Lemma 7.35. Let us write v for v_a $(= v_b)$, for short. Trivially, $pr_u(c)$ is the dual line incident to u and v. In Γ_u we see that $x * pr_u(c)$. Hence $x * v$ and v is incident to all of a, b and c. If $n = 3$, then v is the plane we were looking for. Otherwise, we find such a plane in the projective geometry Γ_v.

Since **IP** is preserved under taking residues (Theorem 6.12), the same can be proved in stars of elements of type $i \leq n-4$. □

Lemma 7.37 *For every dual point u, there is a dual point v such that*

$\sigma_0(u) \cap \sigma_0(v) = \emptyset$. *The same statement (with* $i+1$ *instead of* 0*) holds in stars of elements of* Γ *of type* $i \leq n-3$.

Proof. We work by induction on n. If $n = 2$ the statement is trivial. Let $n \geq 3$ and assume that there is a dual point u such that $\sigma_0(u) \cap \sigma_0(v) \neq \emptyset$ for every dual point v, by way of contradiction.

Given a point $a * u$, by the inductive hypothesis in Γ_a there is a dual point $v * a$ such that $\sigma_1(u) \cap \sigma_1(v) \cap \sigma_1(a) = \emptyset$. Hence $\sigma_0(u) \cap \sigma_0(v) = a$, because Γ_u is a projective geometry and Γ has a good system of lines.

Let us assume that $b \not\perp a$ for some point b. Then $b \notin \sigma_0(u) \cup \sigma_0(v)$, because Γ_u and Γ_v are projective geometries. Let u', v' be the dual points incident to b and dually collinear with u and v respectively (Corollary 7.34) and let us set $x = pr_u(b)$ and $y = pr_v(b)$. We have $b^\perp \cap \sigma_0(u) = \sigma_0(x)$ and $b^\perp \cap \sigma_0(v) = \sigma_0(y)$ by Lemma 7.35. Hence $a \notin \sigma_0(x) \cup \sigma_0(y)$ because $a \not\perp b$. Therefore $\sigma_0(x) \cap \sigma_0(y) = \emptyset$, because $\sigma_0(u) \cap \sigma_0(v) = a$. However, v' and u meet in some point, according to our hypotheses. Let $c \in \sigma_0(v') \cap \sigma_0(u)$. In the projective geometry $\Gamma_{v'}$ we see that $b \perp c$. Hence $c * x = pr_u(b)$, by Lemma 7.35. Let z be the line through b and c. We have $z * v'$ by the property $\mathbf{IP}_0^1$ of Section 6.7.2. Since y is a hyperplane of the projective geometry $\Gamma_{v'}$, z meets y in a point, call it d. We have $d \neq c$ because $c * x$ and $\sigma_0(x) \cap \sigma_0(y) = \emptyset$. Similarly, $d \neq b$. In the projective geometries Γ_u and Γ_v we see that $a \perp c$ and $a \perp d$. As we also have $c \perp d$, there is a plane w incident to all of a, c, d, by the triangular property (Lemma 7.36). As Γ has a good system of planes, the plane w is incident to the line z joining c with d. However, $b * z$. Hence $b * w$ and in the lower residue of w we see that $b \perp a$, contradicting our choice of b.

Therefore we are left with the case of $a \perp b$ for every point b. On the other hand, a has been arbitrarily chosen in $\sigma_0(u)$. Therefore every point of Γ is collinear with all points of u. However, if $b \notin \sigma_0(u)$, then $b^\perp \cap \sigma_0(u) = \sigma_0(pr_u(b))$ by Lemma 7.35 and $\sigma_0(pr_u(b))$ is a proper subset of $\sigma_0(u)$, because $pr_u(b)$ is a hyperplane in the projective geometry Γ_u. Therefore there are no points outside $\sigma_0(u)$. Hence $\sigma_0(v) \subseteq \sigma_0(u)$ for every dual point v. On the other hand, we have proved above that, for every point $a * u$, there is a dual point $v * a$ such that $\sigma_0(u) \cap \sigma_0(v) = a$. Thus, Γ should have just one point. This is evidently an absurdity. The final contradiction is reached.

The same can be done in stars of elements of type $i \leq n-3$, since **IP** and the lemmas we have used above remain true in those stars. □

7.4.3 A characterization of polar spaces

Theorem 7.38 *A* C_n *geometry is a polar space if and only if it satifies the* *intersection property* **IP**.

Proof. We know from Section 6.6 that polar spaces satisfy **IP**.

Conversely, let Γ be a C_n geometry satisfying **IP**. We must check that (i)–(v) of Section 1.3.5 hold in Γ. By the isomorphism property (Section 6.2), the geometry Γ can be identified with its 0-shadow space $\sigma_0(\Gamma)$. Hence Γ can be viewed as a family of subsets of a given set, as required in Section 1.3.5. Axiom (i) of Section 1.3.5 is quite trivial; we only must add $\emptyset$ to $\sigma_0(\Gamma)$ as the minimal element in order to obtain (i), as in the statement of the weak intersection property (Section 6.2). Axiom (ii) is simply the weak intersection property (Section 6.2). We obtain (iii) by Corollary 7.7. Axioms (iv) and (v) are just the statements of Corollary 7.34 and Lemma 7.37, respectively. □

Corollary 7.39 *A C_n geometry with $n \geq 3$ is a polar space if and only if it has good residual systems of lines.*

Trivial, by the previous theorem and Lemma 7.25.

Theorem 7.38 was obtained first by Tits, as a by-product of a 'topological' characterization of buildings as universal objects ([227]; see also Chapter 13 of this book). The proof we have given here is much more elementary than the original proof by Tits.

7.4.4 Non-polar C_n geometries

Many C_n geometries exist that do not satisfy **IP**. We call them *non-polar* C_n geometries, according to Theorem 7.38. We are not going to discuss these geometries in detail here. We only want to give the reader some flavour of what they are.

Many non-polar C_n geometries can be obtained as proper homomorphic images of infinite thick-lined polar spaces (see [227] Example 1.4.1.b; or [170] Example 2.2.ii; or [168] Sections 4 and 5). These images have infinitely many elements. We will examine homomorphic images of polar spaces in Sections 8.4.3 and 11.1.4(1).

Many non-polar C_n geometries also exist where some or all lines are thin ([171], [164]) and many of them can also be obtained as homomorphic images of weak polar spaces. The halved octahedron (Section 2.1.2(9)) is in fact an example of this kind. More examples will be examined in Section 11.1.4.

We also remark that a direct product of C_n geometries (Section 5.3.2) is non-polar if some or all of its factors are non-polar. Thus, we can produce infinitely many non-polar C_n geometries of arbitrarily large rank n, as soon as we are given one non-polar example. All non-polar C_n geometries obtained in this way have some (or all) thin lines.

Only one non-polar finite C_n geometry with thick lines is known, namely the *Alt*(7)-geometry (Section 6.4.2). Thus, the following is quite natural:

Conjecture. *The Alt(7)-geometry is the only non-polar finite C_n geometry with thick lines.*

We are not going to discuss this conjecture here. The reader wanting to know more on this topic may see the survey paper [121]. We will come back to this matter in Chapter 14.

7.5 The diagram D_n

Using Theorem 7.38 we can now describe the geometries belonging to the Coxeter diagram D_n, where 0, 1, ..., $n-3$, +, − are the types:

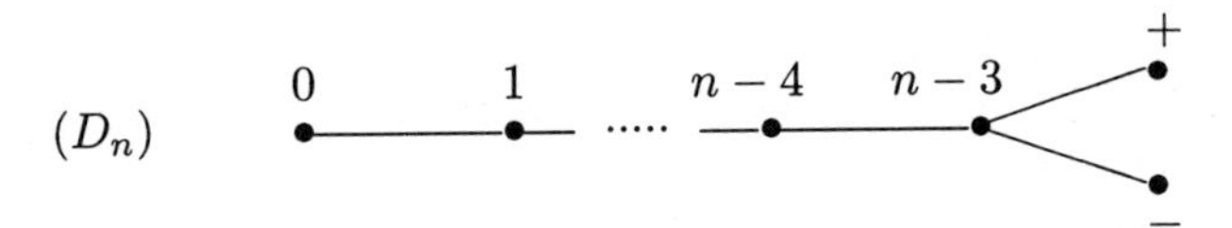

Buildings of type D_n belong to the above diagram. They have been defined in Section 5.3.4 as unfoldings of top-thin polar spaces of rank $n \geq 4$. It is now convenient to extend that definition to the case of $n = 3$. Projective geometries of dimension 3 are precisely unfoldings of rank 3 top-thin polar spaces, by Corollary 5.24. We call them *buildings of type* D_3, taking D_3 as a synonym of A_3, as in Section 3.1.3(5).

We now prove that buildings of type D_n are the only examples for the Coxeter diagram D_n (this result appeared first in [212], where it is attributed to Meixner).

Theorem 7.40 *The diagram D_n characterizes buildings of type D_n.*

Proof. We work by induction on n. Let $n = 3$. The diagram D_3 is simply A_3 drawn in an unusual way and A_3 characterizes 3-dimensional projective geometries (Corollary 7.7), that is buildings of type D_3.

Let $n \geq 4$ and let Γ be a geometry belonging to D_n. We will prove that the 0-Grassmann geometry $Gr_0(\Gamma)$ is a (top-thin) polar space. Having done that, we easily reach the conclusion: Γ is the unfolding of the top-thin polar space $Gr_0(\Gamma)$ (namely, it is a building of type D_n).

By Theorem 7.38, in order to prove that $Gr_0(\Gamma)$ is a polar space we must only prove that **IP** holds in it. By Corollary 7.28, this will be done if we prove that stars of points of $Gr_0(\Gamma)$ are polar spaces and that $Gr_0(\Gamma)$ has a good system of lines.

Let a be a point of $Gr_0(\Gamma)$. Then a is an element of Γ of type 0 and it is easily seen that the star $Gr_0(\Gamma)_a$ of a in $Gr_0(\Gamma)$ is the 1-Grassmann geometry $Gr_1(\Gamma_a)$ of the residue Γ_a of a in Γ, with respect to the initial node 1 of the diagram D_{n-1} of Γ_a (or with respect to the central node 1 of $A_3 = D_3$ when $n = 4$). The residue Γ_a is a building of type D_{n-1}, by the inductive hypothesis. Therefore $Gr_0(\Gamma)_a$ is a polar space.

Let us prove that $Gr_0(\Gamma)$ has a good system of lines. Let a, b be distinct points of $Gr_0(\Gamma)$ and let x, y be lines of $Gr_0(\Gamma)$ incident to both a and b. We must show that $x = y$.

Let $x \neq y$, if possible, and let u be a dual point of $Gr_0(\Gamma)$ incident to x. Then u is an element of Γ of type $+$ or $-$. If $y * u$, then x and y appear as distinct lines both incident to the distinct points a, b in the residue $Gr_0(\Gamma)_u$ of u in $Gr_0(\Gamma)$. However, $Gr_0(\Gamma)_u$ is a projective geometry (Corollary 7.7). Hence the above cannot be.

Therefore, $y \notin \sigma_1(u)$. By Lemma 7.30 in $Gr_0(\Gamma)_a$, there is a dual point v such that $y * v \top u$. Going back to Γ, the above means that v is an element of Γ_a of type $+$ or $-$ and that $y * v * u$ (the elements u and v have opposite types in Γ, as $v * u$ and $v \neq u$). The residue Γ_v of v in Γ is a projective geometry by Corollary 7.7. In this projective geometry we see u as a hyperplane, y as a line and a, b as distinct points incident to both y and u. Therefore, $y * u$. The final contradiction has been reached. □

Exercise 7.14. Let Γ be a geometry belonging to the following diagram of rank $n \geq 3$:

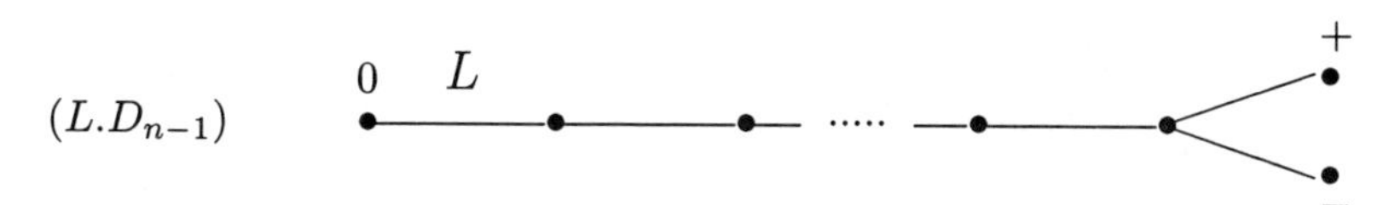

Prove that $\mathbf{IP}_J$ holds in Γ for every nonempty set of types J, except possibly when J contains one of the two types $+$ and $-$.

(Hint: first prove that $\mathbf{IP}_0$ holds in Γ, working with $Gr_0(\Gamma)$ and using the same argument as in the proof of Theorem 7.40, but exploiting Theorem 7.6 instead of Corollary 7.7. Then use Exercise 6.35 to finish.)

(Remark. We have assumed $n > 2$ in the above, but we might also allow $n = 2$, taking $L.D_2$ as a name for the diagram $L^*.L$ of Exercise 6.7.)

7.6 E_6, E_7, E_8 and F_4

We consider the following Coxeter diagrams in this section:

(E_6) $\quad$ 0^+ — 1^+ — C — 1^- — 0^-, with T attached to C

where 0^+, 0^-, 1^+, 1^-, C and T are the types (T and C should remind us of the words 'centre' and 'top').

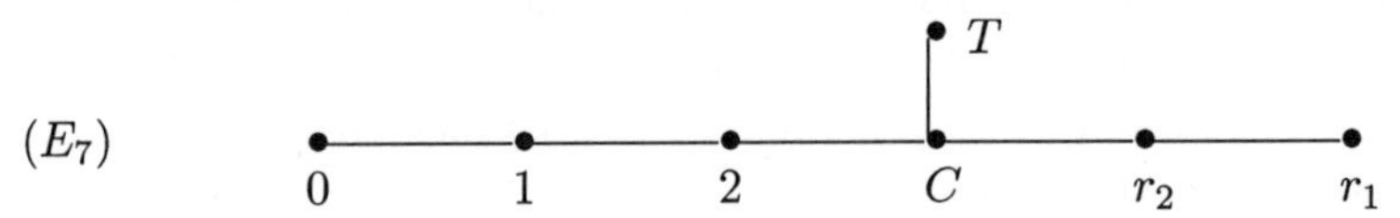

where 0, 1, 2, r_1, r_2, C, T are the types (the letters r_1, r_2 should remind us of the word 'right').

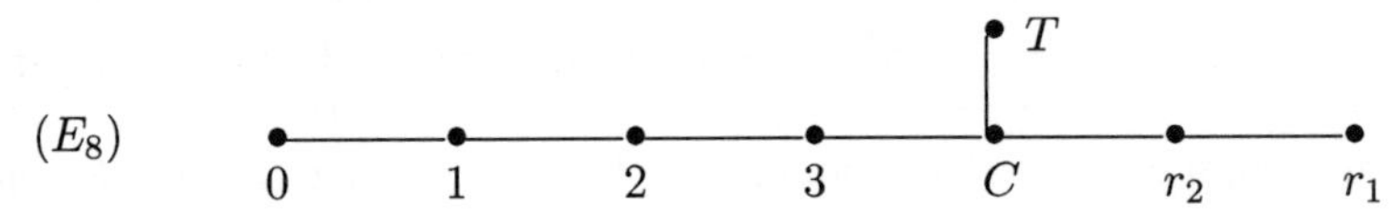

where 0, 1, 2, 3, r_1, r_2, C and T are the types.

These diagrams are special cases of the diagram E_n of Exercise 5.36. We will not state any classification theorem for them. We will only show how Corollary 7.7 and Theorem 7.40 can be used to study properties of geometries belonging to one of these diagrams.

We will also consider the following Coxeter diagram (where 0, 1, 2, 3 are the types)

(F_4) $\quad$ 0 — 1 = 2 — 3

showing how to use the results of Section 7.4 to get some information on geometries belonging to it.

7.6.1 The diagram E_6

In this section Γ is a geometry belonging to E_6. The elements of type 0^+, 1^+, 0^-, 1^-, T and C are called *points*, *lines*, *dual points*, *dual lines*, *maximal subspaces* and *planes* respectively (stating this terminology, we choose one of the two possible 'orientations' for this diagram). The collinearity graph of the point-line system of Γ (with points and lines defined as above) will be called the *collinearity graph* of Γ.

We know the following:

(i) residues of points or dual points are buildings of type D_5 (this follows from Theorem 7.40);

(ii) residues of maximal subspaces are 5-dimensional projective geometries, by Corollary 7.7;

(iii) residues of lines or dual lines are direct sums of two projective geometries, of dimension 1 and 4 respectively, by the direct sum theorem and Corollary 7.7;

(iv) residues of planes are direct sums of a line and two projective planes, by the direct sum theorem.

Proposition 7.41 *Any two points of* Γ *are always incident to a common dual point.*

Proof. Let us consider a configuration a, b, c, x, u where a, b, c are points, x is a line incident to both a and b and u is a dual point incident to both b and c. In the D_5 building Γ_b we find a maximal subspace t of Γ incident to both x and u. Then we find a dual line w of Γ in the D_5 building Γ_u, incident to both c and t. In Γ_x we see that $a * t$. In the projective geometry Γ_t we now find a dual point v incident to both a and w. In Γ_w we see that $c * v$. Therefore v is incident to both a and v. The conclusion now easily follows from the connectedness of the collinearity graph of Γ (Theorem 1.18). □

Corollary 7.42 *The collinearity graph of* Γ *has diameter* $d = 2$.

(Easy, by Proposition 7.41, Theorem 7.40 and because collinearity graphs of polar spaces have diameter $d = 2$).

Proposition 7.43 *The strong intersection property* **IP** *holds in* Γ.

Sketch of the Proof. The first step is to prove that $\mathbf{IP}_i$ holds for $i = 0^+, 0^{-1}, T$. Then we can prove that $\mathbf{IP}_i$ holds for every type i. This is enough to establish **IP** in Γ (see Exercise 6.42). □

7.6.2 The diagram E_7

In this section Γ belongs to the Coxeter diagram E_7. The elements of type 0, 1, r_1 and T are called *points*, *lines*, *dual points* and *maximal subspaces*, respectively. The *collinearity graph* of Γ is the collinearity graph of the point–line system of Γ, with points and lines defined as above.

We know the following on residues:

(i) residues of dual points are buildings of type D_6, by Theorem 7.40;

(ii) residues of maximal subspaces are 6-dimensional projective geometries, by Corollary 7.7;

(iii) residues of lines are direct sums of a line and a building of type D_5, by the direct sum theorem and by Theorem 7.40;

(iv) residues of elements of type 2 are direct sums of a projective plane and a 4-dimensional projective geometry, by the direct sum theorem and Corollary 7.7; similarly, residues of elements of type r_2 are direct sums of a line and a 5-dimensional projective geometry; residues of elements of type C are direct sums of a line, a projective plane and a 3-dimensional projective geometry;

(v) residues of points belong to E_6; some information on them has been given in the previous section.

Lemma 7.44 *Given a point a and a dual point u of Γ, there is a point b incident to u and collinear with a.*

Proof. Suppose there is a point c collinear with a and with a point $d * u$. let x, y be lines incident to a and c and to c and d, respectively. By Proposition 7.41, there is a dual point v in Γ_c incident to both x and y. Similarly, there is a line z in Γ_d incident to v and u. In Γ_v we find a point $b * z$ collinar with a, by Theorem 7.40. The conclusion follows from the connectedness of the collinearity graph of Γ. □

Proposition 7.45 *The collinearity graph of Γ has diameter $d \leq 3$.*

Proof. Given points a and b, let u be a dual point incident to b. By Lemma 7.44, there is a dual point c incident to u and collinear with a. In the D_6 building Γ_u we see that c has distance ≤ 2 from b. □

Lemma 7.46 *The property* $\mathbf{IP}_0$ *holds in the truncation $Tr^-_{r_1}(\Gamma)$.*

Sketch of the Proof. the 0-Grassmann geometry of $Tr^-_{r_1}(\Gamma)$ has the following diagram:

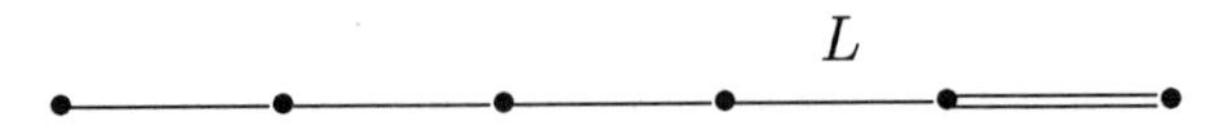

Hence we can use Lemma 7.25. Exploiting the information we have on residues of elements of Γ, we can check that the hypotheses of Lemma 7.25 hold in our case. Then we can use Theorem 6.14 to come back to $Tr^-_{r_1}(\Gamma)$. □

Proposition 7.47 *The strong intersection property* **IP** *holds in* Γ *if and only if the following holds:*

($\mathbf{IP}_0^{1,r_1}$) *given two distinct collinear points* a, b, *let* x *be the line joining them (unique by Lemma 7.46) and let* u *be a dual point; if* $u * a$ *and* $u * b$, *then* $x * u$.

Sketch of the Proof. The 'only if' part if trivial. The first step in the proof of the 'if' claim is to show that $\mathbf{IP}_0^{1,r_1}$ implies $\mathbf{IP}_0^{Red}$. Most of the cases to check are easily handled by Lemma 7.46. The only cases that are not covered by that lemma are those involving an element of type r_1 or a flag of type $\{r_1, T\}$. However, the statement of $\mathbf{IP}_0^{Red}$ is not difficult to prove for those cases, too.

By $\mathbf{IP}_0^{Red}$ we obtain $\mathbf{IP}_0$ (Proposition 6.34). The next step is to prove that $\mathbf{IP}_0$ implies $\mathbf{IP}_{r_1}$ and $\mathbf{IP}_T$. Having done this, we can use the statement of Exercise 6.40 to prove that $\mathbf{IP}_i$ holds for every type i. Then we can work as in Exercise 6.29 to finish the proof. □

Exercise 7.15. Prove that $\mathbf{IP}_C$ holds in Γ, with respect to the central node C of the diagram.

(Hint. By Proposition 6.34, we only need to prove $\mathbf{IP}_C^{Red}$. All cases to consider in $\mathbf{IP}_C^{Red}$ can be dealt with using the information we have on residues of elements of Γ, except when two dual points are involved. However, in that case we find an element of type r_2 such that ...)

7.6.3 The diagram E_8

In this section Γ belongs E_8. The elements of type 0, 1, r_1 and T are called *points*, *lines*, *dual points* and *maximal subspaces*, respectively. The *collinearity graph* of Γ is the collinearity graph of the point–line system of Γ, with points and lines defined as above. We denote the $\{0, r_1\}_+$-truncation of Γ by $\overline{\Gamma}$ and we take points and the dual points of Γ as points and lines of $\overline{\Gamma}$ respectively.

We have the following information on residues:

(i) residues of dual points are buildings of type D_7, by Theorem 7.40;

(ii) residues of maximal subspaces are 7-dimensional projective geometries, by Corollary 7.7;

(iii) the residue of an element of type 2 is the direct sum of a projective plane and a building of type D_5, by the direct sum theorem and by Theorem 7.40;

(iv) by Corollary 7.7 and by the direct sum theorem, the residue of an element of type 3 is the direct sum of two projective geometries, of dimension 3 and 4 respectively; similarly, the residue of an element of type r_2 is the direct sum of a line and a 6-dimensional projective geometry; the residue of an element of type C is the direct sum of a line, a projective plane and a 4-dimensional projective geometry.

(v) residues of points belong to E_7; some information on them has been obtained in the previous section;

(vi) by the direct sum theorem, the residue of a line is the direct sum of a line and a geometry of type E_6; some information on E_6 has been obtained in Section 7.6.1.

Lemma 7.48 *Two points of* Γ *have distance* ≤ 3 *in the collinearity graph of* Γ *if and only if they have distance* ≤ 2 *in the collinearity graph of* $\overline{\Gamma}$.

Proof. Let a, b be points of Γ at distance $d(a,b) \leq 3$ in the collinearity of Γ. We assume $d(a,b) = 3$, leaving the case of $d(a,b) = 2$ for the reader (the case of $d(a,b) = 1$ is trivial). Thus, there are points c, d such that $a \perp b \perp c \perp b$ in Γ. Let x, y, z be lines through a and c, through c and d and through d and b, respectively. Given a dual point u on y, by Lemma 7.44 applied to Γ_c and Γ_d there are lines z_c, z_d incident to c and d respectively and planes v_c, v_d such that $x * v_c * z_c * u * z_d * v_d * z$.

By Theorem 7.40 applied to Γ_u there are planes w_c, w_d incident to u and a point e such that $z_c * w_c * e * w_d * z_d$.

By Proposition 7.41, there are dual points u_c, u_d in Γ_{z_c} and Γ_{z_d} such that $v_c * u_c * w_c$ and $v_d * u_d * w_d$. The configuration a, u_c, e, u_d, b shows that a, b have distance ≤ 2 in the collinearity graph of $\overline{\Gamma}$.

On the other hand, let a, b be points of Γ at distance ≤ 2 in the collinearity graph of $\overline{\Gamma}$. We may assume that a, b have distance 2 in that graph (the case of distance 1 is trivial). Thus, let c be a point and let u, v be dual points such that $a * u * c * v * b$. Given a line x on c in Γ_u, there are a line y and a plane w in Γ_c such that $x * w * y * v$, by Lemma 7.44. In Γ_u and Γ_v we find a point d_x in x collinear with a and a point d_y in y collinear with b. In Γ_w we see that $d_x \perp d_y$. Therefore a, b have distance at most 3 in the collinearity graph of Γ. □

Lemma 7.49 *The collinearity graph of* $\overline{\Gamma}$ *has diameter* ≤ 2.

Proof. Let us consider a configuration of four points a, b, c, d, a line x and two dual points u, v such that $a*x*b*u*c*v*d$. By Lemma 7.44 applied to Γ_b, there are a plane w_b and a line y_b on b such that $x * w_b * y_b * u$. In Γ_u we find a point e_u and a line x_u such that $y_b * e_u * x_u * c$. By Lemma 7.44

applied to Γ_c there are a plane w_c and a line y_c on c such that $x_u * w_c * y_c * v$. In Γ_v we find a point e_v on y_c collinear with d in Γ. In Γ_{w_b} and Γ_{w_c} we see that $a \perp e_u$ and $e_u \perp e_v$ in Γ. Therefore a, b have distance ≤ 3 in the collinearity graph of Γ. By Lemma 7.48, the points a and b have distance ≤ 2 in the collinearity graph of $\overline{\Gamma}$.

The conclusion follows from the connectedness of the collinearity graph of Γ. □

Proposition 7.50 *The collinearity graph of* Γ *has diameter* $d \leq 3$.

(Trivial, by Lemmas 7.48 and 7.49).

Proposition 7.51 *The strong intersection property* **IP** *holds in* Γ *if and only if both the following hold:*

(i) ($\mathbf{IP}_0^{1,r}$) *given two distinct collinear points* a *and* b, *a line* x *joining them and a dual point* u, *if* $u * a$ *and* $u * b$, *then* $x * u$.
(ii) given two distinct points a *and* b, *if there are distinct dual points incident to both* a *and* b, *then* a *and* b *are collinear.*

Sketch of the Proof. The 'only if' claim is trivial. The first step in the proof of the 'if' part is to show that (i) and (ii) imply $\mathbf{IP}_0^{Red}$. This can be done using Theorem 7.40 and Propositions 7.43 and 7.47. Then we can finish as in the proof of Proposition 7.47. □

7.6.4 The diagram F_4

In this section Γ belongs to the Coxeter diagram F_4. We designate by $\mathcal{G}$ the collinearity graph of $Tr^+_{0,3}(\Gamma)$, with the points of Γ taken as points of $Tr^+_{0,3}(\Gamma)$.

Proposition 7.52 *The graph* $\mathcal{G}$ *has diameter* $d \leq 2$.

Proof. Let a, b, c, d be points of Γ with $a \perp b$ in the natural collinearity graph of Γ and b, d adjacent to c in $\mathcal{G}$. Let x be a line through a and b and let u, v be dual points incident to b and c and to c and d, respectively.

By Lemma 7.30 in Γ_b, there are a plane π_b on a and a line y_b on b such that $x * \pi_b * y_b * u$. By the statement of Exercise 7.7 applied to Γ_u, there is a line x_u in Γ_u incident to c and to a point e_u of y_b. By Lemma 7.30 applied to Γ_c, there is a line–plane flag (y_c, π_c) incident to c, x_u and v. By the statement of Exercise 7.7 applied to Γ_v, there is a point e_v of y_c collinear with d in Γ_v. In Γ_{π_b} we find a line z through a and e_u. By the statement of Exercise 7.7 applied to Γ_{e_u}, there are a dual point w and a line y on e_u

such that $z * w * y * \pi_c$. In Γ_{w_c} we find a point e incident to both y and y_c. The configuration a, w, e, v, d shows that a and d have distance ≤ 2 in $\mathcal{G}$.

The conclusion now follows from the connectedness of the collinearity graph of Γ. □

Proposition 7.53 *The strong intersection property* **IP** *holds in* Γ *if and only if all the following hold:*

(i) the geometry Γ *has a good system of lines;*
(ii) ($\mathbf{IP}_0^{1,3}$) *given two distinct collinear points* a *and* b*, a line* x *joining them and a dual point* u*, if* $a * u$ *and* $b * u$*, then* $x * u$*;*
(iii) given two distinct points a *and* b*, if there are distinct dual points incident to both* a *and* b*, then* a *and* b *are collinear in* Γ*.*

Proof. The 'only if' part is trivial. By Corollary 6.39, to prove the 'if' claim we only need to prove that (i)–(iii) imply $\mathbf{IP}_0^{Red}$.

There are six cases to examine in $\mathbf{IP}_0^{Red}$, since Γ has rank 4. However, (i) and (ii) already cover three of them (those where a line is involved). By Lemma 7.29 the $\{0,1,2\}_+$-truncation of Γ belongs to $L.\pi$. Hence **IP** holds in it, by (i) and Lemma 7.25. Thus, only two cases are left to examine, namely those involving a plane and a dual point or two dual points.

Let u be a dual point and v another dual point or a plane such that $\sigma_0(u) \cap \sigma_0(v)$ contains more that one point. If v is a plane, then we can choose a dual point w on v. Otherwise, we set $w = v$. Note that both Γ_u and Γ_w are polar spaces, by (i) and Corollary 7.39. By (iii) on u, w and by (ii) we see that $\sigma_0(u)\cap\sigma_0(v)$ is a singular subspace (Section 1.6.2) both in the polar space Γ_u and in the polar space Γ_w. As polar spaces satisfy the triangular property (Exercise 1.16 or Lemma 7.36), there are planes or lines x, y incident to u and v respectively and such that $\sigma_0(u)\cap\sigma_0(v) = \sigma_0(x) = \sigma_0(y)$ (possibly, $y = v$). However, we have remarked above that the statement of $\mathbf{IP}_0^{Red}$ holds in cases not involving any dual point. Therefore $x = y$.

The proof of $\mathbf{IP}_0^{Red}$ is complete. □

The geometries belonging to F_4 and satisfying **IP** are called *metasymplectic spaces.* Metasymplectic spaces of D_4 buildings, as defined in Section 5.3.6, are metasymplectic spaces in the above meaning too. Indeed **IP** holds in them (see Section 6.3.4(2), Proposition 6.10 and Theorem 6.38).

7.7 $Af.C_{n-1}$ and $Af.D_{n-1}$

From now on we will no longer give complete proofs for the results we mention, except when a few lines are enough. Sometimes, we give a brief outline of the proof. Of course, we always give references.

Even if we are going to state results without proving them, we will only mention those results that can be obtained using elementary techniques, as we promised at the beginning of this chapter (but we warn the reader that an 'elementary' proof need not be easy at all).

7.7.1 The diagram $Af.C_{n-1}$

In this section Γ is a geometry belonging to the following diagram of rank $n \geq 3$, where 0, 1, ..., $n-1$ are the types:

$(Af.C_{n-1})$ $\quad$ 0 Af 1 — 2 ..... $n-3$ — $n-2$ = $n-1$

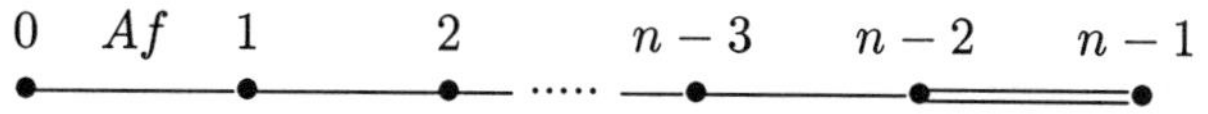

We also assume that Γ satisfies **IP**. By Corollary 7.28 and Theorem 7.38, this is equivalent to assuming that stars of points of Γ are polar spaces and that Γ has a good system of lines.

According to this assumption, we will identify elements of Γ with their 0-shadows, writing $a \notin u$ for $a \notin \sigma_0(u)$, $u \cap v$ for $\sigma_0(u) \cap \sigma_0(v)$, and so on.

We first remark that residues of dual points of Γ are $n-1$-dimensional affine geometries, by Corollary 7.9.

Given a point a and a dual point u not incident to a, let $pr_a(u)$ be the set of dual points on a meeting u in a dual line (i.e., in a hyperplane of the affine geometry Γ_u). Note that $pr_a(u) = \emptyset$ may occur. We define $pr_u(a)$ as follows:

$$pr_u(a) = \{u \cap v \mid v \in pr_a(u)\}.$$

By Lemma 7.30 applied to stars of points of $a^\perp \cap u$, we easily see that

$$a^\perp \cap u = \cup(w \mid w \in pr_u(a)).$$

In particular, we have $a^\perp \cap u \neq \emptyset$ if and only if $pr_u(a) \neq \emptyset$ (if and only if $pr_a(u) \neq \emptyset$).

Let $a^\perp \cap u \neq \emptyset$. By Lemma 7.33 in stars of points of $a^\perp \cap u$, $pr_u(a)$ consists of pairwise parallel hyperplanes of the affine geometry Γ_u (possibly, it consists of just one hyperplane of Γ_u). Furthermore, it is proved in [155] that there is a dual line w on a such that all dual points in $pr_a(u)$ contain w and $w \parallel u \cap v$ in Γ_v, for every $v \in pr_a(u)$. We call $pr_a(u)$, $pr_u(a)$ and w the *projecting bundle* of a onto u, the *projection* of a onto u and the *axis* of the projective bundle, respectively.

It is also proved in [155] that the cardinal number $|pr_a(u)|$ does not depend on the particular choice of a and u, provided that $a \notin u$ and $a^\perp \cap u \neq \emptyset$. We denote this number by γ and we call it the *connection index* of Γ.

Theorem 7.54 *We have $\gamma = 1$ if and only if Γ is an affine polar space.*

Outline of the Proof. The 'if' part is trivial. The 'only if' part has been proved by Buekenhout and Hubaut [26] in the case when Γ has thin lines and (later) by Cohen and Shult [39] in the general case.

The idea of the proof of [39] is quite simple, at least in principle. It is similar to the proof of Theorem 1.4 for affine geometries. We first recover a 'parallelism relation' between the lines of Γ, starting from the parallelism relation in planes of Γ. Next we (re)construct a space at infinity Γ^∞ for Γ, representing the points of Γ^∞ as 'parallelism classes' of lines of Γ (and adding one more point, if necessary). Finally we can check that $\Gamma \cup \Gamma^\infty$ is a polar space and that Γ^∞ is a hyperplane of that polar space.

However, this is only an outline of the proof. The details are not easy at all. The reader should consult [39] for them. We note that Cohen and Shult need to assume thick lines in [39]. However, our hypotheses are a bit stronger than those of [39] and would allow us to use the proof of [39] in the thin-lined case, too. □

The geometry Γ admits orders $s-1$, s, ..., s, t (see Section 3.2.3; of course, $s-1 = s$ when s is infinite).

Trivially, $\gamma \leq min(s, t+1)$. Indeed, if a, u are a point and a dual point with $a \notin u$ and $a^\perp \cap u \neq \emptyset$, then $pr_u(a)$ consists of γ pairwise parallel hyperplanes of the affine geometry Γ_u of order s, whereas $pr_a(u)$ contains γ of the $t+1$ dual points on the axis w of the projecting cone $pr_a(a)$.

The above upper bound for γ is significant if one of s and t is finite. For instance, if stars of points are hyperbolic quadrics, then $t = 1$, hence $\gamma \leq 2$. If stars of points are isomorphic to $\mathcal{S}_{2n-3}(q)$, $\mathcal{Q}^-_{2n-1}(q)$ or $\mathcal{Q}_{2n-2}(q)$, then $s = q$, hence $\gamma \leq q$; if stars of points are isomorphic to $\mathcal{H}_{2n-3}(q^2)$, then $t = q$, hence $\gamma \leq q+1$.

The above upper bound for γ is fairly rough. A stricter upper bound can be determined: if h is the size of a hyperbolic line in the star of a point of Γ, then $\gamma \leq h$ (see Exercise 7.16; see [25] or [166] for the definition and properties of hyperbolic lines in polar spaces; only generalized quadrangles are considered in [166], however what can be said about hyperbolic lines in the rank 2 case is the same as in the general case).

For instance, if stars of points are isomorphic to $\mathcal{H}_m(q^2)$ with $m = 2n-3$ or $2n-2$, then $h = q+1$, hence $\gamma \leq q+1$. If stars of points are isomorphic to $\mathcal{Q}^+_{2n-3}(K)$, $\mathcal{Q}^-_{2n-1}(K)$ or $\mathcal{Q}_{2n-2}(q)$ $(q \neq 2)$, then $h = 2$, hence $\gamma \leq 2$.

The next two propositions have been proved in [155] and [45], respectively. They characterize cases with finite orders and large γ.

Proposition 7.55 ([155]) *Let Γ admit finite orders $s-1$, s, ..., s, t with*

$s \geq 3$. Then we have $\gamma = s$ if and only if Γ is the affine expansion (Section 2.3) of the symplectic variety $\mathcal{S}_{2n-3}(q)$, $q = s$ $(= t)$.

Proposition 7.56 ([45]) *Let the stars of points of Γ be isomorphic to $\mathcal{H}_m(q^2)$, $m = 2n - 3$ or $2n - 2$. Then we have $\gamma = q$ if and only if Γ is the affine expansion of $\mathcal{H}_m(q^2)$.*

Outline of the Proofs. The proofs of propositions 7.55 and 7.56 are quite similar. The 'only if' claim is the only non-trivial part of the statement. To prove it, we must construct an embedding of Γ in an n-dimensional affine geometry $\mathcal{A}$ with the same set of points as Γ, reconstructing that part of $\mathcal{A}$ missing in Γ. The main step in this work is to recover the planes of $\mathcal{A}$ missing in Γ (in the case considered in Proposition 7.56 not all lines of $\mathcal{A}$ are saved by Γ, hence we must recover missing lines, too). The planes of $\mathcal{A}$ missing in Γ give rise to hyperbolic lines in stars of points of Γ. This remark suggests some trick to recover them. Having done this, the conclusion can be obtained using some characterization of affine geometries (such as Proposition 7.12, for instance).

This is only a rough idea of the proofs. The details can be found in [155] and [45]. □

Using universal covers (to be defined in Chapter 12 of this book) much stronger results have been achieved in [46] and [49] (see Section 15.1.1): if $n \geq 4$, then Γ is a homomorphic image of an affine polar space [49]; if $n = 3$, then the same conclusion can be obtained assuming that Γ has finite orders [46].

We are not going to discuss these two theorems in detail here, according to our promise to keep results exploiting non-elementary techniques out of the scope of this chapter. Furthermore, we have not yet defined morphisms of geometries (we will do this in the next chapter), so we could not discuss this matter properly, even if we wanted to. We will come back to this topic in Chapter 15. We only make a few comments now, in order to help the reader to understand how Propositions 7.55 and 7.56 fit with the stronger results of [49] and [46].

The morphisms considered in [46] and [49] are 'nice', in a sense defined in [156] and [49]. As far as we are concerned here, we can take 'nice' as an abbreviation of 'preserving the intersection property' (we remark that many examples are known for $Af.C_{n-1}$ that do not satisfy **IP**; all but one of them are 'non-nice' homomorphic images of affine polar spaces [156]).

We will see in Section 8.4.7 that affine expansions of polar spaces are minimal 'nice' homomorphic images of affine polar spaces of tangent type

[49] (see Section 2.2.2 for the definition of affine polar spaces of tangent or secant type).

If hyperbolic lines in stars of points have finite size h, then the relation $\gamma = h$ characterizes minimal 'nice' homomorphic images of affine polar spaces of secant type [49]. If, furthermore, $h > 2$, then the relation $\gamma + 1 = h$ characterizes affine expansions of polar spaces [49]. Actually, we have $h = s + 1$ in Proposition 7.55 and $h = q + 1$ in Proposition 7.56.

When $h = 2$, then the relation $\gamma = h - 1$ $(= 1)$ only says that we have an affine polar space (Theorem 7.54).

Affine polar spaces of tangent type with $h = 2$ have no 'nice' proper homomorphic images [49]. They are affine expansions of the star of any of their points [49].

We can now understand the case of $\gamma = s = 2$, omitted in Proposition 7.55. When $s = 2$, we have two possibilities for h, namely $h = 3$ $(= s + 1)$ and $h = 2$. According to the above, the relation $\gamma = s$ characterizes the affine expansion of $\mathcal{S}_{2n-3}(2)$ when $h = 3 = s + 1$ (as in the case of $s \geq 3$ considered in Proposition 7.55; see [155]; also [17]). When $h = 2 = s$, then the relation $\gamma = s$ should rather be read as $\gamma = h$, and characterizes minimal 'nice' quotients of affine polar spaces of secant type with stars of points isomorphic to $\mathcal{Q}^+_{2n-3}(2)$ or to $\mathcal{Q}^-_{2n-1}(2)$. Only two examples of this kind exist, corresponding to the two isomorphism types $\mathcal{Q}^+_{2n-3}(2)$ and $\mathcal{Q}^-_{2n-1}(2)$ for stars of points (see [17]; also [155]).

Exercise 7.16. Without using the result of [49] and [46], prove that, if a and u are respectively a point and a dual point of Γ such that $a \not\in u$ and $a^\perp \cap u \neq \emptyset$ and x is a line on u not parallel to the hyperplanes of Γ_u forming $pr_u(a)$, then the lines of Γ joining a to the points of $x \cap a^\perp$ form a subset of a hyperbolic line of the polar space Γ_a (see [25] or [166] for the definition of hyperbolic lines).

(Hint: see [45]; also [49], Lemma 4.5.)

(Remark. As a consequence of the above, if h is the size of a hyperbolic line in the polar space Γ_a, then $\gamma \leq h$.)

Exercise 7.17. Prove that the following holds in the point-line system of Γ: given a non-incident point–line pair (a, x), the point a is collinear with 0, γ or all points of x.

Exercise 7.18. Let a, b be points at distance 2 in the collinearity graph of Γ. Prove that the lines on b that do not meet $a^\perp$ form a hyperplane of the polar space Γ_b.

Exercise 7.19. Without using the result of [49] and [46], prove that the collinearity graph of Γ has diameter $d \leq 3$.

(Hint: use Exercise 7.18.)

Exercise 7.20. Without using Theorem 7.54, prove that the relation $\gamma = 1$ is equivalent to the triangular property.

The geometries considered in the next three exercises do not satisfy **IP**, contrary to the assumptions made on Γ till now. In fact, they are flat. We remark that no flat $Af.C_n$ geometries exist with finite orders and rank $n \geq 5$, by the statement of Exercise 7.11 in stars of points.

Exercise 7.21. (A flat $Af.C_2$ geometry.) Let $\mathcal{G}$ be a coloured graph on a set S of four vertices, with set of colours $\{1, 2, 3\}$ and such that any two distinct vertices are joined by three edges, one for each colour. Take S as set of points and the 18 edges of $\mathcal{G}$ as lines. There are six ways to select one edge for each pair of distinct vertices in such a way that distinct selected edges with a vertex in common have different colours. Each of these choices of edges gives us a complete simple subgraph of $\mathcal{G}$. Take as planes the six subgraphs of $\mathcal{G}$ obtained in this way and define the incidence relation by containment.

Check that this construction gives us a flat geometry belonging to $Af.C_2$ with orders 1, 2, 1.

(Remark: this geometry is actually a 'non-nice' homorphic image of the affine polar space obtained by dropping a secant hyperplane from $\mathcal{Q}_5^+(2)$; see [156].)

Exercise 7.22. (The *Alt*(8)-geometry.) Starting with $AG(3, 2)$, take the points, lines and planes of $AG(3, 2)$ as points, lines and planes of the geometry we want to define, with the incidence relation inherited from $AG(3, 2)$. Let the symmetric group S_8 act on the 8 points of $AG(3, 2)$. The group S_8 transitively permutes the 30 copies of $AG(3, 2)$ that can be built with the 8 points of $AG(3, 2)$. The alternating group A_8 has 2 orbits of size 15 on this set of 30 copies of $AG(3, 2)$. Take one of these orbits as set of dual points, defining the incidence relation between points, lines or planes and these dual points in the natural way, by containment.

The reader may check that a flat $Af.C_3$ geometry is obtained in this way, with orders 1, 2, 2, 2 and stars of points isomorphic to the *Alt*(7)-geometry of Section 6.4.2. We call it the *Alt*(8)-*geometry*.

Prove that the *Alt*(8)-geometry is the only $Af.C_3$ geometry with stars of points isomorphic to the *Alt*(7)-geometry.

(Remark: the *Alt*(8)-geometry was discovered by A. Neumaier [136]; the *Alt*(7)-geometry was obtained in [136] as a by-product, as the isomorphism type of stars of points of the *Alt*(8)-geometry.)

Exercise 7.23. Prove that every flat $Af.C_3$ geometry has a good system of planes (but **IP** does not hold in it; compare with the Remark of Exercise 7.11).

7.7.2 The diagram $Af.D_{n-1}$

We now consider the following diagram of rank $n \geq 3$:

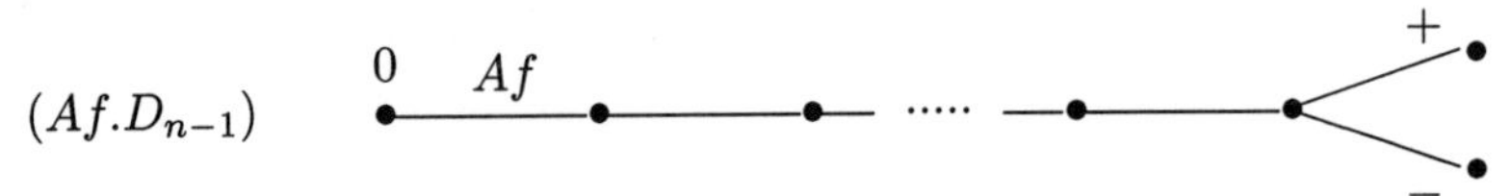

where 0, +, − are types. When $n = 4$, the above diagram consists of a stroke labelled by Af attached to the central node of A_3 ($= D_3$). When $n = 3$, we have the diagram $Af^*.Af$ of Exercise 5.15:

$(Af.D_2 = Af^*.Af)$ 0 — Af — + ; 0 — Af — −

Examples for $Af.D_{n-1}$ can be constructed as follows. Let $\mathcal{P}$ be a top-thin thick-lined polar space of rank $n \geq 3$, and let H be a hyperplane of $\mathcal{P}$. The unfolding $\Gamma(\mathcal{P})$ of $\mathcal{P}$ (Section 5.3.4) is a building of type D_n. Since the affine polar space $\mathcal{P} - H$ is a subgeometry of $\mathcal{P}$, we can also consider the effect of the unfolding construction on $\mathcal{P} - H$. It is not difficult to check that a subgeometry of $\Gamma(\mathcal{P})$ is obtained in this way, which belongs to $Af.D_{n-1}$. We call it the *unfolding* of $\mathcal{P} - H$.

Theorem 7.57 *The diagram $Af.D_{n-1}$ characterizes unfoldings of top-thin affine polar spaces of rank n.*

Proof. Let Γ belong to $Af.C_{n-1}$ and let 0 be the 'initial' node of the diagram, as in the above picture. The 0-Grassmann geometry $Gr_0(\Gamma)$ of Γ belongs to the diagram $Af.C_{n-1}$ and it is top-thin. The strong intersection property **IP** holds in $Gr_0(\Gamma)$ (Exercise 7.14 when $n \geq 4$; Exercise 6.7 when $n = 3$; also Theorem 6.38). Thus, we can compute the connection index γ of $Gr_0(\Gamma)$ (see Section 7.1.1 for the definition of γ).

Let $\gamma \geq 2$. Then we find three distinct dual points u_1, u_2, u_3 of $Gr_0(\Gamma)$ pairwise intersecting in pairwise parallel and distinct dual lines. If we go back to Γ and check what this means, we see that u_1, u_2, u_3 should be distinct pairwise incident elements of Γ of type + or −. However, this is evidently impossible, since distinct incident elements of Γ of type + or − must have opposite types.

Therefore, $\gamma = 1$ and $Gr_0(\Gamma)$ is a (top-thin) affine polar space, by Theorem 7.54. The geometry Γ is the unfolding of this affine polar space. □

7.8 The diagram $L^*.L$

The diagram $Af^*.Af$ (see Section 7.7.2) is a special case of the following:

$$(L^*.L) \qquad \overset{+}{\bullet}\overset{L^*}{\rule{2cm}{0.4pt}}\overset{0}{\bullet}\overset{L}{\rule{2cm}{0.4pt}}\overset{-}{\bullet}$$

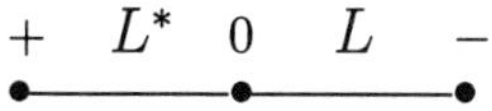

where +, 0, − are the types. This is the diagram we are going to examine now.

7.8.1 Truncated projective geometries

Truncations of projective geometries defined in Section 1.6.1(3) belong to $L^*.L$. More generally, let Γ be a geometry belonging to the following diagram of rank $n \geq 3$, where 0, 1, ..., $n-1$ are the types and $0 < i < n-1$:

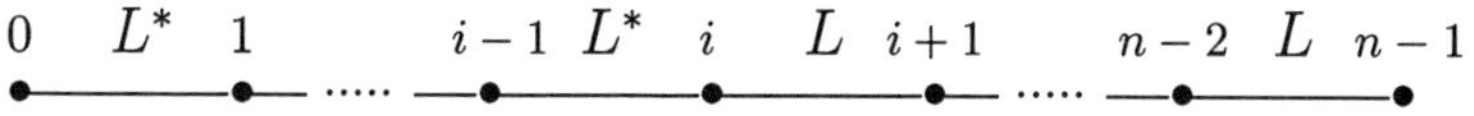

$$\overset{0}{\bullet}\overset{L^*}{\rule{1.5cm}{0.4pt}}\overset{1}{\bullet}\cdots\cdots\overset{i-1}{\bullet}\overset{L^*}{\rule{1.5cm}{0.4pt}}\overset{i}{\bullet}\overset{L}{\rule{1.5cm}{0.4pt}}\overset{i+1}{\bullet}\cdots\cdots\overset{n-2}{\bullet}\overset{L}{\rule{1.5cm}{0.4pt}}\overset{n-1}{\bullet}$$

Then the $\{i-1, i, i+1\}_+$-truncation of Γ belongs to $L^*.L$, by Theorem 7.6 applied to lower residues (or stars) of elements of Γ of type $i+1$ (of type $i-1$). For instance, if Γ is the amalgam of attenuated spaces defined by an m-dimensional subspace of $PG(n, K)$ $(1 < m < n-1)$ (see Section 2.2.1 and Exercise 3.2) and $i = n - m$, then $Tr^+_{\{i-1,i,i+1\}}(\Gamma)$ belongs to $L^*.L$ and residues in $Tr^+_{\{i-1,i,i+1\}}(\Gamma)$ of elements of type $i-1$ or $i+1$ are affine spaces.

Truncations of projective geometries can be characterized among geometries belonging to $L^*.L$, as follows:

Theorem 7.58 (Sprague [197]) *Let Γ be a locally finite geometry belonging to $L^*.L$. Then Γ is a truncation of a (possibly degenerate) projective geometry if and only if* **IP** *holds in it.*

Outline of the Proof. We omit the details of the proof. We only mention its two main steps. First, exploiting **IP** we can prove that linear spaces arising as residues of elements of Γ of type + or − satisfy the Pasch axiom (Section 7.1.4). Thus, those residues are finite projective spaces and we can recover truncated elements of the supposed underlying projective geometry $\overline{\Gamma}$ as subspaces of those projective spaces. However, subspaces in residues of distinct elements of Γ may represent the same element of $\overline{\Gamma}$ (provided that the projective geometry $\overline{\Gamma}$ exists, of course). Thus, the second step of the proof is to identify subspaces of residues of distinct elements of Γ in some consistent way, so as to produce the projective geometry $\overline{\Gamma}$ by them. This can be done, exploiting **IP** once more.

Note that, if we drop the assumption that Γ is locally finite, then it might happen that residues of elements of type $+$ and $-$ are infinite dimensional projective geometries, for each of the types $+$ and $-$. We would get stuck in that case. □

The proof by Sprague [197] can be modified a bit so as to generalize Theorem 7.58 to the following diagram of rank $n \geq 3$:

$(L^*.A_{n-2}.L)$

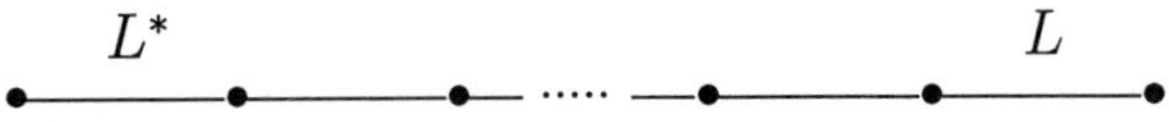

Note that, when $n \geq 3$, there is no need of **IP** to obtain the Pasch axiom in residues of points or dual points; Corollary 7.11 is sufficent for that. However, we still need **IP** to generalize the second part of the proof of Theorem 7.59.

Exercise 7.24. Let Π^+, Π^- be finite thick-lined partial planes with the same set of points P and sets of lines L^+ and L^- respectively. Assume that $|X \cap Y| = 0$ or 2 for every $X \in L^+$ and every $Y \in L^-$. Define a rank 3 geometry $\Gamma(\Pi^+, \Pi^-)$ with types $+$, 0, $-$, where L^+, P, L^- are the sets of elements of type $+$, 0, $-$ respectively, the incidence relation between elements of P and elements of $L^+ \cup L^-$ is the one inherithed from Π^+ and Π^- and elements $X \in L^+, Y \in L^-$ are declared to be incident if $X \cap Y \neq \emptyset$. Check that $\Gamma(\Pi^+, \Pi^-)$ belongs to the following special case of $L^*.L$:

$(c^*.c)$

Exercise 7.25. Let Γ belong to the diagram $c^*.c$ of the the previous exercise and let $\Pi^+ = Tr^+_{\{+,0\}}(\Gamma)$ and $\Pi^- = Tr^-_{\{0,-\}}(\Gamma)$. Prove that Π^+ and Π^- are partial planes as in the previous exercise and that $\Gamma = \Gamma(\Pi^+, \Pi^-)$.

(Hint: use Exercise 4.5.)

(Remark: by Theorem 7.58, **IP** holds in Γ if and only if Γ is a truncation of a simplex.)

Exercise 7.26. Let Γ belong to the following diagram:

$(c.C_2)$

Prove that $\Gamma = Gr_0(\Gamma(\Pi^+, \Pi^-))$ for suitable partial planes Π^+, Π^-, as in Exercise 7.24, if and only if Γ has a good system of lines and the set of planes of Γ can be partitioned in two classes in such a way that every line of Γ is incident to just one plane in each of those two classes.

Exercise 7.27. Let Π^+ be the dual affine plane obtained deleting $(0,0,1)$ and its star from $PG(2,q)$ (q even). Let L^- be the set of conics represented by equations of the following form:

$$ax^2 + by^2 + xy + z^2 = 0$$

and let Π^- be the geometry of rank 2 with the same points as Π^+ and L^- as set of lines. Check that Π^- and Π^- are as in the hypotheses of Exercise 7.24 and that **IP** holds in $\Gamma(\Pi^+, \Pi^-)$ only if $q = 2$.

(Remark: when $q = 2$, then $\Gamma(\Pi^+, \Pi^-)$ is the tetrahedron.)

Exercise 7.28. Let $\mathcal{Q} = \mathcal{Q}_3^-(q)$, defined by a symmetric bilinear form f in $V(4,q)$, $q \equiv 3$ $(mod\ 4)$. The set of points of $PG(3,q)$ not in $\mathcal{Q}$ and represented by vectors of $V(4,q)$ of square (non-square) norm with respect to f will be denoted by P^+ (by P^-, respectively). If X is a line of $PG(3,q)$ tangent to $\mathcal{Q}$, then $X - (\mathcal{Q} \cap X)$ is contained either in P^+ or in P^-. Let $\mathcal{L}$ be the set of lines X of $PG(3,q)$ tangent to $\mathcal{Q}$ and such that $X - (X \cap Q) \subseteq P^+$. Let $\mathcal{P}$ be the set of planes of $PG(3,q)$ of the form $x^\perp$ with $x \in P^-$. Check that P^+, $\mathcal{L}$ and $\mathcal{P}$ form a geometry of rank 3 with respect to the incidence relation inherited from $PG(3,q)$, that this geometry belongs to the diagram $c^*.c$ of Exercise 7.24 and that **IP** holds in it only if $q = 3$.

(Remark: when $q = 3$, then the above geometry is isomorphic to the system of the elements of dimension 1, 2 and 3 of the 5-dimensional simplex.)

Exercise 7.29. Let $\mathcal{Q} = \mathcal{Q}_3^+(q)$, naturally embedded in $\mathcal{P} = PG(3,q)$, $q = 2^h$, $h \geq 2$. Define a geometry Γ over the set of types $\{+, 0, -\}$, as follows. The points (planes) of $\mathcal{P}$ external (secant) to $\mathcal{Q}$ are the elements of Γ of type $+$ (respectively, $-$). The elements of type 0 are the lines of $\mathcal{P}$ external to $\mathcal{Q}$. The incidence relation is the natural one, inherited from $\mathcal{P}$, except that we also require $a^\perp \neq u$ for a point a and a plane u to be incident in Γ. Check that Γ is a geometry of rank 3, with diagram $L^*.L$ and order $(q, q/2 - 1, q)$ and that **IP** fails to hold in Γ.

Exercise 7.30. Let $\mathcal{H} = \mathcal{H}_3(q^2)$ naturally embedded in $\mathcal{P} = PG(3,q^2)$ and let $\mathcal{L}$ be the set of lines of $\mathcal{P}$ tangent to $\mathcal{H}$. Define a geometry Γ over the set of types $\{+, 0, -\}$, as follows. The points (planes) of $\mathcal{P}$ external (secant) to $\mathcal{H}$ are the elements of Γ of type $+$ (respectively, $-$). The elements of type 0 are the lines of $\mathcal{P}$ tangent to $\mathcal{H}$. The incidence relation is the natural one, inherited from $\mathcal{P}$. Check that Γ is a geometry of rank 3, with diagram $L^*.L$ and order $(q^2 - 1, q, q^2 - 1)$ and that **IP** fails to hold in Γ.

7.8.2 Back to $Af^*.Af$

By Theorem 7.57, only two examples exist for $Af^*.Ast$ $(= Af.D_2)$ when we assume finite orders q, $q-1$, q (q is a prime power, by Theorems 7.57 and 1.6):

$$(Af^*.Af) \qquad \underset{q}{\bullet}\overset{Af^*}{\text{———}}\underset{q-1}{\bullet}\overset{Af}{\text{———}}\underset{q}{\bullet}$$

One of these examples is the unfolding of the affine polar space obtained by removing a tangent hyperplane from $\mathcal{Q}_5^+(q)$. We call this geometry $\Gamma_{tg}(q)$. The geometry $\Gamma_{tg}(q)$ is also an amalgam of attenuated spaces (Exercise 5.15).

The other one is the unfolding of the affine polar space obtained by removing a secant hyperplane from $\mathcal{Q}_5^+(q)$ (Exercise 5.16). We call it $\Gamma_{sec}(q)$. The reader may check that every secant hyperplane of $\mathcal{Q}_5^+(q)$ is isomorphic to $\mathcal{Q}_4(q)$ and that $\Gamma_{sec}(q)$ is uniquely determined, up to isomorphisms.

It follows from Theorem 7.58 that **IP** fails to hold in $\Gamma_{tg}(q)$ and $\Gamma_{sec}(q)$ for every $q \geq 3$. On the other hand, $\Gamma_{sec}(2)$ is isomorphic to the system of elements of dimension 1, 2 and 3 of a 5-dimensional simplex (see below, Exercise 7.31) and $\Gamma_{sec}(q) \not\cong \Gamma_{tg}(q)$ for every $q \geq 2$. Therefore **IP** holds in $\Gamma_{sec}(2)$ but it fails to hold in $\Gamma_{tg}(2)$. However, both $\Gamma_{tg}(q)$ and $\Gamma_{sec}(q)$ have good systems of lines, for every $q \geq 2$ (Exercise 4.5).

We have above remarked that $\Gamma_{tg}(q)$ is also an amalgam of attenuated spaces. Considering this, it is not difficult to prove that the following hold in $\Gamma_{tg}(q)$:

(i) given three distinct planes u, v, w, if $|\sigma_0(u) \cap \sigma_0(v) \cap \sigma_0(w)| \geq 2$, then $\sigma_0(w) \supseteq \sigma_0(u) \cap \sigma_0(v)$;

(ii) given a line x and a point $a \notin \sigma_0(x)$, there is just one plane incident to both x and a.

The reader may check that (ii) never holds in $\Gamma_{sec}(q)$, whereas (i) holds in $\Gamma_{sec}(q)$ only when $q = 2$. Actually, (i) and (ii) nearly give us a characterization of $\Gamma_{tg}(q)$:

Proposition 7.59 ([56]) *Let Γ be a geometry belonging to $L^*.L$, with finite orders q, s, q:*

$$(L^*.L) \qquad \underset{q}{\bullet}\overset{L^*}{\text{———}}\underset{s}{\bullet}\overset{L}{\text{———}}\underset{q}{\bullet}$$

Let the above conditions (i) and (ii) hold in Γ. Then either $\Gamma = \Gamma_{tg}(q)$ or $q = (s+1)^3 + (s+1)$.

The proof is a computation with numbers of elements, as in the proofs of Theorems 7.18 and 7.21. Note that once again we encounter the strange relation $q = (s+1)^3 + (s+1)$ (see Theorem 7.21(iv)).

Exercise 7.31. Let $\mathcal{S}$ be the system of elements of dimension 1, 2 and 3 of a 5-dimensional simplex. Prove that $\Gamma_{sec}(2) \cong \mathcal{S}$.

(Hint. $\mathcal{S}$ belongs to $Af^*.Af$ with orders 2, 1, 2. Hence either $\mathcal{S} \cong \Gamma_{tg}(2)$ or $\mathcal{S} \cong \Gamma_{sec}(2)$...)

Exercise 7.32. Let Γ be a geometry belonging to the following diagram:

$$\overset{(g,d_P,d_L)}{\bullet\!\!-\!\!-\!\!-\!\!-\!\!\bullet}\overset{L}{-\!\!-\!\!-\!\!-\!\!\bullet}$$

where $3 \leq g$ and $d_P \leq 4$ (notation as in Section 3.1.2). Prove that, if the above property (ii) holds in Γ, then the triangular property holds in Γ.

(Hint: the only non-trivial step is to prove that Γ has a good system of planes. To this aim, note that, if Π is a partial plane with $d_P \leq 4$, then $a^\perp \cap x \neq \emptyset$ for every point a and every line x of Π.)

Exercise 7.33. Let Γ be as in the previous exercise and let (ii) hold in it. Prove that **IP** holds in Γ if and only if Γ is a matroid (if and only if $g = d_P = 3$).

(Hint: if Γ is not a matroid, then there are points a, b in a plane u at distance 2 in Γ_u. Given a line x on b not in Γ_u, take the plane $v * a, x$ (uniquely determined by (ii)). We have $u \neq v$ ...)

Exercise 7.34. Let Γ be the point–line–plane system of the amalgam of attenuated spaces defined by a line of $PG(m, K)$ ($m \geq 3$; see Section 2.2.1). Check that both (i) and (ii) hold in Γ.

(Remark: if we assume finite orders and dual affine planes, then (ii) and a stronger version of (i) nearly characterize point–line–plane systems of amalgams of attenuated spaces as above (see [56]); we say 'nearly' because of the case $q = (s+1)^3 + (s+1)$ of Proposition 7.59.)

Exercise 7.35. Let Γ be the affine expansion of a linear space $\mathcal{L}$ of order (s, t) realized inside $PG(m, q)$. Check that (i) always holds in Γ and prove that (ii) holds in Γ if and only if $t = (q^m - 1)/(q - 1)$.

Problem. Which triplets of positive integers (x, y, z) with $y \leq x, z$ can be chosen as feasible orders for a geometry belonging to $L^*.L$? In particular, is the case of $q = (s+1)^3 + (s+1)$ of Proposition 7.59 really possible ?.

7.8.3 Pseudo-affine geometries

Recalling that $\mathcal{Q}_5^+(q)$ is the 1-Grassmann geometry of $PG(3,q)$, we see that the construction of $\Gamma_{sec}(q)$ and $\Gamma_{tg}(q)$ (see the previous section) can be generalized as follows.

Let $\mathcal{L}_{m,n}(K)$ and $\mathcal{G}_{m,n}(K)$ be the m-Grassmann point–line system and the m-Grassmann geometry of $PG(n,K)$ ($n \geq 2$, $0 \leq m \leq n-1$; we take dimensions as types for the elements of $PG(n,K)$). **IP** holds in $\mathcal{G}_{m,n}(K)$, as it holds in $PG(n,K)$. Hence the elements of $\mathcal{G}_{m,n}(K)$ can be viewed as subspaces of $\mathcal{L}_{m,n}(K)$ (however, when $1 \leq m < n-2$, not all subspaces of $\mathcal{L}_{m,n}(K)$ represent elements of $\mathcal{G}_{m,n}(K)$). We say that a proper subspace H of $\mathcal{L}_{m,n}(K)$ is a *hyperplane* if every line of $\mathcal{L}_{m,n}(K)$ not contained in H meets H in a point. Let us remove the elements of $PG(n,K)$ that, as elements of $\mathcal{G}_{m,n}$, are contained in H. What is left is a geometry of rank n, which we denote by $AG_H^{m,n}(K)$. Trivially, if $m = 0$ (or $n-1$), then $AG_H^{m,n}(K)$ is an affine geometry (a dual affine geometry).

Let $1 \leq m < n-1$. Then $AG_H^{m,n}(K)$ belongs to the following diagram of rank n:

$$\underset{0}{\bullet}\!-\!\!-\!\!\underset{1}{\bullet}\!-\cdots-\!\underset{m-2}{\bullet}\!-\!\!-\!\!\underset{m-1}{\bullet}\overset{Af^*}{-\!\!-}\underset{m}{\bullet}\overset{Af}{-\!\!-}\underset{m+1}{\bullet}\!-\!\!-\!\!\underset{m+2}{\bullet}\!-\cdots-\!\underset{n-2}{\bullet}\!-\!\!-\!\!\underset{n-1}{\bullet}$$

(we have written types below the nodes of the diagram, this time). We denote this diagram by $Af^*_{m+1}.Af_{n-m}$ and we call $AG_H^{m,n}(K)$ a *pseudo-affine geometry* over K. Note that, as $Af^*_{m+1}.Af_{n-m}$ contains $Af^*.Af$ as a subdiagram, **IP** fails to hold in pseudo-affine geometries, except possibly when $K = GF(2)$ (compare Theorem 5.58; note that **IP** holds in $\Gamma_{sec}(2)$). (Pseudo-affine geometries are given other names in the literature: affine Grassmannians, or symplectic geometries; however, those names might be a bit misleading: $AG_H^{m,n}(K)$ is the unfolding of a Grassmann geometry rather than a Grassmann geometry and it is not a symplectic variety, even if it is sometimes related to symplectic varieties (see [192]; see also Exercise 5.16).)

For instance, let U be a proper nonempty subspace of $PG(n,K)$ and let $m = n-u-1$, with $u = dim(U)$. It is not difficult to check that the set of m-dimensional subspaces of $PG(n,K)$ meeting U in some points is a hyperplane of $\mathcal{L}_{m,n}(K)$. If H is that hyperplane, then $AG_H^{m,n}(K)$ is just an amalgam of attenuated spaces (Section 2.2.1).

If the division ring K is non-commutative, then all pseudo-affine geometries over K are amalgams of attenuated spaces [74]. When K is a field, then hyperplanes of $\mathcal{L}_{m,n}(K)$ can also be obtained from possibly degenerate multilinear alternating forms (see [192] or [47] for the details; we only mention one example here: the set of maximal subspaces of the symplectic variety $\mathcal{S}_{2n-1}(K)$ is a hyperplane of $\mathcal{L}_{n-1,2n-1}(K)$; compare Exercise 5.16 and the construction of $\Gamma_{sec}(q)$).

Clearly, the $\{m+2, m+3, ..., n-1\}_-$-truncation of a pseudo-affine geometry $AG_H^{m,n}(K)$ belongs to the following diagram of rank $m+2$

$(Af^*_{m+1}.L)$ 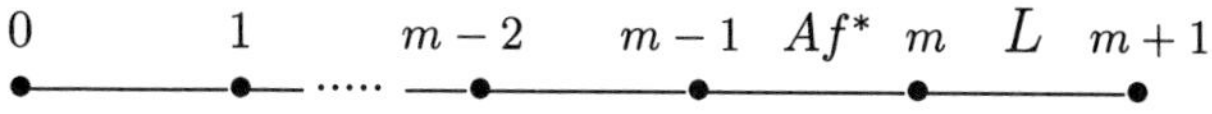

In particular, when $m = 1$ we get the following:

$(Af^*.L)$ 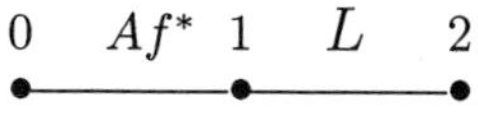

Theorem 7.60 (Cuypers [47], [48]) *All geometries belonging to the diagram* $Af^*_{m+1}.L$ ($m \geq 2$) *and with at least five points on a line are truncations of pseudo-affine geometries.*

Every geometry belonging to $Af^*.L$ *with at least four points on a line is a truncation of a pseudo-affine geometry.*

Therefore, all geometries belonging to the diagram $Af^*_{m+1}.Af_{n-m}$, with at least five points on a line when $m \geq 2$ and with at least four points on a line when $m = 1$, are pseudo-affine geometries.

Geometries belonging to $Af^*.L$ with just three points on every line have been classified by Hall [72], [73]. Not all of them are pseudo-affine geometries. The reader should consult [72] for a detailed description of these examples. We mention only the most obvious of them: the $\{1, 2, 3\}_+$-truncation of an n-dimensional simplex ($n \geq 4$).

Comparing Theorems 7.60 and 7.18 (and using Corollary 7.11), we easily see that that all locally finite geometries belonging to the following diagram of rank $n \geq 4$ (again, we write types below the nodes of the diagram):

$(L^*_{n-1}.L)$ 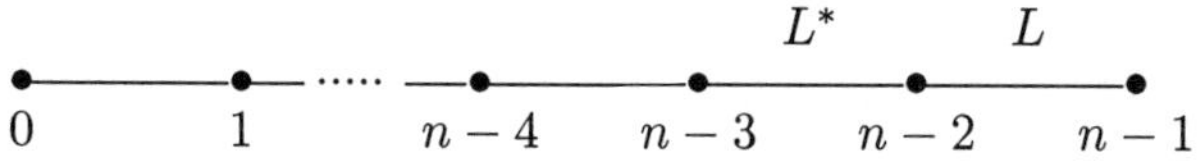

are truncations of projective geometries or of pseudo-affine geometries, except possibly when lines have less than five points or when $n = 4$. When $n = 4$, different examples also exist: for instance, the dual of $S(M_{23})$ and the dual of the $\{0\}_-$-truncation of $S(M_{24})$ (Section 7.2.2, Remark) belong to the above diagram, with orders $(4, 4, 1, 1)$ and $(4, 4, 1, 2)$, respectively:

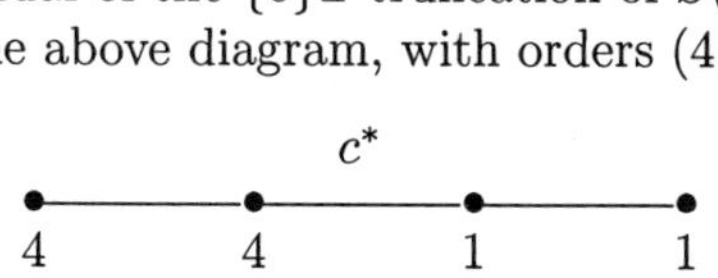

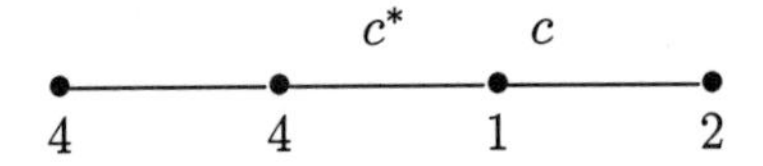

We do not know if these are the only exceptional examples in the case of rank 4. Similarly, all locally finite geometries of rank $n \geq 6$ belonging to the following generalization of $Af^*_{m+1}.Af_{n-m}$:

L^* L

are either projective geometries or pseudo-affine geometries, except possibly when lines have less that 5 points.

7.9 $L.A_{n-2}.L^*$ and $L.A_{n-2}.N^*$

The following diagrams of rank $n \geq 3$ will be denoted by $L.A_{n-2}.L^*$ and $L.A_{n-2}.N^*$, respectively:

$(L.A_{n-2}.L^*)$ L L^*

$(L.A_{n-2}.N^*)$ L N^*

(we recall that the label N^* denotes the class of dual nets). When $n = 3$ we write $L.L^*$ and $L.N^*$ for $L.A_1.L^*$ and $L.A_1.N^*$ respectively.

The above diagrams are special cases of $L_{n-1}.\pi$ (Section 7.3). If we assume finite orders and $n \geq 4$, then we can use Theorems 7.18 and 7.21 to gain control over the class of linear spaces represented by the label L in the above diagram; projective and affine planes are the 'standard' examples here. There are only a few things that might appear besides them in this context, but only when $n = 4$.

On the other hand, if finite orders are not assumed, then the information given us by the label L is too weak for us to say anything definite. Thus, when we do not assume finite orders we need to restrict ourselves to some well-known subclass of the class of linear spaces. Of course, projective and affine planes are the first classes to consider here.

Thus, we will focus on the following special cases ($Af.A_{n-2}.Af^*$ and $A_{n-1}.N^*$) of $L.A_{n-2}.L^*$ and $L.A_{n-2}.N^*$:

$(Af.A_{n-2}.Af^*)$ Af Af^*

$(A_{n-1}.N^*)$ ●——●— ····· —●——●——● (N^*)

7.9.1 The diagram $L.A_{n-2}.L^*$

Bi-affine geometries (Section 2.2.1) belong to the diagram $Af.A_{n-2}.Af^*$. We have remarked that **IP** holds in them (Sections 6.3.4 and 6.7.1). By Corollary 7.27, **IP** is now equivalent to the property of having a good system of lines. Bi-affine geometries are in fact characterized by this property (Lefevre-Percsy [116], Van Nypelseer [233]):

Theorem 7.61 *A geometry belonging to $Af.A_{n-2}.Af^*$ is a bi-affine geometry if and only if it has a good system of lines.*

Examples are known belonging to $Af.A_{n-2}.Af^*$ that do not satisfy **IP**, obtained as proper homomorphic images of bi-affine geometries (see the next chapter, Exercises 8.59–8.61). We do not know if all $Af.A_{n-2}.Af^*$ geometries not satisfying **IP** can be obtained in that way.

We now turn to $L.A_{n-2}.L^*$, but we assume finite orders s, q, ..., q, t:

$(L.A_{n-2}.L^*)$ ●——●——●— ····· —●——●——● (labels L, L^*; orders s, q, q, q, q, t)

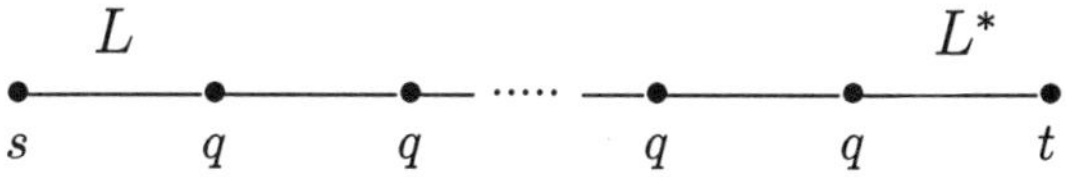

We assume $q > s, t$ in order to keep this diagram distinct from $L.A_{n-1}$ and its dual. We also assume $s \leq t$ so as not to duplicate statements for dual cases.

If $q = t+1 = s+1$, then we are led back to $Af.A_{n-2}.Af^*$. By Theorem 7.21, we have $q \geq s+2$ only if $n \leq 4$.

Many examples are known with $n = 3$ and $q > t \geq s$. For instance: enrichments of biplanes [91] or semibiplanes [86]; truncations of buildings of type D_n as in Section 1.6.1(5). More examples are described in [89] and [165] (Section 8.4). On the other hand, only one example is known with $n = 4$, $q > t \geq s$ and $q \geq s+2$. It can be constructed from the graph for the Higman–Sims sporadic simple group HS. We denote this geometry by $\Gamma(HS)$, just to have a name for it.

We are not going to describe the construction of $\Gamma(HS)$ here. The reader may see [87] for it. We only give some information on this geometry. The system of points and dual points $\Gamma(HS)$ admits a double enrichment (see Exercise 2.10 for the definition of double enrichments) and $\Gamma(HS)$ is the double enrichment of its system of points and dual points. Thus, lines (planes) of $\Gamma(HS)$ are just pairs of points (pairs of dual points). Therefore

$\Gamma(HS)$ has a good system of lines (also a good dual system of lines). Hence **IP** holds in it, by Corollary 7.27.

Residues of dual points and stars of points of $\Gamma(HS)$ are isomorphic to the enrichment $S(M_{22})$ of the Steiner system for the Mathieu group M_{22} and to the dual of $S(M_{22})$, respectively. It is clear from this that $\Gamma(HS)$ has orders 1, 4, 4, 1.

The following has been proved by Hughes in [87] and [88], by a combinatorial analysis of collinearity graphs and of dual collinearity graphs:

Lemma 7.62 *The geometry $\Gamma(HS)$ is the only geometry belonging to the diagram $L.A_2.L^*$ with order $(1,4,4,1)$ and with a good system of lines.*

By this lemma, by Theorems 7.61, 7.18 and 7.21 and Corollary 7.9 we obtain the following:

Theorem 7.63 *Let Γ be a geometry belonging to $L.A_{n-2}.L^*$ $(n \geq 4)$, with a good system of lines and finite orders s, q, ..., q, t, with $q > t \geq s$. Then Γ is one of the following:*

(i) a bi-affine geometry;
(ii) the geometry $\Gamma(HS)$;
(iii) an (unknown) geometry with $n = 4$ and $q = t + 1 = (s + 1)^2$;
(iv) an (unknown) geometry with $n = 4$, $q = (t + 1)^2$ and $s = t \geq 2$;
(v) an (unknown) geometry with $n = 4$, $q = (t + 1)^3 + (t + 1)$ and $s = t$.

One more case might seem to be allowed that is not listed above, namely $q = (t + 1)^2 = (s + 1)^3 + (s + 1)$. However, this case cannot occur, since $(s + 1)^3 + (s + 1)$ is never a square.

Problem. Prove that cases (iii)–(v) of Theorem 7.63 are impossible.

Problem. What can we say if we drop the hypothesis that Γ has a good system of lines ? Note that, even without that hypothesis, we can still obtain that q, s, t are as in one of the following: $q = t + 1 = s + 1$; or $q = (t + 1)^2$ and $s = t$; or as in (iv) or (v) of the previous theorem. Our question is for a completion of Theorem 7.61, dropping the assumption of a good system of lines (possibly assuming finite orders, if it can help); and for a proof of Lemma 7.62 with no use of **IP**.

7.9.2 The diagram $L.A_{n-2}.N^*$

We first consider $A_{n-1}.N^*$. Dual attenuated spaces (Section 2.2.1) belong to this diagram and satisfy **IP** (Sections 6.3.4 and 6.7.1). By Lemma 7.25,

IP is now equivalent to the property of having good residual systems of lines. Actually, this property characterizes dual attenuated spaces:

Theorem 7.64 (Sprague [195], [196]) *A thick-lined geometry belonging to the diagram $A_{n-1}.N^*$ is a dual attenuated space if and only if it has good residual systems of lines.*

We remark that, if a geometry Γ belonging to $A_{n-1}.N^*$ has some thin lines, then the label N^* forces residues of dual points to be thin and we obtain the thin-lined case of C_n (see below, Exercise 7.36). In this case, the property of having good residual systems of lines characterizes thin-lined polar spaces (Corollary 7.39).

We now turn to $L.A_{n-2}.N^*$, but we assume finite orders s, q, ..., q, t:

$(L.A_{n-2}.N^*)$

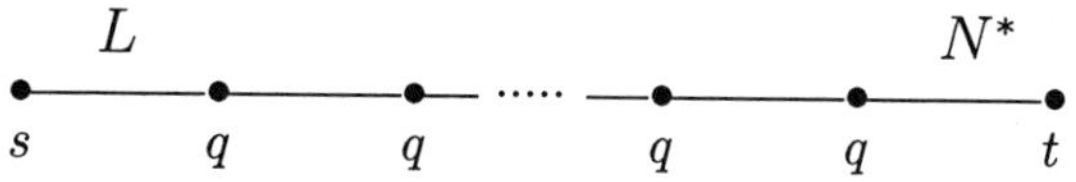

We assume $t \neq q-1$ in order to keep this diagram distinct from $L.A_{n-2}.Af^*$ (hence from $L.A_{n-2}.L^*$). If $n \geq 5$, then $q = s$ or $s+1$, by Theorem 7.18.

When $q = s$, we are led back to $A_{n-1}.N^*$. When $q = s+1$ the diagram is $Af.A_{n-2}.N^*$:

$(Af.A_{n-2}.N^*)$

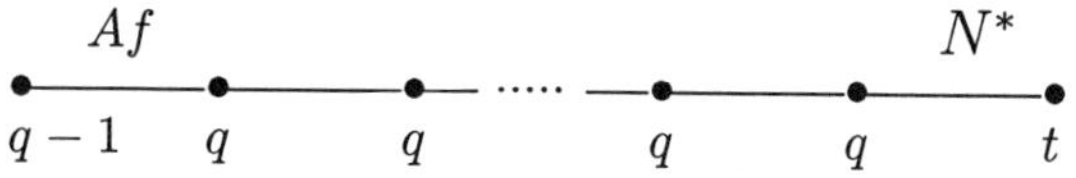

The following result has been proved in [59] for this diagram:

Proposition 7.65 *Let Γ be a geometry belonging to $Af.A_{n-2}.N^*$ ($n \geq 4$), with finite orders $q-1$, q, ..., q, t and $q \geq 4$ and $q \neq 1, t+1$. Then Γ is an affine dual attenuated space if and only if it has good residual systems of lines.*

Many examples are known for the rank 3 case $L.N^*$ of $L.A_{n-2}.N^*$. We only mention enrichments of Laguerre planes besides dual attenuated spaces and affine dual attenuated spaces of rank 3. They belong to the following special case of $L.N^*$:

$(c.N^*)$

The reader may see [82] for more examples.

Exercise 7.36. Let Γ be a geometry belonging to $A_{n-1}.N^*$, with some thin lines. Prove that Γ belongs to the thin-lined case of C_n.

(Hint: some of the dual nets appearing as stars of dual planes of Γ must have some thin lines, by Section 3.2.3(1). By the definition of dual nets, all lines of a dual net are thin if some of its lines are thin ...)

Exercise 7.37 (Laguerre planes). Prove that enrichments of finite Laguerre planes can be characterized as geometries belonging to $c.N^*$ with finite orders 1, q, q, with a good system of lines and satisfying the triangular property.

(Hint: assume the triangular property and finite orders 1, q, q. First check that, if a and u are a point and a plane respectively such that $a \notin u$ but $a^\perp \cap u \neq \emptyset$, then a is collinear with all but one points of u. Then prove that, given a point a and a plane u, we always have $a^\perp \cap u \neq \emptyset$. Use the above to prove that the connected components of the non-collinearity graph are cliques of the same size.)

(Remark: stars of points of enrichments of Laguerre planes are both dual nets and nets. Hence enrichments of Laguerre planes also belong to the diagram $c.N$:

$$(c.N) \qquad \bullet \overset{c}{\text{———}} \bullet \overset{N}{\text{———}} \bullet$$

The previous characterization of enrichments of finite Laguerre planes could be restated for this diagram, too.)

Exercise 7.38 (Minkowski planes). Prove that enrichments of finite Minkowski planes with blocks of size ≥ 5 can be characterized as geometries belonging to the above diagram $c.N$, with finite orders 1, q, $q-1$ ($q \geq 3$), with a good system of lines and satisfying the triangular property.

(Hint: assume the triangular property and finite orders 1, $q+1$, q, $q \geq 2$. First check that, if a and u are a point and a plane respectively such that $a \notin u$ but $a^\perp \cap u \neq \emptyset$, then a is collinear with all but two points of u. Then prove that, for every point a and every plane u, $a^\perp \cap u \neq \emptyset$. Use the above to prove that the non-collinearity graph is the collinearity graph of a grid.)

Exercise 7.39. The diagram $L.N$ with orders 1, 2, 1 is an instance of $Af.C_2$. By Theorem 7.54, there are two geometries belonging to this diagram, with orders 1, 2, 1, with a good system of lines and satisfying the triangular property, namely the two affine polar spaces arising from $\mathcal{Q}_5^+(2)$, of tangent and secant type respectively (Section 7.8.3). Check that the one of tangent type is the enrichment of a Minkoswki plane.

(Remark: it is clear from the above that the restriction $q \geq 3$ is essential for the statement of Exercise 7.38.)

Exercise 7.40. Prove the uniqueness of the Minkowski plane with blocks of size 4.

(Hint: use Exercise 7.39.)

Exercise 7.41. We say that a geometry Γ of rank $n \geq 3$ with string diagram graph satisfies the *residual triangular property* if the triangular property holds in Γ and in the star of every element of Γ of type $i < n-3$.

Let Γ belong to the following diagram of rank $n \geq 3$:

$(c^{n-2}.N)$ ●———●— ····· —●———●———●———● (labels above the last two edges: c, N; orders below: q, $q-1$)

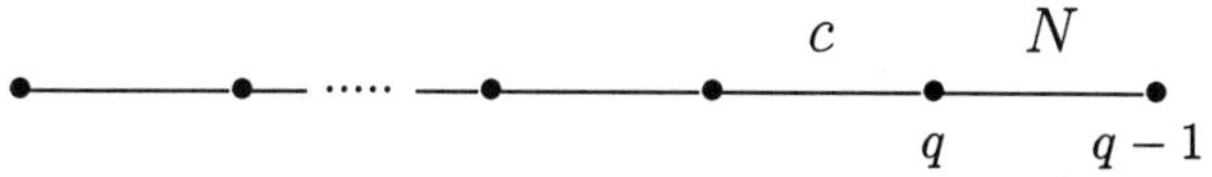

with q, $q-1$ finite orders and $q \geq 3$. Prove that the following are equivalent:

(i) the geometry Γ has good residual systems of lines and satisfies the residual triangular property;
(ii) the geometry Γ is the enrichment of a set of sharply n-transitive permutations of a set of size $n+q-1$, as in Exercise 2.11.

(Hint: to prove that (i) implies (ii) we can work by induction on n, using exercises 7.38 and 2.12 to start the induction.)

Exercise 7.42. Let Γ belong to the following diagram of rank $n \geq 3$:

$(c^{n-2}.N^*)$ ●———●— ····· —●———●———●———● (labels above the last two edges: c, N^*; orders below: q, t)

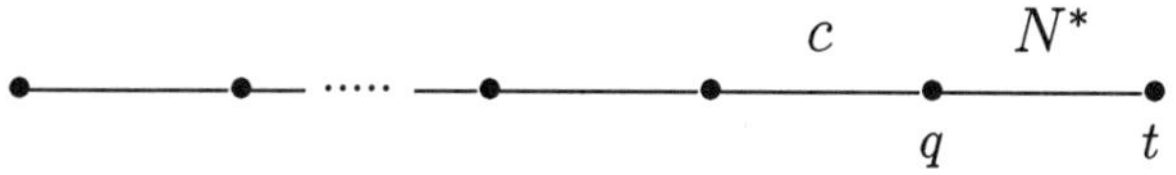

with q, t finite orders. Prove that (i) of Exercise 7.41 holds in Γ if and only if Γ is the enrichment of a set of sharply n-transitive mappings from a set of size $n+q-1$ to a set of size $t+1$, as in Exercise 2.13.

(Remark: using Exercise 7.37, we can obtain the statement of Exercise 2.14 as a special case of the above.)

Exercise 7.43. Let Γ be a thin-lined polar space of rank n admitting finite order t at the last node of the C_n diagram. Prove that Γ is the enrichment of the set of all mappings from a set of size n to a set of size $t+1$.

7.9.3 $N.L$

Let us turn back to Theorem 7.64. We can state it in the form dual to the one we have chosen, substituting $N.A_{n-1}$ for $A_{n-1}.N^*$. We could also state it for the diagram $N.L$:

$(N.L)$ ●———●———● (labels above the edges: N, L)

provided we assume that stars of points are non-degenerate projective spaces of dimension $< n$, for some given positive integer $n \geq 2$: a geometry belonging to $N.L$ with stars of points as above is the point–line–plane system of an attenuated space if and only if **IP** holds in it [196].

Point–line–plane systems of attenuated spaces are only one of the families of geometries belonging to $N.L$. We mention only the following one here: affine expansions of linear spaces of order (r, q) realized in $PG(2, q)$ (these linear spaces are simply complete $r+1$-arcs [81]).

The properties (i) and (ii) of Section 7.8.2 hold in these affine expansions (Exercise 7.35). Actually, these affine expansions can be characterized as geometries belonging to $N.L$, with finite orders $q-1$, r, q and satisfying (i) and (ii) of Section 7.8.2 (see [57]).

The point–line systems of these affine expansions are partial geometries with order $(q-1, (q+1)r)$ and index $\alpha = r$ (see [51]). In particular, when $r = 1$ we obtain the generalized quadrangle $T_2^*(O)$ (Section 2.3).

7.10 Restrictions on labels

All diagrams with finite orders we have considered till now are instances of the following:

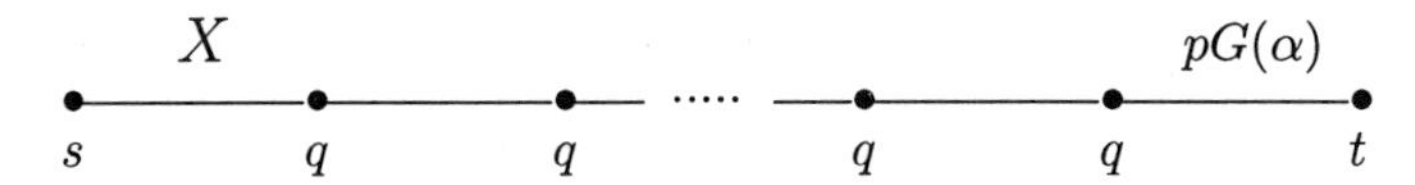

where the label $pG(\alpha)$ denotes the class of partial geometries of index α and X means L or L^*.

However, partial geometries appearing as residues on top are improper in most of the cases we have considered: linear spaces ($\alpha = q+1$); dual linear spaces ($\alpha = t+1$; in particular, dual affine planes); generalized quadrangles ($\alpha = 1$). We have also considered dual nets ($\alpha = q$); these are proper partial geometries, except when $q = t+1$ or $q = 1$ (in these cases we obtain dual affine planes and dual grids, respectively). However, dual nets are certainly not the best representatives of the class of proper partial geometries.

It is natural to ask if some nice class of proper partial geometries can be chosen to label the last stroke of the above diagram. The next theorem shows that the answer is negative if the rank of the diagram is large enough and if the geometry satisfies **IP**.

Theorem 7.66 *Let Γ be a geometry satisfying belonging to the following diagram of rank $n \geq 4$, with finite orders q, ..., q, t and with good residual systems of lines:*

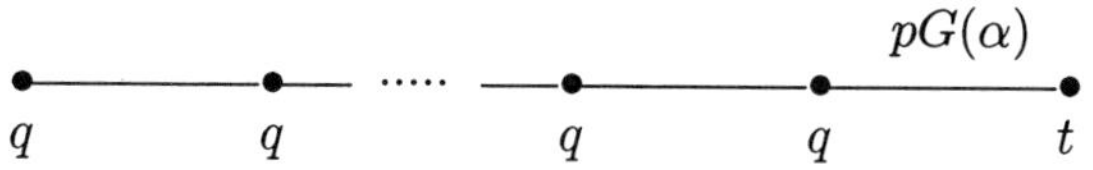

$pG(\alpha)$

q — q ..... q — q — t

Then $\alpha = q + 1$, q *or* 1.

Top residues are linear spaces, dual nets (including dual affine planes) or generalized quadrangles according to whether $\alpha = q + 1$, q or 1. The geometry Γ is a projective geometry or a truncated projective geometry ($\alpha = q + 1$), a dual affine space or a dual attenuated space ($\alpha = q$), or a polar space ($\alpha = 1$), by Corollaries 7.7, 7.11 and 7.9 and Theorems 7.64 and 7.38.

The proof of Theorem 65 is a computation of numbers of elements, as we have done for Lemma 7.16 and Theorems 7.18 and 7.21 (actually, Theorem 7.18 is included in Theorem 65). The reader may find the details of this proof in Section 3 of [58].

Problem. Let us keep the hypotheses of Theorem 7.66, but assume $n = 3$:

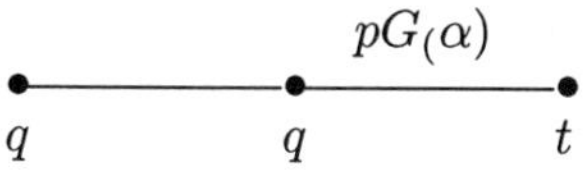

$pG(\alpha)$

q — q — t

Then we can prove that either $\alpha = q + 1$ or α divides q ([58], Proposition 32). Can this result be improved ?

Chapter 8

Morphisms and Quotients

8.1 Morphisms

In this section Γ and Γ' are geometries over sets of types I and I' with type functions t and t', respectively.

A *morphism* $f : \Gamma \longrightarrow \Gamma'$ is a morphism of graphs from the incidence graph of Γ to the incidence graph of Γ'. If $\Gamma' = \Gamma$, then f is said to be an *endomorphism* of Γ.

We say that a morphism $f : \Gamma \longrightarrow \Gamma'$ is *well defined on types* if there is a mapping $\tau_f : I \longrightarrow I'$ such that $\tau_f \cdot t = t' \cdot f$. Trivially, if f is well-defined on types, then the mapping τ_f is uniquely determined by f.

If $I = I'$, if f is well defined on types and if τ_f is the identity mapping on I, then we say that f is *type-preserving* (the term *special* is also used).

Lemma 8.1 *Let $f : \Gamma \longrightarrow \Gamma'$ be a morphism well defined on types and let τ_f be non-injective. Let us set*

$$J = \{i \in I \mid \tau_f(i) = \tau_f(j) \; for some \; j \neq i\}$$

and $J' = \tau_f(J)$. Then f maps $Tr_J^+(\Gamma)$ onto a unique flag F of Γ' of type J'. If furthermore $J \neq I$, then f maps $Tr_J^-(\Gamma)$ into the residue Γ'_F of F.

Proof. Let x, y be distinct incident elements of Γ with $\tau_f(t(x)) = \tau_f(t(y)) = j$, say. Then $f(x) = f(y)$ because $f(x)$ and $f(y)$ are incident and have the same type j. By Theorem 1.18, every element of Γ of type $i \in \tau_f^{-1}(j)$ is mapped onto $f(x)$ $(= f(y))$. □

8.1.1 Chamber morphisms

Lemma 8.2 *The following are equivalent for a morphism $f : \Gamma \longrightarrow \Gamma'$:*

(i) the morphism f maps the chambers of Γ onto chambers of Γ';
(ii) the morphism f is well defined on types and τ_f is surjective.

Proof. Let (i) hold. Then f maps adjacent chambers of Γ onto either the same chamber of Γ' or adjacent chambers of Γ'. By Lemma 1.24, f maps the classes of the type partition of Γ into classes of the type partition of Γ'. Hence (ii) holds.

Conversely, let (ii) hold. Then (i) follows from the surjectivity of τ_f. □

Trivially, if f satisfies (i) of the above lemma, then f induces a morphism from the chamber system $\mathcal{C}(\Gamma)$ of Γ to the chamber system $\mathcal{C}(\Gamma')$ of Γ' (see Section 1.4 for the definition of the chamber system of a geometry). A morphism satisfying (i) (equivalently, (ii)) of Lemma 8.2 will be called a *chamber morphism.*

If Γ and Γ' have the same rank and $f : \Gamma \longrightarrow \Gamma'$ is a chamber morphism, then the mapping $\tau_f : I \longrightarrow I'$ induced by f is a bijection, since it is surjective.

Corollary 8.3 *Let $f : \Gamma \longrightarrow \Gamma'$ be a chamber morphism and let $f(\mathcal{C}(\Gamma)) = \mathcal{C}(\Gamma')$. Then Γ and Γ' have the same rank.*

Trivial, by Lemma 8.1.

8.1.2 Epimorphisms and isomorphisms

Galleries in (chamber systems of) geometries have been defined in Section 1.4. Given a chamber morphism $f : \Gamma \longrightarrow \Gamma'$ and galleries γ and γ' in Γ and Γ' respectively, of the same length, we say that γ' *lifts* to γ through f (also that γ is a *lifting* of γ' through f) if $f(\gamma) = \gamma'$.

Theorem 8.4 *Given a morphism $f : \Gamma \longrightarrow \Gamma'$, the following are equivalent:*

(i) the morphism f is a chamber morphism and $f(\mathcal{C}(\Gamma)_F) = \mathcal{C}(\Gamma')_{f(F)}$ for every flag F of Γ;
(ii) we have $f(\Gamma_F) = \Gamma'_{f(F)}$ for every flag F of Γ;
(iii) for every panel F of Γ, $f(F)$ is a panel of Γ' and $f(\Gamma_F) = \Gamma'_{f(F)}$;
(iv) the morphism f is a chamber morphism and, for every flag F of Γ, every gallery of $\mathcal{C}(\Gamma')_{f(F)}$ lifts through f to a gallery of $\mathcal{C}(\Gamma)_F$.

Proof. Let (i) hold. Then τ_f is bijective, by Corollary 8.3. Let us prove that (ii) holds. Let F be a flag of Γ. If F is a chamber, then $f(F)$ is also a chamber, hence $\Gamma_F = \Gamma_{f(F)} = \emptyset$. The statement of (ii) is trivial in this case. Let F be non-maximal. Every flag of $\Gamma'_{f(F)}$ is contained in a chamber of $\mathcal{C}(\Gamma')_{f(F)}$. Hence f maps the set of flags of Γ_F onto the set of flags of $\Gamma'_{f(F)}$, by (i) and because τ_f is a bijection. Therefore (ii) holds.

Let (ii) hold. By (ii), $\Gamma_F = \emptyset$ if and only if $\Gamma'_{f(F)} = \emptyset$. That is, F is a chamber if and only if $f(F)$ is a chamber. Hence f is a chamber morphism and maps non-maximal flags onto non-maximal flags. If $\tau_f(i) = \tau_f(j)$ for distinct types $i, j \in I$, then a panel of cotype i is mapped onto a chamber, contradicting the above. Therefore τ_f is bijective, Γ and Γ' have the same rank and, given a flag F of Γ, $f(F)$ is a panel if and only if F is a panel. It is now clear that (ii) implies (iii).

Let (iii) hold. As above, f is a chamber morphism and τ_f is bijective. Hence Γ and Γ' have the same rank. We will prove that (iv) holds, by induction on the rank n of Γ and Γ'. If $n = 1$, then there is nothing to prove. Let $n \geq 2$. By the inductive hypothesis, (iv) holds for the restriction of f to Γ_F, for every nonempty flag F. Let $\gamma' = (C'_0, C'_1, ..., C'_m)$ be a gallery in Γ'. As f is a chamber morphism, there is a chamber C_0 of Γ such that $f(C_0) = C'_0$. As τ_f is bijective, $f^{-1}(C'_0 \cap C'_1) \cap C_0$ is a panel in C_0. By (iii), there is some chamber C_1 containing this panel and such that $f(C_1) = C'_1$. We can now repeat the previous argument substituting 1 for 0 and 2 for 1. We obtain a chamber C_2 adjacent to C_1 and such that $f(C_2) = C'_2$. Iterating this process, we eventually obtain a lifting $\gamma = (C_0, C_1, ..., C_m)$ of γ'.

Finally, (iv) implies (i) because (i) is a special case of (iv) (indeed chambers can be viewed as galleries of length 0).

The proof is complete. □

We say that a morphism is *residually surjective* if it satisfies (i) (equivalently, (ii) or (iii) or (iv)) of the previous theorem.

Trivially, a residually surjective morphism $f : \Gamma \longrightarrow \Gamma'$ maps $f(\mathcal{C}(\Gamma))$ onto $\mathcal{C}(\Gamma')$, the set of elements of Γ onto the set of elements of Γ' and the flag-complex $\mathcal{K}(\Gamma)$ of Γ onto the flag-complex $\mathcal{K}(\Gamma')$. The mapping τ_f is bijective and Γ and Γ' have the same rank (Lemma 8.1 and Corollary 8.3).

Given geometries Γ and Γ', if there is a surjective morphism from Γ to Γ' then we say that Γ' is a *homomorphic image* of Γ.

Henceforth we will often use the words *surjective morphism* or *epimorphism* for *residually surjective morphism.* We will use the words *residually surjective* only when we want to be sure to avoid any misunderstanding.

Remark. Surjective morphisms could be defined in a way different from the above, asking only that f maps the set of elements of Γ onto the set of elements of Γ'. Not every morphism that is surjective in this sense is residually surjective (see exercises 8.3-8.7 and 8.20). Actually, this weaker definition of surjectivity might look more natural than the one we have chosen. Furthermore, a morphism f is an epimorphism as defined in category theory if and only if it is surjective in the above weak sense (see Exercise 8.19). Thus, the convention we have previously stated, to write 'surjective' or 'epimorphism' for 'residually surjective', is an abuse. However, morphisms that are surjective in the above weak sense but not residually surjective are quite unusual in diagram geometry. This justifies that abuse.

An *isomorphism* from Γ to Γ' is an isomorphism from the incidence graph of Γ to the incidence graph of Γ'. It is easily seen that a morphism is an isomorphism if and only if it is residually surjective and injective. In particular, isomorphisms are well defined on types. Furthermore, the mapping $\tau_f : I \longrightarrow I'$ induced by an isomorphism $f : \Gamma \longrightarrow \Gamma'$ is a bijection.

A morphism $f : \Gamma \longrightarrow \Gamma'$ is said to be *proper* if it is not an isomorphism.

8.1.3 Exercises

The next exercises are grouped as follows. Exercises 8.1 and 8.2 are just comments on the definition of morphisms. Exercises 8.3–8.7 consider morphisms involving matroids in some way. Exercises 8.8–8.12 deal with rank 2 geometries, focusing on gonalities and diameters. Exercises 8.13–8.18 consider epimorphisms between projective planes and between generalized quadrangles. Exercises 8.19–8.23 deal with epimorphisms and injective morphisms. In the last exercises 8.24–8.30 we choose a 'functorial' point of view.

Exercise 8.1. Given geometries Γ and Γ' over sets of types I and I' respectively and a surjective mapping $\tau : I \longrightarrow I'$, choose a chamber C of Γ' and, for every element x of Γ, let $f(x)$ be the element of C of type $\tau(t(x))$. Check that the mapping f defined in this way is a chamber morphism from Γ to Γ'.

(Remark: if τ is not surjective and if we choose a flag F of Γ' of type $\tau(I)$, then the above gives us a morphism well defined on types. More generally, we may forget about mappings $\tau : I \longrightarrow I'$ and, given an arbitrary nonempty flag F of Γ', we may consider an arbitrary mapping f from the set of elements of Γ to F. Then f is a morphism from Γ to Γ'.)

Exercise 8.2. Let Γ be a geometry over a set of types I with type function

t, let $J \supseteq I$ and $\Gamma' = \oplus_{j \in J} \Gamma_j$. For every element x of Γ, let us pick up an arbitrary element $f(x)$ of $\Gamma_{t(x)}$. Check that the mapping f defined in this way is a morphism from Γ to Γ'.

(Remark: if each of the summands Γ_j ($j \in J$) has rank 1, then the morphism f is well defined on types. If furthermore $J = I$, then f is a chamber morphism.)

Exercise 8.3. Let X be an m-dimensional linear subspace of $V(n, K)$, with $2 \leq m < n$. Let Y be a complement of X and let f be the linear function mapping every vector x onto the projection $f(x) = X \cap (x + Y)$ of x onto X along Y. Check that f induces an endomorphism of $AG(n, K)$, but it is not well defined on types (hence it is not a chamber morphism).

Exercise 8.4. With the notation of the previous exercise, let $\mathcal{A}_m$ be the upper m-truncation of $AG(n, K)$ (see Section 7.1.4). Check that f can also be viewed as a morphism from $\mathcal{A}_m$ to the affine geometry $X = AG(m, K)$, that f maps the set of elements of $\mathcal{A}_m$ onto the set of elements of $X = AG(m, K)$, but f is not well defined on types.

Exercise 8.5. Keeping the notation of Exercise 8.3, let Γ be the geometry obtained from $PG(n-1, K)$ by removing Y together with its lower residue and X together with its star. Check that f defines a morphism from Γ to the projective geometry $PG(m-1, K)$ of all non-trivial proper subspaces of X, that f maps the set of elements of Γ onto the set of elements of $PG(m-1, K)$, but f is not well defined on types.

Exercise 8.6. Let X be a proper nonempty subspace of a matroid $\mathcal{M}$ and let Γ be the geometry obtained from $\mathcal{M}$ by removing X together with its lower residue and all subspaces Y such that $X \cup Y$ spans $\mathcal{M}$. For every element Z of Γ, let $f(Z)$ be the subspace of $\mathcal{M}$ spanned by $X \cup Z$. Check that f is a morphism from Γ to the star of X in $\mathcal{M}$, that f maps the set of elements of Γ onto the star of X, but f is not well defined on types.

Exercise 8.7. Given $\mathcal{M}$ and X as above, let Γ be obtained by deleting X together with its star and all subspaces Y such that $Y \cap X = \emptyset$ and let f be defined by $f(Z) = Z \cap X$, for Z in Γ. Check that f is a morphism from Γ to the lower residue of X in $\mathcal{M}$, that f maps the set of elements of Γ onto the lower residue of X, but f is not well defined on types.

Exercise 8.8. Let Γ be a rank 2 geometry with set of points P and set of lines L, let Γ' be a generalized digon with sets of points P' and set of lines L' and let $f : P \cup L \longrightarrow P' \cup L'$ be a mapping such that $f(P) \subseteq P'$ and $f(L) \subseteq L'$. Check that f is a chamber morphism (compare with Exercise 8.2).

Exercise 8.9. Let Γ be an ordinary n-gon (allowing $n = \infty$) and let Γ' be an ordinary m-gon, where $m < \infty$ and we assume that m is a divisor of n when $n < \infty$. Prove that there are precisely $4m$ epimorphisms from Γ to Γ' ($2m$ of them are type-preserving).

Exercise 8.10. Let Γ be an ordinary m-gon, let Γ' be a rank 2 geometry with gonality $g > m$ and let $f : \Gamma \longrightarrow \Gamma'$ be a chamber morphism. Prove that $f(\Gamma)$ cannot be a geometry.

Exercise 8.11. Let Γ and Γ' be rank 2 geometries with gonalities g and g' respectively and let $f : \Gamma \longrightarrow \Gamma'$ be a non-injective chamber morphism such that, for every element x of Γ, the mapping from Γ_x to $\Gamma'_{f(x)}$ induced by f is injective. Prove that $g' \leq min(g, [d/2])$, where square brackets mean 'integral part of'.

Exercise 8.12. Let Γ and Γ' be rank 2 geometries with diameters d and d' respectively and let $f : \Gamma \longrightarrow \Gamma'$ be an epimorphism. Prove that $d \geq d'$.

Exercise 8.13. We say that an integral domain is a *local ring* if it has just one maximal proper ideal (but we warn the reader that local rings are normally defined in sightly more restrictive ways). Let R be a local ring with maximal proper ideal $P \neq 0$, let R_P be the field of quotients of R and let R/P be the residue field of R. Assume furthermore that R is Noetherian and that the ideal P is principal (for instance, R might be the ring of p-adic integers; in this case P is the ideal generated by the prime p, R_P is the field Q_p of p-adic rational numbers and $R/P = GF(q)$). Prove that there is an epimorphism (that is, a residually surjective morphism) from $PG(2, R_P)$ to $PG(2, R/P)$.

(Hint: under the previous assumptions, R is a principal ideal domain and every proper ideal of R has the form P^n for some positive integer n. Therefore, every point of $PG(2, R_P)$ can be represented by a triple (x, y, z) of homogeneous coordinates such that $x, y, z \in R$ and at most two of x, y, z are in P. If (x, y, z) is such a triple, then $(x + P, y + P, z + P)$ represents a point of $PG(2, R/P)$.)

Exercise 8.14. Let K, K' be fields and $f : PG(2, K) \longrightarrow PG(2, K')$ be a type-preserving epimorphism. Chosen a coordinate system on $PG(2, K)$, let $L = \{(x, x) \mid x \in K \cup \{\infty\}\}$ be the diagonal line with respect to that system of coordinates, let p_∞ be the point at infinity of L and let $O = (0, 0)$ be the origin. We set $L' = f(L)$, $O' = f(O)$, $p'_\infty = f(p_\infty)$, $R = f^{-1}(L' - p'_\infty) \cap L$ and $P = f^{-1}(O') \cap L$. The set $L - p_\infty$ can be identified with K. Thus, R and P can be identified with subsets of K. Prove that the subset R of K is a local ring (in the meaning of Exercise 8.13), that P is the maximal ideal of R, that K is the field of quotients of R and that $K' \cong R/P$.

Exercise 8.15. Let $f : PG(2, K) \longrightarrow PG(2, q)$ be a proper epimorphism. Prove that K is infinite.

Exercise 8.16. Let R be as in Exercise 8.13. Prove that there are epimorphisms from $\mathcal{S}_3(R_P)$ to $\mathcal{S}_3(R/P)$ and from $\mathcal{Q}_4(R_P)$ to $\mathcal{Q}_4(R/P)$.

Exercise 8.17. Let Γ and Γ' be finite degenerate projective planes with $u + 1$ and $v + 1$ points respectively, $u \geq v \geq 3$. Check that there are $v^u - v(v-1)^u$ epimorphisms from Γ to Γ'.

Exercise 8.18. Let Γ and Γ' be finite grids admitting orders $(s, 1)$ and $(t, 1)$ respectively, with $s \geq t$. Check that there are $2[(t+1)^{s+1} - (t+1)t^{s+1}]^2$ epimorphisms from Γ to Γ'.

Exercise 8.19. Let $f : \Gamma \longrightarrow \Gamma'$ be a morphism of geometries. Prove that the following are equivalent:

(i) f maps the set of elements of Γ onto the set of elements of Γ';
(ii) for every geometry Γ'' and for every pair of morphisms h, k from Γ' to Γ'', if $hf = kf$, then $h = k$.

(Hint for (ii)$\Rightarrow$(i): if I' is the set of types of Γ', let Γ'' be a geometry over the set of types I' with totally disconnected diagram graph and let C, D be disjoint chambers of Γ''. Take type-preserving chamber morphisms h and k such that $k(f(\Gamma)) \subseteq h(\Gamma') = C$ and $k(\Gamma' - f(\Gamma)) \subseteq D$.)

Exercise 8.20. Let $f : \Gamma \longrightarrow \Gamma'$ be a morphism of geometries and consider the following conditions:

(i) f maps the set of elements of Γ onto the set of elements of Γ';
(ii) f is a chamber morphism and maps the set of elements of Γ onto the set of elements of Γ';
(iii) f maps the set of edges of the incidence graph of Γ onto the set of edges of the incidence graph of Γ';
(iv) $f(\mathcal{K}(\Gamma)) = \mathcal{K}(\Gamma')$;
(v) f is a chamber morphism and $f(\mathcal{K}(\Gamma)) = \mathcal{K}(\Gamma')$.
(vi) f is a chamber morphism and $f(\mathcal{C}(\Gamma)) = \mathcal{C}(\Gamma')$.
(vii) f is residually surjective.

Prove that the following hold:

(v)$\Rightarrow$(ii)$\Rightarrow$(i),
(v)$\Rightarrow$(iv)$\Rightarrow$(iii)$\Rightarrow$(i),
(vii)$\Rightarrow$(vi)$\Leftrightarrow$(v).

Show by counterexamples that:

(i)$\not\Rightarrow$(ii)$\not\Rightarrow$(iv)$\not\Rightarrow$(v)$\not\Rightarrow$(vii),

(iv)$\not\Rightarrow$(ii)$\not\Rightarrow$(iii)$\not\Rightarrow$(ii),
(i)$\not\Rightarrow$(iii).

Exercise 8.21. Let $f : \Gamma \longrightarrow \Gamma'$ be a morphism of geometries. Prove that the following are equivalent:

(i) f is injective;
(ii) for every geometry Γ'' and every pair of morphisms h, k from Γ'' to Γ, if $fk = fh$, then $h = k$.

(Hint for (ii)$\Rightarrow$(i): choose a rank 1 geometry for Γ''.)

Exercise 8.22. Show by counterexamples that an injective morphism need not be well defined on types.

Exercise 8.23. Let $f : \Gamma \longrightarrow \Gamma'$ be injective and well defined on types. Prove that τ_f is injective.

Exercise 8.24. Let $F_{\mathcal{C}}$ be the functor from the category of geometries and chamber morphisms to the category of graphs and morphisms of graphs, mapping a geometry Γ onto its chamber system $\mathcal{C}(\Gamma)$ and a chamber morphism $f : \Gamma \longrightarrow \Gamma'$ onto the morphism from $\mathcal{C}(\Gamma)$ to $\mathcal{C}(\Gamma')$ induced by f. Prove that, given two geometries Γ and Γ', the functor $F_{\mathcal{C}}$ injectively maps the set of chamber morphisms between Γ and Γ' into the set of morphisms between $\mathcal{C}(\Gamma)$ and $\mathcal{C}(\Gamma')$. Show by counterexamples that this injection need not be a bijection (in the language of category theory: the functor $F_{\mathcal{C}}$ is faithful but not full).

Exercise 8.25. Let Γ_1, Γ_2, ..., Γ_m be geometries over mutually disjoint sets of types. Check that the chamber system $\mathcal{C}(\oplus_{i=1}^m \Gamma_i)$ is the product of the chamber systems $\mathcal{C}(\Gamma_1)$, $\mathcal{C}(\Gamma_2)$, ..., $\mathcal{C}(\Gamma_m)$ in the category of graphs and morphisms of graphs (see Exercise 4.1).

(Remark: however, there is no way to interpret $\oplus_{i=1}^m \Gamma_i$ as a product or a coproduct of Γ_1, Γ_2, ..., Γ_m in the category of geometries and (chamber) morphisms, with respect to Γ_1, Γ_2, ..., Γ_m.)

Exercise 8.26. A morphism from a simplicial complex $\mathcal{K}$ to a simplicial complex $\mathcal{K}'$ is a mapping f from the set of vertices of $\mathcal{K}$ to the set of vertices of $\mathcal{K}'$ such that, for every face X of $\mathcal{K}$, the set $f(X)$ is a face of $\mathcal{K}'$.

Let $F_{\mathcal{K}}$ be the functor from the category of geometries and morphisms of geometries to the category of simplicial complexes and morphisms of simplicial complexes, mapping a geometry Γ onto its flag-complex $\mathcal{K}(\Gamma)$ and a morphism $f : \Gamma \longrightarrow \Gamma'$ onto the morphism induced by f from $\mathcal{K}(\Gamma)$ to $\mathcal{K}(\Gamma')$. Check that $F_{\mathcal{K}}$ bijectively maps the set of morphisms between Γ and Γ' onto the set of morphisms between $\mathcal{K}(\Gamma)$ and $\mathcal{K}(\Gamma')$ (that is, the functor $F_{\mathcal{K}}$ is faithful and full).

Exercise 8.27. Given a finite connected graph $\mathcal{D}$ over a set of vertices I, let Γ and Γ' be geometries over the set of types I with diagram graph $\mathcal{D}$. Given a nonempty subset $J \subseteq I$ let $f : \Gamma \longrightarrow \Gamma'$ be a morphism well defined on types and such that τ_f is an automorphism of the graph $\mathcal{D}$ and $\tau_f(J) = J$. Prove that the morphism f induces a type-preserving morphism $Gr_J(f)$ from $Gr_J(\Gamma)$ to $Gr_J(\Gamma')$.

Exercise 8.28. With the notation of the previous exercise, prove that Gr_J is an epimorphism if and only if f is an epimorphism.

Exercise 8.29. With the notation of Exercise 8.27, prove that $Gr_J(f)$ is an isomorphism if and only if f is an isomorphism.

Exercise 8.30. Let Γ and Γ' be geometries over the same set of types I and let J be a proper subset of I. Check that every type-preserving morphism $f : \Gamma \longrightarrow \Gamma'$ induces a type-preserving morphism from $Tr_{\overline{J}}^{-}(\Gamma)$ to $Tr_{\overline{J}}^{-}(\Gamma')$.

8.2 Quotients

Let $f : \Gamma \longrightarrow \Gamma'$ be an epimorphism (that is, a residually surjective morphism). If I and I' are the sets of types of Γ and Γ' respectively, then $\tau_f : I \longrightarrow I'$ is a bijection, by Corollary 8.3. Thus we can identify I with I' via τ_f. That is, we assume that $I = I'$ and that f is type-preserving.

The sets of elements of Γ and Γ' will be denoted by V and V' respectively. Let $\Phi = \{f^{-1}(x) \mid x \in V'\}$ be the set of fibres of f. We identify the partition Φ of V with the equivalence relation defined by it on V. According to this, given $x \in V$, the class of Φ containing x will be denoted by $[x]\Phi$. Given $F \subseteq V$, we write $[F]\Phi$ for $\{[x]\Phi \mid x \in F\}$. Given $x, y \in V$, we write $x \equiv y$ (Φ) for $f(x) = f(y)$.

The partition Φ is a refinement of the type partition of Γ, since f is type-preserving. Therefore, denoting by t the type function of Γ, we can define the *type* $t(X)$ of a class $X \in \Phi$ as the type of any element of X. By Theorem 8.4(ii), the following holds:

($\mathbf{Q}_1$) Given a flag $F = \{x_1, x_2, ..., x_m\}$ of Γ and a class $X \in \Phi$ of type $t(X) \notin t(F)$, let y_1, y_2, ..., y_m be elements of X such that $\Gamma_{y_i} \cap [x_i]\Phi \neq \emptyset$ for every $i = 1, 2, ..., m$. Then there is an element $y \in X$ such that $y * F$.

By Theorem 8.4(iii) and by the fact that every panel of Γ' is contained in at least two chambers, we also have the following:

($\mathbf{Q}_2$) For every panel F of Γ, its residue Γ_F meets at least two classes of Φ.

Let Γ/Φ be the graph defined as follows: the classes of Φ are the vertices of Γ/Φ and two distinct classes $X, Y \in \Phi$ are adjacent in Γ/Φ if $x * y$ for some elements $x \in X$ and $y \in Y$. By Theorem 8.4(ii), the function mapping every class $X \in \Phi$ onto the element $f(X)$ of Γ' is an isomorphism from Γ/Φ to the incidence graph of Γ'. Therefore Γ/Φ is a geometry. We call this geometry the *quotient* of Γ by Φ.

It is easily seen that the function $p_\Phi : V \longrightarrow \Phi$ mapping $x \in V$ onto $[x]\Phi$ is an epimorphism from Γ to Γ/Φ. We call p_Φ the *projection* of Γ onto Γ/Φ.

We keep I as set of types for Γ/Φ, requiring that the type $t(X)$ of a class $X \in \Phi$ as defined above is the type of the element X of Γ/Φ. The incidence relation of Γ/Φ will still be denoted by $*$. It follows from $\mathbf{Q}_1$ that, given $X, Y \in \Phi$, we have $X * Y$ in Γ/Φ if and only if for every $x \in X$ there is an element $y \in Y$ such that $x * y$.

Quotients can also be defined directly as follows. Let Φ be a refinement of the type partition of Γ satisfying the above properties $\mathbf{Q}_1$ and $\mathbf{Q}_2$. We no longer assume that Φ is the partition of V in fibres of a morphism $f : \Gamma \longrightarrow \Gamma'$. However, we can still define the graph Γ/Φ, just as above. By p_Φ we still denote the function mapping $x \in V$ onto its class $[x]\Phi$. As above, we call the graph Γ/Φ the *quotient* of Γ by Φ and p_Φ is the *projection* of Γ onto Γ/Φ.

Theorem 8.5 *The graph Γ/Φ is a geometry, p_Φ is an epimorphism from Γ to the geometry Γ/Φ and Φ is the set of fibres of p_Φ.*

Proof. Let $n = |I|$ be the rank of Γ and let $\Theta = \{t^{-1}(i) \mid i \in I\}$ be the type partition of Γ. As Φ is a refinement of Θ, the quotient Θ/Φ is an n-partition of the graph Γ/Φ. Furthermore, Γ/Φ is connected, since Γ is connected. We must prove that Γ/Φ is a geometry of rank n with type partition Θ/Φ. We work by induction on n.

Let $n = 1$. Then $\mathbf{Q}_1$ says nothing, whereas $\mathbf{Q}_2$ just says that Φ has at least two classes. In this case Γ/Φ is a trivial graph with at least two vertices. Hence it is a geometry of rank 1 and p_Φ is an epimorphism.

Let $n \geq 2$. Given a class $X \in \Phi$, let $(\Gamma/\Phi)_X$ be the neighbourhood of X in the graph Γ/Φ and, given $x \in X$, let Φ_x be the partition induced by Φ on Γ_x. It is easily seen that, if $n \geq 2$, then the properties $\mathbf{Q}_1$ and $\mathbf{Q}_2$ still hold on Φ_x in Γ_x. Thus, a graph Γ_x/Φ_x can be defined in the same way as Γ/Φ. Using $\mathbf{Q}_1$, it is easy to prove that the graphs $(\Gamma/\Phi)_X$ and Γ_x/Φ_x are isomorphic. However, Γ_x/Φ_x is a geometry, by the inductive hypothesis. Hence $(\Gamma/\Phi)_X \cong \Gamma_x/\Phi_x$ is a geometry. Therefore, Γ/Φ is a geometry. The set I can be taken as set of types for the geometry Γ/Φ, stating that a class $X \in \Phi$ has type $t(x)$, for $x \in X$.

The mapping $p_\Phi : \Gamma \longrightarrow \Gamma/\Phi$ is evidently a type-preserving chamber morphism. By $\mathbf{Q}_1$, the morphism p_Φ satisfies (ii) of Theorem 4.4. Therefore p_Φ is an epimorphism. Trivially, Φ is the set of fibres of p_Φ. $\square$

Thus, homomorphic images and quotients are basically the same things. According to Theorem 8.4, we say that a refinement Φ of the type partition of Γ *defines a quotient* of Γ if it satisfies the properties $\mathbf{Q}_1$ and $\mathbf{Q}_2$. The quotient Γ/Φ is said to be *proper* if p_Φ is a proper morphism, that is if $\Phi \neq \mho$.

Corollary 8.6 *Let Φ be a refinement of the type partition of Γ defining a quotient. Then we have*

$$\mathcal{C}(\Gamma/\Phi) = \{[C]\Phi \mid C \in \mathcal{C}(\Gamma)\}$$
$$\mathcal{K}(\Gamma/\Phi) = \{[F]\Phi \mid F \in \mathcal{K}(\Gamma)\}$$

Proof. Indeed p_Φ, being an epimorphism from Γ to Γ/Φ, induces surjective morphisms from $\mathcal{C}(\Gamma)$ to $\mathcal{C}(\Gamma/\Phi)$ and from $\mathcal{K}(\Gamma)$ to $\mathcal{K}(\Gamma/\Phi)$ (see Exercise 8.20(iv) for this claim concerning flag complexes; see also Exercise 8.26 for the definition of morphisms of simplicial complexes). $\square$

8.3 Covers

In this section Γ and Γ' are geometries over the same set of types I and $n = |I|$ is the rank of Γ and Γ'.

Let m be a positive integer with $m \leq n$. We say that a type-preserving morphism $f : \Gamma \longrightarrow \Gamma'$ is an *m-cover* of Γ' if, for every flag F of Γ of corank m, the morphism f induces an isomorphism from Γ_F to $\Gamma'_{f(F)}$. A quotient Γ/Φ will be called an *m-quotient* if the projection $p_\Phi : \Gamma \longrightarrow \Gamma/\Phi$ is an m-cover.

We will sometimes identify an m-cover $f : \Gamma \longrightarrow \Gamma'$ with its domain Γ, saying that Γ is an *m-cover of* Γ'. Needless to say that this is a linguistic abuse. We will do so only when the m-cover f is uniquely determined by the context or when the particular choice of f is irrelevant for what we want to say.

An m-cover is said to be *proper* if it is not an isomorphism. If there is a proper m-cover $f : \Gamma \longrightarrow \Gamma'$ then we also say that Γ is a *proper* m-cover of Γ'.

An n-cover is just an isomorphism and isomorphisms are m-covers for every $m = 1, 2, ..., n$. An m-cover induces m-covers on residues of flags of corank $h = m+1, m+2, ..., n-1$ and isomorphisms on residues of flags of corank $h < m$. Hence, every m-cover is also a k-cover for every

$k = m-1, m-2, ..., 1$. In particular, all m-covers are 1-covers. By this and by Theorem 8.4(iii) we have the following:

Proposition 8.7 *Every m-cover is an epimorphism.*

Hence, if $f : \Gamma \longrightarrow \Gamma'$ is an m-cover, then every gallery of Γ' lifts through f to some gallery of Γ, by Theorem 8.4(iv). However, we can now say more than this. Indeed:

Proposition 8.8 *Let $f : \Gamma \longrightarrow \Gamma'$ be an m-cover, let C be a chamber of Γ and $C' = f(C)$. Then every gallery of Γ' starting at C' lifts to a unique gallery of Γ starting at C.*

Proof. We already know that the existence claim is true, by (iv) of Theorem 8.4. Let us prove the uniqueness claim.

Let $\gamma = (C_0, C_1, ..., C_h)$ and $\delta = (D_0, D_1, ..., D_k)$ be galleries starting at $C = C_0 = D_0$ and such that $f(\gamma) = f(\delta)$. Since f is also a 1-cover, it induces a bijection from the star of the panel $C_{i-1} \cap C_i$ to the star of the panel $f(C_{i-1} \cap C_i)$, for $i = 1, 2, ..., h$. Therefore $f(C_{i-1}) \neq f(C_i)$ and $f(C_{i-1} \cap C_i) = f(C_{i-1}) \cap f(C_i)$. The same holds for the chambers D_i $(i = 1, 2, ..., k)$. Hence $h = k$, as $f(\gamma) = f(\delta)$. Furthermore, γ and δ have the same type, because f preserves types and $f(\gamma) = f(\delta)$. Therefore $C_0 \cap C_1 = D_0 \cap D_1 = F$, for some panel F of C. As f is a 1-cover, it bijectively maps the star of F onto the star of $f(F)$. Hence $C_1 = D_1$. We can repeat this argument substituting C_1 $(= D_1)$ for C_0 $(= D_0 = C)$. We obtain that $C_2 = D_2$. Continuing by iteration, we eventually obtain $\gamma = \delta$.
□

Let $f : \Gamma \longrightarrow \Gamma'$ be an m-cover, let Φ be the set of fibres of f on the set of elements of Γ, as in Section 8.2, and let

$$\Phi_C = \{f^{-1}(C') \mid C' \in \mathcal{C}(\Gamma')\} = \{[C]\Phi \mid C \in \mathcal{C}(\Gamma)\}$$

be the set of fibres of f on the set $\mathcal{C}(\Gamma)$ of chambers of Γ.

Proposition 8.9 *There is a cardinal number t such that every class of the partition Φ_C of $\mathcal{C}(\Gamma)$ has size t.*

Proof. Let F be a panel of Γ and let C, D be chambers in the cell of F. As f is also a 1-cover, we have $|[C]\Phi| = |[D]\Phi| = |[F]\Phi|$. The proposition now follows from the connectedness of $\mathcal{C}(\Gamma)$. □

Corollary 8.10 *Let t be as in the previous lemma. Then, for every flag F of corank $\leq m$, the class $[F]\Phi$ has size t.*

Proof. Indeed f induces isomorphisms on residues of flags of corank $\leq m$, because it is a k-cover for every $k = 1, 2, ..., m$. □

The common size t of the classes of Φ_C is called the *multiplicity* of f and f is said to be *t-fold*. Trivially, f is an isomorphism if and only if it has multiplicity $t = 1$. If $t = 2$ (or 3 or 4 or...), then we say that f is a *double* m-cover (a *triple* m-cover, or a *quadruple* m-cover or ...). If $m = n-1$ then $|[F]\Phi| = t$ for every nonempty flag F of Γ. In this case all classes of Φ have size t, by Corollary 8.10. In this case we say that the m-quotient Γ/Φ is *t-fold*.

We recall that a topological cover from a simplicial complex $\mathcal{K}$ to a simplicial complex $\mathcal{K}'$ is a morphism of simplicial complexes $f : \mathcal{K} \longrightarrow \mathcal{K}'$ such that, for every vertex x of $\mathcal{K}$, the morphism f induces an isomorphism from the star of x in $\mathcal{K}$ to the star of $f(x)$ in $\mathcal{K}'$ and, for every face X of $\mathcal{K}$, $f(X)$ and X have the same dimension (see Exercise 8.26 for the definition of morphisms of simplicial complexes). It is easily seen that a morphism $f : \Gamma \longrightarrow \Gamma'$ is an $(n-1)$-cover if and only if the morphism induced by f from the simplicial complex $\mathcal{K}(\Gamma)$ to the simplicial complex $\mathcal{K}(\Gamma')$ is a topological cover. Hence the $(n-1)$-covers from Γ to Γ' bijectively correspond to the topological covers from $\mathcal{K}(\Gamma)$ to $\mathcal{K}(\Gamma')$ (see also Exercise 8.26). Because of this, $(n-1)$-covers ($(n-1)$-quotients) will also be called *topological covers* (*topological quotients*). By a linguistic abuse, we also say that the domain Γ of a topological cover $f : \Gamma \longrightarrow \Gamma'$ is a *topological cover* of Γ'.

Topological covers and 2-covers are the epimorphisms most frequently considered in diagram geometry. Trivially, 2-covers of rank 3 geometries are topological covers.

Exercise 8.31. Prove that every epimorphism of thin geometries is a 1-cover.

Exercise 8.32. Prove that a composition of an h-cover and a k-cover is an m-cover, with $m = min(h, k)$.

Exercise 8.33. Let Γ, Γ' and Γ'' be geometries over the same set of types and let $f : \Gamma \longrightarrow \Gamma'$ and $g : \Gamma' \longrightarrow \Gamma''$ be epimorphisms such that gf is an m-cover. Prove that both f and g are m-covers.

Exercise 8.34. Let Γ, Γ' and Γ'' be geometries over the same set of types, let $f : \Gamma \longrightarrow \Gamma'$ be a morphism and $g : \Gamma' \longrightarrow \Gamma''$ be an m-cover such that gf is an m-cover. Prove that f is an m-cover.

Exercise 8.35. Let Γ, Γ' and Γ'' be geometries over the same set of types, let $f : \Gamma \longrightarrow \Gamma'$ be an m-cover and $g : \Gamma' \longrightarrow \Gamma''$ be a morphism such that

gf is an m-cover. Prove that g is an m-cover.

Exercise 8.36. Let $\mathcal{D}$, Γ, Γ' and J be as in Exercise 8.27 and f be an m-cover frm Γ to Γ'. Prove that $Gr_J(f)$ is an m-cover.

Exercise 8.37. With the notation of Exercise 8.30, let $f : \Gamma \longrightarrow \Gamma'$ be an m-cover and let $|J|= k < m < n =|I|$. Check that f induces an $(m-k)$-cover from $Tr_{\bar{J}}^{-}(\Gamma)$ to $Tr_{\bar{J}}^{-}(\Gamma')$.

8.4 Examples

Some examples of morphisms have been given in Exercises 8.1–8.10 and 8.13–8.20. In particular, the geometries considered in Exercises 8.3–8.6 are geometric posets and the morphisms constructed in those exercises are morphisms of posets, that is order-preserving mappings. Trivially, every order preserving mapping from a geometric poset Γ to a geometric poset Γ' is a morphism from the geometry Γ to the geometry Γ'.

A few examples of quotients have also been given in Section 2.1.2, (6)–(9), and in Section 2.1.2(10) (last lines). The reader may check that those quotients are actually topological quotients.

In this section we will study epimorphisms in some special classes of geometries. It will turn out that 'nice' proper epimorphisms are fairly difficult to obtain in many of the classes we will consider.

8.4.1 Epimorphisms of rank 2 geometries

Epimorphisms of Desarguesian projective planes have been examined in Exercises 8.13–8.15. No proper epimorphisms exist between finite Desarguesian projective planes (see Exercise 8.15). This result is due to Klingenberg [114] and it has been generalized to epimorphisms between arbitrary non-degenerate projective planes by Skornjakov [193] (also Hughes [84] and Mortimer [135]). More generally:

Proposition 8.11 ([147]) *Let $f : \Gamma \longrightarrow \Gamma'$ be a proper epimorphism between thick generalized m-gons Γ and Γ', with $3 \leq m < \infty$. Then, for every line L of Γ and every point p of Γ' with $p \in f(L)$, the set $f^{-1}(p) \cap L$ is infinite.*

Corollary 8.12 *There are no proper epimorphisms between finite thick generalized m-gons with $m \geq 3$.*

(Trivial, by Proposition 8.11.) It is worth noticing that proper epimorphisms of generalized m-gons are quite easy to obtain if we drop the thickness hypothesis (see Exercises 8.16 and 8.18).

The previous corollary has the following consequence for geometries belonging to Coxeter diagrams:

Corollary 8.13 *Let Γ and Γ' be thick geometries belonging to the same connected Coxeter diagram of rank $n \geq 2$ and admitting finite orders. Then every epimorphism from Γ to Γ' is a 2-cover.*

The statement of Proposition 8.11 has been generalized by Gevaert and Keppens [70] to the case where Γ' is a generalized quadrangle and Γ is a partial geometry (they allow infinitely many points on a line and infinitely many lines on a point, but they keep the assumption $\alpha < \infty$).

Both in Proposition 8.11 and in the situation considered by Gevaert and Keppens, the gonality and the diameter of Γ are less than or equal to those of Γ'. On the other hand, it is easy to find proper epimorphisms of finite thick geometries where the image has diameter (or gonality) less than the diameter (the gonality) of the domain. For instance, if we map every line (plane) of $AG(3,q)$ onto its point (line) at infinity, then we obtain a proper epimorphism from the line–plane system of $AG(3,q)$ onto $PG(2,q)$. The line–plane system of $AG(3,q)$ has gonality 3, but diameter 4. In the next examples the gonality of the image is less than the gonality of the domain.

Affine generalized quadrangles and their standard quotients. Hyperplanes of thick generalized quadrangles can be defined in the same way as for thick-lined polar spaces of rank $n > 2$ (Section 2.2.2). Given a thick generalized quadrangle $\mathcal{Q}$ and a hyperplane H of $\mathcal{Q}$, let Γ be the geometry obtained removing H from $\mathcal{Q}$. We call Γ an *affine generalized quadrangle.* It is not difficult to check that Γ has gonality 4 and diameter 5 or 6 (the collinearity graph of Γ has diameter 2 or 3, respectively).

Let Φ_0 be the relation defined on points of Γ as follows: given points x, y of Γ we have $x\Phi_0 y$ iff either $x = y$ or $x^\perp \cap H = x^\perp \cap y^\perp$. Then Φ_0 is an equivalence relation and distinct points of Γ correspond in Φ_0 if and only if they have distance 3 in the collinearity graph of Γ. The relation Φ_0 can be extended to the lines of Γ declaring that two lines L, M correspond in Φ_0 when $[L]\Phi_0 = [M]\Phi_0$. The reader may check that we have $L\Phi_0 M$ if and only if $[x]\Phi_0 \cap M \neq \emptyset$ and $[y]\Phi_0 \cap M \neq \emptyset$ for distinct points $x, y \in L$.

It is not difficult to prove that Φ_0 defines a quotient of Γ. The quotient Γ/Φ_0 is actually a 1-quotient. If $\Phi_0 \neq \mho$, then Γ/Φ_0 is a proper quotient of Γ. In this case Γ/Φ_0 has gonality 3 and diameter ≤ 5. The reader may check that we have $\Phi_0 \neq \mho$ if and only if Γ has diameter 6. For instance, this happens if $\mathcal{Q}$ is one of $\mathcal{S}_3(q)$, $\mathcal{H}_3(q^2)$ or $\mathcal{H}_4(q^2)$ and if $H = p^\perp$ for some point p of $\mathcal{Q}$.

If $\Phi_0 \neq \mho$ and Φ is a refinement of Φ_0 defining a quotient, then Γ/Φ is called a *standard* quotient of Γ. We say that Γ/Φ_0 is the *minimal* standard quotient of Γ.

Exercise 8.38. Let Γ be a rank 2 geometry of diameter $d \leq 3$. Prove that there are no proper 1-quotients of Γ.

(Hint: use Exercise 8.11.)

Exercise 8.39. Let Γ be a proper 1-quotient of a generalized quadrangle. Prove that Γ has gonality $g = 2$ and diameter $d \leq 4$.

(Hint: use Exercises 8.10 and 8.11.)

(Remark: see Exercise 8.51 (Remark) for an example of this kind.)

Exercise 8.40. Let Γ be a thick-lined generalized quadrangle, let Γ' be another generalized quadrangle and let $f : \Gamma \longrightarrow \Gamma'$ be a proper epimorphism. Prove that $f(x) = f(y)$ for at least two distinct collinear points x, y of Γ.

(Hint: let $f(x) = f(y)$ for distinct non-collinear points x, y of Γ. Since Γ is thick-lined, there is a line L of Γ not passing through any of x and y. If $x' = x^{\perp} \cap L$ and $y' = y^{\perp} \cap L$, then $f(x') = f(y')$.)

(Remark: the hypothesis that Γ is thick-lined is essential in the above, as the reader can see by dualizing Exercise 8.18.)

Exercise 8.41. Prove that every generalized m-gon can be epimorphically mapped onto an ordinary m-gon.

8.4.2 Matroids and projective geometries

Proposition 8.14 *There are no proper* 1*-quotients of matroids.*

Proof. Let $\mathcal{M}$ be a matroid and let Φ define a 1-quotient of $\mathcal{M}$. Then distinct points of $\mathcal{M}$ never correspond in Φ. Indeed two points of $\mathcal{M}$ are always collinear, whence they are in the residue of some common panel, whereas Φ cannot identify distinct elements in the residue of the same panel, because $\mathcal{M}/\Phi$ is a 1-quotient. On the other hand, if X, Y are subspaces of $\mathcal{M}$ corresponding in Φ, then we have $\{[x]\Phi \mid x \in X\} = \{[y]\Phi \mid y \in Y\}$ by Theorem 8.4(ii) and because the projection p_Φ is an epimorphism. Since no two distinct points of $\mathcal{M}$ are identified by Φ, the above forces X and Y do have the same points. Hence $X = Y$. Therefore $\Phi = \mho$. □

Corollary 8.15 *There are no proper epimorphisms between finite thick projective geometries of dimension $n \geq 2$.*

Proof. This immediately follows from Corollary 8.13 and the above proposition. □

Exercise 8.42. Prove that there are no proper epimorphism of finite affine geometries of dimension $n \geq 3$.

Exercise 8.43. Prove that a simplex has no proper quotients.
(Hint: use Exercise 8.31.)

Exercise 8.44 Let Γ belong to the diagram $L.A_{n-2}.L^*$ (see Section 7.9). Prove that every 1-quotient of Γ is a topological quotient.

8.4.3 Polar spaces and C_n geometries

We know by Corollary 8.12 that there are no proper epimorphisms between finite thick polar spaces of rank 2. We will now consider the case of rank $n \geq 3$.

We first recall a well-known result on dual automorphisms of projective geometries. A *dual automorphism* of a projective geometry Γ (a *duality*, for short) is an isomorphism from Γ to the dual of Γ (dualities of projective geometries are often called *correlations*). A point x of a projective geometry Γ is said to be *absolute* for a duality δ of Γ if $\delta(x) = x$. It is known that all dualities of finite thick projective geometries (of dimension $n \geq 2$) admit some absolute points (see [61], 2.1.16).

Proposition 8.16 *Let $\mathcal{P}$ be a polar space of rank $n \geq 3$ and let Φ define a proper 2-quotient of $\mathcal{P}$. Then there are planes of $\mathcal{P}$ admitting dual automorphisms without any absolute points.*

Proof. Two collinear points of $\mathcal{P}$ are in the residue of some common panel. Therefore distinct collinear points of Γ never correspond in Φ.

Let U, V be distinct maximal subspaces of $\mathcal{P}$ corresponding in Φ. As Φ defines a 2-quotient, we have $dim(U \cap V) \leq n-4$ (possibly $U \cap V = \emptyset$). Hence there is a plane $X \subseteq U$ such that $X \cap V = \emptyset$. The plane X corresponds in Φ to precisely one plane $Y \subseteq V$ because p_Φ is an epimorphism and because Φ induces the identity relation $\mho$ on V, by Proposition 8.14. For the same reason, for every point (line) x of X, there is precisely one point (line) y of Y such that $x \equiv y$ (Φ). Let us denote this point (line) by $\varphi(x)$ and let $\delta(x)$ be the line (point) $X \cap \varphi(x)^\perp$ of X (notation as in Section 1.2.3). It is clear that δ is a dual automorphism of X. Furthermore, δ has no absolute points. Indeed, if x in absolute point of δ, then $x \in \varphi(x)^\perp$. Thus, x and $\varphi(x)$ are distinct collinear points corresponding in Φ, contrary to what we have proved above. □

Corollary 8.17 *Finite thick-lined polar spaces of rank $n \geq 3$ do not admit any proper quotients.*

Proof. Let Γ and Γ' be finite thick-lined C_n geometries with $n \geq 3$ and let $f : \Gamma \longrightarrow \Gamma'$ be a proper epimorphism. We will prove that Γ cannot be a polar space.

Γ and Γ' admit finite orders $(s, s, ..., s, t)$ and $(s', s', ..., s', t')$ respectively, with $s, s' > 1$ (Section 3.2.3(3)). If $t' > 1$, then f is a 2-cover by Corollary 8.13. Let $t' = 1$. The epimorphism f induces 2-covers from residues of dual points of Γ to residues of dual points of Γ', by Corollary 8.13. Hence $s = s'$. Therefore f induces isomorphisms from stars of dual planes of Γ to stars of dual planes of Γ' (Exercise 8.40). In any case, f is a 2-cover.

If Γ is a polar space, then some of its planes admit dual automorphisms without any absolute points, by Proposition 8.16. However, dual automorphisms of non-degenerate finite projective planes always admit some absolute points, as we have remarked at the beginning of this section. Therefore Γ cannot be a polar space. □

Remark. The above corollary is due to Brouwer and Cohen [12], who obtained it as a piece of a more general result on C_n, F_4, E_7 and E_8 (see Propositions 8.26 and 8.28).

Proposition 8.18 *Polar spaces do not admit any proper 2-covers.*

Proof. Let $\mathcal{P}$ be a polar space and let $f : \Gamma \longrightarrow \mathcal{P}$ be a 2-cover. We must prove that f is an isomorphism.

We work by induction on the rank n of $\mathcal{P}$. When $n = 2$ there is nothing to prove. Let $n \geq 3$. For every point a of Γ, f isomorphically maps Γ_a onto $\mathcal{P}_{f(a)}$, by the inductive hypothesis. Let f be not an isomorphism, if possible. Then we have $f(a) = f(b)$ for suitable distinct points a, b of Γ. The points a and b are not collinear, because f is a 2-cover (hence a 1-cover, too). By Lemma 7.31, there is a point c of Γ collinear with both a and c. Let x, y be lines of Γ through c and a and through c and b, respectively. We have $x \neq y$ because $a \not\perp b$. Hence $f(x) \neq f(y)$, since f induces an isomorphism from Γ_c to the polar space $\mathcal{P}_{f(c)}$. Furthermore, $f(c) \neq f(a)$ $(= f(b))$ because $c \perp a$. Therefore the distinct lines $f(x)$, $f(y)$ of $\mathcal{P}$ have two distinct points in common, namely $f(c)$ and $f(a) = f(b)$. However, this cannot be, as $\mathcal{P}$ is a polar space. This is the desired contradiction. □

Corollary 8.19 *No finite thick-lined polar space of rank $n \geq 3$ can be obtained as a proper quotient of a C_n geometry.*

Proof. Let $f : \Gamma \longrightarrow \Gamma'$ be an epimorphism of finite thick-lined C_n geometries with $n \geq 3$ and let Γ' be a polar space. As in the proof of Corollary 8.17 we see that f is a 2-cover. Then f is an isomorphism by Proposition 8.18. □

Exercise 8.45. Prove that every epimorphism between thin geometries of type C_n is a 2-cover.

Exercise 8.46. Prove that the halved octahedron is the only C_3 geometry that can be obtained as a proper quotient of the octahedron.

Exercise 8.47. Prove that the direct sum of an ordinary triangle and a line with 2 points is a proper 1-quotient of the octahedron.

Exercise 8.48. Prove that a thin polar space of rank $n \geq 4$ has just one proper topological quotient, which is the unique almost flat quotient of that polar space (see Section 4.3.2 for the definition of almost flat geometries).

(Hint: exploit the description of thin polar spaces of rank n as systems of cliques of n-partite graphs with classes of size 2, implicit in the final remarks of Section 1.3.5.)

Exercise 8.49. Prove that the thin polar space of rank 4 has just five proper 2-quotients and that all of them are 2-fold (one of them is the topological quotient of Exercise 8.48; the remaining ones are 2-quotients).

Exercise 8.50. Let $\mathcal{P}$ be a polar space of rank $n \geq 3$ and order $(1, 1, ..., 1, t)$ ($t < \infty$). Prove that $\mathcal{P}$ admits an h-fold topological quotient ($h > 1$) if and only if $h \mid t$ and that the t-fold topological quotients of $\mathcal{P}$ (the 'minimal' ones) are characterized by the property of being almost flat.

(Hint: recall that $\mathcal{P}$ is the system of cliques of an n-partite graph with classes of size t, by the final remarks of Section 1.3.5.)

Exercise 8.51. Let $\mathcal{P}$ be the polar space of rank $n \geq 3$ represented by the quadratic form $\sum_{i=1}^{m} x_i^2$ in $V(m, \mathbf{C})$ (where $\mathbf{C}$ is the complex field and $m = 2n$ or $2n+1$). For every vector $x = (x_1, x_2, ..., x_m)$, let $\overline{x} = (\overline{x}_1, \overline{x}_2, ..., \overline{x}_m)$ be the conjugate of x. Identify points of $PG(m-1, \mathbf{C})$ represented by conjugates vectors of $V(m, \mathbf{C})$. Check that a proper toplogical quotient of $\mathcal{P}$ is obtained in this way.

(Remark: if we allow $n = 2$, then we obtain a proper 1-quotient of a thick generalized quadrangle, with diameter 4 and gonality 2.)

8.4.4 Buildings of type D_n

Proposition 8.20 *No proper* 1*-covers exist between geometries belonging to the Coxeter diagram* D_n.

Proof. All geometries belonging to D_n are buildings of type D_n, by Theorem 7.40.

Let $f : \Gamma \longrightarrow \Gamma'$ be a 1-cover of buildings of type D_n. Then f is a 2-cover, because 1-covers between projective planes or between generalized digons are isomorphisms (Exercise 8.38). Let 0 be the initial node of the D_n diagram (or one of the 3 external nodes, when $n = 4$). The 2-cover f induces a 2-cover $Gr_0(f)$ from the polar space $Gr_0(\Gamma)$ to the polar space $Gr_0(\Gamma')$ (Exercise 8.36). The 2-cover $Gr_0(f)$ is an isomorphism, by Proposition 8.19. Therefore f is an isomorphism (see also Exercise 8.29). □

Corollary 8.21 *There are no proper epimorphisms between finite thick buildings of type D_n.*

(Easy, by Corollary 8.13 and Proposition 8.20).

Exercise 8.52. Let $\mathcal{P}$ be a top-thin polar space of rank $n \geq 3$, let $\mathcal{P}/\Phi$ be a proper m-quotient of $\mathcal{P}$, with $m \geq 2$. The relation Φ induces an equivalence relation $\Gamma(\Phi)$ on the set of elements of the unfolding $\Gamma(\mathcal{P})$ of $\mathcal{P}$ (notation as in Section 5.3.4). Nevertheless, $\Gamma(\Phi)$ does not define any quotient of $\Gamma(\mathcal{P})$, as we know by Proposition 8.20. However, such an indirect argument does not explain why $\Gamma(\Phi)$ does not work. Give a direct proof that $\Gamma(\Phi)$ cannot work, without using Proposition 8.20.

(Hint. Let $m = n - 1$, for instance. If X, Y are distinct maximal subspaces of $\mathcal{P}$ corresponding in Φ, then $X \cap Y = \emptyset$ because $m = n - 1$. If n is odd, then X and Y have opposite types in $\Gamma(\mathcal{P})$. Thus, $\Gamma(\Phi)$ mixes the two types corresponding to the two horns of the D_n diagram. Hence $\Gamma(\Phi)$ does not work. Let n be even. Then X and Y have the same type. Let Z be a maximal subspace of $\mathcal{P}$ meeting X in an $(n-2)$-dimensional subspace. Then $x = Z \cap Y$ is a point. We can see that the statement of $\mathbf{Q}_1$ of Section 8.2 fails to hold in $\Gamma(\Phi)$ for the flag $F = \{x_1, x_2\}$ and the elements y_1, y_2 with $x_1 = X$, $x_2 = Z$ and $y_1 = y_2 = x$.)

(Remark: turning back to Exercise 8.36, the above shows that the functor Gr_J need not map the set of m-covers between Γ and Γ' onto the set of m-covers between $Gr_J(\Gamma)$ and $Gr_J(\Gamma')$.)

8.4.5 The diagrams E_6, E_7, E_8 and F_4

Proposition 8.22 *There are no proper* 1*-covers between geometries belonging to the Coxeter diagram E_6.*

Proof. Let Γ, Γ' belong to E_6 and let $f : \Gamma \longrightarrow \Gamma'$ be a 1-cover. Then f is a topological cover by Propositions 8.20 and 8.15. Hence f is an isomorphism if it is injective on the set of points of Γ. On the other hand, any two distinct points of Γ are incident to some common dual point of Γ (Proposition 7.41). Therefore f is injective on the set of points of Γ, by Proposition 8.20. □

Lemma 8.23 *Let Γ belong to the Coxeter diagram E_7 and let Γ/Φ be a proper 1-quotient of Γ. Then distinct points of Γ corresponding in Φ have distance 3 in the collinearity graph of Γ (defined as in Section 7.6.2).*

Proof. The collinearity graph of Γ has diameter $d \leq 3$ (Proposition 7.45). Let a, b, c be distinct points of Γ such that $a \perp b \perp c$ and let x, y be lines on a and b and on b and c, respectively. We have $x \not\equiv y$ (Φ).

Hence $a \not\equiv c$ (Φ), by Lemma 7.46 in Γ/Φ. □

Proposition 8.24 *Let Γ belong to the Coxeter diagram E_7 and let* **IP** *hold in Γ. Then Γ does not admit any proper 2-cover.*

Proof. Let $f : \Gamma' \longrightarrow \Gamma$ be a proper 2-cover of Γ, if possible. By Propositions 8.22, 8.20 and 8.15, f is a topological cover. Therefore f cannot be injective on the set of points of Γ'. That is, $f(a) = f(b)$ for distinct points a, b of Γ'. Let u be a dual point of Γ' incident to b. The point a is not incident to u. Indeed f isomorphically maps Γ'_u onto $\Gamma_{f(u)}$ and we have $f(a) = f(b)$, whereas $a \neq b{*}u$. By Lemma 7.44 there is a point c of Γ' such that $a \perp c{*}u$. Let x be a line of Γ' through a and c. By **IP** in Γ, we have $f(x) * f(u)$ because both $f(x)$ and $f(u)$ are incident to both $f(b)$ $(= f(a))$ and $f(c)$. Hence $x * u$, because f isomorphically maps Γ'_c onto $\Gamma_{f(c)}$. Therefore $a * u$, contrary to what we have said above. □

Proposition 8.25 *Let Γ belong to the Coxeter diagram E_8 and let* **IP** *hold in Γ. Then Γ does not admit any proper 2-cover.*

The proof is quite similar to that of Proposition 8.24. We leave it for the reader (Propositions 8.24, 8.22, 8.20 and 8.15 should be used in this proof; we should also use Lemma 7.49 instead of Lemma 7.44).

Proposition 8.26 (Brouwer–Cohen [12]) *Let Γ be a finite thick geometry belonging to the Coxeter diagram E_7 or E_8 and satisfying* **IP**. *Then Γ does not admit any proper 1-quotient.*

The proof depends on a lemma on automorphisms of certain regular graphs. Using that lemma Brouwer and Cohen obtained the above proposition together with Corollary 8.17 and the next Proposition 8.28, as easy corollaries. We will only consider the case of E_7 here, using an argument more elementary than that of [12].

Proof for the case of E_7. Let Γ be a finite thick geometry belonging to E_7 and satisfying **IP**. Let Φ be an equivalence relation on the set of elements of Γ defining a proper 1-quotient of Γ, if possible. Let X, X' be distinct dual points and Y, Y' distinct maximal subspaces such that $X \equiv X'$ (Φ), $Y \equiv Y'$ (Φ), $X * Y$ and $X' * Y'$. Let $F = \{X, Y\}$ and $F' = \{X', Y'\}$. For every point $a * F$, let $a' = [a]\Phi \cap \Gamma_{F'}$. The points a and a' have distance 3 in the collinearity graph of Γ (Lemma 8.23). Furthermore, $a^\perp \cap \Gamma_X$ is a point (Proposition 7.47; note that we have assumed **IP** in Γ). Let a'' be that point. We have $a'' \notin \Gamma_F$ because a and a' have distance 3. Let $\varphi(a)$ be the dual plane in Γ_F incident to a'' (uniquely determined in the D_6 building Γ_X). As a and a' have distance 3, we have $a \notin \varphi(a)$. The reader may check that φ is a dual automorphism of the projective geometry Γ_F with no absolute points. This is impossible, because Γ_F is finite and thick (see [61], 2.1.16). □

Proposition 8.27 *Let Γ belong to the Coxeter diagram F_4 and let* **IP** *hold in Γ. Then Γ does not admit any proper 2-cover.*

The proof is similar to that of Proposition 8.24. (Hint: use Lemma 7.52.)

Proposition 8.28 *Let Γ be a finite thick geometry belonging to the Coxeter diagram F_4 and satisfying* **IP**. *Then Γ does not admit any proper 2-quotient.*

The reader should consult [12] for the proof.

8.4.6 Affine polar spaces

Proposition 8.29 *Affine polar spaces do not admit any proper 2-covers.*

Proof. Let $f : \Gamma' \longrightarrow \Gamma$ be a 2-cover of an affine polar space Γ. The morphism f is also a topological cover, by Propositions 8.14 and 8.18. Therefore, f is an isomorphism if it is injective on the set of points of Γ'.

Let $f(a) = f(b)$ for distinct points of Γ', by way of contradiction. As f is a 2-cover, the points a and b have distance $d(a, b) \geq 2$ in the collinearity graph of Γ'.

Let $d(a, b) = 2$. Let c be a point of Γ' collinear with both a and b and let x, y be lines of Γ' through a and c and through c and b, respectively. We have $x \neq y$ as $a \not\perp b$. We have $f(x) \neq f(y)$. Indeed f isomorphically maps Γ'_a onto $\Gamma_{f(a)}$ and both x and y are in Γ_a. Trivially, $f(c) \neq f(a)$ $(= f(b))$. Thus, $f(x)$, $f(y)$ are distinct lines of Γ meeting in two distinct points $f(b)$ and $f(c)$. This is impossible, because Γ has a good system of lines, since it is an affine polar space. Therefore $d(a, b) > 2$.

Let $d(a,b) = 3$. Let c, d be points of Γ' such that $a \perp c \perp d \perp b$ and let x, y, z be lines of Γ' through a and c, through c and d and through d and b, respectively. Then $f(x)$, $f(y)$, $f(z)$, $f(b)$ $(= f(a))$, $f(c)$, $f(d)$ form a triangle in Γ. However, Γ is an affine polar space and affine polar spaces satisfy the triangular property (Section 4.3.2). Hence there is a plane u in Γ such that all of $f(x)$, $f(y)$, $f(z)$, $f(b)$, $f(c)$, $f(d)$ are in u. Since f is an epimorphism, there is a plane $u' \in f^{-1}(u) \cap \Gamma'_y$. We have $u' * x$ and $u' * z$ because $u * f(x)$ and $u * f(z)$ and because f isomorphically maps Γ'_c onto $\Gamma_{f(c)}$ and Γ'_d onto $\Gamma_{f(d)}$. Therefore both a and b are in u'. Hence $a \perp b$, contradicting the assumption $d(a,b) = 3$. Therefore $d(a,b) > 3$.

We now prove that **IP** holds in Γ'. As f is a topological cover, stars of points of Γ' are polar spaces. Hence, by Corollary 7.28, we need only prove that Γ' has a good system of lines. Let a, b be distinct points of Γ' and let x, y be lines of Γ' incident to both a, b. The lines $f(x)$ and $f(y)$ are incident to both $f(a)$ and $f(b)$. Since $f(a) \neq f(b)$ and since Γ has a good system of lines, we obtain $f(x) = f(y)$, hence $x = y$ because f is a topological cover. Hence Γ' has a good system of lines, as we wanted to prove.

Using **IP**, it is possible to prove that the collinearity graph of Γ' has diameter $d \leq 3$ (Exercise 7.19; if the reader does not like to refer to Exercise 7.19, he can prove that Γ' has connection index 1, by an argument similar to the one used to prove that $d(a,b) > 3$; thus, Γ' is an affine polar space, by Theorem 7.54; the inequality $d \leq 3$ is straightforward to check in affine polar spaces).

However, we have proved above that $d(a,b) > 3$. This is a final contradiction. □

Corollary 8.30 *There are no proper* 1*-covers between geometries belonging to the diagram* $Af.D_{n-1}$.

The proof is similar to that of Proposition 8.20. We should now use Theorem 7.57 and Proposition 8.21 instead of Theorem 7.40 and Proposition 8.19. We leave the details for the reader.

8.4.7 Standard quotients of affine polar spaces

Let Γ be an affine polar space of rank $n > 2$ and let d be the diameter of the collinearity graph of Γ. It is not hard to check that $d = 2$ or 3.

As for affine generalized quadrangles (Section 8.4.1), the reflexive closure of the relation 'being at distance 3' is an equivalence relation on the set of points of Γ and it can be extended to an equivalence relation Φ_0 between the subspaces of Γ in such a way that Φ_0 defines a quotient of Γ. It is not difficult to check that Γ/Φ_0 is a topological quotient (hence Γ/Φ_0 belongs to the diagram $Af.C_{n-1}$ and it has the same orders as Γ).

Trivially, we have $\Phi_0 = \mho$ if and only if $d = 2$.

Let $d = 3$. The following properties of Γ/Φ_0 are worth a mention (and fairly easy to check). The collinearity graph of Γ/Φ_0 has diameter 1 or 2 and Γ/Φ_0 has a good system of lines (hence **IP** holds in it, by Corollary 7.28). However, the triangular property never holds in Γ/Φ_0 (whereas it holds in Γ). The same can be said for Γ/Φ, for a refinement $\Phi \neq \mho$ of Φ_0 defining a quotient.

If Φ is a refinement of Φ_0 defining a quotient, then the quotient Γ/Φ is called a *standard* quotient of Γ. Standard quotients are topological quotients, since such is Γ/Φ_0 (see Exercise 8.32). We call Γ/Φ_0 the *minimal* standard quotient of Γ.

Note that non-standard topological quotients of affine polar spaces also exist (see Exercise 8.54).

Affine expansions of polar spaces. When the affine polar space Γ arises removing a tangent hyperplane $p^{\perp}$ from a classical polar space $\mathcal{P}$, its minimal standard quotient Γ/Φ_0 can be described in a more 'geometric' way. As $\mathcal{P}$ is classical, $\mathcal{P}$ is the system of non-trivial totally isotropic (or totally singular) subspaces of a vector space V with respect to a non-degenerate (non-singular) sesquilinear (quadratic) form. The hyperplane $p^{\perp}$ spans a hyperplane H of the (possibly infinite dimensional) projective geometry $PG(V)$ of linear subspaces of V (see Section 7.1.4). Every line of $PG(V)$ through p not in H meets $\Gamma = \mathcal{P} - H$ in a point and two distinct points x, y of Γ have distance 3 in the collinearity graph of Γ if and only if the line joining them in $PG(V)$ passes through p (we leave the proofs of these claims for the reader).

Therefore, if Γ has diameter 2 (hence $\Gamma/\Phi_0 = \Gamma$) then every line of $PG(V)$ on p not in H meets Γ in just one point. In this case, Γ is the affine expansion of the residue $\mathcal{P}_p$ of p in $\mathcal{P}$, where $\mathcal{P}_p$ is realized in the star H_p of p in H and the points of Γ are identified with the lines of $PG(V)$ through p not in H. That is, the points of Γ are the points of the affine geometry $PG(V)_p - H_p$ (affine expansions have been defined in Section 2.3 only for geometries realized inside finite dimensional projective geometries; however, it is clear from Section 7.1.4 how that definition can be extended to geometries embedded in infinite dimensional projective geometries).

On the other hand, if Γ has diameter 3 (hence $\Gamma/\Phi_0 \not\cong \Gamma$), then the classes of Φ_0 are the intersections of the lines of $PG(V)$ on p not in H with the set of points of Γ. In this case the affine expansion of the polar space $\mathcal{P}_p$ is Γ/Φ_0.

Conversely, let Γ' be the affine expansion of a classical polar space $\mathcal{P}'$ realized inside the projective geometry $PG(V')$ of linear subspaces of a (possibly infinite dimensional) vector space V' over a division ring K. The vector space V' is a hyperplane of the vector space $V = V' \oplus V(1, K)$ and we can always find a polar space $\mathcal{P}$ in $PG(V)$ and a point p of $\mathcal{P}$ such that $p^{\perp}$ spans V' and $\mathcal{P}_p = \mathcal{P}'$. Let Γ be the affine polar space $\mathcal{P} - p^{\perp}$. As $\mathcal{P}' \subseteq PG(V')$ uniquely determines Γ', we have $\Gamma' = \Gamma/\Phi_0$ ($= \Gamma$ if the collinearity graph of Γ has diameter 2). Therefore, all affine expansions of classical polar spaces are (possibly improper) quotients of affine polar spaces.

Exercise 8.53. Prove that a topological quotient of an affine polar space is a standard quotient if and only if it has a good system of lines.

(Hint: if Γ/Φ has a good system of lines, then any two distinct points in the same class of Φ have distance 3 in the collinearity graph of Γ.)

Exercise 8.54. Let $\Gamma = Af(\mathcal{P})$ be the affine expansion of a classical polar space $\mathcal{P}$ of rank $n \geq 2$ realized in $PG(m, K)$ (we may also allow $\mathcal{P}$ to be a grid). Let p be a point of $PG(m, K)$ not in $\mathcal{P}$ (hence $\mathcal{P}$ is not a symplectic variety in $PG(m, K)$). The planes of $AG(m + 1, K)$ having p as one of the points at infinity form a partition of the set of points of $AG(m+1, K)$, which quite naturally can be extended to a partition Φ_p of the set of elements of Γ. Check that Φ_p defines a topological quotient of Γ (needless to say, the same is true for every refinement Φ of Φ_p defining a quotient).

(Remark: since $\Gamma = Af(\mathcal{P})$ is the minimal (possibly improper) standard quotient of an affine polar space, the above shows how non-standard topological quotients of affine polar spaces can be obtained.)

Exercise 8.55. Check that the quotient Γ/Φ_p of the previous exercise is almost flat (Section 4.3.2).

Exercise 8.56. With the notation of Exercise 8.54, let Φ be a refinement of Φ_p defining a quotient and let $\Phi \neq \mho$. Prove that Γ/Φ does not have a good system of lines (check this directly, without using Exercise 8.53).

Exercise 8.57. Prove that the geometry of Exercise 7.21 is a non-stardard 2-quotient of the affine polar space of rank 3 obtained by dropping a tangent hyperplane from $\mathcal{Q}_5^+(2)$.

Exercise 8.58. Generalize the construction of Exercise 8.54 so as to define topological quotients of the affine expansion $Af(\Gamma)$ of any geometry Γ realized inside a projective geometry $PG(m, K)$, provided that not every point of $PG(m, K)$ is in Γ.

(Remark: all quotients obtained in this way are almost flat.)

Exercise 8.59. Let H and p be a hyperplane and a point respectively in $PG(n,K)$ ($n \geq 2$) and let Γ be the bi-affine geometry obtained by removing H and the star of p from $PG(m,K)$. Let Φ_0 be the equivalence relation defined on the set of elements of Γ as follows: $X\Phi_0 Y$ if $\langle X \cup p\rangle = \langle Y \cup p\rangle$ and $X \cap H = Y \cap H$. Check that the equivalence relation Φ_0 defines a topological quotient of Γ (needless to say, the same is true for every refinement of Φ_0 defining a quotient).

Exercise 8.60. Let Γ and Φ_0 be as in the previous exercise. Prove that the quotient Γ/Φ_0 is proper if and only if either $p \in H$ or $K \neq GF(2)$.

(Remark: when $p \in H$, then Γ is the affine expansion of a dual affine geometry, hence Γ/Φ_0 is just an instance of the construction of Exercise 8.58.)

Exercise 8.61. Let H, p, Γ and Φ_0 be as in Exercise 8.59. Prove that, if $p \in H$, then Γ/Φ_0 is almost flat and that, if $p \notin H$, then the system of points and dual points of Γ/Φ_0 is the system of points and hyperplanes of $PG(n-1,K)$, with $\notin$ as incidence relation.

Exercise 8.62. Prove that bi-affine geometries do not admit any proper 2-covers.

(Hint: we should prove first that the collinearity graph of a geometry belonging to the diagram $Af.A_{n-2}.Af^*$ has diameter $d \leq 2$. Then we can work as in the proof of Proposition 8.18.)

8.4.8 Epimorphisms of thin geometries

Epimorphisms and quotients of thin geometries have already been considered a number of times in previous sections (often in exercises). We only add the next proposition to what we have already said.

Proposition 8.31 *Let Γ, Γ' be thin geometries over the same set of types I, let τ be a permutation of I and let C, C' be chambers of Γ and Γ' respectively. Then there is at most one epimorphism $f : \Gamma \longrightarrow \Gamma'$ such that $f(C) = C'$ and $\tau_f = \tau$.*

Proof. Let $f : \Gamma \longrightarrow \Gamma'$ and $g : \Gamma \longrightarrow \Gamma'$ be epimorphisms such that $f(C) = g(C) = C'$ and $\tau_f = \tau_g = \tau$. Both f and g are 1-covers (see also Exercise 8.31). Therefore $f(D) = g(D)$ for every chamber D adjacent to C. Iterating this argument and using Theorem 1.9 we obtain $f = g$. □

Corollary 8.32 *Let Γ be a thin geometry. Then the identity mapping is the only residually surjective endomorphism of Γ that fixes all elements of some chamber of Γ.*

(Trivial, by the previous proposition).

Exercise 8.63. Let Γ be a simplex or a thin polar space or the unfolding of a thin polar space (as in Section 5.3.4) or the icosahedron or a tessellation of the Euclidean plane as in Sections 2.1.2(5) and 2.1.2(6) or the geometry of white and black triangles of Section 2.1.2(10). Prove that every residually surjective endomorphism of Γ is an automorphism.

Exercise 8.64. Let Γ be the metasymplectic space associated with the thin building of type D_4 (Section 5.3.4). Prove that every proper 1-quotient of the thin polar space of rank 4 defines a proper 1-quotient of Γ.

(Hint: use Exercise 5.28.)

Exercise 8.65. Let $\mathcal{Q}$ be a grid with lines of size 4 and 3 and let Γ be the complement of the collinearity graph of $\mathcal{Q}$. The reader may check that Γ is a rank 3 geometry (the three lines of $\mathcal{Q}$ of size 4 form the type partition of Γ), that Γ is thin and that it belongs to the Coxeter diagram $\widetilde{A}_2$. Prove that Γ is a topological quotient of the geometry Γ of white and black triangles of Section 2.1.2(10).

(Remark: the geometry described above is the smallest example for the diagram $\widetilde{A}_2$.)

(Detail: b) The universal [illegible]position[illegible]

Exercise 8.6.3. Let Γ be a [illegible] or a [illegible] point space or the [illegible] tion polar space in Section 8.4.2. [illegible] the [illegible] the kind [illegible] as done in Sections [illegible] and [illegible] of the geometric [illegible] white and black triangles [illegible] and [illegible] automorphism [illegible]

Exercise 8.6.4. Let Γ be the [illegible] associated with the [illegible] [illegible] type [illegible] Prove that [illegible] [illegible]

[illegible] Exercise [illegible]

Exercise 8.6.5. Let [illegible] with [illegible] and let Γ be the [illegible] [illegible] that Γ is [illegible] [illegible] the [illegible] [illegible] that [illegible] [illegible]

Chapter 9

Automorphisms

9.1 Definitions and notations

9.1.1 The groups $Aut(\Gamma)$ and $Aut_s(\Gamma)$

An *automorphism* of a geometry Γ is an isomorphism from Γ to Γ. The automorphisms of Γ form a group. We denote this group by $Aut(\Gamma)$ and we call it the *full* automorphism group of Γ, also *the* automorphism group of Γ, for short, using the definite article 'the'. When we speak of some (or any, every ...) subgroup of $Aut(\Gamma)$, then we use the indefinite article 'a' (or the words 'some', 'any', 'every' ...).

The type-preserving automorphisms of Γ form a normal subgroup of $Aut(\Gamma)$. We denote this subgroup by $Aut_s(\Gamma)$ and we call it the *special automorphism group* of Γ (we recall that type-preserving morphisms are also called *special* morphisms; see Section 8.1).

Let $S(I)$ be the group of permutations of the set of types I of Γ and let $\tau : Aut(\Gamma) \longrightarrow S(I)$ be the homomorphism that maps $f \in Aut(\Gamma)$ onto τ_f. Trivially, $Aut_s(\Gamma) = Ker(\tau)$ and $Aut(\Gamma)/Aut_s(\Gamma)$ is a subgroup of the automorphism group of the diagram graph $\mathcal{D}(\Gamma)$ of Γ.

Non-type-preserving automorphisms are called *diagram automorphisms.* Let f be a diagram automorphism. If $f^2 \in Aut_s(\Gamma)$ (that is, $\tau_f^2 = 1$), then we call f a *dual automorphism* (also, a *duality*); if $f^3 \in Aut_s(\Gamma)$ (that is, $\tau_f^3 = 1$), then we say that f is a *triality* (these conventions generalize to arbitrary diagrams a terminology sometimes used in the literature for some 'self-dual' diagrams, such as A_n, C_2, F_4, ... and for the diagram D_4).

Given $J \subseteq I$, the automorphisms f of Γ such that $\tau_f(J) = J$ form a subgroup of $Aut(\Gamma)$, denoted by $Aut_J(\Gamma)$. The group $Aut_J(\Gamma)$ contains $Aut_s(\Gamma)$ and it can be viewed as a subgroup of $Aut_s(Gr_J(\Gamma))$ (see Exercise 8.24 for the latter claim). It may happen that $Aut_J(\Gamma)$ is a proper subgroup of $Aut_s(Gr_J(\Gamma))$ (see Exercise 5.41). However, if Γ belongs to a diagram admitting recovery from J-Grassmann geometries (Section 5.4), then $Aut_J(\Gamma) = Aut_s(Gr_J(\Gamma))$.

Turning to the chamber system $\mathcal{C}(\Gamma)$ of Γ, we say that an automorphism φ of the graph $\mathcal{C}(\Gamma)$ is *well defined on types* if there is a mapping $\tau_\varphi : I \longrightarrow I$ such that, for every $i \in I$, if C, D are i-adjacent chambers, then $\varphi(C)$ and $\varphi(D)$ are $\tau_\varphi(i)$-adjacent. Let $Aut(\mathcal{C}(\Gamma), I)$ be the group of automorphisms of $\mathcal{C}(\Gamma)$ that are well-defined on types. The function mapping $\varphi \in Aut(\mathcal{C}(\Gamma), I)$ onto the permutation τ_φ of I is a homomorphism from $Aut(\mathcal{C}(\Gamma), I)$ to the group of permutations of I. We denote the kernel of this homomorphism by $Aut_s(\mathcal{C}(\Gamma), I)$.

Trivially, every automorphism of Γ induces on $\mathcal{C}(\Gamma)$ an automorphism well defined on types. Let $F_\mathcal{C} : Aut(\Gamma) \longrightarrow Aut(\mathcal{C}(\Gamma), I)$ be the function mapping $f \in Aut(\Gamma)$ onto the automorphism of $\mathcal{C}(\Gamma)$ induced by f. By Theorem 1.23, the mapping $F_\mathcal{C}$ is an isomorphism from $Aut(\Gamma)$ to $Aut(\mathcal{C}(\Gamma), I)$. Trivially, f and $F_\mathcal{C}(f)$ induce the same permutation on I, for every $f \in Aut(\Gamma)$. In particular, $F_\mathcal{C}(Aut_s(\Gamma)) = Aut_s(\mathcal{C}(\Gamma), I)$. Thus, we can replace $Aut(\Gamma)$ and $Aut_s(\Gamma)$ by $Aut(\mathcal{C}(\Gamma), I)$ and $Aut_s(\mathcal{C}(\Gamma), I)$ respectively any time we like. For instance, if we want to study the automorphism groups of a tesselation of a surface, we can consider the automorphism group of its barycentric subdivision (see Section 2.1.3), if that is easier to study.

Some of the previous definitions are sometimes extended to isomorphisms between geometries over the same set of types. For instance, given geometries Γ and Γ' over the same set of types I, an isomorphism f from Γ to Γ' is called a *dual isomorphism* if the permutation τ_f induced by f on I is an involution.

9.1.2 Flag transitivity

A subgroup $G \leq Aut_s(\Gamma)$ is said to be *flag-transitive* (also, *chamber-transitive*) if it acts transitively on the set of chambers of Γ (hence, it is transitive on the set of flags of Γ of type J, for every $J \subseteq I$). Trivially, if $G \leq Aut_s(\Gamma)$ is flag-transitive, then $Aut_s(\Gamma)$ is flag-transitive. A geometry Γ is said to be *flag-transitive* if $Aut_s(\Gamma)$ is flag-transitive. If $G \leq Aut_s(\Gamma)$ acts regularly on the set of chambers of Γ, then we say that G is *chamber-regular* (also, *sharply flag-transitive*).

For the rest of this chapter we examine a number of examples of flag-transitive geometries, focusing on thin geometries (Section 9.2), on classical groups (Sections 9.3, 9.4 and 9.5) and on finite matroids (Section 9.6). We leave the general theory of flag-transitive geometries until Chapters 10 and 11.

9.2 The thin case

9.2.1 A bit of theory

In this section Γ is a thin geometry of rank n with set of types I.

Lemma 9.1 *The special automorphism group of Γ acts semi-regularly on the set of chambers of Γ.*

(Easy, by Corollary 8.32.) The next corollaries are trivial consequences of Lemma 9.1.

Corollary 9.2 *Let G be a flag-transitive automorphism group of Γ. Then $G = Aut_s(\Gamma)$.*

Corollary 9.3 *If Γ is flag-transitive, then $Aut_s(\Gamma)$ is chamber-regular.*

Theorem 9.4 *The following are equivalent:*

(i) there is a chamber C of Γ such that, for every type $i \in I$, some element s_i of $Aut_s(\Gamma)$ maps C onto the chamber i-adjacent to C;
(ii) for every chamber C of Γ and for every type $i \in I$ there is an involution $s_i \in Aut_s(\Gamma)$ such that $s_i(C)$ is i-adjacent to C, and $Aut_s(\Gamma) = \langle s_i \rangle_{i \in I}$;
(iii) the geometry Γ is flag-transitive.

Proof. Trivially, each of (ii) and (iii) implies that every chamber satisfies the conditions stated in (i). Let us prove that (i) implies both (ii) and (iii).

Let (i) hold. As Γ is thin, s_i^2 fixes C. Hence $s_i^2 = 1$, by Lemma 9.1. Let us set $G = \langle s_i \rangle_{i \in I}$ ($\leq Aut_s(\Gamma)$).

Let us prove that for every chamber D of Γ there is an element $g \in G$ such that $g(C) = D$. By Theorem 1.9, we can work by induction on the distance m of D from C in $\mathcal{C}(\Gamma)$. If $m = 0$ there is nothing to prove. Let $m > 0$ and let $(C_0, C_1, ..., C_m)$ be a shortest gallery from $C = C_0$ to $D = C_m$. By the inductive hypothesis, there is $f \in G$ such that $f(C) = C_{m-1}$. Hence $f^{-1}(D)$ is adjacent to C. According to the hypotheses of the theorem we want to prove, we have $s_i(C) = f^{-1}(D)$ for some $i \in I$. Hence $D = g(C)$ with $g = fs_i \in G$.

Therefore G is flag-transitive. Also, $Aut_s(\Gamma) = G$ by Corollary 9.2. □

Corollary 9.5 *Let Γ, C and s_i $(i \in I)$ be as in (i) of Theorem 9.4. Let J be a nonempty subset of I, let F be the subflag of C of cotype J and let G_F be the stabilizer of F in $Aut_s(\Gamma)$. Then $G_F = Aut_s(\Gamma_F) = \langle s_j \rangle_{j \in J}$ and G_F is chamber-regular in Γ_F.*

(Trivial, by Theorem 9.4 in Γ_F.)

9.2.2 Examples

(1) **Ordinary m-gons.** Let Γ be an ordinary m-gon with $m < \infty$. We take the symbols P and L as types for Γ, as in Section 1.2. Let $C = (v, e)$ be a chamber of Γ, where v and e are a point (vertex) and a line (edge) of Γ respectively. Let s_L be the type-preserving automorphism of Γ fixing the panel v and interchanging C with the chamber L-adjacent to C. If we represent Γ as a regular m-gon in the Euclidean plane, then s_L can be viewed as a symmetry around an axis through e. Let $s_P \in Aut_s(\Gamma)$ fix the panel e and interchange C with the chamber P-adjacent to C (s_P is a symmetry around an axis through the middle point of e).

Both s_P and s_L are involutions, we have $(s_P s_L)^m = 1$ and s_P, s_L generate the dihedral group D_{2m} of order $2m$. By Theorem 9.4, we have $Aut_s(\Gamma) = D_{2m}$.

Furthermore, Γ admits a dual automorphism δ. Since Γ has rank 2, $Aut_s(\Gamma)$ has index 2 in $Aut(\Gamma)$. Hence $Aut(\Gamma) = Aut_s(\Gamma)\langle\delta\rangle$ ($\approx D_{2m} \cdot \mathbf{2}$; see Chapter 0 for the notation). The flag complex $\mathcal{K}(\Gamma)$ of Γ is an ordinary $2m$-gon (Section 1.4.1) and we can choose δ in such a way as it acts on the $2m$-gon $\mathcal{K}(\Gamma)$ as a symmetry around an axis through the middle point of an edge. Hence we can choose δ of order 2. Thus, $Aut(\Gamma) \approx D_{2m} : \mathbf{2}$.

Let $m = \infty$. As above, the group $Aut_s(\Gamma)$ is generated by two involutions s_P, s_L, but $s_P s_L$ is now aperiodic, that is $\langle s_P s_L\rangle$ is the infinite cyclic group $\mathbf{Z}$. We have $Aut_s(\Gamma) = \mathbf{Z}\langle s\rangle$ where s is an involution such that $sxs = x^{-1}$ for every $x \in \mathbf{Z}$. We can choose any of s_P or s_L for s. An involutory dual automorphism also exists, as in the case of $m < \infty$.

(2) **The n-dimensional simplex.** Let $\mathcal{S}$ be the n-dimensional simplex, $n \geq 2$. We recall that the elements of the geometry $\mathcal{S}$ are the faces of the simplex $\mathcal{S}$. Let V be this set of elements and let $V_0 = \{v_0, v_1, ..., v_n\}$ be the set of elements of $\mathcal{S}$ of type 0 (that is, the set of vertices of the simplex).

It is clear that $Aut_s(\mathcal{S})$ is the group S_{n+1} of all permutations of the elements of V_0. The function $\delta : V \longrightarrow V$ mapping $X \in V$ onto $\delta(X) = V_0 - X$ is a dual automorphism of $\mathcal{S}$. On the other hand, $\mathcal{S}$ belongs to the Coxeter diagram A_n and this diagram has just one symmetry. Furthermore, δ is an involution and commutes with all permutations of V_0. Therefore $Aut(\mathcal{S}) = Aut_s(\mathcal{S}) \times \langle\delta\rangle$ ($= S_{n+1} \times \mathbf{2}$).

Let us set $X_i = \{v_0, v_1, ..., v_i\}$ ($i = 0, 1, ..., n-1$) and $C = \{X_i\}_{i=0}^{n-1}$. Then C is a chamber of $\mathcal{S}$. For every $i = 0, 1, ..., n-1$, let s_i be the transposition (v_i, v_{i+1}). This transposition maps C onto the chamber i-adjacent to C. By Theorem 9.4, we have $S_{n+1} = \langle s_i\rangle_{i=0}^{n-1}$ and $S_{n+1} = Aut_s(\mathcal{S})$ is chamber-regular in $\mathcal{S}$. The involutions s_i ($i = 0, 1, ..., n-1$)

satisfy the following relations:

$$(s_i s_{i+1})^3 = 1 \quad (i = 0, 1, ..., n-2),$$
$$(s_i s_j)^2 = 1 \quad (j > i+1).$$

These relations can easily be checked by direct computations. We can also obtain them by Corollary 9.5 (with $|J| = 2$) and by the description of automorphism groups of ordinary m-gons (see (1) above).

(3) **The thin polar space of rank n.** It is clear from the final remarks of Section 1.3.5 that a thin polar space of rank n is the system of vertices, edges and cliques of a complete n-partite graph, where the classes of the n-partition have size 2. Therefore, there is just one thin polar space of rank n (up to isomorphisms, of course).

Let $\mathcal{G}$ be the complete n-partite graph with classes of size 2, let $\mathcal{A} = \{A_1, A_2, ..., A_n\}$ be the n-partition of $\mathcal{G}$ and let U, V be two disjoint maximal cliques of $\mathcal{G}$. Then $U \cup V$ is the set of vertices of $\mathcal{G}$. Let us set $u_i = U \cap A_i$ and $v_i = V \cap A_i$ for $i = 1, 2, ..., n$. Let $\mathcal{P}$ be the thin polar space of rank n defined by $\mathcal{G}$. We assume $n \geq 3$, since $\mathcal{P}$ is just an ordinary 4-gon when $n = 2$ and we have already considered ordinary m-gons in paragraph (1). Thus, $Aut(\mathcal{P}) = Aut_s(\mathcal{P})$. Trivially, $Aut_s(\mathcal{P})$ $(= Aut(\mathcal{P}))$ is just the automorphism group $Aut(\mathcal{G})$ of the graph $\mathcal{G}$ and it is chamber-regular on $\mathcal{P}$.

The group $G = Aut(\mathcal{G})$ is easy to describe. Let N be the kernel of the action of G on the n-partition $\mathcal{A}$. It is easily seen that $N = \mathbf{2}^n$ (notation as in Chapter 0). Furthermore, N is transitive on the set of maximal cliques of the graph $\mathcal{G}$ and we can identify N with $V(n, 2)$ (the n classes of $\mathcal{A}$ naturally correspond to the vectors of the canonical basis of $V(n, 2)$). By the transitivity of N on the set of maximal cliques of $\mathcal{G}$, stabilizers of maximal cliques of $\mathcal{G}$ form one conjugacy class in G. We take the stabilizer G_U of U as a representative of this conjugacy class. The subgroup G_U acts as the symmetric group S_n on the n vertices belonging to U. For every $g \in G_U$, the action of g on the n vertices belonging to V is uniquely determined by its action in U, since $g(\mathcal{A}) = \mathcal{A}$ and $\mathcal{A}$ defines a bijection between U and V. Trivially, $N \cap G_U = 1$. Hence $G/N \cong S_n = G_U$. Therefore G is the semidirect product of N and G_U. Hence $G \approx \mathbf{2}^n : S_n$. The action of G_U on N is uniquely determined by the action of G_U on $\mathcal{A}$ and by the identification between N and $V(n, 2)$ and between $\mathcal{A}$ and the canonical basis of $V(n, 2)$. This completely determines G.

Let ζ be the element of N interchanging u_i and v_i for all $i = 1, 2, ..., n$. It is easily seen that $Z(G) = \langle \zeta \rangle$. Hence G can also be viewed a central extension of $Z(G)$ by $(N/Z(G))G_U$, that is $G \approx \mathbf{2} \cdot (\mathbf{2}^{n-1} : S_n)$.

Let N^+ be the subgroup of $N = V(n,q)$ represented by vectors of $V(n,2)$ with an even number of non-zero entries. Trivially, $N^+ = \mathbf{2}^{n-1}$ and it is normal in G. Hence G can also be described as follows

$$G \approx (\mathbf{2}^{n-1} : S_n)\langle s\rangle \approx (\mathbf{2}^{n-1} : S_n) : \mathbf{2}$$

where $s \in N - N^+$. We have $\zeta \notin N^+$ if and only if n is odd. In this case we have $G = Z(G) \times N^+G_U \approx \mathbf{2} \times (\mathbf{2}^{n-1} : S_n)$.

It is worth remarking that, when $n = 3$, the thin polar space $\mathcal{P}$ is the octahedron and ζ is the antipodal involution. The halved octahedron is obtained taking as elements the orbits of ζ on the set of elements of $\mathcal{P}$ (cliques of $\mathcal{G}$). Therefore $N^+G_U \approx \mathbf{2}^2 : S_3$ is the automorphism group of the halved octahedron. It is clear from the description given in Section 2.1.2(9) for the halved cube (the dual of the halved octahedron) that the automorphism group of the halved octahedron is the symmetric group S_4. This fits with the above.

We now give a set of involutions generating G. If $U_i = \{u_1, u_2, ..., u_i\}$ for $i = 1, 2, ..., n$, then $C = (U_i)_{i=1}^n$ is a chamber of $\mathcal{P}$. For $i = 1, 2, ..., n-1$, let s_i be the element of $G_U = S_n$ acting as the transposition (u_i, u_{i+1}) on U (that is, s_i acts as $(u_i, u_{i+1})(v_i, v_{i+1})$ on $U \cup V$). Let s_n be the element of N acting as the transposition (u_n, v_n) on $U \cup V$. It is easily seen that s_i maps C onto the chamber i-adjacent to C. Therefore $G = \langle s_i\rangle_{i=1}^n$ by Theorem 9.4. By Corollary 9.5 (or by straightforward computations), the involutions s_i $(i = 1, 2, ..., n)$ satisfy the following relations:

$$\begin{array}{ll} (s_i s_{i+1})^3 = 1 & (i = 1, 2, ..., n-2), \\ (s_{n-1} s_n)^4 = 1, & \\ (s_i s_j)^2 = 1 & (j > i+1). \end{array}$$

(4) **The thin building of type D_n.** Since there is just one thin polar space $\mathcal{P}$ of rank n, there is a unique thin building $\Gamma = \Gamma(\mathcal{P})$ of type D_n $(n \geq 4)$. According to Section 5.3.4, we can describe Γ using the description given for $\mathcal{P}$ in the previous paragraph (3). We keep the notation of paragraph (3) for everything related to $\mathcal{P}$ or to $Aut(\mathcal{P})$.

We take $I = \{1, 2, ..., n-2, -, +\}$ as the set of types for Γ in such a way that the elements of Γ of type $i = 1, 2, ..., n-2$ are the i-cliques of the graph $\mathcal{G}$, the elements of Γ of type $+$ or $-$ are the maximal cliques of $\mathcal{G}$ and two maximal cliques X, Y have the same type if and only if $|X - Y|$ is even. We can always assume that the maximal clique U has type $+$. The incidence relation of Γ between non-maximal cliques or between a non-maximal clique and a maximal clique is the same as in $\mathcal{P}$. Two maximal cliques X, Y are incident in Γ if and only if $|X \cap Y| = n-1$.

It is clear from the above that $Aut(\Gamma) \geq Aut(\mathcal{P})$. On the other hand, if $g \in N - N^+$, then g interchanges the types $+$ and $-$ (g is a duality of Γ). Therefore $Aut_s(\Gamma) = N^+G_U \approx \mathbf{2}^{n-1} : S_n$, the group $Aut_s(\Gamma)$ is chamber-regular on Γ and it is a proper subgroup of $Aut(\mathcal{P})$ ($\leq Aut(\Gamma)$). We have $s_n = (u_n, v_n) \in N - N^+$. Hence $Aut(\mathcal{P}) = Aut_s(\Gamma)\langle s_n \rangle \approx (\mathbf{2}^{n-1} : S_n) : \mathbf{2}$. When n is odd, we have $s_nN^+ = \zeta N^+$ and $Aut(\mathcal{P}) = Aut_s(\Gamma) \times \langle \zeta \rangle \approx (\mathbf{2}^{n-1} : S_n) \times \mathbf{2}$.

When $n \geq 5$, we have $Aut(\Gamma) = Aut(\mathcal{P})$ and $Aut_s(\Gamma)$ has index 2 in $Aut(\Gamma)$. When $n = 4$ the situation is different; we can define a dual automorphism δ of Γ, as follows: for every $h = 1,2,3,4$, we set $\delta(u_i) = (V - v_i) \cup u_i$ and $\delta(v_i) = (U - u_i) \cup v_i$. For every 2-clique $X = \{x, y\}$ of $\mathcal{G}$ we set $\delta(X) = \delta(x) \cap \delta(y)$. For every maximal clique X of type $+$, we define $\delta(X)$ to be the (unique) maximal clique of $\mathcal{G}$ not intersecting X. If X is a clique of type $-$ meeting U in one point, then we set $\delta(X) = X \cap U$. If X is a maximal clique of type $-$ meeting U in 3 points, then we set $\delta(X) = X \cap V$. It is not difficult to check that δ is a dual automorphism of Γ, fixing the types 1 and $+$ and interchanging the types 0 and $-$. On the other hand, s_n is also a duality, it interchanges the types $+$ and $-$ and it fixes 0 and 1. Therefore, s_n and δ induce the full symmetric group S_3 on the three external nodes of the diagram D_4. Hence $Aut(\Gamma)/Aut_s(\Gamma) = S_3$ and $s_n\delta$ is a triality.

Let us come back to the general case and describe a generating set of involutions for $Aut_s(\Gamma)$. Let us set $s_- = s_{n-1}$, $s_+ = (u_n, v_{n-1})(u_{n-1}, v_n) = s_n s_{n-1} s_n$, $U_+ = U$ and $U_- = (U - u_n) \cup v_n$. Then $F = (U_i)_{i \in I}$ is a chamber of Γ and, for every $i \in I = \{1, 2, ..., n-2, -, +\}$, the involution s_i maps F onto the chamber i-adjacent to F. By Theorem 9.4, we have $Aut_s(\Gamma) = \langle s_i \rangle_{i \in I}$ and $Aut_s(\Gamma)$ is chamber-regular in Γ. By Corollary 9.5 (or by straightforward computations), the involutions s_i ($i \in I$) satisfy the following set of relations:

$$
\begin{array}{ll}
(s_i s_{i+1})^3 = 1 & (i = 1, 2, ..., n-3), \\
(s_{n-2} s_-)^3 = (s_{n-2} s_+)^3 = 1, & \\
(s_- s_+)^2 = 1, & \\
(s_i s_-)^2 = (s_i s_+)^2 = 1 & (i = 1, 2, ..., n-3), \\
(s_i s_j)^2 = 1 & (i, j = 1, 2, ..., n-2; j > i+1).
\end{array}
$$

We have assumed $n \geq 4$ till now. When $n = 3$, Γ is the tetrahedron. All the above remain true. We now have $Aut_s(\Gamma) = S_4$.

Exercise 9.1. Let $\mathcal{P}$ be a thin-lined polar space of rank $n \geq 2$ admitting finite order t at the last node of the C_n diagram. Prove that $Aut_s(\mathcal{P}) = S_{t+1} \wr S_n$.

(Remark: when $t = 1$, we obtain $Aut_s(\mathcal{P}) = 2 \wr S_n \approx \mathbf{2}^n : S_n$.)

Exercise 9.2. Let Γ be the metasymplectic space associated with the thin D_4 building (Section 5.3.4). Prove that $Aut_s(\Gamma) \approx (\mathbf{2}^3 : S_4) \cdot S_3$, that $Aut_s(\Gamma)$ has index 2 in $Aut(\Gamma)$, that $Aut_s(\Gamma)$ is chamber-regular in Γ and that it is generated by involutions r_0, r_1, r_2, r_3 satisfying the following relations:

$$\begin{array}{ll} (r_0r_1)^3 = (r_2r_3)^3 = 1, & \\ (r_1r_2)^4 = 1, & \\ (r_ir_j)^2 = 1 & (i+1<j). \end{array}$$

Exercise 9.3. Let $\mathcal{I}$ be the icosahedron. Prove that $Aut_s(\mathcal{I})$ (which is $Aut(\mathcal{I})$) is chamber-regular and that it is generated by three involutions s_1, s_2, s_3 satisfying the following relations:

$$(s_1s_2)^3 = (s_2s_3)^5 = (s_3s_1)^2 = 1.$$

(Remark: we have $Aut(\mathcal{I}) = \langle\zeta\rangle \times A_5 = \mathbf{2} \times A_5$, where ζ is the antipodal involution; the alternating group A_5 is the automorphism group of the halved icosahedron (Section 2.1.2(6)).

Exercise 9.4. Let $\mathcal{T}$ be the tessellation of the Euclidean plane in equilateral triangles. Prove that $Aut_s(\mathcal{T})$ ($= Aut(\mathcal{T})$) is chamber-regular, that it is generated by three involutions s_1, s_2, s_3 satisfying the following relations:

$$(s_1s_2)^3 = (s_2s_3)^6 = (s_3s_1)^2 = 1$$

and that $Aut_s(\mathcal{T}) \approx \mathbf{Z}^2 : D_{12}$.

Exercise 9.5. Let $\mathcal{T}$ be the tessellation of the Euclidean plane in squares. Prove that $Aut_s(\mathcal{T})$ has index 2 in $Aut(\mathcal{T})$, that $Aut_s(\mathcal{T})$ is chamber-regular, that it is generated by three involutions s_1, s_2, s_3 satisfying the following relations:

$$(s_1s_2)^4 = (s_2s_3)^4 = (s_3s_1)^2 = 1$$

and that $Aut_s(\mathcal{T}) \approx \mathbf{Z}^2 : D_8$.

Exercise 9.6. Let Γ be the geometry of white and black triangles of Section 2.1.2(10). Prove that $Aut(\Gamma)/Aut_s(\Gamma) = S_3$.

(Hint: the chamber system $\mathcal{C}(\Gamma)$ can be represented as a tessellation of the Euclidean plane in equilateral triangles (Exercise 2.3).)

Exercise 9.7. Given Γ as in the previous exercise, prove that $Aut_s(\Gamma)$ is chamber-regular, that it is generated by three involutions s_1, s_2, s_3 satisfying the following relations:

$$(s_is_j)^3 = 1 \qquad (i \neq j)$$

and that $Aut_s(\Gamma) \approx \mathbf{Z}^2 : D_6$.

Exercise 9.8. Let $\mathcal{T}$ and Γ be as in exercises 9.4 and 9.6 respectively. Prove that $Aut(\mathcal{T})$ can be realized as a subgroup of $Aut(\Gamma)$ containing $Aut_s(\Gamma)$ as a subgroup of index 2.

(Hint: use Exercise 5.14.)

Exercise 9.9. Given $\mathcal{T}$ and Γ as in Exercises 9.4 and 9.6 respectively, prove that we also have $Aut(\mathcal{T}) \cong Aut(\Gamma)$.

(Hint: use Exercise 2.3.)

(Remark: reedless to say, the above does not contradict the statement of Exercise 9.8. Indeed, we are dealing with infinite groups.)

Exercise 9.10. Let $\mathcal{I}$ be the icosahedron and let V_0 and V_1 be the set of vertices and the set of edges of $\mathcal{I}$, respectively. Set $V = V_0 \times \{+,-\}$ and define an incidence relation on $V \cup V_1$ as follows: given $x \in V_1$, $y \in V_0$ and $i = +$ or $-$, the elements x and (y,i) are incident if x and y are incident in $\mathcal{I}$; given $x, y \in V_0$ and $i, j = +$ or $-$, the elements (x,i) and (y,j) are incident if $i \neq j$ and x, y are adjacent vertices of $\mathcal{I}$. A thin geometry Γ is obtained in this way, belonging to the following Coxeter diagram:

(5) (5)

●———●———●

and we have $Aut_s(\Gamma) = Aut(\mathcal{I})$. Use this information to prove that $Aut(\mathcal{I})$ is also generated by three involutions r_1, r_2, r_3 satisfying the following relations:

$$(r_1r_2)^5 = (r_2r_3)^5 = (r_3r_1)^2 = 1.$$

(Remark: the geometry Γ constructed above is the so-called 'small starred dodecahedron'; equivalently, the so-called 'large dodecahedron'; see [231].

9.3 Thick projective geometries

9.3.1 Classical automorphism groups of $PG(n,K)$

Given a division ring K and an integer $n \geq 2$, by $GL(n+1,K)$ we mean the group of all invertible linear transformations of $V = V(n+1,K)$. Once a basis has been chosen for V, we may identify $GL(n+1,K)$ with the group of all $(n+1) \times (n+1)$ invertible matrices over K.

Let Z be the group of all matrices of the form kI with k a non-zero element of the centre of K and I the $(n+1) \times (n+1)$ identity matrix. Then $PGL(n+1,K)$ denotes the factor group $GL(n+1,K)/Z$. The group

$PGL(n+1,K)$ is called the *n-dimensional (projective) general linear group* over K.

We have $PGL(n+1,K) \leq Aut_s(PG(n,K))$. Let $Aut(K)$ be the automorphism group of the division ring K. Every type-preserving automorphism of $PG(n,K)$ is represented modulo Z by a *semilinear* transformation of V, namely by a composition $f \cdot g_\gamma$ where $f \in GL(n+1,K)$, $\gamma \in Aut(K)$ and $g_\gamma : V \longrightarrow V$ maps $(x_0, x_1, ..., x_n) \in V$ onto $(\gamma(x_0), \gamma(x_1), ..., \gamma(x_n))$. Therefore $Aut_s(PG(n,K))$ is the semidirect product of $PGL(n+1,K)$ and $Aut(K)$.

The semilinear transformations of V form a group, which is denoted by $\Gamma L(n+1,K)$. The symbol $P\Gamma L(n+1,K)$ designates the factor group $\Gamma L(n+1,K)/Z$. Thus:

$$Aut_s(PG(n,K)) \quad = \quad P\Gamma L(n+1,K) \quad \approx \quad PGL(n+1,K) : Aut(K)$$

as we have remarked above.

Trivially, a dual automorphism of $PG(n,K)$ is also a type-preserving isomorphism between $PG(n,K)$ and its dual, which is coordinatized by the dual K^{op} of K. Therefore, $PG(n,K)$ admits dual automorphisms only if K is a field. That is, if K is non-commutative, then $Aut(PG(n,K)) = Aut_s(PG(n,K))$.

Let K be a field and let δ be the function mapping every non-zero vector $a = (a_1, a_2, ..., a_n) \in V$ onto the hyperplane of V represented by the equation $\sum_{i=0}^{n} a_i x_i = 0$. Then δ is a duality of $PG(n,K)$. We have $\delta^2 = 1$ and $\delta f \delta^{-1} = (f^t)^{-1}$ for every $f \in GL(n+1,K)$ (where f^t is the transpose of f) and $\delta g_\gamma = g_\gamma \delta$ for every $\gamma \in Aut(K)$. The full automorphism group $Aut(PG(n,K))$ of $PG(n,K)$ is the semidirect product of $Aut_s(PG(n,K))$ and $\langle \delta \rangle$. Thus, $Aut(PG(n,K)) \approx P\Gamma L(n+1,K) : \mathbf{2}$.

Let K be a field, as above. Then the matrices of determinant 1 form a normal subgroup of $GL(n+1,K)$, denoted by $SL(n+1,K)$. The centre of $SL(n+1,K)$ is $Z \cap SL(n+1,K)$ and $SL(n+1,K)/(SL(n+1,K) \cap Z) \cong (Z \cdot SL(n+1,K))/Z$. Thus, we can identify $SL(n+1,K)/(SL(n+1,K) \cap Z)$ with $(Z \cdot SL(n+1,K))/Z$.

The latter is a normal subgroup of $PGL(n+1,K)$ and it is simple ([35], Section 11.1). It is denoted by $PSL(n+1,K)$ and called the *n-dimensional (projective) special linear group* over K. A subgroup of $P\Gamma L(n+1,K)$ is said to be *classical* if it contains $PSL(n+1,K)$.

When $K = GF(q)$, we write $PGL(n+1,q)$, $P\Gamma L(n+1,q)$ and $L_{n+1}(q)$ for $PGL(n+1,GF(q))$, $P\Gamma L(n+1,GF(q))$ and $PSL(n+1,GF(q))$ respectively.

Remark. We have assumed $n \geq 2$ till now, because in our combinatorial perspective the projective line $PG(1,K)$ is simply a set of $|K|+1$ elements (and $Aut(PG(1,K))$ is the group of all permutations of this set). However, the groups $PGL(n+1,K)$, $P\Gamma L(n+1,K)$ and $PSL(n+1,K)$ can be defined when $n=1$ just as when $n \geq 2$. We also write $L_2(q)$, $PGL(2,q)$ and $P\Gamma L(2,q)$ for $PSL(2,GF(q))$, $PGL(2,GF(q))$ and $P\Gamma L(2,GF(q))$ respectively. The group $L_2(q)$ is simple except when $q = 2$ or 3 (see [35]). We have $L_2(2) \cong S_3$ and $L_2(3) \cong A_4$.

9.3.2 Interlude: apartments and bases

Let $(p_0, p_1, ..., p_n)$ be an ordered basis of a (possibly degenerate) projective geometry Γ of rank $n \geq 2$ (see Section 1.3.2 for the definition of bases of matroids). We use the following notation:

$I = \{0, 1, ..., n-1\}$ and $\overline{I} = I \cup \{n\}$,
$X_J = \langle p_j \rangle_{j \in J}$ for $J \subseteq \overline{I}$,
$X_i = X_{\{0,1,...,i\}}$ $(i \in I)$,
$\mathcal{S} = \{X_J \mid J \subseteq \overline{I}, \emptyset \neq J \neq \overline{I}\}$,
$C = \{X_{\{0,1,...,i\}}\}_{i \in I}$

The subgeometry $\mathcal{S}$ of Γ is an n-dimensional simplex and C is a chamber of Γ contained in $\mathcal{S}$.

The simplex $\mathcal{S}$ is called the *apartment* of Γ relative to the unordered basis $\{p_0, p_1, ..., p_n\}$ (compare Section 1.4.2). It is clear that giving an unordered basis is the same as giving an apartment. Giving an ordered basis is the same as giving an apartment $\mathcal{S}$ and a chamber C in it.

Let C^{op} be the chamber defined by the ordering $(p_n, ..., p_1, p_0)$ of the basis $\{p_0, p_1, ..., p_n\}$, opposite to the ordering $(p_0, p_1, ..., p_n)$ defining C. It is not difficult to prove that the chambers C and C^{op} have maximal distance in $\mathcal{C}(\Gamma)$ (see also Exercise 1.10).

Proposition 9.6 *For every choice of chambers C, D of Γ, there is some apartment of Γ containing both C and D.*

We leave the proof for the reader (the rank 3 case of this statement with C and D at maximal distance has been examined in Section 1.4.2).

9.3.3 Properties of $PGL(n+1,K)$

We now turn back to automorphism groups of $PG(n,K)$. Let $n \geq 2$, as in Section 9.3.1. In this section we assume that K is a field. However, many of the things we will say can be generalized to the case where K is any division ring.

We will only consider $PGL(n+1,K)$, but what we are going to say can be repeated with a few minor changes for any classical subgroup of $P\Gamma L(n+1,K)$. The proofs of most of the properties we will state are nothing but exercises in linear algebra. We leave them for the reader.

Let us write G for $PGL(n+1,K)$, for short. $GL(n+1,K)$ is transitive on the set of ordered bases of $V = V(n+1,K)$. Hence G is transitive on the set of ordered bases of $PG(n,K)$. That is, G is transitive on the set of pairs $(C,\mathcal{S})$ where $\mathcal{S}$ is an apartment of $PG(n,K)$ and C is a chamber in $\mathcal{S}$. In particular, G is flag-transitive.

Given an apartment $\mathcal{S}$ and a chamber C in S, let $(p_0, p_1, ..., p_n)$ be the ordered basis of $PG(n,K)$ associated with the pair $(C,\mathcal{S})$. We choose a representative $e_i \in V$ of p_i for every $i \in \overline{I}$. Thus, the sequence $(e_0, e_1, ..., e_n)$ is an ordered basis of V.

Let B be the stabilizer of C in G and let N be the setwise stabilizer of $\mathcal{S}$ in G (that is, N is the stabilizer of the unordered basis $\{p_0, p_1, ..., p_n\}$ of $PG(n,K)$). Let $H = B \cap N$ be the stabilizer of C and $\mathcal{S}$ in G (that is, H is the stabilizer of the ordered basis $(p_0, p_1, ..., p_n)$ of $PG(n,K)$).

The subgroup B is called the *Borel subgroup* of G relative to the chamber C. It consists of the elements of G represented by upper triangular matrices with respect to the ordered basis $(e_0, e_1, ..., e_n)$ of V. The subgroup H is called the *Cartan subgroup* of G relative to the apartment $\mathcal{S}$. It consists of the elements of G represented by diagonal matrices with respect to $(e_0, e_1, ..., e_n)$. Thus, H is the direct product of n copies of the multiplicative group K^* of K.

By Lemma 9.1, the Cartan subgroup H is the kernel of the action of N on $\mathcal{S}$. Hence $H \trianglelefteq N$. The group $W = N/H$ induced by N on $\mathcal{S}$ is called the *Weyl group* of G. The group W is isomorphic to the special automorphism group S_{n+1} of the simplex $\mathcal{S}$ and it can be represented inside N by the subgroup of all linear transformations permuting the vectors e_0, e_1, ..., e_n. Therefore N is a semidirect product of H and W. Hence $N \approx (K^*)^n : S_{n+1}$.

The upper triangular matrices with all entries 1 on the diagonal form a normal subgroup U of B, called the *unipotent radical* of B. The Borel subgroup B is the semidirect product of U and H. Furthermore, $B = N_G(U)$.

The subgroup U can also be described by means of elations, as we will see in a few lines. Let us define perspectivities before coming to elations. Given a hyperplane X and a point p of $PG(n,K)$ a *perspectivity* of $PG(n,K)$ with *axis* X and *centre* p is an automorphism of $PG(n,K)$ fixing pointwise X and the star of p. A perspectivity f with centre p and axis X is called an *elation* (a *homology*) if $p \in X$ (if $p \notin X$). Let $U_{p,X}$ be the group of perspectivities with centre p and axis X. If $p \in X$, then $U_{p,X} \leq PSL(n+1,K)$ and $U_{p,X}$ is isomorphic to the additive group of K.

If $p \notin X$, then $U_{p,X}$ is isomorphic to the multiplicative group of K and $U_{p,X} \cap PSL(n+1, K)$ is isomorphic to the group of $(n+1)$th-roots of 1 in K.

We have $U = \langle U_{p_i, X_{\overline{I-j}}} \rangle_{0 \leq i < j \leq n}$, where $X_{\overline{I-j}}$ is the hyperplane spanned by $p_0, ..., p_{j-1}, p_{j+1}, ..., p_n$, according to the notation of Section 9.3.2.

Given $J \subseteq I$, we write G_J to denote the stabilizer in G of the subflag of C of type J. The subgroup G_J is called the *parabolic subgroup* of G of *type* $I - J$ relative to the chamber C. A parabolic subgroup G_J of G is said to be *proper* if $\emptyset \neq J \neq I$. We have $G_\emptyset = G$, $G_I = B$ and $G_J = \bigcap_{j \in J} G_j = \langle G_{I-i} \rangle_{i \in I-J}$ for $J \subseteq I$.

Trivially, the stabilizer in G of a flag F of $PG(n, K)$ of type J is a parabolic subgroup of type $I - J$, with respect to any chamber containing F. The parabolic subgroups of type i are stabilizers of panels of cotype i ($i \in I$). They are called *minimal* parabolic subgroups. The parabolic subgroups of type $I - i$ are stabilizers of elements of C ($i \in I$). They are called *maximal* parabolic subgroups.

Since G is flag-transitive, all Borel subgroups are conjugate and all parabolic subgroups of the same type are conjugates. The same holds for stabilizers of apartments and for stabilizers of ordered bases. The reader can check that distinct flags of $PG(n, K)$ have distinct stabilizers (see also Proposition 10.26). Therefore, every parabolic subgroup of G is its own normalizer. In particular $N_G(B) = B$.

Proposition 9.7 (Bruhat decomposition) *We have* $G = BNB$.

Proof. Given $g \in G$, set $C' = g(C)$ and $\mathcal{S}' = g(\mathcal{S})$ and let $\mathcal{S}''$ be an apartment containing C and C' (Proposition 9.6). As G is transitive on the set of ordered bases of $PG(n, K)$, there are an element $b_1 \in B$ mapping $\mathcal{S}$ onto $\mathcal{S}''$, an element $n \in N$ mapping C onto $b_1^{-1}(C')$ and an element $b_2 \in B$ mapping $\mathcal{S}$ onto $n^{-1}b_1^{-1}(\mathcal{S}')$. Trivially, $b_1 n b_2$ maps $(C, \mathcal{S})$ onto $(C', \mathcal{S}')$. Therefore $g^{-1}b_1 n b_2$ fixes both C and $\mathcal{S}$. That is, $g^{-1}b_1 n b_2 \in B \cap N = H$. Hence $g \in BnB$. Therefore $G = BNB$. □

Exercise 9.11. Let Γ be a projective geometry. Prove that every pair of chambers at maximal distance in $\mathcal{C}(\Gamma)$ arises from two opposite orderings of some basis of Γ.

(Hint: use Proposition 9.6.)

Exercise 9.12. With the notation of Sections 9.3.2 and 9.3.3, let B^{op} be the stabilizer of C^{op} in $G = PGL(n+1, K)$. Prove that $G = \langle B, B^{op} \rangle$.

(Hint: we have $N \leq \langle B, B^{op} \rangle$. Hence $G = \langle B, B^{op} \rangle$ by Proposition 9.7.)

Exercise 9.13. With the notation of Section 9.3.3, prove that U contains the commutator subgroup of B and that the commutator subgroup of N has the form $H \cdot A_{n+1}$.

Exercise 9.14. Without using the information that $PSL(n+1, K)$ is simple, prove that $PSL(n+1, K)$ is perfect (namely, it is its own commutator subgroup).

(Hint: let G be the commutator subgroup of $PSL(n+1, K)$ and let $H_s = H \cap PSL(n+1, K)$ be the Cartan subgroup of $PSL(n+1, K)$. Then $B_s = UH_s$ is the Borel subgroup of $PSL(n+1, K)$ and we have $B_s \leq G$ (Exercise 9.13). The statement of Exercise 9.12 also holds for B_s in place of B. The equality $G = PSL(n+1, K)$ follows from that.)

9.3.4 Finite projective geometries

In this section Γ is a flag-transitive n-dimensional finite projective geometry ($n \geq 2$). As Γ is flag-transitive, it admits uniform order q. The thin case ($q = 1$) has been considered in Section 9.2.2(2). Henceforth we assume $q \geq 2$. We recall that Γ is classical if $n \geq 3$ (Theorem 1.3).

We consider the 2-dimensional case, to begin with. We take the symbols P and L as types for points and lines respectively. Given a flag-transitive subgroup G of $Aut_s(\Gamma)$ and a chamber $C = (x_P, x_L)$ of Γ, with x_P a point and x_L a line, we denote by G_P and G_L the stabilizers in G of x_P and x_L respectively. The stabilizer $G_P \cap G_L$ of C will be denoted by B, as in Section 9.3.3.

Two finite classical projective planes are known admitting non-classical flag-transitive automorphism groups. These planes are $\Gamma = PG(2,2)$ and $\Gamma = PG(2,8)$, admitting the Frobenius groups $G = Frob_7^3$ and $G = Frob_{73}^9$ respectively as chamber-regular subgroups of $Aut_s(\Gamma)$ ([61], 4.4.16). We have $G_P \cong G_L \cong \mathbf{3}$ in $G = Frob_7^3$ and $G_P \cong G_L \cong \mathbf{9}$ in $G = Frob_{73}^9$. We have $B = 1$ in both cases. The point x_P (the line x_L) is the only point (the only line) fixed by G_P (by G_L). Furthermore, G_P and G_L belong to the same conjugacy class in G. Hence G_P (respectively, G_L) also fixes a unique line y_L (a unique point y_P). Trivially, $x_P \notin y_L$ and $y_P \notin x_L$. The Frobenius kernel of G acts as a transitive cyclic group both on the set of points and on the set of lines of Γ. $Frob_7^3$ and $Frob_{73}^9$ are the only examples of non-classical flag-transitive subgroups of $P\Gamma L(3, q)$, by the following well-known theorem ([61], 4.4.16):

Theorem 9.8 *Let G be non-classical subgroup of $P\Gamma L(3, q)$, flag-transitive on $PG(2, q)$. Then either $q = 2$ and $G = Frob_7^3$ or $q = 8$ and $G = Frob_{73}^9$.*

Kantor [108] has generalized this theorem to arbitrary finite projective planes and his has later been improved by Feit [65]:

Theorem 9.9 (Kantor–Feit) *Let Γ be a finite projective plane of order $q > 1$ and let $G \leq Aut_s(\Gamma)$ be flag-transitive. Then one of the following holds:*

(i) the number q is a prime power, $\Gamma \cong PG(2, q)$ and G is classical;
(ii) $\Gamma \cong PG(2, 2)$ or $PG(2, 8)$, and $G = Frob_7^3$ or $Frob_{73}^9$ respectively;
(iii) we have $q \equiv 0 \pmod 8$, q is not a prime power, $q + 1 \equiv 0 \pmod 3$, $q^2 + q + 1$ is prime, $G = Frob_{q^2+q+1}^{q+1}$ and G is chamber-regular on Γ.

Furthermore, $q > 14,400,008$ in case (iii) of the above theorem (see [65]). Needless to say, no examples are known for that case.

We have assumed $q > 1$. When $q = 1$, Γ is the ordinary triangle. Its special automorphism group is $S_3 = Frob_3^2$.

We now turn to the case of $n \geq 3$. If $n \geq 3$ then $\Gamma = PG(n, q)$ by Theorem 1.3. The following proposition could be obtained as a trivial consequence from a stronger result of Higman [80] (see Theorem 9.11). However, we will give a straightforward proof of it, to offer the reader an example of how to work with finite groups.

Proposition 9.10 *Let $G \leq P\Gamma L(4, q)$ be flag-transitive on $\Gamma = PG(3, q)$. Then, for every plane u of Γ, the stabilizer G_u of u in G acts as a classical group in the residue Γ_u of u.*

Proof. Let G be a counterexample to the above statement. By Theorem 9.8, we have $q = 2$ or 8. We will only consider the case of $q = 8$, leaving the other one for the reader.

Considering the stabilizer G_F of a point–plane flag F and the action of G_F on Γ_F and applying Theorem 9.8, we see that G_F acts as the cyclic group **9** on the nine lines of Γ_F. By this and by the flag-transitivity of G, for every point or plane x, the stabilizer G_x of x in G acts as $Frob_{73}^9$ in Γ_x. The kernel K_x of that action is the only subgroup of G_x fixing some chamber containing x, since $Frob_{73}^9$ is chamber-regular in $PG(2, 8)$.

Let $g \in K_x$ and let x be a point, to fix ideas. Let y be any plane incident to x, let G_y be the stabilizer of y in G and let K_y be the kernel of the action of G_y in Γ_y. We have $g \in K_y$ because g fixes the chambers of Γ_y containing x and because K_y is the only subgroup of G_y fixing some chamber containing y. Iterating this argument (and exploiting the connectedness of Γ) we see that g fixes all chambers of Γ, that is $g = 1$. Therefore, for every point or plane x we have $K_x = 1$ and $G_x = Frob_{73}^9$, chamber-regular in Γ_x. Hence G is chamber-regular in Γ.

For every point (plane) x, let $F_x \approx Z_{73}$ be the Frobenius kernel of G_x. If F_x fixes another point (plane) $y \neq x$, then it also fixes the line through x and y. However, we see this to be impossible when we consider the action of F_x in Γ_x. Therefore F_x acts semi-regularly on the $8 \cdot 73$ points (planes) of Γ other than x, with 8 orbits of size 73 on this set of points (planes). Since 73 does not divide the total number of planes (points) of Γ, the group F_x must also fix some plane (point). By the same argument as above, F_x cannot fix two distinct planes (points). Hence it fixes just one plane (point), let us call it $\delta(x)$. On the other hand, F_x does not fix anything in $\Gamma_{\delta(x)}$. Therefore x and $\delta(x)$ are not incident. Furthermore, $F_{\delta(x)} = F_x$, hence $\delta(\delta(x)) = x$.

Let us again assume that x is a point and let H be a complement of F_x in G_x. We have $H \approx \mathbf{9}$. Considering the action of H in Γ_x we also see that H fixes one line v_H on x and one plane y_H on x but not on v_H. Since H normalizes F_x, it must also fix the unique plane $\delta(x)$ fixed by F_x. Hence H fixes the point $v_H \cap \delta(x)$. Thus, seven points remain on v_H that can be moved by H. Since H is cyclic of order 9, there is a subgroup K of H of order 3 fixing all points of v_H. If z is any point of v_H, the group K appears as a subgroup of order 3 in the stabilizer G_z of z. Hence it normalizes F_z and fixes $\delta(z)$. Therefore K also fixes the line $y_H \cap \delta(z)$. However, $K \leq H \leq G_{y_H}$, where G_{y_H} is the stabilizer of y_H in G. We see from the action of G_{y_H} in Γ_{y_H} that K fixes just one line w_H on y_H, which is the unique line on y_H fixed by H. Hence $y_H \cap \delta(z) = w_H$ for all points $z \in v_H$. Therefore, $\delta(z) \supseteq w_H$ for all points $z \in v_H$.

However, we can substitute any other line for v_H in the above argument, by the flag-transitivity of G. Therefore the function δ mapping a point or a plane x onto $\delta(x)$ is a polarity of $PG(2,8)$. On the other hand, we have remarked that $x \notin \delta(x)$ for every point x. Hence δ should be a polarity of $PG(2,8)$ without any absolute points. This cannot be (see [61], 2.1.16). The contradiction is achieved. □

As we have remarked above, a statement much stronger than Proposition 9.10 holds:

Theorem 9.11 (Higman [80]) *Let G be a flag-transitive automorphism group of $PG(n,q)$, where $n \geq 3$. Then either G is classical or $q = 2$, $n = 3$ and $G \cong A_7$.*

The action of A_7 on $PG(3,2)$. The flag-transitive action of the alternating group A_7 on $PG(3,2)$ (Theorem 9.11) can be described using the following model of $PG(3,2)$.

We have remarked in Section 6.4.2 that there are 30 distinct ways to draw a model of $PG(2,2)$ on a given set S of seven points and that A_7 has two orbits of size 15 on that set of 30 models. Let X and Y be those two orbits. We take X as set of points, Y as set of planes and the 3-subsets of S as lines. Given a point or a plane $z \in X \cup Y$ and a line u, we say that u and z are incident if the 3-set u is one of the lines of the model z of $PG(2,2)$. Given a point $x \in X$ and a plane $y \in Y$, we say that x and y are incident when there is a line incident to both of them. The reader may check that the above clauses define a geometry Γ belonging to the Coxeter diagram A_3 with uniform order 2. We have $\Gamma = PG(3,2)$, by Corollary 7.7 (and Theorem 1.3). The stabilizer in S_7 of a model of $PG(2,2)$ is $L_3(2)$. However, $L_3(2)$ is simple. Hence it is contained in A_7. Therefore A_7 acts flag-transitively in Γ, with point-stabilizers and plane-stabilizers isomorphic to $L_3(2)$.

The symmetric group S_7 permutes X and Y. Hence it is obtained adding a polarity of $PG(3,2)$ to A_7.

The above also shows that the point–line system of $PG(3,2)$ is just the dual point–line system of the $Alt(7)$-geometry.

The reader may see [136] for another description of the action of A_7 in $PG(3,2)$, exploiting a construction of the exceptional isomorphism $A_8 \cong L_4(2)$ inside the Mathieu group M_{24}.

9.4 Classical polar spaces

9.4.1 Classical automorphism groups of polar spaces

We will only consider classical polar spaces defined by finite dimensional vector spaces over fields.

Let $\mathcal{P}$ be a classical polar space defined by a trace valued non-degenerate (σ,ε)-sesquilinear form φ of Witt index $n \geq 2$ in $V = V(m,K)$ (Chapter 1, Appendix I). We now assume that the division ring K is commutative (that is, it is a field). When φ is a symmetric bilinear form and $m = 2n$, then we also assume $n \geq 3$, in order to avoid grids (which are not classical polar spaces, according to a convention stated in Chapter 1).

$Aut_s(\mathcal{P})$ is the group of those elements of $P\Gamma L(m,K)$ represented by a semilinear transformation $f \in \Gamma L(m,K)$ such that, for every choice of vectors $x, y \in V$, we have $\varphi(x,y) = 0$ if and only if $\varphi(f(x), f(y)) = 0$. If $n \geq 3$, then $Aut_s(\mathcal{P}) = Aut(\mathcal{P})$. If $n = 2$, then either $Aut(\mathcal{P}) = Aut_s(\mathcal{P})$ or $Aut_s(\mathcal{P})$ has index 2 in $Aut(\mathcal{P})$ (this happens for $\mathcal{S}_3(q)$ with q even, for instance; see Section 1.2.5).

By $PGO(\varphi)$ (by $P\Gamma O(\varphi)$) we denote the normal subgroup of $Aut_s(\mathcal{P})$ consisting of those elements represented by a linear (respectively, semilinear) transformation f such that $\varphi(x,y) = \varphi(f(x), f(y))$ for every choice of vectors $x, y \in V$. We set $PSO(\varphi) = PGO(\varphi) \cap PSL(m, K)$. The commutator subgroup of $PSO(\varphi)$ is denoted by $P\Omega(\varphi)$. Note that $P\Omega(\varphi)$ might be a proper subgroup of $PSO(\varphi)$. Also, $P\Gamma O(\varphi)$ may be a proper subgroup of $Aut_s(\mathcal{P})$. For instance, all semilinear transformations f that map φ onto a form proportional to φ represent elements of $Aut_s(\mathcal{P})$, but they need not represent elements of $P\Gamma O(\varphi)$. To give a more precise description of $Aut_s(\mathcal{P})$ we should define dominant embeddings of polar spaces in projective spaces ([222], Chapter 8), but we are not going so far.

A subgroup of $Aut_s(\mathcal{P})$ is said to be *classical* if it contains $P\Omega(\varphi)$.

If φ is a symmetric bilinear form then the groups $PGO(\varphi)$, $PSO(\varpi)$ and $P\Omega(\varphi)$ are called *orthogonal groups.*

If φ is a symmetric bilinear form of Witt index n and $m = 2n+1$, then we write $PGO(2n+1, K)$, $PSO(2n+1, K)$ and $P\Omega(2n+1, K)$ for $PGO(\varphi)$, $PSO(\varphi)$ and $P\Omega(\varphi)$. If furthermore $K = GF(q)$, then we substitute q for K in the above symbols and we write $O_{2n+1}(q)$ for $P\Omega(2n+1, q)$.

When φ is a symmetric bilinear form of Witt index $n \geq 3$ and $m = 2n$, then we use the symbols $PGO^+(2n, K)$, $PSO^+(2n, K)$ and $P\Omega^+(2n, K)$ instead of $PGO(\varphi)$, $PSO(\varphi)$ and $P\Omega(\varphi)$. If $K = GF(q)$, then we substitute q for K in the above symbols and we write $O^+_{2n}(q)$ for $P\Omega^+(2n, q)$.

When φ is a symmetric bilinear form of Witt index n in $V(2n+2, q)$, then we write $PGO^-(2n+2, q)$, $PSO^-(2n+2, q)$ and $P\Omega^-(2n+2, q)$ (also $O^-_{2n+2}(q)$) for $PGO(\varphi)$, $PSO(\varphi)$ and $P\Omega(\varphi)$ respectively.

If φ is an alternating bilinear form of Witt index n (hence $m = 2n$), then $PGO(\varphi) = PSO(\varphi) = P\Omega(\varphi)$ (see [10]). We write $PSp(2n, K)$ for $PGO(\varphi)$. When $K = GF(q)$, then we write $S_{2n}(q)$ for $PSp(2n, GF(q))$. The group $PSp(2n, K)$ is called the *symplectic group* of rank n over K.

If $K = GF(q^2)$ and φ is a Hermitian form of Witt index n ($m = 2n$ or $2n+1$), then $PSO(\varphi) = P\Omega(\varphi)$. We now write $PGU(m, q^2)$ and $U_m(q)$ for $PGO(\varphi)$ and $PSO(\varphi)$, respectively. The groups $PGU(m, q^2)$ and $U_m(q)$ are called *unitary groups.*

Definition and notation as above can also be stated in the case when $\mathcal{P}$ is defined by a non-singular (σ, ε)-quadratic form φ. In this case the definition of $PGO(\varphi)$ must be modified requiring that $\varphi(x) = \varphi(f(x))$ instead of $\varphi(x, y) = \varphi(f(x), f(y))$. An element of $P\Gamma L(m, K)$ belongs to $Aut_s(\mathcal{P})$ if and only if it is represented by a semilinear transformation f such that both the following hold:

(i) for every vector $x \in V$, we have $\varphi(x) = 0$ iff $\varphi(f(x)) = 0$;
(ii) given any two vectors $x, y \in V$, we have $\Phi(x, y) = 0$ if and only if $\Phi(f(x), f(y)) = 0$, where Φ is the sesquilinearization of φ.

In particular, we need to refer to quadratic forms instead of bilinear forms to define the orthogonal groups $O_{2n+1}(q)$, $O^+_{2n}(q)$ and $O^-_{2n+2}(q)$ with q even.

We notice that, if q is even, then $O_{2n+1}(q) \cong S_{2n}(q)$, in accordance with the fact that the polar spaces $\mathcal{Q}_{2n}(q)$ and $\mathcal{S}_{2n-1}(q)$ are isomorphic in this case (see Section 1.3.4; also [62], [63]). Furthermore, $O_5(q) \cong S_4(q)$ for every prime power q, in accordance with the fact that the generalized quadrangles $\mathcal{Q}_4(q)$ and $\mathcal{S}_3(q)$ are dually isomorphic (Section 1.2.5; also [62], [63]).

The groups $O_{2n+1}(q)$, $O^+_{2n}(q)$ $(n \geq 3)$, $O^-_{2n+2}(q)$, $S_{2n}(q)$, $U_{2n}(q)$ and $U_{2n+1}(q)$ are simple, except for $S_4(2)$ $(= U_5(2))$ (see [35]). Actually, $S_4(2)$ is isomorphic to the symmetric group S_6 on 6 elements (see the discussion of case (i) of Theorem 9.16 in Section 9.4.5).

We have assumed $n \geq 3$ in $O^+_{2n}(q)$ because we do not want to consider grids. Note that the automorphism group of $\mathcal{Q}^+_3(q)$ is a wreath product $S_{q+1} \wr 2$.

9.4.2 Interlude: apartments and frames

Let $\mathcal{P}$ be a (possibly non-classical or even weak) polar space of rank $n \geq 2$. Given a maximal subspace X of $\mathcal{P}$, let $\mathcal{A} = (a_1, a_2, ..., a_n)$ be an ordered basis of X. Set $I = \{1, 2, ..., n\}$ and $X_J = \langle a_j \rangle_{j \in J}$ for $J \subseteq I$. Given another maximal subspace X' of $\mathcal{P}$ not intersecting X (see Section 1.3.5(v)), let a'_i be the point $X^{\perp}_{I-i} \cap X'$, for $i = 1, 2, ..., n$. The n-tuple $\mathcal{A}' = (a'_1, a'_2, ..., a'_n)$ is an ordered basis of X'. The pair of n-tuples $\mathcal{F} = ((a_1, a_2, ..., a_n), (a'_1, a'_2, ..., a'_n))$ is called an *ordered frame* for $\mathcal{P}$. Let us set $A = \{a_1, a_2, ..., a_n\}$, $A' = \{a'_1, a'_2, ..., a'_n\}$ and $F = A \cup A'$. The set F is called the *unordered frame* of $\mathcal{P}$ associated with the ordered frame $\mathcal{F}$.

The collinearity graph of $\mathcal{P}$ induces a complete n-partite graph $\mathcal{G}_F$ on F with classes $\{a_i, a'_i\}$ $(i \in I)$. The cliques of this graph form a thin polar space $\mathcal{P}_F$ of rank n and can be identified with the subspaces of $\mathcal{P}$ which they span. Thus, $\mathcal{P}_F$ can be viewed as a subgeometry of $\mathcal{P}$. We call $\mathcal{P}_F$ the *apartment* of $\mathcal{P}$ defined by the unordered frame F.

The chamber $C = \{X_{\{1,2,...,i\}}\}_{i=1}^{n}$ of $\mathcal{P}$ is also a chamber of the apartment $\mathcal{P}_F$. Thus, giving an ordered frame of $\mathcal{P}$ is the same as giving a pair $(C, \mathcal{P}_F)$ where $\mathcal{P}_F$ is an apartment and C is a chamber of $\mathcal{P}_F$. It is clear from the way in which frames have been defined that every chamber of $\mathcal{P}$ is contained in some apartment. Moreover:

Proposition 9.12 *Given any two chambers C, D of $\mathcal{P}$, there is some apartment containing both C and D.*

The rank 3 case of the above statement with C, D at maximal distance has been considered in Exercise 1.11. We leave the proof of the general case for the reader (the reader may also consult Chapter 7 of [222]).

When $\mathcal{P}$ is classical, giving an ordered frame for $\mathcal{P}$ is essentially the same as giving a matrix for the sesquilinear form representing $\mathcal{P}$ or associated with the quadratic form representing $\mathcal{P}$. Indeed, let $\mathcal{P}$ be classical, defined by a non-degenerate trace valued (σ, ε)-sesquilinear form φ of Witt index n in $V = V(m, K)$ (the quadratic form case can be dealt with in a similar way). Let $((p_1, p_2, ..., p_n), (p'_1, p'_2, ..., p'_n))$ be an ordered frame of $\mathcal{P}$. Then we can choose representative vectors e_i and e'_i of p_i and p'_i respectively $(i = 1, 2, ..., n)$ and vectors e''_1, e''_2, ..., e''_k (if $k = m - 2n > 0$) in such a way that $\mathcal{E} = (e_1, ..., e_n, e'_1, ..., e'_n, e''_1, ..., e''_k)$ is a basis of V and the form φ is represented with respect to $\mathcal{E}$ by an $m \times m$ matrix as follows:

$$\begin{pmatrix} O_n & I_n & O_{n,k} \\ I_n & O_n & O_{n,k} \\ O_{k,n} & O_{k,n} & D \end{pmatrix}$$

The symbols O_n and I_n here denote the $n \times n$ null and identity matrices respectively, $O_{n,k}$ and $O_{k,n}$ are the $n \times k$ and the $k \times n$ null matrices and D is an invertible diagonal $k \times k$ matrix (needless to say that the matrices $O_{n,k}$, $O_{k,n}$, D appear only if $m > 2n$). Conversely, given a basis $\mathcal{E}$ of V with respect to which φ is represented by a matrix as above, the first $2n$ vectors of $\mathcal{E}$ give us an ordered frame of $\mathcal{P}$ (the reader may notice that Exercise 1.13 dealt with a particular case of the above).

9.4.3 Properties of $PGO(\varphi)$ (the thick case)

Let $\mathcal{P}$, φ, n be as in Section 9.4.1. We now assume that $\mathcal{P}$ is thick. We will only consider $PGO(\varphi)$, but what we will say can be repeated with a few minor changes for any classical subgroup of $P\Gamma O(\varphi)$ containing $PSO(\varphi)$. Classical subgroups of $P\Gamma O(\varphi)$ not containing $PSO(\varphi)$ (in particular, $P\Omega(\varphi)$ when $P\Omega(\varphi) \neq PSO(\varphi)$) require some more work, but we are not going to deal with them here. The reader may consult the books of Dieudonné [62] and [63] for more information on those subgroups.

Let us write G for $PGO(\varphi)$, for short. The group G is transitive on the set of ordered frames of $\mathcal{P}$. That is, it is transitive on the set of pairs $(C, \mathcal{P}_F)$, where $\mathcal{P}_F$ is an apartment and C is a chamber in it. In particular, G is flag-transitive.

Given a pair $(C, \mathcal{P}_F)$ as above, let B be the stabilizer in G of the chamber C of $\mathcal{P}$ and let N be the stabilizer of the apartment $\mathcal{P}_F$. The subgroup B is called the *Borel* subgroup of G relative to the chamber C. The subgroup $H = B \cap N$ is the stabilizer of the ordered frame $\mathcal{F}$ associated with the pair $(C, \mathcal{P}_F)$. It is called the *Cartan subgroup* of $PGO(\varphi)$ relative to the apartment $\mathcal{P}_F$. We have $H \trianglelefteq N$ and $W = N/H$ is the full automorphism group of the thin polar space $\mathcal{P}_F$. Hence $W \approx \mathbf{2}^n : S_n$ (see Section 9.2.2(3)). The factor group W is called the *Weyl group* of G.

Let us take $I = \{1, 2, ..., n\}$ as set of types for $\mathcal{P}$ (that is, the type of an element of $\mathcal{P}$ is its dimension as a linear subspace of V). The stabilizer in G of the subflag F_J of C of type $J \subseteq I$ will be denoted by G_J. We call it the *parabolic subgroup* of type $I - J$ relative to the chamber C. We have $G_\emptyset = G$, $G_I = B$ and $G_J = \bigcap_{j \in J} G_j = \langle G_{I-i} \rangle_{i \in I-J}$ for $J \subseteq I$. If $\emptyset \neq J \neq I$, then the parabolic subgroup G_J is said to be *proper*. The parabolic subgroups of type $I - i$ ($i \in I$) are stabilizers of elements of $\mathcal{P}$ of type i. They are called *maximal* parabolic subgroups. The parabolic subgroups of type $i \in I$ are stabilizers of panels of cotype i. They are called *minimal* parabolic subgroups.

Proposition 9.13 (Bruhat decomposition) *We have* $G = BNB$.

Proof. The proof is just the same as that of Proposition 9.7, except for using Proposition 9.12 instead of Proposition 9.6. □

The reader may check that distinct flags have distinct stabilizers (see also Proposition 10.26). Therefore, every parabolic subgroup is its own normalizer. In particular, $B = N_G(B)$.

9.4.4 Properties of $PGO(\varphi)$ (the top-thin case)

In this section $\mathcal{P}$ is a top-thin classical polar space of rank $n \geq 4$. The D_n building obtained by unfolding $\mathcal{P}$ (Proposition 5.23) will be denoted by Γ.

Proposition 9.14 (Tits [222], Section 8.4.3) *We have* $\mathcal{P} = \mathcal{Q}^+_{2n-1}(K)$ *for some field* K.

Hence $PGO(\varphi) = PGO^+(2n, K)$. All that we said for $PGO(\varphi)$ in Section 9.4.3 also holds for $PGO^+(2n, K)$, except for the very last claims of Section 9.4.3. Indeed, as $\mathcal{P}$ is now top-thin, every panel of $\mathcal{P}$ of cotype n is in precisely two chambers. Therefore, if C and C' are the two chambers intersecting in a panel of cotype n, if G_{I-n} denotes the stabilizer of the panel $C \cap C'$ and B is the stabilizer of C, then B has index 2 in G_{I-n}. Hence $B \trianglelefteq G_{I-n}$ and B also stabilizes C'. Actually, G_{I-n} is the normalizer of B and two distinct chambers of $\mathcal{P}$ are stabilized by the same subgroup if and only if they meet in a panel of cotype n (that is, they are n-adjacent).

When $n \neq 4$, we set $G_0 = Aut(\Gamma)$. If $n = 4$, then G_0 will denote the subgroup of $Aut(\Gamma)$ fixing the type of Γ corresponding to the points of $\mathcal{P}$. Since the diagram D_n admits recovery from Grassmann geometries with respect to its initial node (Section 5.4), we have $Aut(\mathcal{P}) = G_0$.

However, as $Aut(\mathcal{P})$ is flag-transitive in $\mathcal{P}$, it also contains some element f interchanging two n-adjacent chambers C and C' of $\mathcal{P}$. The maximal subspaces of $\mathcal{P}$ contained in C and in C' have opposite types in Γ and f interchanges them. Hence f cannot be type-preserving when viewed as an automorphism of Γ.

Therefore $Aut_s(\Gamma)$ has index 2 in $Aut(\mathcal{P})$. Hence $Aut_s(\Gamma)$ is a normal proper subgroup of $Aut(\mathcal{P})$ and it contains the commutator subgroup of $Aut(\mathcal{P})$. Similarly, if G is any flag-transitive subgroup of $Aut(\mathcal{P})$, then the subgroup $G \cap Aut_s(\Gamma)$ has index 2 in G and it contains the commutator subgroup of G. Therefore, the subgroup $Aut_s(\Gamma)$ of $Aut(\mathcal{P})$ is not flag-transitive on $\mathcal{P}$ and, for every subgroup G of $Aut(\mathcal{P})$, the commutator subgroup of G cannot be flag-transitive on $\mathcal{P}$. In particular, $P\Omega^+(2n, K)$ is not flag-transitive on $\mathcal{P}$. We will now prove that, nevertheless, $P\Omega^+(2n, K)$ is flag-transitive on Γ.

Given an apartment $\mathcal{P}_F$ of $\mathcal{P}$ and a chamber C of $\mathcal{P}_F$, let N and B be the stabilizers in $PGO^+(2n, K)$ of $\mathcal{P}_F$ and C. Set $H = B \cap N$ and $W = N/H$, as in Section 9.4.3. Let X be the maximal subspace of $\mathcal{P}$ in C and let G_n denote the stabilizer of X. We have $G_n \leq Aut_s(\Gamma)$, because G_n cannot interchange the two horns of the D_n diagram, since it fixes X. Let N' and G'_n be the commutator subgroups of N and G_n respectively. The reader may check that, for a suitable basis of $V(2n, K)$ (see Section 9.4.2), the elements of G_n are represented by matrices of the following form:

$$\begin{pmatrix} M & MA \\ O_{n,n} & (M^t)^{-1} \end{pmatrix}$$

where $M \in GL(n, K)$, M^t is the transpose of M and A is an $n \times n$ antisymmetric matrix. Therefore $G_n \leq PSO^+(2n, K)$ and the elements of G'_n are represented by matrices as above, except that we must now take M in $SL(n, K)$ instead of in $GL(n, K)$.

The Weyl group $W \approx \mathbf{2}^n : S_n$ is the automorphism group of the thin C_n building $\mathcal{P}_F$. Using the information given in Section 9.2.2, (3) and (4), the reader can check that the commutator subgroup W' of W is the special automorphism group $Aut_s(\Gamma_F) \approx \mathbf{2}^{n-1} : S_n$ of the thin D_n building Γ_F obtained by unfolding $\mathcal{P}_F$. The group W' $(= Aut_s(\Gamma_F))$ is chamber-regular on Γ_F (see Section 9.2.2(4)).

The commutator subgroup G'_n of G_n is transitive on the set of apartments of $\mathcal{P}$ containing X (the reader may check this by a direct computation, exploiting the information on G_n we have given above). Hence G'_n is transitive on the set of apartments of $\mathcal{P}$ containing C. As $PGO^+(2n, K) = BNB$ and $B \leq G_n$, we also have $PGO^+(2n, K) = \langle G_n, N \rangle$. Therefore the group $P\Omega^+(2n, K)$ (which contains $\langle G'_n, N' \rangle$) is flag-transitive on Γ. Actually, it is the minimal flag-transitive subgroup of $Aut_s(\Gamma)$.

As for automorphism groups of polar spaces, a subgroup of $Aut_s(\Gamma)$ is said to be *classical* if it contains $P\Omega^+(2n, K)$.

Remark. We have assumed above $n \geq 4$, but all we have said remains true in the case of $n = 3$. In that case, we have $\Gamma = PG(3, K)$ by Proposition 5.23, where K is a field because $\mathcal{P}$ is assumed to be classical. We have $\mathcal{P} = \mathcal{Q}_5^+(K)$ (see Section 1.3.6 or [222], 8.4.3). Therefore $P\Omega^+(6, K) = PSL(4, K)$.

Exercise 9.15. Prove that, if $char(K) \neq 2$ and n is even, then we have $PSO^+(2n, K) = PGO^+(2n, K) \cap Aut_s(\Gamma)$.

Exercise 9.16. Describe the subgroups B, N and H of $PGO^+(2n, K)$ as groups of matrices.

Exercise 9.17. Prove that a parabolic subgroup of $PGO^+(2n, K)$ of type J is contained in $PSO^+(2n, K)$ if and only if $n \notin J$.

Exercise 9.18. Let Γ be the unfolding of $\mathcal{Q}_{2n-1}^+(K)$. Prove that distinct flags of Γ have distinct stabilizers in $PSO^+(2n, K)$ (that is, stabilizers in $PSO^+(2n, K)$ of flags of Γ are their own normalizers in $PSO^+(2n, K)$).

Exercise 9.19. Let $G = PSO(2n + 1, K)$ with $n \geq 2$ and $char(K) \neq 2$. Prove that G contains a subgroup $A \approx \mathbf{2}^{2n} : A_{2n+1}$.

(Hint: the subgroup N of $PSL(2n + 1, K)$ is contained in G.)

Exercise 9.20. Let $A \approx \mathbf{2}^4 : A_5$ be a subgroup of $U_5(3)$ ($= S_4(3)$) as in the previous exercise. Prove that A is flag-transitive on $\mathcal{Q}_4(3)$ (on $\mathcal{S}_3(3)$).

(Hint. The stabilizer in A of a chamber of $\mathcal{Q}_4(3)$ has index 160 in A. There are 160 chambers in $\mathcal{Q}_4(3)$, hence ...)

Exercise 9.21. Affine polar spaces. Let Γ be an affine polar space obtained by removing a hyperplane H from a classical polar space $\mathcal{P}$ of rank $n \geq 3$ (Section 2.2.2). Let G_H be the stabilizer of H in $Aut(\mathcal{P})$. Prove that $G_H = Aut(\Gamma)$ and that G_H is flag-transitive on Γ.

9.4.5 The finite case

All finite thick polar spaces of rank $n \geq 3$ are classical (Theorem 1.6). Furthermore:

Theorem 9.15 (Seitz [187]) *All flag-transitive automorphism groups of finite thick polar spaces of rank $n \geq 3$ are classical.*

In the rank 2 case we have the following:

Theorem 9.16 (Seitz [187]) *Let Γ be a finite thick classical generalized quadrangle and let G be a flag-transitive subgroup of $Aut_s(\Gamma)$. Then either G is classical or one of the following occurs:*

(i) $\Gamma = \mathcal{S}_3(2)$ $(= \mathcal{Q}_4(2))$ *and* $G = A_6$;
(ii) $\Gamma = \mathcal{S}_3(3)$ *or* $\mathcal{Q}_4(3)$ *and* G *is one of* $\mathbf{2}^4 : A_5$, $\mathbf{2}^4 : S_5$ *or* $\mathbf{2}^4 : Frob_5^4$;
(iii) $\Gamma = \mathcal{Q}_5^-(3)$ *or* $\mathcal{H}_3(3^2)$ *and* $G \approx L_3(4) \cdot \mathbf{2}$ *or* $G \approx L_3(4) : \mathbf{2}^2$.

We warn that (iii) and the $\mathbf{2}^4 : Frob_5^4$ case of (ii) are missing in [187]. Seitz wrote a correction soon after his paper [187] had been published, but he never published it. However, corrected versions of his theorem are available in the literature (Theorem C.7.1 of [106], for instance; also [125] (Lemma 4.1) and [128] (2.1 and 2.2)).

As for (i) of Theorem 9.16, we remark that $S_4(2)$ is not simple. Indeed we have $S_4(2) \cong S_6$ (see [41]), hence A_6 is the commutator subgroup of $S_4(2)$. The isomorphism $S_4(2) \cong S_6$ becomes evident if we remark that we can realize the generalized quadrangle $\mathcal{S}_3(2)$ inside $PG(2,4)$ as follows. Let O be a hyperoval of $PG(2,4)$, consisting of a non-degenerate conic plus its nucleus [81]. Take $PG(2,4) - \mathcal{O}$ as set of points. As lines we take the lines of $PG(2,4)$ meeting $\mathcal{O}$. The incidence relation is the one inherited from $PG(2,4)$. It is straightforward to check that a generalized quadrangle Γ of order $(2,2)$ is obtained in this way, and we have $\Gamma = \mathcal{S}_3(2)$ (by [166], Chapter 6). The stabilizer of $\mathcal{O}$ in $P\Gamma L(3,4)$ is S_6 and it is an automorphism group for Γ. However, the groups S_6 and $S_4(2) = Aut_s(\Gamma)$ have the same order. Hence $S_6 = S_4(2)$.

Let us turn to (ii) of Theorem 9.16. The following model of $\mathcal{S}_3(3)$ can explain how things go here. We take $PG(3,4) - \mathcal{H}_3(2^2)$ as set of points and we take as lines the bases of $PG(3,4)$ consisting of four non-isotropic mutually orthogonal points (with respect to the Hermitian form defining $\mathcal{H}_3(2^2)$). The reader may check that a generalized quadrangle Γ of order $(3,3)$ is obained in this way. We have $\Gamma = \mathcal{S}_3(3)$ ([166], Chapter 6). Trivially, $U_4(2) \leq Aut_s(\Gamma) = S_4(3)$. Comparing orders, we see that $U_4(2) = S_4(3)$. The group $\mathbf{2}^4 : L_2(4)$ is the stabilizer in $U_4(2)$ of a line of $\mathcal{H}_3(2^2)$. Hence it is a subgroup of $S_4(3)$. Furthermore, $A_5 = L_2(4)$ (see [41]). Hence $\mathbf{2}^4 : A_5 = \mathbf{2}^4 : L_2(4)$. Considering the action of this group in Γ, we can see that $\mathbf{2}^4 : A_5$ is indeed flag-transitive on $\mathcal{S}_3(3)$.

The group $\mathbf{2}^4 : S_5$ is the stabilizer in $PGU(4,2^2)$ of a line of $\mathcal{H}_3(2^2)$ and it contains $\mathbf{2}^4 : Frob_5^4$. To discuss the case of $\mathbf{2}^4 : Frob_5^4$ and case (iii) of Theorem 9.16 would involve too many technical details, and hence this is omitted. We only remark that the group $G \approx L_3(4) \cdot 2$ of (iii) can be any of the two groups called $L_3(4) \cdot 2_2$ and $L_3(4) \cdot 2_3$ on page 23 of [41].

Only classical generalized quadrangles are considered in Theorem 9.16. Actually, just two finite thick flag-transitive non-classical generalized quadrangles are known. They are of type $T_2^*(O)$ (see Section 2.3), with orders $(3,5)$ and $(15,17)$ respectively. We remark that, if Γ is a generalized quadrangle of type $T_2^*(O)$ obtained from a hyperoval O of the plane at infinity

$PG(2,q)$ of $AG(3,q)$ ($q = 2^h$, $h \geq 2$), then Γ admits an automorphism group $G \approx V(3,q) : G_O$, where $V(3,q)$ is the translation group of $AG(3,q)$ and G_O is the stabilizer of O on $\Gamma L(3,q)$. It is clear that G is flag-transitive in Γ if and only if O has been chosen in such a way that G_O acts transitively on the $q+2$ points of O. Such a choice of O is possible when $q = 4$ or 16 ([81], p. 177; actually, $q = 2$ would also work, but we have assumed $q \geq 4$, in order to avoid the non-thick case). Every hyperoval O of $PG(2,4)$ is obtained by adding the nucleus to a non-degenerate conic, and G_O acts as S_6 on O (see [81]). The Lunelli–Sce hyperoval is the unique (up to isomorphism) hyperoval O of $PG(2,16)$ such that G_0 is transitive on O (see [81]).

We have not considered the top-thin case in Theorem 9.15 because $PSO^+(2n,q)$ is not flag-transitive on $\mathcal{Q}^+_{2n-1}(q)$ (see Section 9.4.4). Instead of $\mathcal{Q}^+_{2n-1}(q)$ we should consider the building of type D_n obtained by unfolding $\mathcal{Q}^+_{2n-1}(q)$. We recall that all thick buildings of type D_n are unfoldings of hyperbolic quadrics (Proposition 9.14).

Theorem 9.17 (Seitz [187]) *Every flag-transitive automorphism group of a finite thick building of type D_n ($n \geq 4$) is classical.*

The next corollary is an easy consequence of Theorem 9.17 and Proposition 5.23.

Corollary 9.18 *Let $\mathcal{P}$ be a finite thick-lined top-thin polar space of rank $n \geq 4$. Then all flag-transitive automorphism groups of $\mathcal{P}$ are classical.*

When $\mathcal{P}$ has rank 3, using Theorem 9.11 instead of Theorem 9.17 we obtain the following:

Corollary 9.19 *Let $\mathcal{P}$ be a finite thick-lined top-thin polar space of rank 3 and let $G \leq Aut(\mathcal{P})$ be flag-transitive. Then either G is classical or $\mathcal{P} = \mathcal{Q}^+_5(2)$ and $G = S_7$.*

(Note that S_7 contains a polarity of $PG(3,2)$; see the description of the action of A_7 on $PG(3,2)$ at the end of Section 9.3.5.)

9.4.6 The $Alt(7)$-geometry

We finish this section with a few remarks on the $Alt(7)$-geometry, even if it is not a polar space. Let Γ be the $Alt(7)$-geometry. It is clear from Section 6.4.2 that the alternating group A_7 is the full automorphism group of Γ. It is transitive on the set of planes of Γ with stabilizers of planes isomorphic to $L_3(2)$ (see the description of the action of A_7 on $PG(3,2)$, in Section

9.3.4). Hence A_7 is flag-transitive on Γ. As Γ has seven points, stabilizers in A_7 of points of Γ have index 7 in A_7. Hence they are isomorphic to $S_4(2)$ (see [41]).

The alternating group A_7 is the unique flag-transitive automorphism group of Γ. Indeed, let G be a flag-transitive automorphism group of Γ. If $G \neq A_7$, then the stabilizer G_u in G of a plane u of Γ is a proper subgroup of $L_3(2)$, flag-transitive on $\Gamma_u = PG(2,2)$. By Theorem 9.8, we have $G_u = Frob_7^3$. Therefore G has index 8 in A_7. However, A_7 has no subgroup of index 8 (see [41]). Hence $G = A_7$.

9.5 Affine geometries

9.5.1 Classical automorphism groups of $AG(n, K)$

Let $V = V(n, K)$, with K a division ring and $n \geq 2$, let $AG(n, K)$ be the affine geometry of V (Section 1.3.1) and $G = Aut(AG(n, K))$.

Given a vector $v \in V$, the *translation* of $AG(n, K)$ defined by v is the automorphism of $AG(n, K)$ mapping x onto $x + v$, for $x \in V$. Trivially, the translations of $AG(n, K)$ form a subgroup of G, isomorphic to the additive group of V. By abuse of notation, we use the letter V to designate the group of translations of $AG(n, K)$, too. The group V is transitive (even regular) on the set of points of $AG(n, K)$. Therefore, stabilizers in G of points of $AG(n, K)$ form one conjugacy class in G. Thus, if we deal with a point-stabilizer in G, we may always assume that the point we consider is the zero vector 0 of V. The group G is the semidirect product of V and of the stabilizer G_0 of 0 in G.

The geometry $AG(n, K)$ can be viewed as the complement of a hyperplane H of $PG(n, K)$, where points and lines of H correspond to the parallelism classes of lines and planes of $AG(n, K)$, respectively (Section 1.3.1). Hence G is the stabilizer of H in $P\Gamma L(n+1, K)$. Therefore $G_0 = \Gamma L(n, K)$ and $P\Gamma L(n, K)$ is the action of G_0 on the star of 0. Hence G is the group of all functions from V to V mapping $x \in V$ onto $f(x) + a$, for some choice of $f \in \Gamma L(n, K)$ and $a \in V$. Since G_0 acts flag-transitively on the star of 0 and V is transitive on the set of points of $AG(n, K)$, the group G is flag-transitive. We denote it by $A\Gamma L(n, K)$.

Let K be field. The subgroup $V : SL(n, K)$ of $A\Gamma L(n, K)$ is denoted by $ASL(n, K)$. It is flag-transitive on $AG(n, K)$. The subgroups of $A\Gamma L(n, K)$ containing $ASL(n, K)$ are said to be *classical.*

When $K = GF(q)$, we write $A\Gamma L(n, q)$, $AGL(n, q)$ and $ASL(n, q)$ for $A\Gamma L(n, K)$, $AGL(n, K)$ and $ASL(n, K)$ respectively.

Let $G_0 \leq \Gamma L(n, K)$ be flag-transitive on $PG(n-1, K)$ (note that, when

$n = 2$, flag-transitivity on $PG(1, K)$ is just transitivity on the $|K|+1$ points of $PG(1, K)$). The group $V : G_0$ is flag-transitive on $AG(n+1, K)$. In particular, $A\Gamma L(n, K)$, $AGL(n, K)$ and $ASL(n, K)$ (when K is a field) are flag-transitive on $AG(n, K)$.

Remark. We have assumed $n \geq 2$. However, we can also define $A\Gamma L(1, K)$ and $AGL(1, K)$. The group $AGL(1, K)$ consists of all functions from K to K mapping $x \in K$ onto $ax + b$ for some choice of $a, b \in K$ with $a \neq 0$. The elements of $A\Gamma L(1, K)$ are obtained by composing elements of $AGL(1, K)$ with automorphisms of K.

When $K = GF(q)$, we write $A\Gamma L(1, q)$ and $AGL(1, q)$ for $A\Gamma L(1, K)$ and $AGL(1, K)$ respectively.

9.5.2 Finite affine geometries

Let G be a flag-transitive subgroup of $A\Gamma L(n, q))$ ($n \geq 2$). It follows from [61] (Section 4.4.22) that $G = V : G_0$, where V is the group of translations of $AG(n, q)$, as in the previous section, and $G_0/Z(G_0)$ is flag-transitive on $PG(n-1, q)$. Thus, by Theorems 9.8 and 9.11 (and 1.4) we obtain the following classification of flag-transitive automorphism groups of finite affine geometries of dimension $n \geq 3$.

Theorem 9.20 *Let G be a flag-transitive automorphism group of $AG(n, q)$, with $n \geq 3$. Then one of the following holds:*

(i) the group G is classical;
(ii) we have $n = 4$, $q = 2$ and $G \approx \mathbf{2}^4 : A_7$;
(iii) we have $n = 3$, $q = 2$ and $G \approx \mathbf{2}^3 : Frob_7^3$;
(iv) we have $n = 3$, $q = 8$ and $\mathbf{8}^3 : Frob_{73}^9 \leq G \leq \mathbf{8}^3 : (\mathbf{7} \cdot Frob_{73}^9)$ (the factor $\mathbf{7}$ is the centre of $GL(3, 8)$).

More degrees of freedom are left for flag-transitive automorphism groups of finite affine planes. In particular, many non-classical flag-transitive finite affine planes are known [122] whereas no non-classical flag-transitive finite projective plane is known (apart from the ordinary triangle, of course). We remark that a finite affine plane may be flag-transitive even if the corresponding projective plane is not flag-transitive. Indeed, the flag-transitivity of an automorphism group G of a finite affine plane Γ only implies that Γ is a translation plane [122], that G contains all translations of Γ ([61], 4.4.22) and that the stabilizer in G of a point of Γ acts transitively on the bundle of lines on that point.

We are not going to discuss translation planes here. The reader wanting to know more on them is referred to [122] (or to [61], Chapter 4).

Exercise 9.22. Prove that $AGL(n, K)$ acts regularly on the set of ordered bases of the matroid $AG(n, K)$.

Exercise 9.23. Prove that the symmetric group S_{n+1} can be embedded in $PGL(n, K)$.

(Hint: if $n \geq 2$, then the stabilizer in $AGL(n, K)$ of a basis of $AG(n, K)$ is S_{n+1} and it does not contain any non-trivial translation.)

Exercise 9.24. Affine–dual-affine geometries. Let Γ be a bi-affine geometry obtained dropping from $PG(n, K)$ a hyperplane H and the star of a point p. Let $G_{H,p}$ be the stabilizer of H and p in $P\Gamma L(n+1, K)$. Prove that $Aut_s(\Gamma) = G_{H,p}$ and that $G_{H,p}$ is flag-transitive on Γ.

Exercise 9.25. The $Alt(8)$-geometry. Let Γ be the $Alt(8)$-geometry (Exercise 7.22). Check that Γ is flag-transitive, with A_8 as the unique flag-transitive automorphism group, and that stabilizers in A_8 of dual points of Γ are isomorphic to $ASL(3, 2)$ (needless to say, stabilizers in A_8 of points of Γ are isomorphic to A_7).

9.6 Finite matroids

9.6.1 Finite linear spaces

Flag-transitive finite projective or affine planes are finite flag-transitive linear spaces. We have already discussed them in Sections 9.3.4 and 9.5.2. The point–line system of $PG(n, q)$ $(n \geq 3)$ has the same automorphism group as $PG(n, q)$. Trivially, it is flag-transitive. The point-line system of $AG(n, q)$ $(n \geq 3,\ q \geq 3)$ is also a flag-transitive linear space, with the same automorphism group as $AG(n, q)$.

Every circular space with v points is flag-transitive, with special automorphism group S_v, which is also the full automorphism group when $v \geq 4$ (when $v = 3$, the circular space is a triangle, its special automorphism group is S_3 and its full automorphism group is $S_3 \times 2$). The chambers of a circular space can be viewed as ordered pairs of points, a chamber $\{x, \{x, y\}\}$ being represented by the pair (x, y). Thus, if $\mathcal{C}$ is a circular space with set of points S, then every 2-transitive group of permutations of S is a flag-transitive automorphism group of $\mathcal{C}$.

We remark that the point–line system of $AG(n, 2)$ is just the circular space with $v = 2^n$ points. Hence its full automorphism group is S_v. Note that S_v properly contains $AGL(n, 2)$ when $n \geq 3$. On the other hand, $S_4 = AGL(2, 2)$.

The geometries mentioned above are the most obvious examples of finite flag-transitive linear spaces, but more examples are known. We now describe some of them.

(1) **Hermitian unitals.** The Hermitian variety $H = \mathcal{H}_2(q^2)$ consists of q^3+1 points of $PG(2,q^2)$ with the property that, if a line L of $PG(2,q^2)$ meets H in at least two points, then $|L \cap H| = q+1$. Therefore, the lines of $PG(2,q^2)$ meeting H in at least two points form a linear space with order (q, q^2-1). We denote this linear space by $U_H(q)$ and we call it the *Hermitian unital* of *order* q.

The Hermitian unital $U_H(q)$ admits $U_3(q)$ as a flag-transitive automorphism group. We also remark that $U_3(q)$ is simple, except when $q=2$. As for this latter case, $U_3(2)$ is a subgroup of $ASL(2,3)$ with the following structure: $U_3(2) \approx \mathbf{3}^2 : \mathbf{2}^{1+2}$ (indeed we have $U_H(2) \cong AG(2,3)$, since the linear space $U_H(2)$ has order $(2,3)$).

(2) **Ree unitals.** Some knowledge of simple groups of twisted Lie type is needed to understand what follows. The reader who has never heard of these groups may skip this paragraph and go to the next one, or consult [35].

Given an odd positive integer h, let $q = 3^h$ and let G be the Ree group ${}^2G_2(q)$. The group G has q^3+1 Borel subgroups, forming one conjugacy class $\mathcal{B}$. Given two Borel subgroups $B_1, B_2 \in \mathcal{B}$, there is just one involution $i \in B_1 \cap B_2$ and i belongs to precisely $q+1$ Borel subgroups (including B_1 and B_2 among them, of course). Let $[B_1, B_2]$ be the set of Borel subgroups containing i. We take $\mathcal{B}$ as set of points and the sets $[B_1, B_2]$ as lines $(B_1, B_2 \in \mathcal{B},\ B_1 \neq B_2)$. We obtain a linear space of order (q, q^2-1). We denote this linear space by $U_R(q)$ and we call it the *Ree unital* of *order* q. The Ree group G is a flag-transitive automorphism group of $U_R(q)$ and we have $Aut(U_R(q)) = Aut(G)$.

We remark that G is simple if $h > 1$. If $h = 1$, then $G \cong P\Gamma L(2,8)$.

(3) **Witt–Bose–Shrikande spaces.** Let $q = 2^h$ $(h \geq 2)$ and let O be a hyperoval of $PG(2,q)$, consisting of a non-degenerate conic and its nucleus. We take $PG(2,q) - O$ as the set of 'lines'. The 'lines' of $PG(2,q)$ not meeting O are taken as 'points'. The incidence relation is the natural one, inherited from $PG(2,q)$. We obtain a flag-transitive linear space of order $(q/2-1, q)$. This linear space is often denoted by $\mathcal{W}(q)$ and it is called the *Witt–Bose–Shrikande space* of *order* q.

$\mathcal{W}(4)$ is just the circular space on 6 points. Hence $Aut(\mathcal{W}(4)) = S_6$. When $q \geq 8$, $Aut(\mathcal{W}(q)) = P\Gamma L(2,q)$, but $L_2(q)$ also acts flag-transitively on $\mathcal{W}(q)$.

(4) **Hering spaces.** The reader is referred to [79] for the description of these spaces. We only remark that they have order $(8, 90)$.

The following classification of finite flag-transitive linear spaces has been obtained by Buekenhout, Delandtsheer, Doyen, Kleidman, Liebeck and Saxl ([30] and [31]):

Theorem 9.21 *Let Γ be a finite linear space admitting a flag-transitive automorphism group G. Then one of the following occurs:*

(i) the linear space Γ is a flag-transitive projective or affine plane, or the point–line system of $PG(m,q)$ or of $AG(m,q)$, or a circular space;
(ii) the linear space Γ is one of the above examples (1)–(4);
(iii) we have $G \leq A\Gamma L(1,q)$ for some prime power q, and Γ has q points.

Note that (i) and (iii) are not mutually exclusive. For instance, all flag-transitive affine planes of order $q = p^m$ (p prime, $m \geq 1$) with full automorphism group not contained in $AGL(2m,p)$ belong to case (iii) (see [31]). If $\Gamma = PG(2,2)$ or $PG(2,8)$ and $G = Frob_7^3$ or $Frob_{73}^9$, then $G \leq AGL(1,p)$ with $p = 7$ or 73, respectively. Thus, we are in case (iii). Note also that $AGL(1,q)$ is 2-transitive on the q points of the affine line $AG(1,q)$, for every prime power q. Therefore, a circular space Γ with q points is included in (iii) if we choose $G \leq Aut(\Gamma)$ between $AGL(1,q)$ and $A\Gamma L(1,q)$.

Let us describe another class of examples for (iii). Let q be a prime power with $q \equiv 7 \ (mod\ 12)$. Hence $q \equiv 1 \ (mod\ 3)$ and the equation $x^3 = 1$ has three distinct roots in $GF(q)$. Let K be the set of cubic roots of 1 in $GF(q)$, let $P\Sigma L(2,q)$ be the extension of $L_2(q)$ by the automorphism group of $GF(q)$ and let $P\Sigma L(2,q)_\infty$ be the stabilizer in $P\Sigma L(2,q)$ of the point ∞ of $PG(1,q)$. Let us take $GF(q) = PG(1,q) - \infty$ as set of points and the images of K under $P\Sigma L(2,q)_\infty$ as lines. We obtain a linear space of order $(2,(q-3)/2)$ where $P\Sigma L(2,q)_\infty$ acts faithfully as a flag-transitive automorphism group. This linear space is called a *Netto triple system* and it is denoted by $N(q)$. Note that $N(7) = PG(2,2)$ and $P\Sigma L(2,7)_\infty = Frob_7^3$. On the other hand, when $q > 7$ then $N(q)$ has nothing to do with (i) of Theorem 9.21. In this case $Aut(N(q)) = P\Sigma L(2,q)_\infty$ (see [172]).

More examples for (iii) of Theorem 9.21 are given in [31] and in [111].

9.6.2 The case of rank $n \geq 3$

We first list the known examples of finite flag-transitive matroids of dimension $n \geq 3$.

(1) The n-dimensional simplex is a flag-transitive. The symmetric group S_{n+1} is its special automorphism group (Section 2.2.2(2)). The chambers of an n-dimensional simplex $\mathcal{S}$ can be identified with the $(n+1)!$ ways of ordering the $n+1$ vertices of $\mathcal{S}$ (a chamber $(v_0, \{v_0, v_1\}, ..., \{v_0, ..., v_{n-1}\})$ corresponds to the ordering $(v_0, ..., v_{n-1}, v_n)$). Therefore, a flag-transitive automorphism group of $\mathcal{S}$ must act $(n+1)$-transitively on the $n+1$ vertices of $\mathcal{S}$. Hence S_{n+1} is the unique flag-transitive automorphism group of $\mathcal{S}$.

Let Γ the upper n-truncation of an m-dimensional simplex $\mathcal{S}$ with $m > n$. Trivially, Γ is a flag-transitive matroid with the same special automorphism group S_{m+1} as $\mathcal{S}$. The chambers of Γ can be identified with the $(m+1)!/(m+1-n)!$ ways of choosing n of the $m+1$ vertices of $\mathcal{S}$ and ordering them. Therefore, any n-transitive group of permutations of the $m+1$ vertices of $\mathcal{S}$ is flag-transitive in Γ. For instance, A_{m+1} is flag-transitive in Γ (but not in $\mathcal{S}$). When $n \geq 6$, when $n = 5$ but $m \neq 11$ or 23 and when $n = 4$ but $m \neq 10$ or 22, then A_{m+1} and S_{m+1} are the only flag-transitive automorphism groups of Γ.

(2) The flag-transitive automorphism groups of $PG(n,q)$ have been classified in Theorem 9.11. The upper n-truncation of $PG(m,q)$ $(m > n)$ is also flag-transitive, with the same special automorphism group as $PG(m,q)$.

(3) The flag-transitive automorphism groups of $AG(n,q)$ have been classified in Theorem 9.20. The upper n-truncation of $AG(m,q)$ $(m > n)$ is also flag-transitive, with the same full automorphism group as $AG(m,q)$.

(4) The matroids $S(M_{22})$, $S(M_{23})$, $S(M_{24})$ of Theorem 7.23(iii) and the matroids $S(M_{11})$, $S(M_{12})$ of Theorem 7.24(iii) are flag-transitive. The automorphism group of $S(M_{22})$ is $Aut(M_{22})$ $(\approx M_{22} \cdot 2)$, but M_{22} also acts flag-transitively $S(M_{22})$. The Mathieu groups M_{23} and M_{24} are the unique flag-transitive automorphism groups of $S(M_{23})$ and $S(M_{24})$ respectively. The Mathieu groups M_{11} and M_{12} are the unique flag-transitive automorphism groups of $S(M_{11})$ and $S(M_{12})$ respectively.

(5) Let $\mathcal{D}$ be the 3-$(q^d+1, q+1, 1)$-design obtained by taking the projective line $PG(1,q^d)$ as set of points (with $q \geq 3$ and $d \geq 2$), choosing a copy X of $PG(1,q)$ inside $PG(1,q^d)$ and taking as blocks all images of X under the action of $P\Gamma L(2,q^d)$. The design $\mathcal{D}$ admits an enrichment $E(\mathcal{D})$ of rank 3, which is a flag-transitive matroid with automorphism group $Aut(E(\mathcal{D})) = P\Gamma L(2,q^d)$. Stabilizers in $Aut(E(\mathcal{D}))$ of blocks of $\mathcal{D}$ are isomorphic to $P\Gamma L(2,q) \times Z_d$. The subgroup $PGL(2,q^d) \leq Aut(E(\mathcal{D}))$ is also flag-transitive on Γ, with block stabilizers isomorphic to $PGL(2,q)$.

When $d = 2$, the matroid $E(\mathcal{D})$ belongs to the diagram $c.Af$ and $\mathcal{D}$ is a Mőbius plane (Exercise 7.3). Actually, it is the classical Mőbius plane obtained from $\mathcal{Q}_3^-(q)$ (see Exercise 10.25).

(6) Let $\mathcal{D}$ be the 3-$(q+1,4,1)$-design (with $q \equiv 7$ $(mod\ 12)$) obtained by taking the projective line $PG(1,q) = \infty \cup GF(q)$ as set of points and the images of $\infty \cup K$ under $P\Sigma L(2,q)$ as blocks, where K is the set of solutions of the equation $x^3 = 1$ in $GF(q)$ and $P\Sigma L(2,q)$ is the extension of $L_2(q)$ by the group of field automorphisms of $GF(q)$. This design admits an enrichment $E(\mathcal{D})$ of rank 3, which is a flag-transitive matroid belonging to the diagram $Af.L$ of Section 7.1.5 with order $(1,2,(q-3)/2)$. Residues of points of $E(\mathcal{D})$ are Netto triple systems and we have $Aut(E(\mathcal{D})) = P\Sigma L(2,q)$.

Theorem 9.22 (Delandtsheer) *The examples mentioned above are the only flag-transitive finite matroids of dimension $n \geq 3$.*

Outline of the Proof. The strategy of the proof is quite easy to explain, but the details are fairly technical. The reader should consult [52] and [53] for them.

Let Γ be as in the hypotheses of the theorem with $G \leq Aut_s(\Gamma)$ flag-transitive. We can work by induction on n.

Let $n = 3$, let $F = (a,u)$ be a point–plane flag of Γ and let G_a, G_u, G_F be the stabilizers in G of a, u, F respectively. By Theorem 9.21 we know the feasible isomorphism types of the residues Γ_a and Γ_u and feasible actions of G_a and G_u in Γ_a and Γ_u. We can now describe feasible actions of G_F in Γ_F. We can do this in two ways, either by considering G_F as the stabilizer of u in G_a and exploiting the information we have on feasible actions of G_a in Γ_a, or by considering G_F as the stabilizer of a in G_u and using what we know on feasible actions of G_u in Γ_u. The information we get on G_F from G_a must fit with that from G_u.

A case-by-case analysis shows that this happens only if Γ belongs to one of the following diagrams: $A_2.L$, $Af.L$ with lines of size ≥ 4, $L.A_2$ with order $(1,4,4)$, $Af.L$ with lines of size 3 and $c.L$ (that is, $Af.L$ with lines of size 2).

In the first three cases the results of Sections 7.1.3, 7.1.4, 7.2.2 and 7.2.4 can be applied to prove that Γ is one of the 3-dimensional examples arising in (1)–(4).

The geometries belonging to $Af.L$ with lines of size 3 are Hall–Steiner systems (Section 7.1.5). It is proved in [76] that upper 3-truncations of affine geometries of order 3 are the only flag-transitive Hall–Steiner systems.

Finally, (5), (6) and upper 3-truncations of affine geometries of order 2 are the only possibilities that can arise in the case of $c.L$ (see [52]).

Let $n \geq 4$. We can consider a point–hyperplane flag $F = (a, u)$. Feasible isomorphism types of Γ_a and Γ_u are known by the inductive hypothesis. Theorems 7.23 and 7.24 are sufficent to obtain the conclusion, except for a bit of work when $n = 4$, to rule out the cases of order $(1, 4, 4, t)$ $(t > 4)$ and $(1, q-1, q, t)$ $(t > q+1)$. We remark that (iv) of Theorem 7.24 cannot occur here, because in that case stars of lines are non-Desarguesian affine planes, whereas the affine planes arising as stars of points in (5) are Desarguesian.
□

The technique used in the first part of the above proof is quite standard when working on the classification of some class of flag-transitive geometries of rank 3. We call it *comparison of feasible induced actions.*

Chapter 10

Parabolic Systems

10.1 Coset geometries

10.1.1 Preliminaries

Let Γ be a geometry of rank $n \geq 2$ over a set of types I and let $G \leq Aut_s(\Gamma)$ be a group of type-preserving automorphisms of Γ. Given a chamber C_0 of Γ, let B be the stabilizer of C_0 in G. For every type $i \in I$, let x_i be the element of C_0 of type i and let G_i be the stabilizer of x_i in G. For every subset $J \subseteq I$, let F_J be the subflag of C_0 of type J. Thus, the subgroup $G_J = \bigcap_{j \in J} G_j$ is the stabilizer of F_J in G. In particular, $B = G_I$, $G = G_\emptyset$ and G_{I-i} is the stabilizer in G of the panel of C_0 of cotype i ($i \in I$). We set

$$G^i = G_{I-i}, \qquad G^J = \langle G^j \rangle_{j \in J}, \qquad G^\emptyset = B$$

We also write $G_{i,j}$ and $G^{i,j}$ for $G_{\{i,j\}}$ and $G^{\{i,j\}}$, for short. Needless to say that, if $C = g(G_0)$ ($g \in G$) is another chamber in the same orbit of G containing C_0 and if we substitute C for C_0 in the above, then the subgroups G_i, G_J etc. should be replaced by their conjugates gG_ig^{-1}, gG_Jg^{-1} etc.

Lemma 10.1 *The group G is flag-transitive if and only if, for every $i \in I$, the subgroup G^i is transitive on the set of chambers containing the panel F_{I-i}.*

Proof. The 'only if' part is trivial. Let us prove the 'if' claim. Let G^i be transitive on the set of chambers containing F_{I-i}, for every $i \in I$. Given a chamber C, let $C_0, C_1, ..., C_m = C$ be a gallery from C_0 to C. We prove by induction on m that there is an element $g \in G$ mapping C_0 onto C. If

$m = 0$ there is nothing to prove. Let $m \geq 1$. By the inductive hypothesis, there is an element $g_1 \in G$ mapping C_0 onto C_{m-1}. Let i be the cotype of the panel $C_{m-1} \cap C_m$. The chambers $C_0 = g_1^{-1}(C_{m-1})$ and $g_1^{-1}(C_m)$ are i-adjacent. Therefore, there is an element $g_2 \in G^i$ mapping C_0 onto $g_1^{-1}(C_m)$, by assumption. Hence $g_1 g_2$ maps C_0 onto C. □

Note that the equivalence of (i) and (iii) of Theorem 9.4 is simply a special case of the previous lemma.

10.1.2 From geometries to parabolic systems

From now on we assume that $G \leq Aut_s(\Gamma)$ is flag-transitive. We denote the family $(G_i)_{i \in I}$ by $\mathcal{G}(\Gamma, G)$ and we call it the *(concrete) parabolic system* of Γ and G, relative to the chamber C_0. Since G is flag-transitive, if we subsitute another chamber for C_0, then we should only replace $\mathcal{G}(\Gamma, G)$ by a system conjugate to $\mathcal{G}(\Gamma, C)$. In this sense, $\mathcal{G}(\Gamma, G)$ does not essentially depend on the choice of C_0.

The subgroups G_J (with $J \subseteq I$) are called the *parabolic* subgroups of G (relative to C_0). A parabolic subgroup G_J is said to be *proper* if $\emptyset \neq J \neq I$; note that, by the flag-transitivity of G and since Γ has at least two elements of each type, a proper parabolic subgroup of G is also a proper subgroup of G in the usual sense. The subgroups B, G_i and G^i ($i \in I$) are called the *Borel* subgroup, *maximal parabolic* subgroups and *minimal parabolic* subgroups, respectively (compare with Sections 9.3.3 and 9.4.3). We will often use the word 'parabolic' as a noun rather than an adjective, thus writing 'parabolic' instead of 'parabolic subgroup', for short.

The word 'parabolic' ('Borel subgroup') is often used in the literature to denote any conjugate of a subgroup G_J (of B, respectively). We did this in Sections 9.3.3 and 9.4.3. When this convention is used, the subgroups B and G_J ($J \subseteq I$) are distinguished among all their conjugates by being called the *fundamental* Borel subgroup and the *fundamental* parabolics, respectively. However, from now on we will use the words 'Borel' and 'parabolic' only for the fundamental Borel subgroup and for fundamental parabolic subgroups, contrary to what we did in Sections 9.3.3 and 9.4.3. A terminology as in Sections 9.3.3 and 9.4.3 is preferred when the function mapping every flag onto its stabilizer is injective (see Section 10.3). However, that function is not injective in general (see Section 10.3).

Lemma 10.2 *For every $J \subseteq I$, we have $G_J = G^{I-J}$.*

Proof. Given a flag $F_J \subseteq C_0$ and a chamber $C \supseteq F_J$, by Theorem 1.19 there is a gallery C_0, C_1, ..., $C_m = C$ from C_0 to C in the cell of F_J. Let

$i_h \in I - J$ be the cotype of the panel $C_{h-1} \cap C_h$, for $h = 1, 2, ..., m$. By the same argument used to prove Lemma 10.1, there are elements $g_1 \in G^{i_1}$, $g_2 \in G^{i_2}$, ..., $g_m \in G^{i_m}$ such that the element $f_h = g_1 g_2 ... g_h \in G^{I-J}$ maps C_0 onto C_h, for $h = 1, 2, ..., m$. In particular, $f_m(C_0) = C$. Let g be an element of G_J mapping C_0 onto C. We have $g^{-1} f_m \in B \subseteq G^{I-J}$. Hence $g \in G^{I-J}$, since $f_m \in G^{I-J}$. The inclusion $G_J \leq G^{I-J}$ is proved. The converse inclusion is trivial. Therefore $G_J = G^{I-J}$. □

Corollary 10.3 *We have $G = G_\emptyset = G^I$ and $G_i = G^{I-i}$ (for $i \in I$).*

(Trivial, by the previous lemma.)

Note that Lemma 10.2 is just a group-theoretic translation of Corollary 1.21.

By the flag-transitivity of G, the chambers of Γ bijectively correspond to the right cosets of B: the right coset gB corresponds to the chamber $g(C_0)$. Note that gB is the set of elements of G mapping C_0 onto $g(C_0)$. We denote the set of right cosets of B by $\mathcal{C}(\mathcal{G})$, where we write $\mathcal{G}$ for $\mathcal{G}(\Gamma, G)$, for short.

For every $i \in I$, let Φ^i be the i-adjacency relation in $\mathcal{C}(\Gamma)$, as in Section 1.4. We define an equivalence relation Ψ^i on $\mathcal{C}(\mathcal{G})$ by the following clause: given $f, g \in G$, the cosets gB and fB correspond in Ψ^i if and only if $g^{-1} f \in G^i$ (that is, if and only if gB and fB belong to the same right coset of G^i). Let us set

$$\Psi^J = \bigvee_{j \in J} \Psi^j \quad (\text{for } J \subseteq I),$$

$$\Psi_i = \Psi^{I-i} \quad (\text{for } i \in I),$$

$$\Psi_J = \bigcap_{j \in J} \Psi_j \quad (\text{for } J \subseteq I).$$

The reader may compare the above definitions with the definitions of Φ^J, Φ_i and Φ_J in Section 1.5. The system of equivalence relations $(\Psi^i)_{i \in I}$ can also be viewed as a coloured graph with set of colours I, just as in the case of $(\Phi^i)_{i \in I}$ (Sections 1.4 and 1.5.2).

Lemma 10.4 *The coloured graphs $(\Psi^i)_{i \in I}$ and $(\Phi^i)_{i \in I}$ are isomorphic.*

Proof. We need only show that, given $f, g \in G$ and $i \in I$, the chambers $g(C_0)$ and $f(C_0)$ are i-adjacent if and only if $g^{-1} f \in G^i$. The chambers $g(C_0)$ and $f(C_0)$ are i-adjacent if and only if C_0 and $g^{-1} f(C_0)$ are i-adjacent. By Lemma 10.1, the chambers C_0 and $g^{-1} f(C_0)$ are i-adjacent if and only if $g^{-1}f(C_0) = h(G_0)$ for some $h \in G^i$. That is, if and only if $g^{-1}f \in G^i$. □

Lemma 10.5 *Given $J \subseteq I$ and $f, g \in G$, we have $gB \equiv fB(\Psi^J)$ if and only if $g^{-1}f \in G^J$ (that is, if and only if f and g belong to the same right coset of G^J).*

Proof. Let $gB \equiv fB(\Psi^J)$. Then there are types $i_1, i_2, ..., i_m \in J$ and elements $g_h \in G$ $(h = 0, 1, ..., m)$ such that $g_0 = g$, $g_m = f$ and $g_{h-1}^{-1}g_h \in G^{i_h}$ for $h = 1, 2, ..., m$. Hence: $g^{-1}f = g_0^{-1}g_1g_1^{-1}g_2g_2^{-1}g_3...g_{m-1}^{-1}g_m \in G^J$.

Conversely, let $g^{-1}f \in G^J$. Then there are types $i_1, i_2, ..., i_m \in J$ and elements $f_h \in G^{i_h}$ $(h = 1, 2, ..., m)$ such that $g^{-1}f = f_1f_2...f_m$. Let us define $g_0 = g$ and $g_h = g_{h-1}f_h$ $(h = 1, 2, ..., m)$. Then $g_m = f$ and $g_{h-1}^{-1}g_h \in G^{i_h}$ for $h = 1, 2, ..., m$. Hence $gB \equiv fB(\Psi^J)$. □

Corollary 10.6 *Given $J \subseteq I$ and $f, g \in G$, we have $gB \equiv fB(\Psi^J)$ if and only if f and g belong to the same right coset of G_{I-J}.*

(Trivial, by Lemmas 10.2 and 10.5.)

Let $\Gamma(\mathcal{G})$ be the graph having the right cosets gG_i as vertices $(g \in G, i \in I)$, two distinct vertices gG_i, fG_j being adjacent when $gG_i \cap fG_j \neq \emptyset$ (that is, when $gG_i \cap fG_j$ contains some left coset of B). Furthermore, let $\mathcal{K}(\mathcal{G})$ be the set of cosets gG_J $(g \in G, \emptyset \neq J \subseteq I)$, ordered by reverse inclusion (we remark that gG_J is the set of elements of G mapping F_J onto $g(F_J)$). The next theorem easily follows from Lemma 10.4, Corollary 10.6, Theorem 1.23 and Lemma 1.22:

Theorem 10.7 *We have $\Gamma(\mathcal{G}) \cong \Gamma$ and $\mathcal{K}(\mathcal{G}) \cong \mathcal{K}(\Gamma)$.*

In particular, $\Gamma(\mathcal{G})$ is a geometry. A type function is naturally defined on $\Gamma(\mathcal{G})$ giving gG_i the type i (for $g \in G$ and $i \in I$). Thus, the isomorphism $\Gamma \cong \Gamma(\mathcal{G})$ mapping $g(x_i)$ onto gG_i $(i \in I)$ is type-preserving (we recall that x_i is the element of C_0 of type i).

Corollary 10.8 *We have $\bigcap_{g \in G} gBg^{-1} = 1$.*

Proof. Let $f \in \bigcap_{g \in G} gBg^{-1}$. Then $fgB = gB$ for every $g \in G$. Hence f fixes all chambers of Γ, by Theorem 10.7. Therefore $f = 1$. □

Lemma 10.4 is actually the main step in the proof of Theorem 10.7. It also shows that the equalities $G_i = G^{I-i}$ $(i \in I)$ are the essential part of the statement of Lemma 10.2. Indeed, the rest of that statement can also be obtained by these equalities, by Lemmas 10.4 and 10.5 (whose proofs do not use Lemma 10.2) and Corollary 1.21.

Remark. We might also define parabolic systems starting from the system $(G^i)_{i\in I}$ of minimal parabolics rather than from maximal parabolics. If we did so, we would first define G^i as the stabilizer of the panel F_{I-i} of C_0 of cotype i, take the equality $G_J = G^{I-J}$ as a definition of G_J, then show that G_i (defined as $G_i = G^{I-i}$) is in fact the stabilizer of the element x_i of C_0 (the flag-transitivity of G is essential to prove this) and, finally, we would obtain that G_J is the stabilizer of F_J, as a consequence of Corollary 1.21. A number of authors indeed prefer this approach, which is most natural when focusing on chamber systems rather than on geometries (see Section 12.3).

Remark. Many authors use the expressions 'parabolic subgroup' and 'Borel subgroup' in the meaning stated in this section. However, only a few of them use the expression 'parabolic system' and almost nobody uses it in our sense.

For instance, Timmesfeld ([212], [213], [214], [216], [217], [218]) has defined parabolic systems in connection with locally classical locally finite flag-transitive geometries for Coxeter diagrams. Another definition of parabolic systems has been stated by Ronan and Stroth [183], where the maximal parabolics are p-local subgroups. Parabolic systems as defined by Timmesfeld or by Ronan and Stroth form a proper subclass of the class of parabolic systems as defined here.

Some authors use the word 'amalgam' with the meaning we have stated for 'parabolic system'. However, this word is used in a more general sense in group theory, and normally in situations where we look for a group generated by a family of subgroups with prescribed isomorphism types and prescribed mutual intersections. Thus, the word 'amalgam' suggests that we are facing an existence problem. From this point of view, calling parabolic systems 'amalgams' might be a bit misleading.

Thus, we have preferred to avoid this word in this chapter and we have borrowed the expression 'parabolic system' from Timmesfeld and from [183], stating a more general meaning for it. We will define amalgams in the next chapter, in connection with amalgamated products.

10.1.3 Parabolic systems: an abstract definition

Parabolic systems can also be defined in a purely group-theoretic way, without mentioning geometries at all. We give such an abstract definition in this section. We should also show that this definition is indeed equivalent to the 'geometric' definition given in the previous section. We will do this in Section 10.1.4.

Given a group G and a family $\mathcal{G} = (G_i)_{i\in I}$ of subgroups of G, with $2 \leq |I| < \infty$, we will use the following notation:

$$\begin{array}{ll} G_J = \bigcap_{j\in J} G_j, & \text{for } J \subseteq I \quad \text{(in particular, } G_\emptyset = G\text{)},\\ B = G_I, & \\ G^i = G_{I-i}, & \text{for } i \in I,\\ G^J = \langle G^j \rangle_{j\in J}, & \text{for } J \subseteq I,\ J \neq \emptyset,\\ G^\emptyset = B. & \end{array}$$

We say that $\mathcal{G}$ is an (*abstract*) *parabolic system* of *rank* $n = |I|$ over the set of *types* I in the group G, if the following hold:

($\mathbf{P}_1$) we have $B \neq G^i$ for every $i \in I$;

($\mathbf{P}_2$) we have $G_J = G^{I-J}$ for every $J \subseteq I$;

($\mathbf{P}_3$) given distinct types $i, j \in I$, a subset J of $I - \{i, j\}$ and an element $g \in G_J$, if $G_i \cap gG_j \neq \emptyset$, then $G_J \cap G_i \cap gG_j \neq \emptyset$;

($\mathbf{P}_4$) we have $\bigcap_{g\in G} gBg^{-1} = 1$ (that is, B does not contain any non-trivial normal subgroup of G).

This definition is consistent with the one we have given in the previous section. Indeed:

Theorem 10.9 *Every concrete parabolic system is an abstract parabolic system.*

Proof. Let $\mathcal{G}(\Gamma, G)$ be a concrete parabolic system. Condition $\mathbf{P}_1$ follows from the flag-transitivity of G and from the fact that every panel is contained in at least two chambers. Condition $\mathbf{P}_2$ is just the statement of Lemma 10.2. Condition $\mathbf{P}_3$ is an easy consequence of Theorem 10.7. Condition $\mathbf{P}_4$ has been obtained in Corollary 10.8. □

10.1.4 From parabolic systems to geometries

In this section $\mathcal{G} = (G_i)_{i\in I}$ is an abstract parabolic system of rank $n = |I|$ over the set of types I.

We will invert Theorem 10.9, constructing a geometry from $\mathcal{G}$ and showing that $\mathcal{G}$ can be identified with a concrete parabolic system for that geometry. This will show that abstract and concrete parabolic systems are just the same things. Thus, we may omit the words 'concrete' and 'abstract' when speaking of parabolic systems, and speak of parabolics and Borel subgroups in abstract parabolic systems just as in concrete parabolic systems.

Lemma 10.10 *Let J, K be distinct subsets of I. Then $G_J \neq G_K$.*

Proof. Let $j \in K - J$. Then $G^j \leq G^{I-J}$ by $\mathbf{P}_2$. If $G_J = G_K$, then $G_J \leq G_j$, as $G_K \leq G_j$. Hence $G^j \cap G_j = G_j$. On the other hand, we have $G^j = G_{I-j}$ by $\mathbf{P}_2$. Therefore $G^j \cap G_j = G_I = B$. We obtain $G^j = B$, contradicting $\mathbf{P}_1$. □

Lemma 10.11 *Given a nonempty subset $J \subseteq I$, let $(g_j)_{j\in J}$ be a family of elements of G such that $g_iG_i \cap g_jG_j \neq \emptyset$ for every choice of $i, j \in J$. Then $\bigcap_{j\in J} g_jG_j = gG_J$ for some $g \in G$.*

Proof. We work by induction on $m =|J|$. If $m \leq 2$, then the statement is trivial. Let $m \geq 3$. Let h, k be distinct types in J and let us set $K = J - \{h, k\}$. By the inductive hypothesis, there are elements $f_h, f_k, f \in G$ such that $f_hG_{K\cup h} = g_hG_h \cap (\bigcap_{i\in K} g_iG_i)$, $f_kG_{K\cup k} = g_kG_k \cap (\bigcap_{i\in K} g_iG_i)$ and $fG_K = \bigcap_{i\in K} g_iG_i$.

Therefore we have $g_hG_h \cap fG_K \neq \emptyset$ and $g_kG_k \cap fG_K \neq \emptyset$. That is, $f^{-1}g_hG_h \cap G_K \neq \emptyset$ and $f^{-1}g_kG_k \cap G_K \neq \emptyset$. Since $g_hG_h \cap g_kG_k \neq \emptyset$ by assumption, we also have $f^{-1}g_hG_h \cap f^{-1}g_kG_k \neq \emptyset$.

Then $f^{-1}g_hG_h \cap f^{-1}g_kG_k \cap G_K \neq \emptyset$, by $\mathbf{P}_3$.

That is, $g_hG_h \cap g_kG_k \cap fG_K \neq \emptyset$.

Let $g \in g_hG_h \cap g_kG_k \cap fG_K$.

Then $gG_h = g_hG_h$, $gG_k = g_kG_k$ and $gG_K = fG_K$ $(= \bigcap_{i\in K} g_iG_i)$.

Therefore $gG_J = g_hG_h \cap g_kG_k \cap fG_K = \bigcap_{j\in J} g_jG_j$. □

We can now define a graph $\Gamma(\mathcal{G})$ by taking the cosets gG_i as vertices ($g \in G$, $i \in I$) and stating that two distinct vertices gG_i, fG_j are adjacent if and only if $gG_i \cap fG_j \neq \emptyset$.

The cliques of $\Gamma(\mathcal{G})$ bijectively correspond to the cosets gG_J ($g \in G$, $J \subseteq I$). Indeed, a coset gG_J uniquely determines the set of types J, by Lemma 10.10. Hence it uniquely determines the clique $(gG_j)_{j\in J}$ of $\Gamma(\mathcal{G})$. On the other hand, every nonempty clique of $\Gamma(\mathcal{G})$ arises in this way, by Lemma 10.11, whereas G corresponds to the empty clique $\emptyset$.

The maximal cliques of $\Gamma(\mathcal{G})$ are represented by cosets of B and they have size n. By $\mathbf{P}_2$, the cliques of size $n-1$ are represented by the cosets gG^i ($g \in G$, $i \in I$).

Lemma 10.12 *The graph $\Gamma(\mathcal{G})$ is connected.*

Proof. We will prove that, given two maximal cliques C_0 and C of $\Gamma(\mathcal{G})$, there is a sequence $C_0, C_1, ..., C_m = C$ of maximal cliques of $\Gamma(\mathcal{G})$ such that $|C_{h-1} \cap C_h| = n-1$ for every $h = 1, 2, ..., m$, thus proving also that all maximal cliques have size n. The statement of the lemma follows from this.

We may assume without loss of generality that C_0 is the maximal clique represented by B. Let C be represented by gB. By $\mathbf{P}_2$, there are types $i_1, i_2, ..., i_m \in I$ and elements $g_h \in G^{i_h}$ $(h = 1, 2, ..., m)$ such that $g = g_1 g_2 ... g_m$.

We set $f_0 = 1$ and $f_h = g_1 g_2 ... g_h = f_{h-1} g_h$, for $h = 1, 2, ..., m$. Hence $f_m = g$. Let C_h be the maximal clique represented by $f_h B$ $(h = 0, 1, ..., m)$. As $f_h = f_{h-1} g_h$, we have $f_h B \cup f_{h-1} B \subseteq f_{h-1} G^{i_h}$ $(h = 1, 2, ..., m)$. Hence $|C_{h-1} \cap C_h| = n-1$ $(h = 1, 2, ..., m)$. The sequence C_0, C_1, ..., C_m has the required properties. □

Let us set $T_i = \{gG_i\}_{g \in G}$ for $i \in I$ and $\Theta = (T_i)_{i \in I}$. The following is an easy consequence of Lemma 10.10:

Lemma 10.13 *The family Θ is an n-partition of the graph $\Gamma(\mathcal{G})$.*

Let F_J be the clique of $\Gamma(\mathcal{G})$ represented by G_J. Set $m = |I - J|$ and

$$K_J = \bigcap_{g \in G_J} gBg^{-1}, \quad \mathcal{G}_J = (G_{J \cup i}/K_J)_{i \in I - J}$$

If $m = 1$, then $\mathcal{G}_J$ consists of one group, actually a factor group of B. When $m \geq 2$ we have the following:

Lemma 10.14 *Let $m \geq 2$. Then the family $\mathcal{G}_J$ is a parabolic system of rank m in G_J/K_J over the set of types $I - J$.*

Proof. For every $i \in I - J$ we have

$$\begin{aligned} G^i &= \textstyle\bigcap_{j \in I - i} G_j = (\bigcap_{j \in I - (J \cup i)} G_j) \cap (\bigcap_{j \in J} G_j) \\ &= (\textstyle\bigcap_{j \in I - (J \cup i)} G_j) \cap G_J = \bigcap_{j \in I - (J \cup i)} G_{J \cup i}) \end{aligned}$$

Thus, the minimal parabolic subgroups G^i of G with $i \in I - J$ play the role of minimal parabolic subgroups in $\mathcal{G}_J$. By $\mathbf{P}_2$ in $\mathcal{G}$, we have $\langle G^i \rangle_{i \in I - J} = G_J$ and, for every proper subset $K \subseteq I - J$, we have the following:

$$\begin{aligned} \langle G^k \rangle_{k \in K} &= \textstyle\bigcap_{i \in I - K} G_i = (\bigcap_{i \in I - (K \cup J)} G_i) \cap (\bigcap_{j \in J} G_j) \\ &= (\textstyle\bigcap_{i \in I - (K \cup J)} G_i) \cap G_J = \bigcap_{i \in (I - J) - K} G_{J \cup i} \end{aligned}$$

The above shows that $\mathbf{P}_2$ holds in $\mathcal{G}_J$. Since the minimal parabolics of $\mathcal{G}_J$ are minimal parabolics in $\mathcal{G}$ too, condition $\mathbf{P}_1$ holds in $\mathcal{G}_J$ because it holds in $\mathcal{G}$. Condition $\mathbf{P}_3$ holds in $\mathcal{G}_J$ by Lemma 10.11. Finally, $\mathbf{P}_4$ holds, because we are working inside G_J/K_J. □

Therefore, when $m \geq 2$ we can define a graph $\Gamma(\mathcal{G}_J)$ for the parabolic system $\mathcal{G}_J$ in the same way as we have defined $\Gamma(\mathcal{G})$ for $\mathcal{G}$. When $J = I - i$ for some $i \in I$ (that is, when $m = 1$), $\Gamma(\mathcal{G}_J)$ will denote the set of right cosets of B in G^i (that is, $\Gamma(\mathcal{G}_J)$ is now a trivial graph). Since $B \neq G^i$ by $\mathbf{P}_1$, this set has at least two elements.

Let Γ_J be the neighbourhood of the clique F_J in $\Gamma(\mathcal{G})$.

Lemma 10.15 *We have* $\Gamma_J \cong \Gamma(\mathcal{G}_J)$.

Proof. By Lemma 10.11, the neighbourhood Γ_J of F_J consists of those cosets gG_i $(i \in I - J)$ such that $gG_i \cap G_J = fG_{J\cup i}$ for some $f \in G_J$.

Let $\alpha : \Gamma_J \longrightarrow \Gamma(\mathcal{G}_J)$ be the function mapping gG_i onto $f(G_{J\cup i}/K_J)$ with i, g, f as above. Trivially, α is well defined and surjective. It is also injective, by Lemma 10.10. By Lemma 10.11, the bijection α is an isomorphism. □

Corollary 10.16 *Let F be a non-maximal clique of $\Gamma(\mathcal{G})$, represented by gG_J for some $g \in G$ and some proper subset J of I. Then the neighbourhood of F in $\Gamma(\mathcal{G})$ is isomorphic to $\Gamma(\mathcal{G}_J)$.*

Proof. Indeed, the left multiplication by g gives us an isomorphism from the neighbourhood of F_J to the neighbourhood of F. The statement follows from Lemma 10.15. □

We are now ready to prove the converse of Theorem 10.9:

Theorem 10.17 *The graph $\Gamma(\mathcal{G})$ is a geometry of rank n over the set of types I, with type function mapping gG_i onto i, for $i \in I$, $g \in G$. The group G, acting faithfully on $\Gamma(\mathcal{G})$ by left multiplication, is a flag-transitive group of type-preserving automorphisms of the geometry $\Gamma(\mathcal{G})$ and $\mathcal{G}$ is the (concrete) parabolic system of G and $\Gamma(\mathcal{G})$.*

Proof. We work by induction on n. By Corollary 10.16, for every vertex gG_i of $\Gamma(\mathcal{G})$, the neighbourhood of gG_i in $\Gamma(\mathcal{G})$ is isomorphic to $\Gamma(\mathcal{G}_i)$. The latter is a geometry of rank $n-1$ (by the inductive hypothesis when $n \geq 3$; when $n = 2$, then $\Gamma(\mathcal{G}_i)$ is a trivial graph with at least two vertices, hence it is a geometry of rank 1). Lemmas 10.12 and 10.13 now give us the conclusion: $\Gamma(\mathcal{G})$ is a geometry.

By $\mathbf{P}_4$ and Lemma 10.11, the action of G on $\Gamma(\mathcal{G})$ is faithful. The rest of the statement is trivial. □

Exercise 10.1. The statement of Lemma 10.11 is equivalent to $\mathbf{P}_3$. Check that this equivalence does not depend on any of the properties $\mathbf{P}_1$, $\mathbf{P}_2$ and $\mathbf{P}_4$.

Exercise 10.2. Without using any of $\mathbf{P}_1$, $\mathbf{P}_2$ or $\mathbf{P}_4$, prove that each of the following is equivalent to $\mathbf{P}_3$:

($\mathbf{P}_{3,1}$) for every choice of subsets J, H, K of I and of elements f, g, h of G, if the cosets fG_J, gG_H, hG_K have pairwise nonempty intersections, then $fG_J \cap gG_H \cap hG_K \neq \emptyset$;

($\mathbf{P}_{3,2}$) the following holds for every choice of $J, H, K \subseteq I$:
$$(G_J \cap G_H)(G_J \cap G_K) = G_J \cap (G_H G_K);$$

($\mathbf{P}_{3,3}$) the following holds for every choice of $J, H, K \subseteq I$:
$$(G_J G_H) \cap (G_J G_K) = G_J(G_H \cap G_K).$$

Exercise 10.3. Without using any of $\mathbf{P}_1$, $\mathbf{P}_3$ or $\mathbf{P}_4$, prove that $\mathbf{P}_2$ is equivalent to the following:

($\mathbf{P}'_2$) we have $G^J \cap G^K = G^{J \cap K}$ for any two subsets J, K of I.

Exercise 10.4. Let $\mathcal{G} = (G_i)_{i \in I}$ be a finite family of subgroups of a group G. Define $G_J = \bigcap_{j \in J} G_j$ for $J \subset I$ (in particular, $G_\emptyset = G$). Assume that, for every nonempty proper subset J of I and every $j \in J$, G_J is a maximal proper subgroup of G_{J-j}. Prove that $\mathbf{P}_1$ and $\mathbf{P}_2$ hold in $\mathcal{G}$.

Exercise 10.5 (Truncations). Let $\mathcal{G} = (G_i)_{i \in I}$ be a parabolic system and let $J \subseteq I$ with $|J| \geq 2$. Define $N_J = \bigcap_{g \in G} gG_J g^{-1}$ and $Tr^+_J(\mathcal{G}) = (G_j/N_J)_{j \in J}$. Prove that $Tr^+_J(\mathcal{G})$ is a parabolic system in G/N_J over the set of types J and that $\Gamma(Tr^+_J(\mathcal{G})) = Tr^+_J(\Gamma(\mathcal{G}))$.

10.1.5 Residues, kernels and actions

Turning back to the conventions stated before Lemma 10.14, it is now clear that K_J is the kernel of the action of G_J in the residue Γ_J of F_J in $\Gamma(\mathcal{G})$. We call K_J the *kernel* of G_J.

We can substitute any flag F of type J for F_J in the above, except that we must replace suitable conjugates of G_J and K_J for G_J and K_J respectively. Indeed, let F be a flag of type J. Then $F = g(F_J)$ for some $g \in G$, as G is flag-transitive in $\Gamma(\mathcal{G})$; that is, F is represented by the coset gG_J of G_J. Let G_F be the stabilizer of F in G and let K_F be the kernel of the action of G_F in the residue Γ_F of F. It is clear that $G_F = gG_J g^{-1}$ and $K_F = gK_J g^{-1}$. Hence $G_F/K_F \cong G_J/K_J$. Furthermore, left multiplication by g is an isomorphism from the residue Γ_J of F_J to Γ_F, by Corollary 10.16.

Therefore, when speaking of stabilizers of flags of type J and of kernels and actions of these stabilizers, we can always refer to F_J, Γ_J, G_J, K_J and G_J/K_J, without loss of generality.

10.1.6 The rank 2 case

Let $\mathcal{G} = (G_P, G_L)$ be a parabolic system of rank 2 in a group G, over the set of types $\{P, L\}$. According to the notation of Section 10.1.3, we set $B = G_P \cap G_L$. Condition $\mathbf{P}_3$ now says nothing, $\mathbf{P}_1$ amounts to say that $G_P \neq G_L$ and $\mathbf{P}_2$ only says that $G = \langle G_P, G_L \rangle$. Indeed, the rest of $\mathbf{P}_2$ now only reminds us of the conventions $G^P = G_L$, $G^L = G_P$ and $G^{\emptyset} = B$.

Let $\Gamma(G_P, G_L)$ be the geometry defined by $\mathcal{G} = (G_P, G_L)$. According to the notation of Section 1.2, we denote the P-diameter and the L-diameter of $\Gamma(G_P, G_L)$ by d_P and d_L respectively. We also use the following conventions (compare with Section 1.4.1):

$$G_0^{P,L} = G_0^{L,P} = B,$$
$$G_{m+1}^{P,L} = G^P G_m^{L,P} = G^P G^L G^P ... \quad (m+1 \text{ factors})$$
$$G_{m+1}^{L,P} = G^L G_m^{P,L} = G^L G^P G^L ... \quad (m+1 \text{ factors})$$

We warn the reader that the subsets $G_m^{P,L}$ and $G_m^{L,P}$ of G need not be subgroups of G. When $m \geq 2$, the set $G_m^{P,L}$ is a subgroup of G if and only if $G_m^{P,L} = G$. If $G_m^{P,L} = G$, then $G_{m+1}^{P,L} = G_{m+1}^{L,P} = G$. If $G_m^{P,L} = G$ with m even, then $G_m^{L,P} = G$. And similarly, interchanging P and L.

By Section 10.1, the chambers of $\Gamma(G_P, G_L)$ are the right cosets of B and the P- and L-adjacency relations are the partitions of this set of cosets induced by the cosets of G^P and G^L, respectively. Therefore, Proposition 1.11 and Corollaries 1.13 and 1.14 can now be rephrased as follows:

Proposition 10.18 *We have $m = d_P$ if and only if $G_m^{L,P} = G \neq G_{m-1}^{L,P}$. Similarly for $G_m^{P,L}$ and d_L.*

Corollary 10.19 *The geometry $\Gamma(G_P, G_L)$ is a generalized digon if and only if $G_P G_L = G_L G_P$ $(= G)$.*

Corollary 10.20 *We have $d_P = d_L = m$ if and only if $G_m^{L,P} = G_m^{P,L}$, but $G_{m-1}^{L,P} \neq G_m^{L,P}$ and $G_{m-1}^{P,L} \neq G_m^{P,L}$.*

Exercise 10.6. (Gonalities). Prove that the gonality of $\Gamma(G_P, G_L)$ is the minimal positive integer m such that:

$$G_m^{L,P} \cap G_m^{P,L} \neq G_{m-1}^{L,P} \cup G_{m-1}^{P,L}.$$

Exercise 10.7. Prove that $\Gamma(G_P, G_L)$ is a projective plane if and only if all the following hold:

$$G_P G_L G_P = G_L G_P G_L,$$
$$G_P G_L \neq G_L G_P,$$
$$G_P G_L \cap G_L G_P = G_P \cup G_L.$$

Exercise 10.8. Prove that $\Gamma(G_P, G_L)$ is a linear space if and only if all the following hold:

$$G_P G_L G_P = G_L G_P G_L G_P \neq G_L G_P G_L,$$
$$G_P G_L \cap G_L G_P = G_P \cup G_L.$$

Exercise 10.9. Prove that $\Gamma(G_P, G_L)$ is a generalized m-gon if and only if all the following hold:

(i) $G_m^{L,P} = G_m^{P,L}$;
(ii) $G_{m-1}^{L,P} \neq G_m^{L,P}$ and $G_{m-1}^{P,L} \neq G_m^{P,L}$;
(iii) $G_k^{L,P} \cap G_k^{P,L} = G_{k-1}^{P,L} \cup G_{k-1}^{L,P}$ for every $k = 2, 3, ..., m-1$.

Exercise 10.10. Prove that, if m is odd, then we can substitute (ii) of Exercise 10.9 with the following:

(ii') $G_{m-1}^{L,P} \neq G_{m-1}^{P,L}$.

Exercise 10.11. Let Γ be a flag-transitive geometry of rank 2, let $G = Aut_s(\Gamma)$ and let (G_P, G_L) be the parabolic system in G relative to a given chamber of Γ. Prove that the flag complex $\mathcal{K}(\Gamma)$ is flag-transitive if and only if Γ admits a duality interchanging G_P and G_L.

Exercise 10.12. With the notation of the previous exercise, let δ be a duality of Γ interchanging G_P and G_L, hence normalizing B. Let us set $G'_P = G_P$ or G_L and $G'_L = B\langle\delta\rangle$. Check that (G'_P, G'_L) is a parabolic system for $\mathcal{K}(\Gamma)$ and use this information to give a group-theoretic proof of the fact that the diameter (the gonality) of $\mathcal{K}(\Gamma)$ is twice the diameter (the gonality) of Γ (compare with Section 1.4.1).

10.1.7 Diagram graphs

Let $\mathcal{G} = (G_i)_{i \in I}$ be a parabolic system and let $\Gamma = \Gamma(\mathcal{G})$ be the geometry defined by $\mathcal{G}$. Let $\mathcal{D}(\Gamma)$ be the diagram graph of Γ. Given distinct types $i, j \in I$, the symbols $K^{i,j}$ and $\mathcal{G}^{i,j}$ will denote the kernel $K_{I-\{i,j\}}$ of $G^{i,j}$ and the parabolic system $(G^i/K^{i,j}, G^j/K^{i,j})$, respectively. Thus, $\Gamma(\mathcal{G}^{i,j})$ is isomorphic to the residues of Γ of type $\{i, j\}$.

By Corollaries 10.16 and 10.19, two distinct types $i, j \in I$ are joined in $\mathcal{D}(\Gamma)$ if and only if $(G^i/K^{i,j})(G^j K^{i,j}) \neq (G^j/K^{i,j})(G^i/K^{i,j})$. Therefore:

Proposition 10.21 *Two distinct types $i, j \in I$ are joined in $\mathcal{D}(\Gamma)$ if and only if $G^iG^j \neq G^jG^i$.*

Exercise 10.13. (The direct sum theorem). Let J, K be two nonempty disjoint subsets of I such that $J \cup K = I$. Prove the following group-theoretic version of the direct sum theorem:

$\Gamma = Tr_J^+(\Gamma) \oplus Tr_K^+(\Gamma)$ if and only if $G^J G^K = G^K G^J$.

Exercise 10.14. Let $\mathcal{G} = (G_1, G_2, G_3)$ a triple of subgroups of a group G satisfying $\mathbf{P}_1$ and $\mathbf{P}_2$ and such that $G^1G^3 = G^3G^1$, where $G^i = G_j \cap G_k$ ($\{i, j, k\} = \{1, 2, 3\}$), according to the notation of Section 10.1. Prove that $\mathbf{P}_3$ holds in $\mathcal{G}$.

Exercise 10.15. Let $\mathcal{G} = (G_i)_{i=1}^n$ be a system of subgroups of a group G satisfying $\mathbf{P}_1$ and $\mathbf{P}_2$ and such that $G^iG^j = G^jG^i$ for every choice of $i, j = 1, 2, ..., n$ with $i + 1 < j$ (where $G^i = \bigcap_{j \neq i} G_j$, according to the notation of Section 10.1). Prove that $\mathbf{P}_3$ holds in $\mathcal{G}$.

(Remark: the above is due to Meixner and Timmesfeld [134]. Trivially, if we also assume that $\mathbf{P}_4$ holds for $\mathcal{G}$ in addition to the previous hypotheses, then $\mathcal{G}$ is a parabolic system and the diagram graph of $\Gamma(\mathcal{G})$ is a string or a join of strings or of isolated nodes.)

10.2 Examples

10.2.1 The thin case

In this section $\mathcal{G} = (G_i)_{i \in I}$ is a parabolic system in a group G, defining a thin geometry $\Gamma(\mathcal{G})$. According to the notation of Section 10.1, $B = \bigcap_{i \in I} G_i$ is the Borel subgroup of $\mathcal{G}$, the subgroups G^i ($i \in I$) are the minimal parabolic subgroups and Γ_J is the residue of the flag F_J of type J represented by G_J, for $J \subseteq I$.

By Theorem 9.4, the group G is the special automorphism group of $\Gamma(\mathcal{G})$, is chamber-regular in $\Gamma(\mathcal{G})$ (that is, $B = 1$) and the minimal parabolic subgroups have order 2; that is, there are involutions s_i ($i \in I$) such that $G^i = \langle s_i \rangle$ for every $i \in I$. Hence $G_J = \langle s_i \rangle_{i \in I - J}$ for every $J \subseteq I$, by $\mathbf{P}_2$. In particular, G is generated by the family of involutions $(s_i)_{i \in I}$.

As $B = 1$, the chambers of $\Gamma(\mathcal{G})$ are represented by the elements of G.

Since $\Gamma(\mathcal{G})$ is flag-transitive and thin, it belongs to a Coxeter diagram (Proposition 3.2). Let $M = (m_{ij})_{i,j \in I}$ be the Coxeter matrix of that Coxeter diagram (we recall that $m_{ij} = \infty$ is allowed and that $m_{ii} = 1$ for all $i \in I$). For every choice of distinct types $i, j \in I$, the residues of Γ of type

$\{i,j\}$ are ordinary m_{ij}-gons. Furthermore, G^{ij} acts faithfully on $\Gamma_{I-\{i,j\}}$ because $B = 1$.

Therefore $G^{i,j}$ is the dihedral group $D_{2m_{ij}}$ or the infinite cyclic group $\mathbf{Z}$, according to whether $m_{ij} < \infty$ or $m_{ij} = \infty$ (see Section 9.2.2(1)). That is, $G^{i,j}$ is the group presented by the following relations (Section 9.2.2(1)):

$$(1)_{ij} \quad s_i^2 = s_j^2 = 1$$

$$(2)_{ij} \quad (s_i s_j)^{m_{ij}} = 1$$

The group G satisfies the following set of relations:

$$(3) \quad (s_i s_j)^{m_{ij}} = 1 \ (i, j \in I)$$

which includes both the relations $(2)_{ij}$ and the relations $(1)_{ij}$ for all $i, j \in I$ (the reader may notice that, if we choose $i = j$ in (3), then we just obtain $s_i^2 = 1$, because $m_{ii} = 1$ for all $i \in I$). We call (3) the set of *Coxeter relations* for $\mathcal{G}$ (or for $\Gamma(\mathcal{G})$).

Note that, even if the subgroups $G^{i,j}$ are presented by the relations $(1)_{ij}$ and $(2)_{ij}$, the set of relations (3) need not be a presentation for the group G. However, in all the examples described in Section 9.2.2 (including Exercises 9.2–9.9) the group is actually presented by the Coxeter set of relations of its parabolic system. This is not the case in the example of Exercise 9.10. Indeed the automorphism group of that example is $A_5 \times \mathbf{2}$, whereas the Coxeter relations considered there define an infinite group (see Section 11.3.1). We will prove in Chapter 11 that a flag-transitive thin geometry Γ is a proper 2-quotient of another thin geometry if and only $Aut_s(\Gamma)$ is not presented by the Coxeter relations of Γ.

10.2.2 Examples from Chapter 9

(1) **Classical groups.** Parabolic systems in classical automorphism groups of $PG(n, K)$, of classical polar spaces and of thick D_n buildings have been described in Sections 9.3.3, 9.4.3 and 9.4.4. Let us turn to $AG(n, K)$.

If $(G_i)_{i=0}^{n-1}$ is the parabolic system in $AGL(n, K)$ defining $AG(n, K)$ $(n \geq 2)$ and Z is the centre of $GL(n, K)$, then $G_{n-1} = AG(n-1, K)$ and $(G_i/Z)_{i=0}^{n-2}$ is the parabolic system defining $PG(n-1, K)$ in $PGL(n, K)$. The Borel subgroup is the group of all elements of $GL(n, K)$ represented by triangular matrices (with respect to a suitable basis of $V(n, K)$). When $n = 2$, then $G_0 = GL(2, K)$ and $G_1 = AGL(1, K)$ form the parabolic system defining $AG(2, K)$ in $AGL(2, n)$.

Parabolic systems in other classical automorphism groups of $AG(n, K)$ admit a description similar to the above.

(2) **The exceptional cases of Theorems 9.9 and 9.11.** Let Γ, G, q be as in (ii) or (iii) of Theorem 9.9 and $p = q^2 + q + 1$ (we recall that p is prime). It is quite evident that the parabolic system defining Γ in G consists of two distinct copies of Z_{q+1}. On the other hand, let N be the normalizer in the full symmetric group S_p of the Frobenius kernel Z_p of G. We have $N \geq AGL(1,p)$ and the latter is 2-transitive on the set of points of Γ, now viewed as points of the affine line $AG(1,q)$. Therefore N acts 2-transitively on the set of copies of Z_{q+1} in G, by conjugation. Hence any two such copies form a parabolic system defining a projective plane isomorphic to Γ (note also that S_p permutes models of Γ drawn on the set of points of Γ).

In particular, any two distinct copies of Z_3 in $Frob_7^3$ form a parabolic system defining a model of $PG(2,2)$ and any two distinct copies of Z_9 in $Frob_{73}^9$ form a parabolic system defining a model of $PG(2,8)$.

The flag-transitive action of A_7 in $PG(3,2)$ has been described in Section 9.3.4. Let (G_0, G_1, G_2) be the parabolic system defined in A_7 by that action, where the indices 0, 1, 2 are the types and correspond to points, lines and planes, as usual. The reader may check the following:

$$
\begin{aligned}
&G_0 \cong G_2 \cong L_3(2),\\
&G_1 \approx (A_4 \times \mathbf{3}) : \mathbf{2} \approx \mathbf{2}^2 : (\mathbf{3}^2 : \mathbf{2}),\\
&G^0 = G_1 \cap G_2 \cong G^2 = G_0 \cap G_1 \cong S_4 \approx \mathbf{2}^2 : S_3,\\
&G^1 = G_0 \cap G_2 \cong S_4 \approx \mathbf{2}^2 : S_3,\\
&B = G_0 \cap G_1 \cap C_2 \cong D_8.
\end{aligned}
$$

The maximal parabolics G_0, G_2 are not conjugate in A_7 (but they are conjugate in S_7).

(3) **The exceptional cases of Theorem 9.16.** The flag-transitive action of A_6 on $\mathcal{S}_3(2)$ has been described in Section 9.4.5. The reader may check that the parabolic system arising from that action consists of two non-conjugate copies of S_4 intersecting in a Sylow 2-subgroup $\mathbf{2}^2 : \mathbf{2}$ of S_4.

Let (G_P, G_L) be the parabolic system defining $\mathcal{S}_3(3)$ in $\mathbf{2}^4 : A_5$ (Theorem 9.16(ii)), with G_P and G_L stabilizers of a and u respectively, for some point a of $\mathcal{S}_3(3)$ and some line u through a. Using the description of the flag-transitive action of $\mathbf{2}^4 : A_5$ on $\mathcal{S}_3(3)$ given in Section 9.4.5, the reader can check that $G_P \approx \mathbf{2}{\cdot}A_4$ with kernel $K_P = \mathbf{2}$ (hence G_P acts as A_4 on the four lines through a), $G_L \approx \mathbf{3} : D_8$ with $K_L = \mathbf{3}$ (hence G_L acts on the 4 points of u as D_8 on the 4 points of a square) and $B = G_P \cap G_L = \mathbf{2} \times \mathbf{3} = K_P \times K_L$. Note that G_P has just one Sylow 2-subgroup, isomorphic to Q_8 and acting as $\mathbf{2}^2$ on the 4 lines through a.

Similarly, if (G_P, G_L) is the parabolic system defining $\mathcal{S}_3(3)$ in $\mathbf{2}^4 : S_5$, then $G_P \approx \mathbf{2} \cdot S_4$ with $K_P = \mathbf{2}$, $G_L \approx S_3 : D_8$ with $K_L = S_3$ and $B = \mathbf{2} \times S_3 = K_P \times K_L$. Sylow 2-subgroups of G_P are now central non-split extensions of Z_2 by the Sylow 2-subgroup D_8 of S_4, whereas Sylow 2-subgroups of G_L are direct products $\mathbf{2} \times D_8$.

We have not described the flag-transitive action of $\mathbf{2}^4 : Frob_5^4$ on $\mathcal{S}_3(3)$ in Section 9.4.5. However, we can compute the parabolic system (G_P, G_L) arising from that action, using some indirect arguments and the information we have given above on $\mathbf{2}^4 : A_5$ and $\mathbf{2}^4 : S_3$. This will be sufficent to uniquely determine that action. We write G for $\mathbf{2}^4 : Frob_5^4$ and $\overline{G}$ for $G \cap (\mathbf{2}^4 : A_5)$, for short. The subgroup $\overline{G}$ has index 2 in G, as $G \not\leq \mathbf{2}^4 : A_5$. The Borel subgroup B of G has order 2, by the flag-transitivity of G and because $|G|= 320$ and $\mathcal{S}_3(3)$ has 160 chambers. On the other hand, $\overline{G}$ is not flag-transitive on $\mathcal{S}_3(3)$, by Theorem 9.16(ii). Therefore it has two orbits of size 80 on the set of chambers of $\mathcal{S}_3(3)$. Hence $B \leq \overline{G}$. The group G has index 6 in $\mathbf{2}^4 : S_4$ and G_L is the stabilizer of a line u. Hence G_L has index 2 in a Sylow 2-subgroup of the stabilizer of u in $\mathbf{2}^4 : S_4$. Furthermore, G_u is transitive on the four points of u. Therefore $G_L = D_8$, acting faithfully on the four points of u. Hence $G_L \cap \overline{G}$ acts transitively as $\mathbf{2}^2$ on the four points of u. The subgroup G_P fixes a point $a \in u$, it is transitive on the four lines through a and it has index 2 in a Sylow 2-subgroup of the stabilizer of a in $\mathbf{2}^4 : S_4$. The latter has been described in the previous paragraph. By that description we see that either G_P is the Sylow 2-subgroup Q_8 of the stabilizer of a in $\mathbf{2}^4 : A_5$, or $G_P = Z_8$. In the first case $G_P \leq \mathbf{2}^4 : A_5$ is transitive on the four lines through a. As $G_L \leq \mathbf{2}^4 : A_5$ is also transitive on the four points of u, we would obtain that $\overline{G}$ is flag-transitive on $\mathcal{S}_3(3)$, which is not true. Therefore $G_P = Z_8$.

Let us turn to (iii) of Theorem 9.16. We firstly consider the cases of $G = L_3(4) \cdot \mathbf{2}_i$ ($i = 2$ or 3; notation as in [41]). If G is viewed as an automorphism group of $\mathcal{Q}_5^-(3)$, then G_L is the stabilizer in G of a line u of $\mathcal{Q}_5^-(3)$ and G_P is the stabilizer of a point a of u. The quadric $\mathcal{Q}_5^-(3)$ has 112 points and 280 lines. Thus, G_P and G_L have indices 112 and 280 in G, respectively. Consulting [41] we see that $G_P = L_2(9)$, acting faithfully on the 10 lines through a because $L_2(9)$ is simple, and that $G_L \approx \mathbf{3}^2 : (Q_8 \cdot \mathbf{2})$. The Borel subgroup B can easily be computed inside G_P and we see that $B = \mathbf{3}^2 : \mathbf{4} = ASL(1,9)$. On the other hand, the kernel K_L of G_L is a subgroup of B, which is contained in G_L. It is now clear that $K_L \approx \mathbf{3}^2 : \mathbf{2}$ and that G_L acts as the Sylow 2-subgroup D_8 of S_4 on the four points of u. Hence $G_L \approx (\mathbf{3}^2 : \mathbf{2}) \cdot D_8$.

Let now $G = L_3(4) \cdot \mathbf{2}^2$. The group G contains both the groups $L_3(4) \cdot \mathbf{2}_2$ and $L_3(4) \cdot \mathbf{2}_3$ considered above and the group denoted by $L_3(4) \cdot \mathbf{2}_1$ in [41]. The group $L_3(4)$ has two conjugacy classes of 56 subgroups isomorphic to $L_2(9)$. These two classes form one conjugacy class of size 112 in $L_3(4) \cdot \mathbf{2}_i$

($i = 2$ or 3), whereas in $L_3(4)\cdot\mathbf{2}_1$ they give rise to two conjugacy classes of 56 subgroups isomorphic to M_{10} (also denoted by $L_2(9)\cdot 2_3$ in [41]). These 112 ($= 56 + 56$) copies of M_{10} form one conjugacy class of G, because $L_3(4)\cdot\mathbf{2}_i$ is contained in G ($i = 2, 3$). Therefore $G_P = M_{10}$, acting faithfully on the ten lines through a.

Hence $B \approx \mathbf{3}^2 : D_8$ and $G_L \approx (\mathbf{3}^2 : \mathbf{2}^2)\cdot D_8$.

(4) **A model of $\mathcal{H}_3(2^2)$ in $PG(4,3)$.** We saw in Section 9.4.5 how to realize $\mathcal{S}_3(3)$ in $PG(3,4)$. The following construction is the converse of that one. We know that $O_5(3) = S_4(3) = U_4(2)$ (see Section 9.4.5). Let us write G for $O_5(3)$ or $U_4(2)$, for short. We have remarked in Section 9.4.5 that the stabilizer G_L in G of a line of $\mathcal{H}_3(2^2)$ has the following form: $G_L \approx \mathbf{2}^4 : A_5$.

Let f be the bilinear symmetric form defining $\mathcal{Q}_4(3)$ in $PG(4,3)$. The subgroup G_L is the stabilizer in G of a basis of $PG(4,3)$ formed by points represented by mutually orthogonal vectors of norm 1 with respect to f. Indeed the stabilizer in G of such a basis acts as A_5 on the five points of the basis, with kernel $\mathbf{2}^4$ consisting of elements represented by diagonal matrices. Thus, it has the form $\mathbf{2}^4 : A_5$. On the other hand, there is just one conjugacy class of subgroups of that form in G (see [41]). Therefore G_L is one of those stabilizers.

We can prove in a similar way that stabilizers of points of $PG(4,3)$ represented by vectors of norm 1 with respect to f and stabilizers of points of $\mathcal{H}_3(2^2)$ form the same conjugacy class in G, and that if we choose a basis X of $PG(4,3)$ of the kind considered above and a point $a \in X$, then the stabilizer of X and a in G is the stabilizer in G of a chamber of $\mathcal{H}_3(2^2)$.

Thus, we can form a model of $\mathcal{H}_3(2^2)$ as follows. The points of $PG(4,3)$ represented by vectors of norm 1 with respect to f are taken as points of that model. As lines we take the bases of $PG(4,3)$ formed by points represented by mutually orthogonal vectors of norm 1 with respect to f. We obtain a geometry Γ of rank 2 admitting G as a flag-transitive automorphism group. It is clear from the above that Γ and $\mathcal{H}_3(2^2)$ define the same parabolic system in G. Therefore $\Gamma \cong \mathcal{H}_3(2^2)$.

Exercise 10.16. Check that, in the thin case, condition (i) of Exercise 10.9 is nothing but $(2)_{ij}$ and that (ii) and (iii) of Exercise 10.9 now just say that the group $G^{i,j}$ is presented by the relations $(1)_{ij}$ and $(2)_{ij}$.

Exercise 10.17. All affine polar spaces obtained from classical polar spaces are flag-transitive (Exercise 9.22). Choose one of them, the one you like most, and describe its parabolic system.

Exercise 10.18. Compute the parabolic system in the alternating group

A_7 for the $Alt(7)$-geometry (see Section 9.4.6).

Exercise 10.19. Compute the parabolic system in the alternating group A_8 for the $Alt(8)$-geometry (see Exercise 9.25).

Exercise 10.20. Compute a parabolic system for the Hermitian unital and for the Witt–Bose–Shrikande space.

Exercise 10.21. Compute the parabolic systems for $S(M_{22})$, $S(M_{23})$ and $S(M_{24})$ in M_{22}, M_{23} and M_{24} respectively (see Section 9.6.2(4)).

(Hint. Use the information given in Chapter 8 of [91] on $S(3,6,22)$, $S(4,7,23)$ and $S(5,8,24)$. Otherwise, use [41] to recognize suitable subgroups of right index. For instance, $S(M_{22})$ has 22 points, 77 planes and 231 lines. Thus, we obtain the following parabolic system in M_{22}:
$G_0 = L_3(4)$, $G_1 \approx \mathbf{2}^4 : S_5$, $G_2 \approx \mathbf{2}^4 : A_6$
$G_{0,1} \cong G_{0,2} \cong \mathbf{2}^4 : A_5$, $G_{1,2} \approx \mathbf{2}^4 : S_4$
$B \approx \mathbf{2}^4 : A_4$)

Exercise 10.22. Compute the parabolic systems defining $S(M_{11})$ and $S(M_{12})$ in M_{11} and M_{12} respectively (see Section 9.6.2(4)).

(Hint. Use the information given in Section 4.4 of [91] on the actions of M_{11} and M_{12} on their Steiner systems. Otherwise, use [41] to recognize suitable subgroups of M_{11} and M_{12} with the right indices.)

Exercise 10.23. Check that A_7 admits a parabolic system $\mathcal{G} = (G_i)_{i=1}^3$ with $G_i \cong Frob_7^3$ for $i = 1,2,3$, $G^i \cong \mathbf{3}$ for $i,j = 1,2,3$, $i \neq j$ and $B = 1$. Also check that $\Gamma(\mathcal{G})$ belongs to the Coxeter diagram $\widetilde{A}_2$ (see Section 3.1.3(6.e)) with uniform order 2.

(Remark. This geometry has been discovered by Ronan [177].)

Exercise 10.24. Prove that we can pick up three distinct copies G_1, G_2, G_3 of $Frob_7^3$ inside $L_3(2)$ in such a way that $G_1 \cap G_2 \cap G_3 = 1$. Check that $\mathbf{P}_1$, $\mathbf{P}_2$ and $\mathbf{P}_4$ (but not $\mathbf{P}_3$) hold for that triple of subgroups of $L_3(2)$.

Exercise 10.25. Let Γ be the geometry described in Section 9.6.2(5). Prove that Γ is the enrichment of the classical Mőbius plane arising from $\mathcal{Q}_3^-(q)$.

(Hint. We have $L_2(q^2) = O_4^-(q)$ (see [41]) and $PGL(2,q^2)$, viewed as an overgroup of $O_4^-(q)$, acts flag-transitively on the enrichment of the Mőbius plane obtained from $\mathcal{Q}_3^-(q)$, with the same parabolics as in Γ. A parabolic system uniquely determines the geometry which it comes from ...)

Exercise 10.26. Prove that the existence of a sharply flag-transitive projective plane of order q (Theorem 9.9, (ii) and (iii)) is equivalent to the existence of a Frobenius group of order $(q+1)(q^2+q+1)$ with Frobenius kernel of order q^2+q+1 and containing two distinct copies G_P, G_L of Z_{q+1} such that $G_PG_L \cap G_LG_P = G_P \cup G_L$.

(Hint: use Exercise 9.6.)

Exercise 10.27. Let $q = s^2+s+1$ be a prime power. Prove that $AGL(1,q)$ contains a Frobenius subgroup F of order $q(s+1)$ and that, if s is even, then we can take F inside $ASL(1,q)$.

10.2.3 Sporadic geometries

Many nice flag-transitive geometries arise in connection with sporadic simple groups or with exceptional isomorphisms between non-sporadic small simple groups. They are often called *sporadic geometries* (needless to say, this is not a formal definition).

For instance, the $Alt(7)$-geometry is related to the exceptional action of A_7 on $PG(3,2)$, which is in turn related to the exceptional isomorphism $A_8 \cong L_4(2)$. The geometries $S(M_{22})$, $S(M_{23})$, $S(M_{24})$ and $S(M_{11})$, $S(M_{12})$ of theorems 7.23(iii) and 7.24(iii) arise from the Mathieu groups. The geometry $\Gamma(HS)$ of Theorem 7.63(iii) is defined by a parabolic system in the Higman–Sims group HS (we will describe this parabolic system very soon).

We are not going to describe all known sporadic geometries. The reader can find a catalogue of sporadic geometries in [22]. Some years have passed since that list was compiled and a number of new examples have been discovered in the meantime. Nevertheless, that catalogue is still quite useful nowadays (also because nobody till now has found the courage to write a new updated catalogue).

We will only examine a few examples here.

(1) **The geometry $\Gamma(HS)$.** Some information on this geometry was given in Section 7.9.1. We complete it here. The simple group HS is a flag-transitive automorphism group of $\Gamma(HS)$ and $Aut(HS) \approx HS \cdot \mathbf{2}$ is the special automorphism group of $\Gamma(HS)$, whereas $Aut(HS) \times \mathbf{2}$ is the full automorphism group of $\Gamma(HS)$ (see [159]). We gave the orders and a diagram of $\Gamma(HS)$ in Section 7.9.1. The geometry $\Gamma(HS)$ has 100 points and 100 dual points (see [87]) and residues of dual points (stars of points) are isomorphic to $S(M_{22})$ (to the dual of $S(M_{22})$). Hence $\Gamma(HS)$ has 3850 lines and 3850 planes. By the above, by the information given on HS in [41] and on flag-transitive automorphism groups of $S(M_{22})$ in [91], we find that the maximal parabolics G_0, G_1, G_2, G_3 stabilizing in HS the elements x_0, x_1, x_2, x_3 of a chamber of $\Gamma(HS)$ are as follows:

$$G_0 \cong G_3 \cong M_{22}, \quad G_1 \cong G_2 \cong \mathbf{2}^4 \cdot S_6$$

The subgroups G_0 and G_3 are conjugate in HS and act faithfully on the residues of x_0 and x_2 respectively, that is they have trivial kernels ($K_0 = K_3 = 1$). The groups G_1 and G_2 also belong to the same conjugacy class. We have $K_1 \cong K_2 \cong \mathbf{2}^4$. Hence G_1 (respectively G_2) acts as S_6 on the star of x_1 (on the lower residue of x_2) and permutes the two elements of type 0 (of type 3) incident to x_1 (to x_2). As $K_0 = K_3 = 1$, we can compute the parabolics $G_{i,j}$, $G_{i,j,k}$ and the Borel subgroup B inside $G_i = M_{22}$, for $i = 0$ or 3, $j \neq i$, $k \neq i, j$ (see the hint given for Exercise 10.21)). The subgroup $G_{1,2}$ can be computed inside G_1 or G_2, considering that G_1 (respectively, G_2) acts as S_6 on the six dual points (on the six points) incident to the line x_1 (to the plane x_2). Thus, $G_{1,2} \approx \mathbf{2}^4 : (\mathbf{2} \times S_4)$.

(2) **Two geometries related to the isomorphism $S_4(3) \cong U_4(2)$.** We have seen in Section 9.4.5 how $\mathcal{S}_3(3)$ can be realized inside the complement of $\mathcal{H}_3(2^2)$ in $PG(3,4)$. We can imitate that construction in $PG(4,4)$ as follows. We take the complement of $\mathcal{H}_4(2^2)$ in $PG(4,4)$ as set of points. As blocks we take the bases of $PG(4,4)$ consisting of five points represented by mutually orthogonal vectors non isotropic for the form defining $\mathcal{H}_4(2^2)$. The reader can check that we obtain a rank 2 geometry admitting an enrichment of rank 3. We denote that enrichment by $\Gamma(U_5(2))$. It is evident that the star of a point of $\Gamma(U_5(2))$ is (isomorphic to) the model of $\mathcal{S}_3(3)$ constructed in Section 9.4.5. Therefore $\Gamma(U_4(2))$ has diagram and orders as follows:

$(c.C_2)$ ●—c—●═══● with orders 3 3

Trivially, $U_5(2)$ is an automorphism group for $\Gamma(U_5(2))$ and it is not hard to check that $U_5(2)$ is flag-transitive on $\Gamma(U_5(2))$, with parabolic system as follows:

$$G_0 = \mathbf{3} \times S_4(3), \quad G_1 \approx (\mathbf{3}^{2+2} : \mathbf{2}) \cdot (\mathbf{2} \times A_5), \quad G_2 \approx \mathbf{3}^4 : S_5,$$
$$G^0 \approx (\mathbf{3}^4 : \mathbf{6}) : \mathbf{2}, \quad G^1 \approx \mathbf{3}^4 : S_4, \quad G^2 \approx (\mathbf{3}^{2+2} : \mathbf{2}) \cdot A_4,$$
$$B \approx \mathbf{3}^4 : S_3.$$

The Hermitian variety $\mathcal{H}_3(2^2)$ can be realized inside the set of points of $PG(4,3)$ represented by vectors of norm 1 with respect to the form defining $\mathcal{Q}_4(3)$ (see Section 9.2.2). We can imitate that construction in $PG(5,3)$ as follows. The points of $PG(5,3)$ represented by vectors of norm 1 with respect to the form defining $\mathcal{Q}_5^-(3)$ are taken as points. As blocks we take the bases of $PG(5,3)$ consisting of six points represented by mutually orthogonal vectors of norm 1 with respect to the form defining $\mathcal{Q}_5^-(3)$. The reader can check that we obtain a rank 2 geometry admitting an enrichment of rank 3. We denote that enrichment by $\Gamma(O_6^-(3))$. It is quite evident that stars of points of $\Gamma(O_6^-(3))$ are (isomorphic to) the model of $\mathcal{H}_3(2^2)$ constructed in Section 10.2.2(4). Therefore $\Gamma(O_6^-(3))$ has diagram and orders as follows:

$$(c.C_2) \qquad \bullet \overset{c}{\text{———}} \underset{4}{\bullet} \text{===} \underset{2}{\bullet}$$

Trivially, $O_6^-(3)$ is an automorphism group for $\Gamma(U_5(2))$ and it is not hard to check that $O_6^-(3)$ is flag-transitive on $\Gamma(O_6^-(3))$, with parabolic system as follows:

$$G_0 = U_4(2), \quad G_1 \approx SO^+(4,3) : \mathbf{2}, \quad G_2 \approx \mathbf{2}^4 : A_6$$
$$G^0 \approx \mathbf{2}^4 : S_4, \quad G^1 \approx \mathbf{2}^4 : A_5, \quad G^2 \approx (\mathbf{2} \cdot (A_4 \times \mathbf{2}^2) \cdot S_3$$
$$B \approx \mathbf{2}^4 : A_4$$

The geometry $\Gamma(O_6^-(3))$ also admits a flag-transitive 3-fold 2-cover (see [60], [158] or [241]; also [13], Section 13.2.C). We denote it by $3 \cdot \Gamma(O_6^-(3))$.

(3) **Three geometries for** McL, Suz **and** $Aut(HS)$**.** A geometry is known with diagram and orders as follows, admitting the McLaughlin group McL as a flag-transitive automorphism group ([26] or [22], Example 22; actually, McL is the minimal flag-transitive automorphism group of this geometry):

$$(c.C_2) \qquad \bullet \overset{c}{\text{———}} \underset{3}{\bullet} \text{===} \underset{9}{\bullet}$$

We denote this geometry by $\Gamma(McL)$. Stars of points of $\Gamma(McL)$ are isomorphic to $\mathcal{Q}_5^-(3)$. The parabolic system for $\Gamma(McL)$ in McL is as follows ([158] or [241]):

$$G_0 = O_6^-(3), \quad G_1 \approx \mathbf{3}^4 : M_{10}, \quad G_2 \approx (\mathbf{3}^{1+4} : \mathbf{2}) \cdot S_5$$
$$G^0 \approx (\mathbf{3}^{1+4} : \mathbf{2}) \cdot (S_3 \times \mathbf{2}), \quad G^1 \approx (\mathbf{3}^{1+4} : \mathbf{2}) \cdot S_4, \quad G^2 \approx \mathbf{3}^4 : L_2(9)$$
$$B \approx (\mathbf{3}^{1+4} : \mathbf{2}) \cdot S_3$$

We have $Aut(\Gamma(McL)) = Aut(McL) \approx McL \cdot \mathbf{2}$.

A geometry is known with diagram and orders as follows, admitting the Suzuki group Suz as (minimal) flag-transitive automorphism group ([22], Example 6):

$(c.C_2)$ — diagram: c — 9 — 3

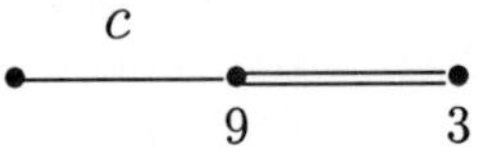

We denote this geometry by $\Gamma(Suz)$. Stars of points of $\Gamma(Suz)$ are isomorphic to $\mathcal{H}_3(3^2)$. The parabolic system for $\Gamma(Suz)$ in Suz can be described as follows ([158] or [241]):

$$G_0 \approx \mathbf{3}\cdot(U_4(3)\cdot\mathbf{2}_3, \quad G_1 \approx \mathbf{3}^{2+4} : (2\cdot(S_4\times\mathbf{2}^2)), \quad G_2 \approx \mathbf{3}^5 : M_{11}$$
$$G^0 \approx (\mathbf{3}^{5+2} : Q_8)\cdot\mathbf{2}, \quad G^1 \approx \mathbf{3}^5 : M_{10}, \quad G^2 \approx (\mathbf{3}^{2+4}\cdot\mathbf{4})\cdot S_4$$
$$B \approx \mathbf{3}^{2+4} : (\mathbf{2}^{1+1}\cdot S_3)$$

We have $Aut(\Gamma(Suz)) = Aut(Suz) \approx Suz\cdot\mathbf{2}$.

A geometry is known with diagram and orders as follows and admitting $Aut(HS)$ as its (unique) flag-transitive automorphism group [240]:

$(c.C_2)$ — diagram: c — 9 — 3

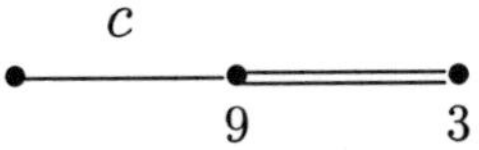

We call this geometry the *Yoshiara geometry* after its discoverer [240]. Stars of points of the Yoshiara geometry are isomorphic to $\mathcal{H}_3(3^2)$. The parabolic system admits the following description ([238], [241] or [159]):

$$G_0 = L_3(4)\cdot\mathbf{2}^2, \quad G_1 \approx (((\mathbf{3}^2 : Q_8)\cdot\mathbf{2})\times\mathbf{2})\cdot\mathbf{2}, \quad G_2 = M_{11}$$
$$G^0 \approx (\mathbf{3}^2 : Q_8)\cdot\mathbf{2}, \quad G^1 = M_{10}, \quad G^2 \approx (\mathbf{3}^2 : \mathbf{2}^2) : D_8$$
$$B \approx \mathbf{3}^2 : Q_8$$

(4) **Three geometries for Fi_{22}, Fi_{23} and Fi_{24}.** A geometry is known with diagram and orders as follows, admitting the Fischer group Fi_{22} as (minimal) flag-transitive automorphism group (see [26] or Example 46 of [22]):

$(c.C_3)$ — diagram: c — 4 — 4 — 2

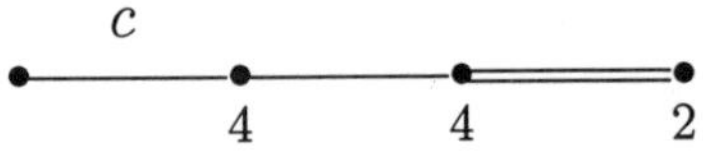

We denote this geometry by $\Gamma(Fi_{22})$. Stars of points of $\Gamma(Fi_{22})$ are isomorphic to $\mathcal{H}_5(2^2)$ and residues of dual points are isomorphic to $S(M_{22})$. The maximal parabolics and the Borel subgroup for $\Gamma(Fi_{22})$ in Fi_{22} are as follows ([22]):

$$G_0 \approx \mathbf{2}\cdot U_6(2)$$
$$G_1 \approx (\mathbf{2}\times\mathbf{2}^{1+8})\cdot(U_4(2)\cdot\mathbf{2})$$

$G_2 \approx \mathbf{2}^{5+8} \cdot (S_3 \times A_6)$
$G_3 \approx \mathbf{2}^{10} : M_{22}$
$B \approx \mathbf{2}^{10} : (\mathbf{2}^4 : A_4)$

We have $Aut(\Gamma(Fi_{22})) = Aut(Fi_{22}) \approx Fi_{22} \cdot \mathbf{2}$. The geometry $\Gamma(Fi_{22})$ is isomorphic to stars of points of a larger geometry with the following diagram and with the Fischer group Fi_{23} as (unique) flag-transitive automorphism group ([22], Example 47):

$(c^2.C_3)$ [diagram: •—•—•^c—•—4—•=2=•; labels c, 4, 4, 2]

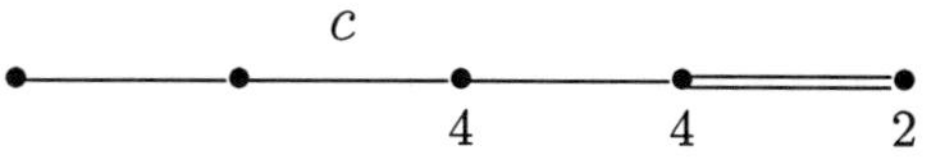

We denote this rank 5 geometry by $\Gamma(Fi_{23})$. Residues of dual points of $\Gamma(Fi_{23})$ are isomorphic to $S(M_{23})$. The geometry $\Gamma(Fi_{23})$ is in turn isomorphic to stars of points of a geometry of rank 6 with the following diagram and with the Fischer group Fi_{24} as flag-transitive (full) automorphism group ([22], Example 48):

$(c^3.C_3)$ [diagram: •—•—•—•^c—•—4—•=2=•; labels c, 4, 4, 2]

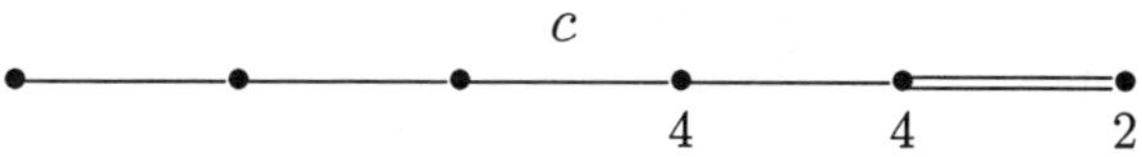

We denote this latter geometry by $\Gamma(Fi_{24})$. Residues of dual points of $\Gamma(Fi_{24})$ are isomorphic to $S(M_{24})$. The commutator subgroup Fi'_{24} of Fi_{24} is the minimal flag-transitive automorphism group of $\Gamma(Fi_{24})$. Furthermore, $\Gamma(Fi_{24})$ admits a flag-transitive 3-fold 2-cover [174]. We denote it by $3 \cdot \Gamma(Fi_{24})$.

The reader wanting to know more on the geometries $\Gamma(Fi_{22})$, $\Gamma(Fi_{23})$, $\Gamma(Fi_{24})$ and on their parabolic systems may consult [26], [22], [232] or [129].

10.3 Conjugacy systems

We have shown in Section 10.1 how to recover a flag-transitive geometry Γ from right cosets of the stabilizers of the elements of a given chamber C_0 of Γ in a given flag-transitive subgroup $G \leq Aut_s(\Gamma)$. However, cosets are not very handy to work with, except when using computers. Futhermore, the representation of Γ by means of cosets depends on the choice of the chamber C_0. Actually, that choice has no influence on the isomorphism type of the parabolic system we obtain, since parabolic systems obtained from different chambers are conjugate in G. However, a feeling of arbitrariness remains marring this way of representing geometries.

On the other hand, stabilizers in G of flags of Γ of a given type form one conjugacy class in G, because G is flag-transitive. Therefore, if we can identify elements or flags of Γ with their stabilizers, then we can work with conjugacy classes instead of cosets. Thus, we would also obtain a representation of Γ that need not pivot on the choice of a chamber C_0.

Unfortunately, there are cases where this cannot be done. In this section we examine conditions that allow us to identify elements or flags with their stabilizers.

10.3.1 Preliminaries

We keep the notation of Section 10.1. Thus, Γ is a flag-transitive geometry and $G \leq Aut_s(\Gamma)$ is flag-transitive. $\mathcal{G} = (G_i)_{i \in I}$ is the parabolic system for Γ in G, relative to a given chamber $C_0 = (x_i)_{i \in I}$. The Borel subgroup of $\mathcal{G}$ is denoted by B and, given $J \subseteq I$, the subflag of C_0 of type J is denoted by F_J, as in Section 10.1.

Trivially, if $F = g(F_J)$ for some $g \in G$, then gG_Jg^{-1} is the stabilizer in G of the flag F. We also denote it by G_F, as usual (note that $G_J = G_{F_J}$).

It may happen that G_J and G_K are conjugate in G for some distinct subsets $J, K \subseteq I$ (some examples of this kind will be mentioned in Section 10.3.5). If that happens, then the function mapping a flag F of Γ onto its stabilizer G_F is not injective and we cannot replace flags by their stabilizers. This shows that the object we should associate with a flag F of Γ is not just its stabilizer G_F in G, but rather the pair $(G_F, t(F))$, where t is the type function of Γ. Thus, let γ be the function mapping every flag F of Γ onto the pair $(G_F, t(F))$.

Lemma 10.22 *The mapping γ is injective if and only if $N_G(G_J) = G_J$ for every $J \subseteq I$.*

Proof. The mapping γ is injective if and only if the following holds for any two elements f, g of G and for every subset J of I: we have $gG_Jg^{-1} = fG_Jf^{-1}$ if and only if $g(F_J) = f(F_J)$ (that is, if and only if $f^{-1}g \in G_J$). The statement is now evident. □

10.3.2 Strong conjugacy systems

Obviously, given flags $X = g(F_J)$ and $Y = f(F_K)$ of Γ of types J and K respectively with $J \subseteq K$, we have $X \subseteq Y$ in Γ if and only if there is some $h \in G$ such that $hG_Jh^{-1} = gG_Jg^{-1}$ $(= G_X)$ and $hG_Kh^{-1} = fG_Kf^{-1}$ $(= G_Y)$. However, this way of representing inclusions of flags is a bit clumsy. The following way would be better.

Define a poset $\mathcal{K}_{cong}(\Gamma)$ by taking the pairs $(G_F, t(F))$ as elements and stating that $(G_X, t(X)) \leq (G_Y, t(Y))$ in this poset precisely when $G_Y \leq$

G_X and $t(X) \subseteq t(Y)$. Trivially, γ is a morphism of posets from the flag-complex $\mathcal{K}(\Gamma)$ of Γ to $\mathcal{K}_{cong}(\Gamma)$. Assume that γ is injective. If γ is an isomorphism, namely if $G_Y \leq G_X$ and $t(X) \subseteq t(Y)$ imply $X \subseteq Y$, then we can substitute $\mathcal{K}_{cong}(\Gamma)$ for $\mathcal{K}(\Gamma)$ and we have fully recovered Γ as a system of conjugates.

We say that $\mathcal{G}$ is a *strong conjugacy system* if γ is an isomorphism from $\mathcal{K}(\Gamma)$ to $\mathcal{K}_{cong}(\Gamma)$.

Theorem 10.23 *The parabolic system $\mathcal{G}$ is a strong conjugacy system if and only if the following holds:*

(**SCS**) *for every $i \in I$ and $g \in G$, if $gBg^{-1} \leq G_i$, then $g \in G_i$*

Proof. Let $\mathcal{G}$ be a strong conjugacy system. Let $gBg^{-1} \leq G_i$. Then the chamber $g(C_0)$ contains the element $x_i \in C_0$ of type i. Hence $g(x_i) = x_i$. That is, $g \in G_i$.

On the other hand, let **SCS** hold. Given $J \subseteq I$ and $g \in N_G(G_J)$, we have $gBg^{-1} \leq G_J \leq G_j$ for every $j \in J$. Hence $g \in G_J$, by **SCS**. Therefore $N_G(G_J) = G_J$. By Lemma 10.22, the mapping γ is injective.

Let now X, Y be flags with $J = t(X) \subseteq t(Y) = K$ and $G_Y \leq G_X$. By the flag-transitivity of G, we can always assume that $X = F_J$. Hence $G_X = G_J$. We have $Y = g(F_K)$ and $G_Y = gG_Kg^{-1}$ for some $g \in G$. Since $gG_Kg^{-1} = G_Y \leq G_X = G_J$, we have $gBg^{-1} \leq G_j$ for every $j \in J$. Hence $g \in G_J$, by **SCS**. Therefore $X = F_J = g(F_J) \subseteq g(F_K) = Y$. □

Corollary 10.24 *The following holds in every strong conjugacy system:*

(**SCS**$'$) *for every $J \subseteq I$ and every $g \in G$, if $gBg^{-1} \leq G_J$, then $g \in G_J$.*

(Trivial, by the previous theorem and because $G_J = \bigcap_{j \in J} G_j$.)

Corollary 10.25 *Let $\mathcal{G}$ be a strong conjugacy system and let J and K be distinct subsets of I. Then G_J and G_K belong to distinct conjugacy classes in G.*

Proof. Let $G_J = gG_Kg^{-1}$ for some $g \in G$. Then $gBg^{-1} \leq G_J$. Hence $g \in G_J$ by **SCS**$'$ of Corollary 10.24. Therefore $G_K = G_J$ and $K = J$. □

Thus, if $\mathcal{G}$ is a strong conjugacy system, distinct flags never have the same stabilizers. Therefore the definition of γ can be simplified in strong conjugacy systems, just setting $\gamma(F) = G_F$ for every flag F of Γ.

10.3.3 Examples and counterexamples

Let Γ be $PG(n, K)$ $(n \geq 2)$, as in Section 9.3.3, or a classical thick polar space of rank $n \geq 2$, as in Section 9.4.3. Let G be a classical subgroup of $Aut_s(\Gamma)$. Given a chamber C, let $\mathcal{G}$ be the parabolic system defined by C in G.

Proposition 10.26 *The parabolic system $\mathcal{G}$ is a strong conjugacy system.*

Proof. We need to prove that **SCS** of Theorem 10.23 holds in $\mathcal{G}$. That is, given $x \in C$ and $g \in G$ such that gBg^{-1} fixes x, we have $g(x) = x$.

Let $g(x) \neq x$, if possible. Let $y = g^{-1}(x)$. As $g(x) \neq x$, we have $y \neq x$. Let $\gamma = (C_0, C_1, ..., C_m)$ be a gallery of Γ such that $C_0 = C$ and $y \in C_m$. We can always assume to have chosen γ of minimal length compatible with the above conditions. Therefore $y \notin C_{m-1}$. Since Γ is thick, there is a chamber C' different from both C_{m-1} and C_m and containing the panel $X = C_{m-1} \cap C_m$. Trivially, $y \notin C'$. By Propositions 9.6 and 9.12, there is an apartment $\mathcal{A}$ containing C and C_m and an apartment $\mathcal{A}'$ containing C and C'.

It is possible to prove that, if an apartment contains a chamber C and a flag X, then it contains all galleries $\gamma = (C_0, C_1, ..., C_m)$ with $C_0 = C$ and $X \subseteq C_m$ and such that m is minimal with respect to those two conditions.

We do not prove this claim here. The reader can obtain it as a special case of (iv) of Section 13.1.1. Otherwise, a straightforward proof can be constructed starting from cases of low rank ($n = 2$ or 3), where that claim is fairly evident (see also Section 1.4.2), then generalizing to higher rank cases.

Set $\delta = (C_0, C_1, ..., C_{m-1})$ and let δ' be a gallery of length $m' < m - 1$ going from C to a chamber containing X, if possible. Then C_m is not a chamber of δ', as we have chosen γ of minimal length with respect to the property of starting at C and ending at a chamber containing y. We can add C_m to δ', thus obtaining a gallery γ' from C to C_m of length $m' + 1 < m$. This contradicts our choice of γ. Therefore every gallery from C to a chamber containing the panel X has length $\geq m - 1$ and δ is a shortest gallery from C to X. By the above property of apartments, both $\mathcal{A}$ and $\mathcal{A}'$ contain δ. Hence C', C_{m-1} are the two chambers of $\mathcal{A}'$ on the panel X.

We have remarked in Sections 9.3.3 and 9.4.3 that the group G is transitive on the set of pairs consisting of an apartment and a chamber of that apartment. Hence there is an element f of G fixing C and mapping $\mathcal{A}$ onto $\mathcal{A}'$. As δ is contained in both $\mathcal{A}$ and $\mathcal{A}'$ and both $\mathcal{A}$ and $\mathcal{A}'$ are thin, f fixes δ. In particular, it fixes C_{m-1}. Hence it maps C_m onto C'. Therefore $f(y) \in C'$. As $y \notin C'$, we have $f(y) \neq y$. On the other hand, we have

$f(y) = y$ because $y = g^{-1}(x)$ and $f \in B \leq g^{-1}G_xg$. A contradiction has been reached. □

In Sections 9.3.3 and 9.4.3 we encouraged the reader to check that distinct flags of Γ have distinct stabilizers in G. This claim is a consequence of the previous proposition and of Corollary 10.25.

By arguments similar to those used in the proof of Proposition 10.26 it is possible to prove also that the parabolic system in a classical automorphism group of a building of type D_n is a strong conjugacy system.

On the other hand, the parabolic system in $PGO^+(2n, K)$ for the polar space $\mathcal{Q}^+_{2n-1}(K)$ examined in Section 9.4.4 is not a strong conjugacy system. Indeed the Borel subgroup B now has index 2 in the minimal parabolic subgroup G^{n-1} corresponding to the last node $n-1$ of the diagram. Hence $B \trianglelefteq G^{n-1}$ and γ is not injective (Lemma 10.22). The same happens in any automorphism group of $\mathcal{Q}^+_{2n-1}(K)$ containing $PGO^+(2n, K)$.

More generally, we have the following:

Proposition 10.27 *Let $\mathcal{G}$ be a parabolic system such that $\Gamma(\mathcal{G})$ is thin at one type. Then $\mathcal{G}$ is not a strong conjugacy system.*

Proof. Let $\Gamma = \Gamma(\mathcal{G})$ be thin at $0 \in I$. Then B has index 2 in G_0. Hence $B \trianglelefteq G_0$ and γ is not injective, by Lemma 10.22. □

Therefore, parabolic systems of flag-transitive thin geometries are never strong conjugacy systems. This can also be proved in a slightly different way, as follows: we have $B = 1$ in thin geometries. Hence $N_G(B) = G \neq B$ and γ is not injective. More generally, parabolic systems with $B = 1$ are never strong conjugacy systems.

Exercise 10.28. Let $\mathcal{B}$, $\mathcal{C}$ be two conjugacy classes of subgroups of a group G such that some member B of $\mathcal{B}$ is contained in some member X of $\mathcal{C}$. Prove that the following are equivalent:

(i) $\{g \in G \mid gBg^{-1} \leq X\} \subseteq X$;
(ii) every member of $\mathcal{B}$ is contained in just one member of $\mathcal{C}$.

(Remark: the property **SCS** of Theorem 10.23 just states that (i) holds with B the Borel subgroup and X any maximal parabolic.)

Exercise 10.29. Let $\mathcal{B}$, $\mathcal{C}$ be as in the previous exercise, satisfying (i) (equivalently (ii)). Prove that $N_G(X) = X$ for every $X \in \mathcal{C}$.

Exercise 10.30. Let G be a group, let $\mathcal{B}$ be a conjugacy class of proper subgroups of G and let $(\mathcal{C}_i)_{i \in I}$ be a finite family of conjugacy classes of proper subgroups of G such that for every $i \in I$ some member of $\mathcal{C}_i$ contains some member of $\mathcal{B}$ and (i) (equivalently (ii)) of Exercise 10.28 holds for $\mathcal{B}$ and $\mathcal{C}_i$. Hence, given $B \in \mathcal{B}$, there is a unique member $G_i \in \mathcal{C}_i$ containing B.

Assume that $B = \bigcap_{i\in I} G_i$, that $G = \langle G_i \rangle_{i\in I}$ and that $\mathbf{P}_1$, $\mathbf{P}_2$ and $\mathbf{P}_4$ of Section 10.1 hold in $(G_i)_{i\in I}$. Assume furthermore that, for every $J \subseteq I$, for every choice of distinct indices $i, j \in I - J$ and for every member $X \in \mathcal{C}_j$, if both $G_J \cap X$ and $G_i \cap X$ contain some members of $\mathcal{B}$, then $G_J \cap G_i \cap X$ contains a member of $\mathcal{B}$.

Prove that under the previous assumptions $(G_i)_{i\in I}$ is a strong conjugacy system.

10.3.4 Conjugacy systems

We have defined strong conjugacy systems in Section 10.3.2 with flag complexes in mind. We will now focus on incidence graphs. Let Γ, $C_0 = (x_i)_{i\in I}$, G, $\mathcal{G} = (G_i)_{i\in I}$ and γ be as in Section 10.3.1 and let γ_{vert} be the mapping induced by γ on the set of elements of Γ. The same argument used for Lemma 10.22 proves the following:

Lemma 10.28 *The mapping γ_{vert} is injective if and only if $N_G(G_i) = G_i$ for every $i \in I$.*

Corollary 10.29 *If γ_{vert} is injective, then $Z(G) = 1$.*

Proof. Indeed, $Z(G) \leq \bigcap_{i\in I} G_i = B$ by Lemma 10.28. By property $\mathbf{P}_4$ of parabolic systems, $Z(G) = 1$. □

Let γ_{vert} be injective. Then two elements $x = g(x_i)$ and $y = f(x_j)$ of Γ (with $g, f \in G$, $i, j \in I$, $i \neq j$) are incident in Γ if and only if there is an element $h \in G$ such that $hG_ih^{-1} = gG_ig^{-1}$ $(= G_x)$ and $hG_jh^{-1} = fG_jf^{-1}$ $(= G_y)$. However, this way of representing the incidence relation of Γ is a bit clumsy. The following would be better.

We remark that, if x and y are incident in Γ, then $gG_ig^{-1} \cap fG_jf^{-1} \geq hG_{i,j}h^{-1}$ for some $h \in G$. Thus, we define a graph $\Gamma_{cong}(\mathcal{G})$ taking the pairs (gG_ig^{-1}, i) as vertices (with $g \in G$, $i \in I$) and stating that (gG_ig^{-1}, i) and (fG_jf^{-1}, j) $(i \neq j)$ are adjacent in this graph when $gG_ig^{-1} \cap fG_jf^{-1} \geq hG_{i,j}h^{-1}$ for some element $h \in G$. We say that $\mathcal{G}$ is a *conjugacy system* if γ_{vert} is an isomorphism from the incidence graph of Γ to $\Gamma_{cong}(\mathcal{G})$.

Theorem 10.30 *The parabolic system $\mathcal{G}$ is a conjugacy system if and only if both the following hold:*

($\mathbf{CS}_1$) we have $N_G(G_i) = G_i$ for every $i \in I$;
($\mathbf{CS}_2$) we have $N_G(G_{i,j}) \subseteq G_iG_j$ for any two distinct types $i, j \in I$.

Proof. Let $\mathcal{G}$ be a conjugacy system. That is, γ_{vert} is an isomorphism. Hence $\mathbf{CS}_1$ holds, by Lemma 10.28. We first prove the following:

$$\{g \in G \mid gG_{i,j}g^{-1} \leq G_i\} = G_iN_G(G_{i,j}) \text{ for any two } i, j \in I.$$

Trivially $\{g \in G \mid gG_{i,j}g^{-1} \leq G_i\} \supseteq G_iN_G(G_{i,j})$. Let us prove the converse inclusion. Let $gG_{i,j}g^{-1} \leq G_i$. Then $gG_{i,j}g^{-1} \leq G_i \cap gG_jg^{-1}$. Therefore $hG_ih^{-1} = G_i$ and $hG_jh^{-1} = gG_jg^{-1}$ for some $h \in G$, as γ_{vert} is an isomorphism. By $\mathbf{CS}_1$, we have $h \in G_i$. Furthermore, $hG_{i,j}h^{-1} = hG_ih^{-1} \cap hG_jh^{-1} = G_i \cap gG_jg^{-1} \geq gG_{i,j}g^{-1}$ because $gG_{ij}g^{-1} \leq G_i$, gG_jg^{-1}. Hence $fG_{i,j}h^{-1} = gG_{ij}g^{-1}$. That is, $h^{-1}g \in N_G(G_{i,j})$. Since $h \in G_i$, we obtain $g \in G_iN_G(G_{i,j})$, as we wanted.

We can now prove $\mathbf{CS}_2$. Let $g = N_G(G_{i,j})$. Then $G_{i,j} \leq g^{-1}G_ig \cap G_j$. As γ_{vert} is an isomorphism, there is an element $k \in G$ such that $kG_ik^{-1} = g^{-1}G_ig$ and $kG_jk^{-1} = G_j$. By $\mathbf{CS}_1$ we have $gk \in G_i$ and $k \in G_j$. Therefore $g \in G_iG_j$.

On the other hand, let $\mathbf{CS}_1$ and $\mathbf{CS}_2$ hold. Then γ_{vert} is injective, by Lemma 10.28. Let (gG_ig^{-1}, i), (fG_jf^{-1}, j) be adjacent in $\Gamma_{cong}(\mathcal{G})$. Then $hG_{i,j}h^{-1} \leq gG_ig^{-1} \cap fG_jf^{-1}$ for some $h \in G$. Hence $g^{-1}hG_{i,j}h^{-1}g \leq G_i$ and $f^{-1}hG_{i,j}h^{-1}f \leq G_j$. By $\mathbf{CS}_1$, there are elements $a_i \in G_i$, $a_j \in G_j$ and $b_i, b_j \in N_G(G_{i,j})$ such that $g^{-1}h = a_ib_i$ and $f^{-1}h = a_jb_j$. Therefore, $g^{-1}f \in G_iN_G(G_{i,j})G_j$. By $\mathbf{CS}_2$ we have $g^{-1}f = c_ic_j$ for suitable elements $c_i \in G_i$ and $c_j \in G_j$. Therefore $gG_ig^{-1} = kG_ik^{-1}$ and $fG_jf^{-1} = kG_jk^{-1}$, with $k = gc_i$. That is, the elements $g(x_i)$ and $f(x_j)$ of Γ are incident. Hence γ_{vert} is an isomorphism. □

Theorem 10.31 *Let $\mathcal{G}$ be a conjugacy system. Then maximal parabolics of $\mathcal{G}$ of distinct types belong to distinct conjugacy classes.*

Proof. Let $\mathcal{G}$ be a conjugacy system and let $G_i = fG_jf^{-1}$ for some $f \in G$ and some distinct types i, j, if possible. Then $G_i \cap fG_jf^{-1} = G_i \geq G_{i,j}$. Therefore $G_i \cap fG_j \neq \emptyset$, because γ_{vert} is an isomorphism. Hence $f = g_ig_j$ for suitable $g_i \in G_i$ and $g_j \in G_j$. By this and by the relation $G_i = fG_jf^{-1}$ we get the equality $G_i = G_j$, which is impossible. □

Thus, if $\mathcal{G}$ is a conjugacy system in the meaning stated above, distinct elements never have the same stabilizers. Therefore the definition of γ_{vert} can be simplified in conjugacy systems by setting $\gamma(x) = G_x$ for every element x of Γ.

10.3.5 Examples and counterexamples

A strong conjugacy system is also a conjugacy system in the meaning of Section 10.3.4. Indeed, the incidence graph of Γ is the skeleton of the simplicial complex $\mathcal{K}(\Gamma)$ and, if $\mathcal{G}$ is a strong conjugacy system, then γ is an isomorphism, the graph $\Gamma_{cong}(\mathcal{G})$ is the skeleton of the simplicial complex $\mathcal{K}_{cong}(\mathcal{G})$ and γ_{vert} is the isomorphism of graphs induced by the isomorphism of simplicial complexes γ. However, there are many conjugacy systems that are not strong conjugacy systems. For instance, the parabolic system of $P\Gamma O^+(2n, K)$ in $O^+_{2n-1}(K)$, which is not a strong conjugacy system, is nevertheless a conjugacy system (we leave the proof of this claim for the reader).

It is easily seen that, if $\mathcal{G}$ has rank 2 and Borel subgroup $B = 1$, then $\Gamma_{cong}(\mathcal{G})$ is a complete bipartite graph. In this case $\mathcal{G}$ is not a conjugacy system, except possibly when $\Gamma(\mathcal{G})$ is a generalized digon. For instance, parabolic systems defining ordinary polygons are not conjugacy systems. The parabolic systems in $Frob^3_7$ and $Frob^9_{73}$ for $PG(2,2)$ and $PG(2,8)$ respectively are not conjugacy systems.

Note also that the stabilizer of a point of $PG(2,2)$ (of $PG(2,8)$) in $Frob^3_7$ (in $Frob^9_{73}$) also stabilizes a line, whereas by Theorem 10.31 this could not happen if we had a conjugacy system. The same happens in automorphism groups of ordinary m-gons with m odd: the stabilizer of a vertex x also stabilizes the edge opposite to x. Similarly, the parabolic system defined by $\Gamma(HS)$ in HS (see Section 10.2.3) is not a conjugacy system, since there are parabolics of distinct types in it that are conjugate in the group HS, contrary to what Theorem 10.31 states for conjugacy systems. Using the same criterion as above, we immediately see that the n-dimensional simplex does not define a conjugacy system in S_{n+1}, even if $n \geq 3$. Indeed, if x is an i-dimensional face of the n-dimensional simplex ($i = 0, 1, ..., n-1$) and y is the $(n-1-i)$-dimensional face opposite to it, then x and y have the same stabilizer in S_{n+1}.

On the other hand, many parabolic systems exist that satisfy the conclusion of Theorem 10.31 but are not conjugacy systems. Parabolic systems defining ordinary m-gons with m even and ≥ 4 are obvious examples of this kind. Many less trivial examples exist, too. For instance, if $\mathcal{P}$ is a thin polar space, then elements of $\mathcal{P}$ of distinct types have distinct stabilizers in $Aut_s(\mathcal{P})$. However, for every type i and for every element x of type i, the stabilizer of x in $Aut_s(\mathcal{P})$ also stabilizes another element of type i, hence γ_{vert} is not injective.

Truncations. Trivially, a (strong) conjugacy system $\mathcal{G}$ induces (strong) conjugacy systems on all truncations of $\Gamma(\mathcal{G})$ of rank ≥ 2 (see Exercise 10.5

for parabolic systems induced on truncations).

On the other hand, there are parabolic systems that are not conjugacy systems and nevertheless induce conjugacy systems on suitable truncations. Let $\mathcal{G}$ be a parabolic system of this kind, inducing a conjugacy system $\mathcal{G}'$ on a truncation Γ' of the geometry $\Gamma(\mathcal{G})$ defined by $\mathcal{G}$. If we can recover $\Gamma(\mathcal{G})$ from Γ', then we can still handle $\Gamma(\mathcal{G})$ by conjugates instead of cosets, replacing $\mathcal{G}$ by $\mathcal{G}'$.

For instance, let $G = AGL(n,2)$, with $n \geq 4$. Since the stabilizer in G of a hyperplane u of $AG(n,2)$ also fixes the (unique) hyperplane parallel to u, the mapping γ_{vert} is not injective, hence the parabolic system $(G_i)_{i=0}^{n-1}$ defined by $AG(n,2)$ in G is not a conjugacy system. However, the parabolic system (G_0, G_1, G_2) induced in the upper 3-truncation of $AG(n,2)$ is a (non-strong) conjugacy system. We can recover $AG(n,2)$ from its upper 3-truncation. Hence we can work with the conjugacy system (G_0, G_1, G_2) instead of the parabolic system $(G_i)_{i=0}^{n-1}$.

Residues. It is easy to prove that a strong conjugacy system induces strong conjugacy systems in residues. On the other hand, a conjugacy system might not induce conjugacy systems on residues. For instance, let Γ be the upper 3-truncation of $AG(n,2)$ ($n \geq 4$). Then Γ is obtained from a (non-strong) conjugacy system (see above). On the other hand, residues of planes of Γ are circular spaces on four points. A circular space on four points admits only two flag-transitive automorphism groups, namely A_4 and S_4. In both of them the stabilizer of a line also fixes another line. Therefore there are no conjugacy systems for the circular space on four points.

Exercise 10.31. Let Γ be a circular space with $s+2 \geq 5$ points and let $G = S_{s+2}$ or $G = A_{s+2}$. Prove that the parabolic system of Γ in G is a strong conjugacy system.

Exercise 10.32. Let Γ be the bi-affine geometry obtained by removing a plane H and the star of a point p from $PG(3,q)$ (Section 2.2.1). Let $\mathcal{G}$ be the parabolic system of Γ in $Aut_s(\Gamma)$. Prove that $\mathcal{G}$ is a conjugacy system if and only if $q \neq 2$. Also prove that, if $\mathcal{G}$ is a conjugacy system, then it is a strong conjugacy system.

Exercise 10.33. Let $\Gamma = AG(3,8)$ and $G \approx \mathbf{2}^9 : Frob_{73}^9$, flag-transitive on Γ. Check that the parabolic system of Γ in G is not a conjugacy system.

Exercise 10.34. Let Γ be the $Alt(8)$-geometry and let $\mathcal{G}$ be the parabolic system of Γ in A_8 (Exercise 10.19). Check that $\mathcal{G}$ is not a conjugacy system. Prove that, on the other hand, $\mathcal{G}$ induces strong conjugacy systems in residues of points.

Exercise 10.35. Let $\mathcal{G}$ be the parabolic system in A_7 defined in Exercise 10.23. Check that $\mathcal{G}$ is not a conjugacy system.

Exercise 10.36. Let Γ be the thin geometry of type $\widetilde{A}_2$ constructed in Exercise 8.65. This geometry is flag-transitive with special automorphism group $Aut_s(\Gamma) = S_4$ (and $Aut(\Gamma) = S_4 \times S_3$). Check that the parabolic system of Γ in S_4 is not a conjugacy system.

Exercise 10.37. Find a conjugacy system defining a generalized digon and with Borel subgroup $B = 1$.

(Hint: take A_4 in its natural action on four objects; take those four objects as points and the three involutions of A_4 as lines ...)

Exercise 10.38. Given a parabolic system $\mathcal{G}$ in a group G for a geometry Γ, define a coloured graph $\mathcal{C}_{cong}(\mathcal{G})$ as follows (possibly with multiple edges and loops): the conjugates of the Borel subgroup B are the vertices; a triplet (gBg^{-1}, fBf^{-1}, i) is an edge of colour $i \in I$ and vertices gBg^{-1}, fBf^{-1} if there is some $h \in G$ such that $gBg^{-1}, fBf^{-1} \subseteq hG^ih^{-1}$. Let γ_{cham} be the function mapping every chamber $g(C_0)$ of Γ onto its stabilizer gBg^{-1} and every pair $(g(C_0), f(C_0))$ of i-adjacent chambers onto the edge (gBg^{-1}, fBf^{-1}, i) of $\mathcal{C}_{cong}(\mathcal{G})$. Prove that γ_{cham} is an isomorphism from the coloured graph $\mathcal{C}(\Gamma)$ to $\mathcal{C}_{cong}(\mathcal{G})$ if and only if both the following hold:

(i) $N_G(B) = B$;

(ii) $B \subseteq (G^iG^j) \cap (G^jG^i)$ for any two distinct types $i, j \in I$.

Problem. We have set $\gamma_{vert}(x) = (G_x, t(x))$ (and $\gamma(F) = (G_F, t(F))$, more generally) because we wanted a definition that could be applied to cases where there are parabolics of distinct types that are conjugate in G, as in the parabolic system for $\Gamma(HS)$ in HS. However, that precaution has been useless (Theorem 10.31). This failure depends on having defined the adjacency relation of the graph $\Gamma_{cong}(\mathcal{G})$ by means of the inclusion $gG_ig^{-1} \cap fG_jf^{-1} \geq hG_{i,j}h^{-1}$ instead of the equality $gG_ig^{-1} \cap fG_jf^{-1} = hG_{i,j}h^{-1}$. Defining the adjacency relation by the equality $gG_ig^{-1} \cap fG_jf^{-1} = hG_{i,j}h^{-1}$, we obtain a subgraph Γ' of $\Gamma_{cong}(\mathcal{G})$. Thus, we can state a weaker definition of conjugacy systems, requiring only that γ_{vert} is an isomorphism from the incidence graph of Γ to Γ'. This definition covers such cases as that of $\Gamma(HS)$. However, to take profit from it in practice, we need a characterization in the same style of Theorems 10.23 and 10.30 (by means of normalizing conditions) for those parabolic systems that are conjugacy systems in the above weaker sense. Try to obtain such a characterization.

10.4 From $Aut(G)$ to $Aut(\Gamma)$

As in Sections 10.1 and 10.3, $\mathcal{G}$ denotes a parabolic system in a group G and $\Gamma = \Gamma(\mathcal{G})$ is the geometry defined by $\mathcal{G}$. B and G_i ($i \in I$) are the Borel subgroup and the maximal parabolics of $\mathcal{G}$. The chamber of Γ represented by B will be denoted by C_0.

We have $G \leq Aut_s(\Gamma)$ by Theorem 10.17. Henceforth we also assume $G \trianglelefteq Aut(\Gamma)$. However, what we are going to say holds in the general case as well, provided we substitute $N_{Aut(\Gamma)}(G)$ for $Aut(\Gamma)$ everywhere in what follows.

It is easily seen that $\mathbf{P}_4$ of Section 10.1.3 forces $B \cap Z(G) = 1$. Therefore B can be viewed as a subgroup of the inner automorphism group $Inn(G)$ of G.

Given $f \in Aut(\Gamma)$ and $g \in G$, the automorphism f maps the chamber $g(C_0)$ of Γ onto the chamber $f(g(C_0)) = fgf^{-1}(f(C_0))$. Let τ_f be the permutation induced by f on I and let g_f be an element of G such that $f(C_0) = g_f(C_0)$ (note that g_f is determined by f modulo B). The action of f on Γ can be represented on cosets of maximal parabolics as follows:

$$(1) \qquad f(gG_i) = fgf^{-1}g_fG_{\tau_f(i)}$$

for all $i \in I$ and all $g \in G$ (note that $fgf^{-1} \in G$ because we have assumed $G \trianglelefteq Aut(\Gamma)$).

10.4.1 The automorphism group of $\mathcal{G}$

An *automorphism* of $\mathcal{G}$ is an automorphism φ of G such that $\varphi(G_i) = G_{\tau_\varphi(i)}$ for all $i \in I$, for some permutation τ_φ of I (trivially, φ uniquely determines τ_φ). The automorphisms of $\mathcal{G}$ form a subgroup of $Aut(G)$. We denote this subgroup by $Aut(\mathcal{G})$.

Trivially, we have $\varphi(G_J) = G_{\tau_\varphi}(J)$ for every $J \subseteq I$ and for every $\varphi \in Aut(\mathcal{G})$. In particular, $\varphi(B) = B$ for all $\varphi \in Aut(\mathcal{G})$. We have above remarked that $B \leq Inn(G)$. It is clear from the definition of $Aut(\mathcal{G})$ that $B \leq Aut(\mathcal{G}) \cap Inn(G)$ and that $B \trianglelefteq Aut(\mathcal{G})$.

Let $\overline{B}$ be the stabilizer of the chamber C_0 in $Aut(\Gamma)$, that is the group of those $f \in Aut(\Gamma)$ such that $g_f \in B$. Note that $\overline{B} \leq N_{Aut(\Gamma)}(B)$, but it might happen that $\overline{B} \neq N_{Aut(\Gamma)}(B)$.

For every $\varphi \in Aut(\mathcal{G})$, let $\alpha(\varphi)$ be the function mapping the element gG_i of Γ ($g \in G$, $i \in I$) onto the element $\varphi(gG_i) = \varphi(g)G_{\tau_\varphi(i)}$. It is easily seen that $\alpha(\varphi) \in \overline{B}$ and that the function α mapping $\varphi \in Aut(\mathcal{G})$ onto $\alpha(\varphi)$ is a homomorphism from $Aut(\mathcal{G})$ to $\overline{B}$.

Given $f \in \overline{B}$, let $\beta(f)$ be the automorphism of G mapping $g \in G$ onto fgf^{-1}. It is not hard to prove that $\beta(f) \in Aut(\mathcal{G})$ and that the function

β mapping $f \in \overline{B}$ onto $\beta(f)$ is a homormorphism from $\overline{B}$ to $Aut(\mathcal{G})$. The next theorem can easily be obtained by the description (1) of the action of automorphisms of Γ on cosets of maximal parabolics.

Theorem 10.32 *We have* $\alpha\beta = 1_{\overline{B}}$, *where* $1_{\overline{B}}$ *is the identity automorphism of* $\overline{B}$.

Therefore α is surjective and β is injective.

Corollary 10.33 *We have* $Aut(\mathcal{G}) \approx Ker(\alpha) : \overline{B}$.

Proof. By Theorem 10.32, $\beta(\overline{B}) \cap Ker(\alpha) = 1$ and α induces on $\beta(\overline{B})$ an isomorphism to $\overline{B}$. Therefore $Aut(\mathcal{G})$ is a semidirect product of $Ker(\alpha)$ and $\beta(\overline{B}) \cong \overline{B}$. □

Given $\varphi \in Aut(\mathcal{G})$, let $\widehat{\varphi}$ be the function mapping $g \in G$ onto $\varphi(g)g^{-1}$. It is clear from the definition of α that the subgroup $Ker(\alpha)$ can be described as follows:

$$(2) \qquad Ker(\alpha) = \{\varphi \in Aut(\mathcal{G}) \mid \widehat{\varphi}(G) \subseteq B\}$$

Given $f \in Aut(\Gamma)$, let $g_f \in G$ be as in (1). Then $g_f^{-1}f \in \overline{B}$. Hence $Aut(\Gamma) = G\overline{B}$. Trivially, $\overline{B} \cap G = B$ and $B \trianglelefteq \overline{B}$, since we have assumed $G \trianglelefteq Aut(\Gamma)$. Therefore

$$(3) \qquad Aut(\Gamma)/G \cong \overline{B}/B$$

Lemma 10.34 *We have* $\langle Ker(\alpha), B\rangle = Ker(\alpha) \times B$.

Proof. Let $b \in B \cap Ker(\alpha)$. Then $bgb^{-1}g^{-1} \in B$ for all $g \in G$, by (2). Hence $b \in \bigcap_{g\in G} gBg^{-1}$. Therefore $b = 1$, by $\mathbf{P}_4$. Thus, $Ker(\alpha) \cap B = 1$. This gives us the conclusion, since both $Ker(\alpha)$ and B are normal in $Aut(\mathcal{G})$. □

Theorem 10.35 *We have the following isomorphism:*

$$(4) \qquad \frac{Aut(\Gamma)}{G} \cong \frac{Aut(\mathcal{G})}{Ker(\alpha) \times B}$$

(Easy, by Lemma 10.34 and (3); note that $Ker(\alpha) \times B \trianglelefteq Aut(\mathcal{G})$, because both $Ker(\alpha)$ and B are normal in $Aut(\mathcal{G})$.)

Corollary 10.36 *Let* $B = 1$. *Then* $Aut(\Gamma)/G \cong Aut(\mathcal{G})$.

Proof. If $B = 1$, then $Ker(\alpha) = 1$, by (2). The conclusion now follows from the previous theorem. □

10.4.2 The group $Aut_{\mathcal{G}}(G)$

Let $Aut_{\mathcal{G}}(G)$ be the subgroup of the semidirect product $G : Aut(G)$ consisting of those pairs (g, ψ) ($g \in G$, $\psi \in Aut(G)$) such that $\psi(G_i) = gG_{\tau_{\psi,g}(i)}g^{-1}$ for all $i \in I$, for a suitable permutation $\tau_{\psi,g}$ of I. Note that $\tau_{\psi,g}$ is uniquely determined by ψ and g.

It is clear that, given $g \in G$ and denoted by $Inn(g)$ the conjugation by g, then $(g, Inn(g)) \in Aut_{\mathcal{G}}(G)$. Thus, G can be viewed as a subgroup of $Aut_{\mathcal{G}}(G)$, that is $G \leq Aut_{\mathcal{G}}(G)$. Furthermore, it is easily seen that G is normal in $Aut_{\mathcal{G}}(G)$.

Every element $\varphi \in Aut(\mathcal{G})$ can be identified with the pair $(1, \varphi) \in Aut_{\mathcal{G}}(G)$. Thus, we also have $Aut(\mathcal{G}) \leq Aut_{\mathcal{G}}(G)$ and $Aut(\mathcal{G})$ normalizes the subgroup B of $Aut_{\mathcal{G}}(G)$. It is also easy to prove that $Aut_{\mathcal{G}}(G)$ is the semidirect product of G and $Aut(\mathcal{G})$:

$$(5) \qquad Aut_{\mathcal{G}}(G) = G : Aut(\mathcal{G})$$

Given $(g, \psi) \in Aut_{\mathcal{G}}(G)$, let $\alpha(g, \psi)$ be the function mapping xG_i onto $\psi(x)gG_{\tau_{\psi,g}(i)} = \psi(xG_i)g$ (for every $x \in G$ and $i \in I$). It is straightforward to check that $\alpha(g, \psi)$ is an automorphism of Γ. The function $\overline{\alpha}$ mapping $(g, \psi) \in Aut_{\mathcal{G}}(G)$ onto $\alpha(g, \psi) \in Aut(\Gamma)$ is a homomorphism from $Aut_{\mathcal{G}}(G)$ to $Aut(\Gamma)$ and the restriction of $\overline{\alpha}$ to $Aut(\mathcal{G})$ is the homomorphism α from $Aut(\mathcal{G})$ to $\overline{B}$ considered in the previous section.

Theorem 10.37 *The homomorphism $\overline{\alpha}$ is surjective and we have*

$$(6) \qquad Ker(\overline{\alpha}) = \{(b, \varphi) \mid b \in B, \varphi \in Ker(\alpha)\}$$

Proof. The surjectivity of $\overline{\alpha}$ is an easy consequence of (1). The relation (6) follows from definition of $\overline{\alpha}$ and from (2). □

Corollary 10.38 *The following hold:*

$$(7) \qquad \langle Ker(\overline{\alpha}), G \rangle = Ker(\overline{\alpha}) \times G$$

$$(8) \qquad Ker(\overline{\alpha}) \cong Ker(\alpha) \times B$$

Proof. Let $(g, Inn(g)) \in Ker(\overline{\alpha})$. Then $g \in Z(G)$. On the other hand, $g \in B$ by (6). Therefore $g = 1$, by $\mathbf{P}_4$ of Section 10.1.3. Hence $Ker(\overline{\alpha}) \cap G = 1$. The equality (7) follows from this and from the fact that both $Ker(\overline{\alpha})$ and G are normal in $Aut_{\mathcal{G}}(G)$. The isomorphism (8) follows from (5) and (6) and (7). □

Note that $Ker(\overline{\alpha}) \cap Aut(\mathcal{G}) = Ker(\alpha)$. Hence $Ker(\overline{\alpha})$ is not a subgroup of $Aut(\mathcal{G})$ (the isomorphism (8) is not an identity).

It is now evident that (4) of Theorem 10.35 could also be obtained as a consequence of Theorem 10.37 and of (5) and that $Aut_{\mathcal{G}}(G)$ admits the following description:

$$(9) \qquad Aut_{\mathcal{G}}(G) \approx (G \times Ker(\alpha) \times B) \cdot Aut(\Gamma)/G$$

10.4.3 Two examples

Let $\Gamma = PG(n,K)$ $(n \geq 2)$ and $G = P\Gamma L(n,K) = Aut_s(\Gamma)$. Let K be a field. The reader can check that the function δ mapping every non-zero vector $(v_0, v_1, ..., v_n)$ of $V(n+1,K)$ onto the hyperplane of $V(n+1,K)$ represented by the equation $\sum_{i=0}^{n} v_i x_i = 0$ induces a polarity in $PG(n,K)$ and determines an outer automorphism of $GL(n,K)$, mapping a matrix $M \in GL(n,K)$ onto the transpose of the inverse of M. Therefore, δ also induces an outer automorphism of G. We have $Aut(\Gamma) = Aut(G) = G \cdot \mathbf{2}$, $Ker(\alpha) = 1$ and $Aut(\mathcal{G}) = B \cdot \mathbf{2}$.

On the other hand, if K is non-commutative, then $Aut(\Gamma) = G = Aut(G)$ and $Aut\mathcal{G} = B$.

The n-dimensional simplex gives us a quite different example. Let Γ be the n-dimensional simplex and $G = Aut_s(\Gamma) = S_{n+1}$. We have seen in Section 9.2.2(2) that Γ admits an involutory dual automorphism δ mapping every i-dimensional face onto the $(n-i-1)$-dimensional face opposite to it and that $Aut(\Gamma) = G \times \langle\delta\rangle = S_{n+1} \times \mathbf{2}$. On the other hand, if $\mathcal{G}$ is the parabolic system for Γ in S_{n+1} (see Section 10.2.1), we have $Aut(\Gamma)/G \cong Aut(\mathcal{G})$ by Corollary 10.36, because $B = 1$. Hence $Aut(\mathcal{G}) = \mathbf{2}$. If we think of the previous example, we might expect to see that the non-trivial element of $Aut(\mathcal{G})$ is contributed by some outer automorphism of G. However, this is not the case at all, because $Aut(G) = G$. Actually, the non-trivial element of $Aut(\mathcal{G})$ is an inner automorphism of G, obtained by conjugation from the unique (involutory) element $w_0 \in G$ mapping the chamber C_0 from which $\mathcal{G}$ is defined onto the chamber at maximal distance from C_0.

We will turn back to this example in Section 11.3.5

Exercise 10.39. Prove that $Ker(\alpha) \cap Inn(G) = 1$.

(Hint: if $fgf^{-1}g^{-1} \in B$ for all $g \in G$, then $f \in B$ and we obtain $f = 1$ by $\mathbf{P}_4$ of Section 10.1.3.)

Exercise 10.40. Let $N_G(G_i) = G_i$ for all $i \in I$ and let π be the projection of $G : Aut(G)$ onto $Aut(G)$, mapping (g, ψ) onto $\psi^{\bullet}$ $(g \in G, \psi \in Aut(G))$. Prove that $Ker(\pi) \cap Aut_{\mathcal{G}}(G)$ $(\leq G)$ is an extension $B \cdot K_G$ of (a copy of the Borel subgroup) B by a suitable group K_G of permutations of I.

(Hint. Under the previous hypotheses, for every $\psi \in Aut(G)$ and every permutation τ of I there is at most one coset gB such that $(g, \psi) \in Aut_{\mathcal{G}}(G)$

and $\tau_{\psi,g} = \tau$. Therefore ...)

Exercise 10.41. Define an equivalence relation $\equiv$ on I by the following clause: given $i, j \in I$, we set $i \equiv j$ if G_i and G_j are conjugate in G. With the hypotheses and the notation of the previous exercise, prove that the permutations of I corresponding to elements of K_G fix all classes of $\equiv$ (hence $K_G = 1$ if $\mathcal{G}$ is a conjugacy system).

Exercise 10.42. Keeping the hypotheses and the notation of Exercise 10.40, prove that $Aut(\mathcal{G}) \cap Inn(G) \cong Ker(\pi)$.

Exercise 10.43. With the hypotheses and the notation of Exercise 10.40, prove that $Aut(\Gamma)$ contains a normal subgroup $K_\Gamma \cong K_G$, centralizing G and such that $K_\Gamma \cap G = 1$.

Exercise 10.44. With the notation and the hypotheses of Exercises 10.40 and 10.43, prove that the isomorphism stated in (4) of Theorem 10.35 maps K_Γ onto $(Aut(\mathcal{G}) \cap Inn(G))/B$.

Chapter 11

Amalgams

We will define amalgams in Section 11.2 of this chapter, aiming at a construction of universal objects in categories of parabolic systems (we postpone the more general problem of constructing universal objects in categories of geometries to the next chapter). Then we apply amalgams to the investigation of flag-transitive thin geometries (Section 11.3).

However, before coming to amalgams, we need to complete the theory developed in Sections 8.2 and 8.3 for quotients and covers, showing how certain subgroups of $Aut_s(\Gamma)$ can be used to define quotients (in particular, m-quotients) of a geometry Γ.

11.1 Back to quotients

Let $f : \Gamma \longrightarrow \Gamma'$ be an epimorphism of geometries. An automorphism g of Γ is called a *deck transformation* of f if $fg = f$. Since τ_f is a bijection from the set of types of Γ to the set of types of Γ', the deck transformations of f belong to $Aut_s(\Gamma)$. Hence they form a subgroup of $Aut_s(\Gamma)$. We call this subgroup the *deck group* of f and we denote it by $Aut(f)$.

The orbits of $Aut(f)$ on the set of elements of Γ are subsets of fibres of f. Hence they form a refinement of the type partition of Γ and we can define types for them, stating that the type $t(X)$ of an orbit X of $Aut(f)$ is the type $t(x)$ of any element $x \in X$.

The case when the orbits of $Aut(f)$ are precisely the fibres of f is particularly interesting. In this case properties $\mathbf{Q}_1$ and $\mathbf{Q}_2$ of Section 8.2 can be rephrased as follows, with $G = Aut(f)$:

($\mathbf{Q}_1'$) Given a flag $F = \{x_1, x_2, ..., x_k\}$ of Γ and an orbit X of G of type

$t(X) \notin t(F)$, let $y_1, y_2, ..., y_k$ be elements of X such that $y_i * x_i$ for $i = 1, 2, ..., k$. Then there is an element $y \in X$ such that $y * F$.

($\mathbf{Q}_2'$) For every type i, the group G has at least two orbits on the set of elements of Γ of type i and, for every panel F of Γ of cotype i, the residue Γ_F meets at least two of those orbits.

We remark that, if we rephrase $\mathbf{Q}_1$ for G word-by-word, then we should state $\mathbf{Q}_1'$ for any m-tuple of elements $y_1, y_2, ..., y_k$ of X such $\Gamma_{y_i} \cap G(x_i) \neq \emptyset$ for every $i = 1, 2, ..., k$. However, the statement we obtain in this way is just equivalent to $\mathbf{Q}_1'$, even if it looks stronger. Indeed, if $g_i(x_i) * y_i$ for some $g_i \in G$, then we can substitute $g_i^{-1}(y_i)$ for y_i and we are back to the hypotheses of $\mathbf{Q}_1'$.

Conversely, let G be a subgroup of $Aut_s(\Gamma)$ satisfying $\mathbf{Q}_1'$ and $\mathbf{Q}_2'$. Then the orbits of G on the set of elements of Γ form a refinement Φ of the type partition of Γ satisfying $\mathbf{Q}_1$ and $\mathbf{Q}_2$ of Section 8.2. Hence Φ defines a quotient of Γ (Theorem 8.5). We denote this quotient by Γ/G and we call it the *quotient* of Γ by G. Thus, if $G \leq Aut_s(\Gamma)$ satisfies $\mathbf{Q}_1'$ and $\mathbf{Q}_2'$, then we say that G *defines a quotient* of Γ.

Let $G \leq Aut_s(\Gamma)$ define a quotient of Γ and let p_G be the projection of Γ onto Γ/G. Then $G \leq Aut(p_G)$, the deck group $Aut(p_G)$ of p_G defines a quotient of Γ and we have $\Gamma/G = \Gamma/Aut(p_G)$. Trivially, $Aut(p_G)$ is the largest subgroup of $Aut_s(\Gamma)$ that defines the same quotient as G.

11.1.1 Groups residually defining quotients

Let $G \leq Aut_s(\Gamma)$ define a quotient of Γ, let F be a non-maximal flag of Γ and let G_F be the stabilizer of F in G. Let $\overline{G}_F \leq Aut_s(\Gamma_F)$ be the action of G_F in Γ_F. It is easily seen that $\overline{G}_F$ defines a quotient $\Gamma_F/\overline{G}_F$ of Γ_F. Furthermore, the function $p_{G,F}$ mapping $\overline{G}_F(x)$ onto $G(x)$ for $x \in \Gamma_F$ is an epimorphism from $\Gamma_F/\overline{G}_F$ to the residue $(\Gamma/G)_{G(F)}$ of the flag $G(F) = \{G(x) \mid x \in F\}$ in Γ/G. Trivially, $p_{G,F}$ is an isomorphism if and only if, for every orbit X of G, $G_F \cap X$ is an orbit of G_F.

Theorem 11.1 *Let $G \leq Aut_s(\Gamma)$ define a quotient of Γ. Then the following are equivalent:*

(i) the epimorphism $p_{G,F} : \Gamma_F/\overline{G}_F \longrightarrow (\Gamma/G)_{G(F)}$ is an isomorphism for every non-maximal flag F of Γ;
(ii) for every flag F of Γ and every orbit X of G meeting Γ_F, the set $X \cap \Gamma_F$ is an orbit of G_F;
(iii) given flags $F = \{x_1, x_2, ..., x_k\}$ and $F' = \{y_1, y_2, ..., y_k\}$ with $y_i \in G(x_i)$ for $i = 1, 2, ..., k$, there is an element $g \in G$ such that $g(F) = F'$.

Proof. We have already remarked that the statements of (i) and (ii) are equivalent. Let us prove that (ii) implies (iii).

Let (ii) hold. The property (ii) is preserved under taking residues and stabilizers of flags. Hence we can work by induction on the rank n of Γ.

If $n = 1$ there is nothing to prove. Let $n > 1$ and let F and F' be as in the hypotheses of (iii). If $k = 1$ there is nothing to prove. Let $k > 1$ and let $g_1 \in G$ map x_1 onto y_1. By (ii) we have $y_i \in G_{y_1}(g_1(x_i))$ for $i = 2, 3, ..., k$. By the inductive hypothesis in Γ_{y_1} applied to G_{y_1} and to the flags $F_1 = g_1(F - x_1)$ and $F'_1 = F' - y_1$, there is an element $g_2 \in G_{y_1}$ mapping F_1 onto F'_1. Hence $g = g_2 g_1$ maps F onto F'. We are done.

Finally, we prove that (iii) implies (ii). Let (iii) hold. Given a flag F and elements $x, y \in \Gamma_F$ in the same orbit of G, by (iii) there is an element $g \in G$ mapping $F \cup x$ onto $F \cup y$. Trivially, $g(F) = F$. Thus, (ii) holds. □

We say that a subgroup G of $Aut_s(\Gamma)$ *residually defines quotients* in Γ if it defines a quotient of Γ and satisfies any of the equivalent conditions (i), (ii) or (iii) of the previous theorem.

Exercise 11.1. Note that we have not used the hypothesis that G defines a quotient in the proof of the equivalence of (ii) and (iii) of Theorem 11.1. In view of this, let $G \leq Aut_s(\Gamma)$ satisfy (ii) or (iii) of Theorem 11.1. Prove that G satisfies $\mathbf{Q}'_1$ if and only if the following holds in G:

($\mathbf{Q}''_1$) given a flag F, elements x', y' of Γ_F and elements $x \in G(x')$, $y \in G(y')$ with $x * y$, there is an element $g \in G$ such that $g(x), g(y) \in \Gamma_F$.

(Remark. Tits [227] uses $\mathbf{Q}''_1$ to define quotients by groups. We have preferred to use $\mathbf{Q}'_1$ instead of $\mathbf{Q}''_1$ because $\mathbf{Q}'_1$ is just the group-theoretic translation of $\mathbf{Q}_1$ of Chapter 8, which is a merely geometric property, whereas we do not see how to rephrase $\mathbf{Q}''_1$ without mentioning groups at all.)

Exercise 11.2. Let Γ have rank ≥ 2, let $G \leq Aut_s(\Gamma)$ define a quotient of Γ and assume that, for every element x of Γ and for every element $g \in G$ with $g \neq 1$, the elements x and $g(x)$ have distance ≥ 3 in the incidence graph of Γ. Prove that G residually defines quotients in Γ and that Γ/G is a topological quotient of Γ (Section 8.3).

Exercise 11.3. Let Γ have rank ≥ 2 and let $G \leq Aut_s(\Gamma)$ be such that, for every element x of Γ and for every element $g \in G$ with $g \neq 1$, the elements x and $g(x)$ have distance ≥ 4 in the incidence graph of Γ. Prove that G residually defines quotients in Γ and that Γ/G is a topological quotient of Γ.

Exercise 11.4. Let $G_0 \leq Aut_s(\Gamma)$ define a quotient of Γ and let Γ/G_0 be a topological quotient. Prove that G_0 residually defines quotients in Γ, that the same holds for every subgroup of G_0 and that Γ/G is a topological quotient for every subgroup $G \leq G_0$.

Exercise 11.5. Let $G \leq Aut_s(\Gamma)$ define a quotient of Γ and let Γ/G be an m-quotient, for some $m \geq 1$. Prove, if (iii) of Theorem 11.1 holds for flags of corank $\leq m$, then G residually defines quotients in Γ.

11.1.2 Liftings and projections of automorphisms

Let $G \leq Aut_s(\Gamma)$ define a quotient of Γ and let p_G be the projection of Γ onto Γ/G.

Given $f \in Aut(\Gamma)$ and $f' \in Aut(\Gamma/G)$, if $p_G f = f' p_G$ then we say that f is a *lifting* of f' via p_G and that f' is a *projection* of f onto Γ/G. Trivially, every automorphism of Γ has at most one projection onto Γ/G. If $f \in Aut(\Gamma)$ has a projection onto Γ/G, then this projection will be denoted by $p_G(f)$.

The identity automorphism of Γ is a lifting of the identity automorphism of Γ/G. If f and g admit projections onto Γ/G, then gf and f^{-1} also admit projections and we have $p_G(fg) = p_G(f)p_G(g)$ and $p_G(f^{-1}) = (p_G(f))^{-1}$. Therefore, if $L(G)$ denotes the set of automorphisms of Γ admitting projections onto Γ/G, then $L(G)$ is subgroup of $Aut(\Gamma)$ and the function $p_G : L(G) \longrightarrow Aut(\Gamma/G)$, mapping every element $f \in L(G)$ onto its projection $p_G(f)$, is a (possibly non-surjective) homomorphism. The kernel of this homomorphism is just the deck group $Aut(p_G)$ of p_G.

Trivially, if A is a subgroup of $Aut(\Gamma)$, then $p_G^{-1}(A)$ is a subgroup of $L(G)$. In particular, we have $p_G^{-1}(Aut_s(\Gamma/G)) = L(G) \cap Aut_s(\Gamma)$. We will write $L_s(G)$ for $L(G) \cap Aut_s(\Gamma)$.

Let $N(Aut(p_G))$ and $N_s(Aut(p_G))$ be the normalizers of $Aut(p_G)$ in $Aut(\Gamma)$ and in $Aut_s(\Gamma)$, respectively.

Theorem 11.2 *We have* $L(G) = N(Aut(p_G))$ *and* $L_s(G) = N_s(Aut(p_G))$.

Proof. The second equality is a trivial consequence of the first one. Let us prove the first one.

Since $Aut(p_G)$ is the kernel of the homomorphism $p_G : L(G) \longrightarrow \Gamma/G$, we have $L(G) \leq N(Aut(p_G))$. Let us prove the converse inclusion. Let $f \in N(Aut(p_G))$. We define a mapping $p_f : \Gamma/G \longrightarrow \Gamma/G$ by the following clause: $p_f(p_G(x)) = p_G(f(x))$, for every element x of Γ. It is not difficult to check that p_f is an endomorphism of Γ/G, provided that it is well defined.

To prove that p_f is well defined we must show that, if $p_G(x) = p_G(y)$ for distinct elements x, y of Γ, then $p_G(f(x)) = p_G(f(y))$. Let $p_G(x) = p_G(y)$.

Then $y = g(x)$ for some $g \in G$. Hence $f(y) = f(g(x)) = g'f(x)$ for some $g' \in Aut(p_G)$, because $G \leq Aut(p_G)$ and $f \in N(Aut(p_G))$. Hence $p_G(f(y)) = p_G(f(x))$. Therefore p_f is well defined.

We can also define $p_{f^{-1}}$ and it is easily seen that $p_{f^{-1}}p_f = p_f p_{f^{-1}} = 1$. Hence p_f is bijective and $p_f^{-1} = p_{f^{-1}}$. Therefore $p_f \in Aut(\Gamma/G)$ and f is a lifting of p_f. Hence $f \in L(G)$. The equality $L(G) = N(Aut(p_G))$ is proved. □

Corollary 11.3 *Let $Aut(p_G) \trianglelefteq Aut_s(\Gamma)$ and let Γ be flag-transitive. Then Γ/G is flag-transitive.*

Proof. We have $Aut_s(\Gamma) \leq L(G)$, by Theorem 11.2. The group $Aut_s(\Gamma)$ is flag-transitive by assumption. It is projected by p_G onto a flag-transitive subgroup of $Aut_s(\Gamma)$. □

11.1.3 Deck groups of covers

Lemma 11.4 *Let $f : \Gamma \longrightarrow \Gamma'$ be an m-cover. Then $Aut(f)$ acts semi-regularly on the set of flags of Γ of corank $\leq m$.*

Proof. Given a flag F of Γ of corank $k \leq m$ and an element g of $Aut(f)$ fixing F, let C be a chamber of Γ containing F. We have $g(C) = C$ because f is an m-cover. Given any other chamber D of Γ, let γ be a gallery from C to D. Then $g(\gamma)$ is a gallery from $C = g(C)$ to $g(D)$. Since $g \in Aut(f)$, the galleries γ and $g(\gamma)$ are mapped onto the same gallery of Γ'. Hence $\gamma = g(\gamma)$, by Proposition 8.8. In particular, $g(D) = D$. This shows that $g = 1$. □

Lemma 11.5 *Let $G \leq Aut_s(\Gamma)$ residually define quotients in Γ and let Γ/G be an m-quotient of Γ. Let p_G^m be the mapping induced by p_G from the set of flags of Γ of corank $\leq m$ to the set of flags of Γ/G of corank $\leq m$. Then G stabilizes each of the fibres of p_G^m and acts regularly in each of them.*

Proof. Trivially, $G(X) = X$ for every fibre X of p_G^m. Let $F, F' \in X$. Then $F' = g(F)$ by (iii) of Theorem 11.1, because G residually defines quotients. Therefore, G acts transitively in X. If X consists of chambers, then G acts regularly in X by Lemma 11.4.

Let X consist of non-maximal flags and let $F \in X$. By (i) of Theorem 11.1, we have $\Gamma_F/\overline{G}_F \cong (\Gamma/G)_{G(F)}$. However, $(\Gamma/G)_{G(F)} \cong \Gamma_F$ because p_G is an m-cover and F has corank $\leq m$. Hence $\overline{G}_F = 1$ and G_F fixes all chambers containing F. Hence $G_F = 1$, because G acts semi-regularly on the set of chambers of Γ. □

Theorem 11.6 *Let $G \leq Aut_s(\Gamma)$ residually define quotients in Γ and let Γ/G be an m-quotient of Γ. Then we have $G = Aut(p_G)$.*

Proof. Given a fibre X of the morphism induced by p_G from $\mathcal{C}(\Gamma)$ to $\mathcal{C}(\Gamma/G)$, both G and $Aut(p_G)$ stabilize X and act semi-regularly on X, by Lemma 11.4. On the other hand, G also acts regularly on X, by Lemma 11.5. Hence $Aut(p_G)$ cannot be larger than G. □

11.1.4 Examples

Many examples of quotients given in Chapter 8 are actually quotients by groups. We will now recall a few of them.

(1) **Quotients of polar spaces.** Let $\mathcal{P}$ be a polar space of rank n and let $G \leq Aut_s(\mathcal{P})$. The reader can prove that G residually defines quotients in $\mathcal{P}$ if and only if the following hold:

($\mathbf{PQ}_1$) we have $x^\perp \cap G(x) = x$ for every point x;

($\mathbf{PQ}_2$) we have $x^\perp \cap G(y) = G_x(y)$ for every choice of distinct collinear points x, y of $\mathcal{P}$;

($\mathbf{PQ}_3$) for every $(n-2)$-dimensional subspace X of $\mathcal{P}$ the stabilizer G_X of X in G is not transitive on the star of X.

The condition $\mathbf{PQ}_1$ just states that the orbits of G are cocliques of the collinearity graph of $\mathcal{P}$. That is, $X \cap g(X) = \emptyset$ for every maximal subspace X of $\mathcal{P}$ and for every non-identity element $g \in G$. If furthermore we want $\mathcal{P}/G$ to be an m-quotient, then we need only assume the following (which entails $\mathbf{PQ}_3$):

($\mathbf{PQ}_{3,m}$) for every $(n-m-1)$-dimensional subspace X of $\mathcal{P}$, the stabilizer G_X of X in G acts trivially in the star of X (hence $G_X = 1$, by $\mathbf{PQ}_1$ and Lemma 11.4).

In particular, $\mathcal{P}/G$ is a topological quotient if and only if G acts semi-regularly on the set of points of $\mathcal{P}$.

It is worth remarking that, if $\mathbf{PQ}_1$ or $\mathbf{PQ}_3$ (or $\mathbf{PQ}_{3,m}$) hold in a subgroup $G \leq Aut_s(\mathcal{P})$, then they hold in all subgroups of G. Therefore, if G residually defines quotients in $\mathcal{P}$, then every subgroup $H \leq G$ satisfying $\mathbf{PQ}_2$ residually defines quotients in $\mathcal{P}$ (and, if $\mathcal{P}/G$ is an m-quotient, then $\mathcal{P}/H$ is an m-quotient).

Furthermore, if G residually defines quotients in $\mathcal{P}$ and if $\mathcal{P}/G$ is a topological quotient, then every subgroup $H \leq G$ residually defines quotients in $\mathcal{P}$ and $\mathcal{P}/H$ is a topological quotient of $\mathcal{P}$ (see Exercise 11.4).

A quotient $\mathcal{P}/G$ is flag-transitive if G is normalized by a flag-transitive subgroup of $Aut_s(\mathcal{P})$. We will prove in Chapter 12 (Theorem 12.59 and Proposition 12.48) that an m-quotient $\mathcal{P}/G$ ($m = 2, 3, ..., n-1$) is flag-transitive only if the normalizer of G is flag-transitive in $\mathcal{P}$. We will also prove (Corollary 12.57 and Proposition 12.48) that every m-quotient of a polar space $\mathcal{P}$ (m as above) arises as a quotient by a subgroup $G \leq Aut_s(\mathcal{P})$ satisfying $\mathbf{PQ}_1$, $\mathbf{PQ}_2$ and $\mathbf{PQ}_3$.

In particular, classical polar spaces defined over infinite division rings often admit infinitely many proper quotients, but these quotients are never flag-transitive. Indeed, groups such as $P\Gamma O(f)$, $PGO(f)$, $PSO(f)$ or $P\Omega(f)$ never admit non-trivial proper normal subgroups 'small' enough to define proper quotients (see for instance [92], 27.5 or [35], 11.1 and 14.4).

For instance, if $\mathcal{P}$ is as in Exercise 8.51 and γ is the involutory semilinear transformation mapping a vector $x = (x_1, x_2, ..., x_m)$ onto its conjugate $\gamma(x) = \overline{x} = (\overline{x}_1, \overline{x}_2, ..., \overline{x}_m)$, then $\langle\gamma\rangle$ defines a topological 2-fold quotient of $\mathcal{P}$, but this quotient is not flag-transitive. It is clear that many other examples of m-quotients of classical polar spaces can be constructed with $m \geq 2$, choosing suitably small subgroups of the automorphism group of the polar space. The reader can find some of them in [170], [191] or [168]. As we remarked above, none of these quotients is flag-transitive. Furthermore, all of them arise from polar spaces defined over infinite division rings, according to Corollary 8.17.

It is worth mentioning that some flat C_3 geometries can also be obtained as quotients as above, starting from classical polar spaces of rank 3 defined over ordered fields (see [170]).

Now let $\mathcal{P}$ be thin. Then $\mathcal{P}$ can be realized as the system of cliques of a complete n-partite graph with classes A_1, A_2, ..., A_n of size 2. We know from Section 9.2.2(3) that, if $e_i \in Aut_s(\mathcal{P})$ is the involution permuting the two points in A_i and fixing all remaining points, then e_1, e_1, ..., e_n form a basis for $V(n, 2)$ and $2^n = V(n, 2)$ is a normal subgroup of $Aut_s(\mathcal{P})$.

It is not hard to check that all m-quotients of $\mathcal{P}$ arise as quotients by subgroups $G \leq V(n, 2)$ such that all non-zero vectors in G have at most $n - m - 1$ entries equal to 0. Furthermore, if G defines an m-quotient of $\mathcal{P}$, then it residually defines quotients into $\mathcal{P}$. By Lemma 11.5, the m-quotient $\mathcal{P}/G$ is t-fold with $t = |G|$.

For instance, let x be a vector of $V(n, 2)$ with at most $n - m - 1$ entries equal to 0. Then the group $G = \langle x \rangle$ produces a 2-fold m-quotient of $\mathcal{P}$. However, other choices are often possible for the group G. For instance, given a proper divisor d of n and a partition of the set of indices $\{1, 2, ..., n\}$

in d classes $I_1, I_2, ..., I_d$ of size n/d, the group $G = \langle \sum_{j \in I_i} e_j \rangle_{i=1}^{d}$ produces a 2^d-fold $(n/d - 1)$-quotient of $\mathcal{P}$.

By Corollary 11.3, a quotient $\mathcal{P}/G$ is flag-transitive if G is normal in $Aut_s(\mathcal{P})$. Actually, a partial converse of this statement is also true: an m-quotient $\mathcal{P}/G$ $(m = 2, 3, ..., n-1)$ is flag-transitive only if $G \trianglelefteq Aut_s(\mathcal{P})$ (see Section 11.3.4(2)).

Problem. Something like a classification could perhaps be achieved for some classes of m-quotients of classical polar spaces, if not for all of them.

(2) **Affine polar spaces.** Let $\mathcal{P}$ be a classical polar space defined by a sesquilinear or quadratic form φ in $PG(m, K)$, let $\overline{H}$ be a hyperplane of $PG(m, K)$ and let Γ be the affine polar space obtained by deleting the hyperplane section H of $\mathcal{P}$ defined by $\overline{H}$ (Section 2.2.2). Let $p \in PG(m, K)$ be the polar point of $\overline{H}$ with respect to φ and let $U_{\overline{H},p}$ be the group of all perspectivities of $PG(m, K)$ with centre p and axis $\overline{H}$. Then $G_0 = U_{p,\overline{H}} \cap Aut_s(\mathcal{P})$ is the elementwise stabilizer of H in $Aut_s(\mathcal{P})$ $(= Aut(\mathcal{P}))$ and it can be viewed as a subgroup of $Aut_s(\Gamma)$ $(= Aut(\Gamma))$. We have $G_0 \trianglelefteq Aut_s(\Gamma)$.

It is not hard to prove that G_0 residually defines quotients in Γ and that Γ/G_0 is the minimal standard quotient Γ/Φ_0 of Γ defined in Section 8.4.5. Since G_0 is normal in the stabilizer G_H of H in $Aut_s(\mathcal{P})$ and $G_H = Aut_s(\Gamma)$ (Exercise 9.22), the quotient Γ/G_0 is flag-transitive, by Corollary 11.3.

Furthermore, every subgroup $G \leq G_0$ residually defines quotients in Γ and Γ/G is a standard quotient of Γ (see Exercise 11.4). A quotient Γ/G is standard if and only if $G \leq G_0$ (see [156]).

It will follow from Theorem 12.56 and Proposition 12.50 of that all k-quotients of Γ $(k = 2, 3, ..., n-1)$ can be obtained from subgroups of $Aut_s(\Gamma)$. In particular, all standard quotients of Γ arise from subgroups of G_0.

(3) **Tessellations.** All quotients of tessellations of surfaces considered in Section 2.1.2 were actually quotients by groups. For instance, pairs of antipodal elements of the icosahedron are orbits of the unique non-trivial central element of the automorphism group $\mathbf{2} \times A_5$ of the icosahedron (see Exercise 9.3). Thus, the halved incosahedron is obtained factorizing the icosahedron by the centre $\mathbf{2}$ of $\mathbf{2} \times A_5$.

(4) **Affine–dual-affine geometries.** Let Γ be a bi-affine geometry of rank $n \geq 3$, obtained by removing from $PG(n, K)$ a hyperplane H and the star

of a point p. Let $U_{H,p}$ be the group of all perspectivities of $PG(n,K)$ with centre p and axis H. It is not difficult to check that $U_{H,p}$ defines the quotient Γ/Φ_0 of Γ considered in Exercise 8.59 and that every subgroup of $U_{H,p}$ defines a quotient of Γ.

All 1-quotients of Γ are topological quotients (see Exercise 8.44) and all topological quotients of Γ arise from subgroups of $U_{H,p}$ (Theorem 12.56 and Exercise 12.31). In particular, if $p \notin H$ and $K = GF(2)$, then Γ does not admit any proper 1-quotient (compare with Exercise 8.60)).

Exercise 11.6. Let $\mathcal{P}$ be the rank 3 polar space of order $(1,1,3)$. Prove that, up to isomorphisms, $\mathcal{P}$ admits just two flat 2-quotients and just one 2-fold 2-quotient.

(Hint: there are just three conjugacy classes of subgroups of S_4 acting semi-regularly on the four points being permuted.)

Exercise 11.7. Let $\mathcal{P}$ be a polar space of order $(1,1,...,1,t)$, $t < \infty$. Prove that $\mathcal{P}$ admits a flag-transitive d-fold topological quotient for every divisor d of t.

Exercise 11.8. Let R, P, R_P and R/P be as in Exercise 8.13. For $n \geq 2$, let H be the set of elements of $PGL(n+1,R_P)$ represented by matrices of the form $I+M$, where I is the $(n+1)\times(n+1)$ identity matrix and M is an $(n+1)\times(n+1)$ matrix with entries in P. Prove that H is a subgroup of $PGL(n+1,R_P)$, that it residually defines quotients in $PG(n,R_P)$ and that $PG(n,R_P)/H \cong PG(n,R/P)$ (needless to say, this quotient is not a 1-quotient).

(Remark. The epimorphism $f: PG(2,R_P) \longrightarrow PG(2,R/P)$ considered in the hint of Exercise 8.13 is just an example of the above.)

Exercise 11.9. With the notation of Exercise 11.8, let p_H be the projection of $PG(n,R_P)$ onto $PG(n,R_P)/H = PG(n,R/P)$. Prove that $H = Aut(p_H)$.

Exercise 11.10. With the notation of Exercise 11.8, let $G(n+1,R)$ be the set of elements of $PGL(n+1,R_P)$ represented by matrices with entries in R and determinant in $R-P$. Prove that $G(n+1,R)$ is a subgroup of $PGL(n+1,R_P)$ and that $G(n+1,R) = L(H)\cap PGL(n+1,R_P)$, where $L(H)$ is the group of those automorphisms of $PG(n,R_P)$ that admit projections onto $PG(n,R_P)/H$, as in Section 11.1.2 (hence $H \trianglelefteq G(n+1,R)$ and H is the kernel of the homomorphism $p_H : G(n+1,R) \longrightarrow PGL(n,R/P)$, which is now surjective).

Exercise 11.11. Let R, P, R_P and R/P be as in Exercise 8.13. Prove that there are epimorphisms from $\mathcal{Q}^+_{2n-1}(R_P)$ to $\mathcal{Q}^+_{2n-1}(R/P)$ and from

$\mathcal{S}_{2n-1}(R_P)$ to $\mathcal{S}_{2n-1}(R/P)$ (note that these epimorphisms are not 1-covers).

11.2 Amalgams

11.2.1 Morphisms of parabolic systems

Let $\mathcal{G} = (G_i)_{i \in I}$ and $\mathcal{G}' = (G'_i)_{i \in I}$ be parabolic systems of rank n over the same set of types I in groups G and G' respectively (see Section 10.1.3).

A homomorphism $\varphi : G \longrightarrow G'$ is called a *morphism* from $\mathcal{G}$ to $\mathcal{G}'$ if $\varphi(G_J) = G'_J$ for every $J \subseteq I$. In particular, if φ is a morphism from $\mathcal{G}$ to $\mathcal{G}'$, then $\varphi(G_\emptyset) = G'_\emptyset$, that is $\varphi(G) = G'$ (that is, φ is surjective).

We write $\varphi : \mathcal{G} \longrightarrow \mathcal{G}'$ to mean that φ is a morphism from $\mathcal{G}$ to $\mathcal{G}'$. As in Section 10.1, $\Gamma(\mathcal{G})$ and $\Gamma(\mathcal{G}')$ denote the geometries defined by $\mathcal{G}$ and $\mathcal{G}'$, respectively.

Lemma 11.7 *Every morphism $\varphi : \mathcal{G} \longrightarrow \mathcal{G}'$ uniquely determines a type-preserving epimorphism $f_\varphi : \Gamma(\mathcal{G}) \longrightarrow \Gamma(\mathcal{G}')$. Furthermore, $Aut(f_\varphi) \cap G = Ker(\varphi)$.*

Proof. Given gG_i of $\Gamma(\mathcal{G})$ ($g \in G$, $i \in I$), define $f_\varphi(gG_i) = \varphi(gG_i) = \varphi(g)G'_i$. It is straightforward to check that the mapping f_φ defined in this way is a type-preserving morphism from $\Gamma(\mathcal{G})$ to $\Gamma(\mathcal{G}')$. Using Lemma 10.15, we can easily check that (iii) of Theorem 8.4 holds. Hence f_φ is an epimorphism. Trivially, $Ker(\varphi) \leq Aut(f_\varphi) \cap G$. Let us prove the converse inequality.

Let $h \in Aut(f_\varphi) \cap G$. Then we have $\varphi(h)\varphi(g)G'_i = \varphi(g)G'_i$ for every $i \in I$ and every $g \in G$. As $\varphi : G \longrightarrow G'$ is surjective, the above can be restated as follows: $\varphi(h) \in xG'_ix^{-1}$ for all $i \in I$ and $x \in G'$. Hence $\varphi(h) \in \bigcap_{x \in G'} xB'x^{-1}$, where B' is the Borel subgroup of $\mathcal{G}'$. On the other hand, we have $\bigcap_{x \in G'} xB'x^{-1} = 1$ by $\mathbf{P}_4$. Therefore $\varphi(h) = 1$. □

Corollary 11.8 *Let $\varphi : \mathcal{G} \longrightarrow \mathcal{G}'$ be a morphism. Then $Ker(\varphi)$ residually defines quotients in $\Gamma(\mathcal{G})$ and $\Gamma(\mathcal{G}') \cong \Gamma(\mathcal{G})/Ker(\varphi)$.*

Proof. Let $f_\varphi : \Gamma(\mathcal{G}) \longrightarrow \Gamma(\mathcal{G}')$ be the epimorphism induced by φ, defined by the clause $f_\varphi(gG_i) = \varphi(g)G'_i$ ($g \in G$, $i \in I$), as in the proof of Lemma 11.7. We have $f_\varphi(gG_i) = f_\varphi(G_i)$ if and only if $g \in Ker(\varphi)G_i$, by the definition of f_φ. Hence the orbits of $Ker(\varphi)$ in $\Gamma(\mathcal{G})$ are just the fibres of f_φ. Therefore $Ker(\varphi)$ defines a quotient and $\Gamma(\mathcal{G}') \cong \Gamma(\mathcal{G})/Ker(\varphi)$. We must still prove that $Ker(\varphi)$ residually defines quotients.

Given $J \subseteq I$, let φ_J be the surjective homomorphism from G_J to G'_J induced by φ. Trivially, $Ker(\varphi_J) = Ker(\varphi) \cap G_J$. Let Γ_J be the residue of the flag F_J of $\Gamma(\mathcal{G})$ of type J represented by G_J. Let $i \in I-J$. The elements of Γ_J of type i are the cosets gG_i with $g \in G_J$. Let $f_\varphi(G_i) = f_\varphi(gG_i)$ for some $g \in G_J$. Then $g \in Ker(\varphi)G_i$ (see above). Hence $g \in G_J \cap Ker(\varphi)G_i$. However, $\varphi(G_J \cap K(\varphi)G_i) = G'_{J\cup i} = \varphi(G_{J\cup i})$. Therefore we have $g^{-1}g' \in Ker(\varphi) \cap G_J = Ker(\varphi_J)$ for some $g' \in G_{J\cup i}$.

It follows from the above that the orbits of $Ker(\varphi)$ on the set of cosets $\{gG_i\}_{g\in G_J}$ (are the fibres of $Aut(f_\varphi)$ contained in that set and) bijectively correspond to the orbits of $Ker(\varphi_J)$ on the set of cosets $\{gG_{J\cup i}\}_{g\in G_J}$. On the other hand, we can also represent the elements of Γ_J of type i by the cosets $gG_{J\cup i}$ $(g \in G_J)$, as in Lemma 10.15.

It is now clear that $Ker(\varphi)$ satisfies (i) of Theorem 11.1. Hence, it residually defines quotients. □

Given a morphism $\varphi : \mathcal{G} \longrightarrow \mathcal{G}'$, let $f_\varphi : \Gamma(\mathcal{G}) \longrightarrow \Gamma(\mathcal{G}')$ be the epimorphism induced by φ. Given $J \subseteq I$, let φ_J be the surjective homomorphism from G_J to G'_J induced by φ. Let $m = 1, 2, ...$ or $n-1$.

Lemma 11.9 *The epimorphism $f_\varphi : \Gamma(\mathcal{G}) \longrightarrow \Gamma(\mathcal{G}')$ is an m-cover if and only if $\varphi_J : G_J \longrightarrow G'_J$ is an isomorphism for every $J \subseteq I$ with $|I-J|= m$.*

Proof. The 'if' part of this statement follows from the fact that $Ker(\varphi)$ residually defines quotients (Corollary 11.8). Let us prove the 'only if' part. Let f_φ be an m-cover. Then $Ker(\varphi)$ acts semi-regularly on the set of flags of $\Gamma(\mathcal{G})$ of corank m, by Lemmas 11.7 and 11.4. Hence $Ker(\varphi) \cap G_J = 1$ for every $J \subseteq I$ with $|I - J|= m$. □

Corollary 11.10 *If $Ker(\varphi) \cap G^J = 1$ for every $J \subseteq I$ with $|J|= m$, then $Ker(\varphi) = Aut(f_\varphi)$.*

Proof. Easy, by Corollary 11.8, Lemma 11.9 and Theorem 11.6. □

11.2.2 Amalgams

Let $\mathcal{G} = (G_i)_{i\in I}$ be a parabolic system of rank n over the set of types I in a group G.

For every $m = 1, 2, ..., n-1$, let $\mathbf{G}_m$ be the family of subgroups G_J with $|J|= n-m$ $(J \subseteq I)$ and let G_m be the amalgamated product of the family $\mathbf{G}_m$, with amalgamation of the subgroups G_K with $|K|> n-m$ (see Section 0.8.3). Let $\varphi_m : G_m \longrightarrow G$ be the canonical homomorphism from G_m to G. The homomorphism φ_m is surjective, as $G = G^I = \langle G^i\rangle_{i\in I}$.

For every $K \subseteq I$ with $|K|\geq n-m$, the subgroup $G_K \leq G$ is contained in some subgroups of the family $\mathbf{G}_m$, hence it is naturally identified via φ_m with a subgroup $\widehat{G}_K \leq G_m$ and the homomorphism φ_m induces an isomorphism from $\widehat{G}_K$ to G_K. Therefore, if we set

$$\widehat{G}^i = \widehat{G}_{I-i} \quad (i \in I)$$

$$\widehat{G}^J = \langle \widehat{G}^j \rangle_{j \in J} \quad (J \subseteq I)$$

then we have

$$(1) \quad \widehat{G}_K = \widehat{G}^{I-K}$$

for every $K \subseteq I$ with $|K| \geq n - m$, and

$$(2) \quad \widehat{G}^I = G_m$$

because $G_m = \langle \widehat{G}_J \mid J \subseteq I, \, |J| = n - m \rangle$. Set

$$\widehat{G}_i = \widehat{G}^{I-i} \quad (i \in I)$$

$$\mathcal{G}_m = (\widehat{G}_i)_{i \in I}$$

We call $\mathcal{G}_m$ the *m-amalgam* of $\mathcal{G}$. We say that $\mathcal{G}_m$ is *tight* if $\widehat{G}^i = \bigcap_{j \in I-i} \widehat{G}_j$ for every $i \in I$. Thus, if $\mathcal{G}_m$ is a parabolic system and it is tight, then the minimal parabolic subgroups of $\mathcal{G}_m$ are just the subgroups $\widehat{G}^i$ $(i \in I)$ of G_m and, for every $K \subseteq I$ with $|K| \geq n - m$, we have $\widehat{G}_K = \bigcap_{k \in K} \widehat{G}_k$ (by (1) and because $\mathbf{P}_2$ holds in $\mathcal{G}_m$) and $\widehat{G}_K$ is a parabolic in $\mathcal{G}_m$.

Remark. As we remarked at the end of Section 10.1.2, many authors use the word 'amalgam' with a meaning more general than the one we have stated above, for any parabolic system or even any family of subgroups of a given group with prescribed intersections and generating that group.

The next theorem is an easy consequence of Lemma 11.9:

Theorem 11.11 *If the m-amalgam $\mathcal{G}_m$ of $\mathcal{G}$ is a parabolic system and is tight, then φ_m is a morphism of parabolic systems from $\mathcal{G}_m$ to $\mathcal{G}$ and it induces an m-cover from $\Gamma(\mathcal{G}_m)$ to $\Gamma(\mathcal{G})$.*

Unfortunately, we cannot prove that the hypotheses of this lemma hold in general. However, we can do it when $m = n - 1$:

Theorem 11.12 *The $(n-1)$-amalgam of $\mathcal{G}$ is a parabolic system and it is tight.*

Proof. The $(n-1)$-amalgam $\mathcal{G}_{n-1}$ is tight because $\widehat{G}^J$ is an isomorphic copy of G^J for every proper subset $J \subset I$. For the same reason, $\mathbf{P}_1$ and $\mathbf{P}_2$ hold in $\mathcal{G}_{n-1}$.

Let us prove that $\mathbf{P}_3$ holds in $\mathcal{G}_{n-1}$. Given distinct types $i, j \in I$, a subset $J \subseteq I - \{i,j\}$ and an element $g \in \widehat{G}_J$ such that $\widehat{G}_i \cap g\widehat{G}_j \neq \emptyset$, we have $G_i \cap \varphi_{n-1}(g)G_j \neq \emptyset$. Hence $G_J \cap G_i \cap \varphi_{n-1}(g)G_j \neq \emptyset$ by $\mathbf{P}_3$ in $\mathcal{G}$.

Let $a \in G_J \cap G_i \cap \varphi_{n-1}(g)G_j$. As φ_{n-1} induces isomorphisms on $\widehat{G}_J$ and $\widehat{G}_i$, elements $g_1 \in \widehat{G}_J \cap g\widehat{G}_j$ and $g_2 \in \widehat{G}_i \cap g\widehat{G}_j$ are uniquely determined in $\varphi_{n-1}^{-1}(a)$. Furthermore, φ_{n-1} also induces a bijection from $g\widehat{G}_j$ to $\varphi_{n-1}(g)G_j$. Hence $g_1 = g_2$, because $g_1, g_2 \in \varphi_{n-1}^{-1}(a) \cap g\widehat{G}_j$. Therefore $g_1 = g_2 \in \widehat{G}_J \cap \widehat{G}_i \cap g\widehat{G}_j$. Thus, $\mathbf{P}_3$ is proved.

We must still prove $\mathbf{P}_4$. Let us set $\widehat{B} = \widehat{G}_I$, according to the notation of Section 10.1. We have $\bigcap_{a\in G} aBa^{-1} = 1$ by $\mathbf{P}_4$ in $\mathcal{G}$. Hence $\bigcap_{g\in\widehat{G}} g\widehat{B}g^{-1} \leq Ker(\varphi_{n-1}) \cap \widehat{B}$. However, φ_{n-1} induces an isomorphism on $\widehat{B}$. Hence $\bigcap_{g\in\widehat{G}} g\widehat{B}g^{-1} = 1$. □

Exercise 11.12. Let $\mathcal{G}$ and $\overline{\mathcal{G}}$ be parabolic systems of rank n over the same set of types I in groups G and $\overline{G}$ respectively and let us set $B = G_I$, $\overline{B} = \overline{G}_I$, $G^i = G_{I-i}$ and $\overline{G}^i = \overline{G}_{I-i}$, according to the notation of Section 10.1. Prove that a homomorphism $\varphi : G \longrightarrow \overline{G}$ is a morphism from $\mathcal{G}$ to $\overline{\mathcal{G}}$ if and only if $\varphi(B) = \overline{B}$ and $\varphi(G^i) = \overline{G}^i$ for all $i \in I$.

Exercise 11.13. Let Γ, Γ' be flag-transitive thin geometries over the same set of types and belonging to the same Coxeter diagram. Let us set $G = Aut_s(\Gamma)$, $G' = Aut_s(\Gamma')$ and let $(s_i)_{i\in I}$ and $(s'_i)_{i\in I}$ be the sets of generating involutions of G and G' respectively, according to (3) of Section 10.2.1. Let $\varphi : G \longrightarrow G'$ be a homomorphism mapping s_i onto s'_i, for every $i \in I$. Prove that φ induces a 2-cover from Γ to Γ'.

Exercise 11.14. Let $\mathcal{G}$ be a parabolic system of rank $n = 2$. Prove that the geometry $\Gamma(\mathcal{G}_1)$ of its 1-amalgam (Theorem 11.12) is an infinite tree without terminal nodes.

Exercise 11.15. Let $\mathcal{G}$ be a parabolic system of rank n defining a thin geometry. Prove that the following hold:

(i) the 1-amalgam $\mathcal{G}_1$ is a parabolic system and is tight;
(ii) the group G_1 is freely generated by n involutions;
(iii) the chamber system of $\Gamma(\mathcal{G}_1)$ is an infinite tree where every node has valency n.

Problem. Find some nice conditions sufficent for an m-amalgam with $2 \leq m \leq n-2$ to be a parabolic system and tight. (The case $m = n-1$ has been considered in Theorem 11.12. It will follow from Theorem 12.28

and Exercise 12.4 that all 1-amalgams are parabolic systems and tight.)

11.3 Coxeter systems

11.3.1 Definition and notation

Given a Coxeter matrix $M = (m_{ij})_{i,j \in I}$ (Section 3.1.2) and a symbol s_i for every index $i \in I$, the following set of relations is called the set of *Coxeter relations* of *type* M:

$$(\mathbf{M}) \quad (s_i s_j)^{m_{ij}} = 1 \quad (i, j \in I)$$

The group presented by the set of relations $\mathbf{M}$ (Section 0.8.2) is called the *Coxeter group* of *type* M. We denote it by $W(M)$.

We recall that, since $m_{ii} = 1$ for every $i \in I$, the set of relations $\mathbf{M}$ also contains the relation $s_i^2 = 1$ for all $i \in I$. That is, all symbols s_i $(i \in I)$ denote involutions. We call them the *Coxeter involutions* of $\mathbf{M}$ (or relative to M).

When the Coxeter diagram represented by M already has a name, such as A_n, C_n, D_n etc., then we say that $W(M)$ has type ... (name of that Coxeter diagram). We use the same convention for Coxeter relations. We also identify the Coxeter matrix M with the Coxeter diagram it represents, thus speaking of the Coxeter diagram M, although this a linguistic abuse.

Since we have given $W(M)$ by means of its presentation $\mathbf{M}$, the symbol $W(M)$ should mean not just the group $W(M)$, but rather $W(M)$ together with its presentation $\mathbf{M}$, or the pair $(W(M), (s_i)_{i \in I})$, where $(s_i)_{i \in I}$ is the set of Coxeter involutions (or any set of involutions obtained from it by an automorphism of $W(M)$). We will give $W(M)$ this meaning, taking the symbol $W(M)$ as a shortening of $(W(M), (s_i)_{i \in I})$ (or of $(W(M), \mathbf{M})$), and using the word 'group' by abuse, since we are properly speaking of pairs consisting of a group and of a presentation of that group.

The following is also worth noticing: it may happen that the same group W can be generated by different sets of involutions in such a way that W is a Coxeter group with respect to one of these sets but it is not a Coxeter group with respect to another one. For instance, the group $A_5 \times \mathbf{2}$ is generated by three involutions satisfying the Coxeter relations of type H_3 (Exercise 9.3 and Section 3.1.3(6.c)). It is known that $A_5 \times \mathbf{2}$ is in fact the Coxeter group of type H_3 ([43], Table 2). Nevertheless, $A_5 \times \mathbf{2}$ is also generated by three involutions satisfying relations of another Coxeter type (Exercise 9.10). We will see in Section 11.3.7 that, with respect to this set of generating involutions, the group $A_5 \times \mathbf{2}$ is not a Coxeter group.

Given a Coxeter matrix $M = (m_{ij})_{i,j \in I}$ and a nonempty subset J of I, we define $M_J = (m_{ij})_{i,j \in J}$. Trivially, M_J is still a Coxeter matrix. According to the above notation, $\mathbf{M}_J$ and $W(M_J)$ are the Coxeter set of relations of type M_J and the Coxeter group of type M_J, respectively. When $J = \{i, j\}$, then we write M_{ij} and $\mathbf{M}_{ij}$ instead of $M_{\{i,j\}}$ and $\mathbf{M}_{\{i,j\}}$, for short.

If s_i ($i \in I$) are the Coxeter involutions of $W(M)$, then we denote by W^J the subgroup $\langle s_j \rangle_{j \in J}$ of $W(M)$. When $J = \{i, j\}$, we write $W^{i,j}$ for $W^{\{i,j\}}$. We also set $W^{\emptyset} = 1$.

The following lemma is well known [11]

Lemma 11.13 *$W^J = W(M_J)$ for every nonempty subset J of I.*

Corollary 11.14 *Given distinct types $i, j \in I$, we have $W^{i,j} = D_{2m_{ij}}$ or $W^{i,j} = \mathbf{Z}$, according to whether $m_{ij} < \infty$ or $m_{ij} = \infty$.*

Proof. This follows from the previous lemma, remarking that $\mathbf{M}_{ij}$ is a presentation of the dihedral group $D_{2m_{ij}}$ or of $\mathbf{Z}$, according to whether $m_{ij} < \infty$ or $m_{ij} = \infty$. □

11.3.2 Coxeter systems and Coxeter complexes

With the notation of the previous section, set $W_i = W^{I-i}$ for every $i \in I$ and $W_J = \bigcap_{j \in J} W_j$ for every $J \subseteq I$. In particular, $W_{\emptyset} = W(M)$. Define $\mathcal{W}(M) = (W_i)_{i \in I}$. Henceforth we assume $n = |I| \geq 2$.

The following two lemmas are well known [11]:

Lemma 11.15 *For every $J \subseteq I$, we have $W^J = W_{I-J}$.*

Lemma 11.16 *We have $W^J(W^H \cap W^K) = (W^J W^H) \cap (W^J W^K)$ for every choice of subsets J, H, K of I.*

Theorem 11.17 *The family $\mathcal{W}(M)$ is a parabolic system in $W(M)$, with minimal parabolics $W^i = \langle s_i \rangle$ ($i \in I$) and Borel subgroup $B = 1$. The geometry $\Gamma(\mathcal{W}(M))$ is thin and belongs to the Coxeter diagram M.*

Proof. By Lemma 11.15, the properties $\mathbf{P}_1$ and $\mathbf{P}_2$ hold in $\mathcal{M}$ and we have $B = G_I = 1$ and $W_{I-i} = W^i = \langle s_i \rangle$, for all $i \in I$. It is not hard to prove that the property stated in Lemma 11.16 implies $\mathbf{P}_3$ (see also Exercise 10.1). Trivially, $\mathbf{P}_1$ holds because $B = 1$. Therefore $\mathcal{W}(M)$ is a parabolic system.

The geometry $\Gamma(\mathcal{W}(M))$ is thin because the minimal parabolics have order 2. It belongs to the Coxeter diagram M, by Corollary 11.14 (see also Section 10.2.1). □

We call $\mathcal{W}(M)$ and $\Gamma(\mathcal{W}(M))$ the *Coxeter system of type* M and the *Coxeter complex of type* M, respectively. We write $\Gamma(M)$ for $\Gamma(\mathcal{W}(M))$, for short.

Needless to say, all thin geometries of rank 2 are Coxeter complexes.

The following is an easy consequence of Lemma 11.13:

Corollary 11.18 *Let F be a flag of $\Gamma(M)$ of cotype $J \neq \emptyset$. Then the residue of F in $\Gamma(M)$ is (isomorphic to) the Coxeter complex of type M_J.*

11.3.3 Flag-transitive thin geometries

We can now finish the investigation of flag-transitive thin geometries, begun in Section 10.2.1.

Let $\mathcal{G}$ be a parabolic system in a group G defining a thin geometry $\Gamma = \Gamma(\mathcal{G})$ with Coxeter diagram M. According to the notation of Section 11.2.2, let G_2 be the amalgamated product of the family $\mathbf{G}_2 = (G^J \mid J \subseteq I, |J| = 2)$ of parabolics of $\mathcal{G}$, let $\mathcal{G}_2$ be the 2-amalgam of $\mathcal{G}$ and let φ_2 be the canonical homomorphism from G_2 to G.

The relations $(1)_{ij}$ and $(2)_{ij}$ of Section 10.2.1 give us a presentation of the parabolic subgroup $\langle s_i, s_j \rangle$ of $\mathcal{G}$ ($i, j \in I, i \neq j$). Hence $G_2 = W(M)$ and φ_2 is the canonical homomorphism from $W(M)$ to G. Furthermore, $\mathcal{G}_2 = \mathcal{W}(M)$. Therefore $\mathcal{G}_2$ is a parabolic system and is tight (Theorem 11.17) and we have $\Gamma(\mathcal{G}_2) = \Gamma(M)$. By Theorem 11.11 we have the following:

Theorem 11.19 *The canonical homomorphism $\varphi_2 : W(M) \longrightarrow G$ induces a 2-cover from $\Gamma(M)$ to Γ.*

Thus, $\Gamma = \Gamma(M)/Ker(\varphi_2)$. Furthermore, $Ker(\varphi_2)$ is the group of deck transformations of the 2-cover induced by φ_2, by Corollary 11.10. Hence

Corollary 11.20 *If Γ has no proper 2-covers, then $G = W(M)$.*

Proof. If $G \neq W(M)$, then $Ker(\varphi_2) \neq 1$, hence $\Gamma = \Gamma(M)/Ker(\varphi_2)$ is a proper 2-quotient of $\Gamma(M)$. □

11.3.4 Examples

(1) The n-dimensional simplex is the unique thin geometry of type A_n (Corollary 7.8). Hence it is a Coxeter complex (Corollary 11.20) and its special automorphism group S_{n+1} is the Coxeter group of type A_n.

(2) The thin polar space $\mathcal{P}$ of rank n does not admit any proper 2-cover (Proposition 8.18). Therefore $\mathcal{P}$ is the Coxeter complex of type C_n by Corollary 11.20 and $Aut_s(\mathcal{P})$ ($\approx \mathbf{2}^n : S_n$) is the Coxeter group of type C_n.

Furthermore, by Theorem 11.19 all flag-transitive proper 2-quotients of $\mathcal{P}$ arise as quotients by a non-trivial normal subgroup $K \trianglelefteq Aut_s(\mathcal{P})$ such that $K \cap \langle s_i, s_j \rangle = 1$ for $1 \leq i < j \leq n$ (s_1, s_2, ..., s_n are the Coxeter involutions of $Aut_s(\mathcal{P})$, as in Section 9.2.2(3)). It is clear from the structure of $Aut_s(\mathcal{P})$ as described in Section 9.2.2(3) that the centre $\langle \zeta \rangle$ of $Aut_s(\mathcal{P})$ is the the only possible choice for K (provided that $n \geq 3$, of course). Therefore, the 2-fold topological quotient of $\mathcal{P}$ (Exercises 8.46 and 8.48) is the only flag-transitive proper 2-quotient of $\mathcal{P}$. That is, the thin polar space of rank n and its 2-fold topological quotient are the only two flag-transitive thin geometries of type C_n.

(3) The thin building of type D_n is the unique thin geometry of type D_n (Theorem 7.40). Hence it is the Coxeter complex of type D_n and its special automorphism group $\mathbf{2}^{n-1} : S_n$ is the Coxeter group of type D_n.

(4) There are no proper 2-covers between geometries belonging to the Coxeter diagram E_6 (Proposition 8.22). Therefore the Coxeter complex of type E_6 is the unique flag-transitive thin geometry of type E_6, by Corollary 11.20.

We will see later (Theorem 13.25) that more than the above is true: the Coxeter complex of type E_6 is the unique thin geometry of type E_6.

Geometries belonging to the Coxeter diagrams E_7, E_8 and F_4 and satisfying the intersection property **IP** do not admit any proper 2-covers (Propositions 8.24, 8.25 and 8.28). Therefore, the Coxeter complexes of type E_7, E_8 and F_4 are the only flag-transitive thin geometries of type E_7, E_8 and F_4 that satisfy **IP**, by Corollary 11.20.

A result stronger than the above will be obtained in Chapter 13: the Coxeter complexes of type E_7, E_8 and F_4 are the only thin geometries of type E_7, E_7 and E_8 that satisfy **IP** (Theorems 13.26 and 13.27).

(5) We have remarked in Section 11.3.1 that the automorphism group $A_5 \times \mathbf{2}$ of the icosahedron is the Coxeter group of type H_3. Hence the icosahedron is the Coxeter complex of type H_3. The halved icosahedron is obtained by factorizing by the centre $\mathbf{2}$ of the Coxeter group $A_5 \times \mathbf{2}$ and admits A_5 as a flag-transitive automorphism group (Section 11.1.4(3)). Since the centre is the only non-trivial proper normal subgroup of the group $A_5 \times \mathbf{2}$, the icosahedron and the halved icosahedron are the only flag-transitive thin geometries of type H_3 (needless to say, we are not distinguishing any more between a geometry and its dual). **IP** holds in these two geometries (see Section 6.3.4).

(6) The tessellation of the plane into squares is the Coxeter complex of type $\widetilde{B}_2$ ([43]; see Section 3.1.3(6.d) for the diagram and Exercise 9.5 for the relations).

The tessellation of the plane by equilateral triangles (or by regular hexagons, if the reader prefers hexagons to triangles) is the Coxeter complex of type $\widetilde{G}_2$ ([43]; see Section 3.1.3(6.d) for the diagram and Exercise 9.4 for the relations).

The geometry of black and white triangles of Section 2.1.2(10) is the Coxeter complex of type $\widetilde{A}_2$ (see [43], where the flag-complex of this geometry is represented as a tessellation of the plane into equilateral triangles; see Section 3.1.3(6.e) for the diagram and Exercise 9.7 for the relations).

11.3.5 Diagram automorphisms

Given a Coxeter diagram $M = (m_{ij})_{i,j \in I}$, an *automorphism* of M is a permutation p of I such that $m_{i,j} = m_{p(i),p(j)}$ for every choice of $i, j \in I$. Trivially, the automorphisms of M form a group. We denote this group by $Aut(M)$.

Automorphisms of M can also be viewed as permutations of the Coxeter involutions of $\mathbf{M}$ preserving the set $\mathbf{M}$ of Coxeter relations. Therefore, every automorphism of M uniquely determines an automorphism of the Coxeter group $W(M)$, which is also an automorphism of the parabolic system $\mathcal{W}(M)$, in the meaning of Section 10.4.1. Conversely, every automorphism of $\mathcal{W}(M)$ is induced by a uniquely determined automorphism of M. Therefore $Aut(M) \cong Aut(\mathcal{W}(M))$. Hence, by Corollary 10.36 we obtain the following (recall that $W(M) = Aut_s(\Gamma(M))$):

Theorem 11.21 $Aut(\Gamma(M))/W(M) \cong Aut(M)$

For instance, let $M = A_n$ with $n \geq 2$. Recall that $\Gamma(M)$ is the n-dmensional simplex (Section 11.3.4(1)). The diagram M has a symmetry σ, which gives us a duality δ_σ of the n-dimensional simplex $\Gamma(M)$. However, the duality δ_σ obtained in this way is not the same duality δ that we have considered in Sections 9.2.2(2) and 10.4.3. Indeed, δ acts trivially on $W(M)$ $(= S_{n+1})$ whereas δ_σ acts as an involution on $W(M)$. The product $w_0 = \delta_g \delta$ belongs to $W(M)$. It maps the chamber C_0 represented by B $(= 1)$ onto the chamber at maximal distance from C_0 (see Section 10.4.3) and it acts by conjugaction on S_{n+1} just as σ. Thus, the automorphism of $\mathcal{W}(M)$ induced by σ is in fact the inner automorphism $Inn(w_0)$ (notation as in Section 11.4), as we have remarked in Section 11.4.3.

11.3.6 Reducibility

A Coxeter diagram is said to be *reducible* if its underlying diagram graph is not connected. Otherwise, it is said to be *irreducible*. If M is a reducible Coxeter diagram and M_1, M_2, ..., M_m are the Coxeter diagrams induced on the connected components of the underlying diagram graph of M, then M_1, M_2, ..., M_m are called the *irreducible components* of M. We extend this terminology to the case when M is irreducible, saying that an irreducible Coxeter diagram is its (unique) irreducible component.

If I is the set of types of M and $\{J, K\}$ is a partition of I, then we write $M = M_J + M_K$ to mean that the sets of types J and K are not connected by any path in the underlying diagram graph of M (the symbols M_J and M_K denote the Coxeter diagrams induced by M on J and K respectively, as in Section 11.3.1). Thus, if M_1, M_2, ..., M_m are the irreducible components of M, then $M = M_1 + M_2 + ... + M_m$.

The symbols W^J and W^K in the next theorem have the meaning stated in Section 11.3.1.

Theorem 11.22 (Reducibility Theorem) *Let I be the set of types of the Coxeter diagram M and let J, K be disjoint nonempty subsets of I such that $J \cup K = I$. Then the following are equivalent:*

(i) $\Gamma(M) = \Gamma(M_J) \oplus \Gamma(M_K)$;
(ii) $M = M_J + M_K$;
(iii) $W(M) = W^J \times W^K$;
(iv) $W^J W^K = W^K W^J$.

Proof. The equivalence of (i) and (ii) is just the direct sum theorem. Trivially, (iii) implies both (ii) and (iv).

Since J and K are disjoint, $W^J \cap W^K = 1$ by Lemma 11.15. Hence (ii) implies (iii). The implication from (iv) to (ii) holds in general (see Exercise 10.13). Thus, all the above are equivalent. □

11.3.7 Finiteness

Let $M = M_1 + M_2 + ... + M_m$ be the decomposition of M into its irreducible components. By Theorem 11.22, the Coxeter group $W(M)$ (the Coxeter complex $\Gamma(M)$) is finite if and only if all of $W(M_1)$, $W(M_2)$, ..., $W(M_m)$ (all of $\Gamma(M_1)$, $\Gamma(M_2)$, ..., $\Gamma(M_m)$) are finite.

A Coxeter diagram M is said to be of *spherical type* (also *spherical*, for short) when $W(M)$ is finite; that is, when $\Gamma(M)$ is finite. The name 'spherical' is due to the fact that $\Gamma(M)$ is finite if and only if its flag complex can be realized as a tessellation of a hypersphere (see [43] or [11]). If M

is of spherical type, then we also say that the complex $\Gamma(M)$, the system $\mathcal{W}(M)$ and the group $W(M)$ are of *spherical type.*

The irreducible Coxeter diagrams of spherical type are the following ones (see [43] or [11]):

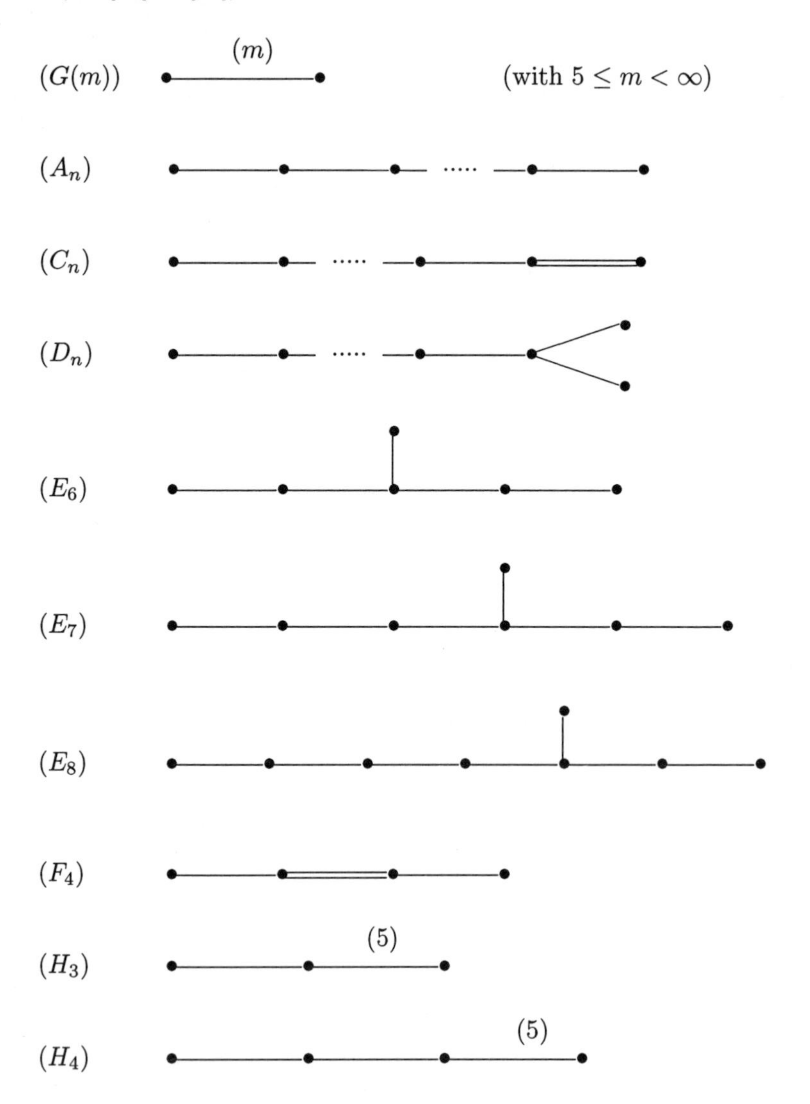

Thus, the irreducible Coxeter diagrams of rank 3 and of spherical type are A_3, C_3 and H_3, which describe the tetrahedron, the octahedron and the icosahedron respectively (see also Exercise 2.6).

The diagrams $G(6)$, A_n, C_n, D_n, E_6, E_7, E_8 and F_4 are also called *Dynkin diagrams.*

The Coxeter complexes belonging to Dynkin diagrams arise in the context of simple Lie algebras (the reader interested in this may consult [35]). In that context the diagram C_n describes two distinct structures of simple Lie algebras, denoted by B_n and C_n respectively. Thus, some authors use the notation B_n in diagram geometry too, as a synonym of C_n. Some authors also include the diagram $G(8)$ among Dynkin diagrams, so as to have a Dynkin diagram for every class of finite simple groups of Lie type (see [35]).

The irreducible Coxeter diagrams describing Coxeter complexes whose flag complexes can be realized as tessellations of an affine space over the field of real numbers are called *affine diagrams.* The reader may find the complete list of affine diagrams in [11] (Chapter 6, Section 4.3, Theorem 4). We will only mention the affine diagrams of rank ≤ 3 here (we have already encountered them quite often in this book).

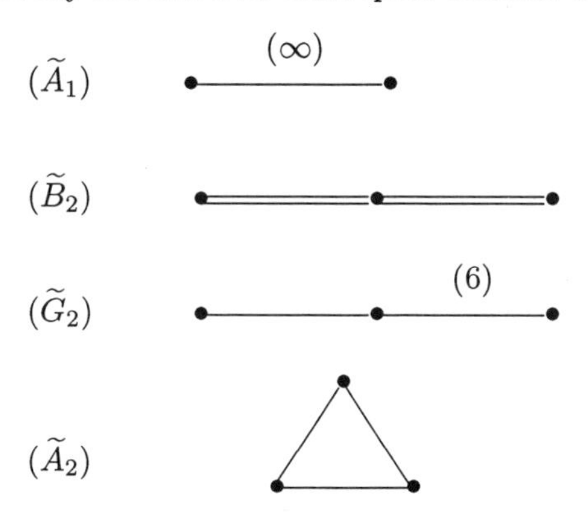

Chapter 12

Universal Covers

12.1 Definition

Given geometries Γ and $\widehat{\Gamma}$ over the same set of types I, let $f : \widehat{\Gamma} \longrightarrow \Gamma$ be an m-cover $(1 \leq m < n = |I|)$.

We say that f is *universal* if, for every m-cover $g : \Gamma' \longrightarrow \Gamma$, there is just one morphism $h_g : \widehat{\Gamma} \longrightarrow \Gamma'$ such that $gh_g = f$. The morphism h_g is actually an m-cover. We leave the (easy) proof of this claim for the reader. It is also easy to prove that universal m-covers are uniquely determined up to isomorphisms, when they exist. That is, given two universal m-covers $f_1 : \widehat{\Gamma}_1 \longrightarrow \Gamma$ and $f_2 : \widehat{\Gamma}_2 \longrightarrow \Gamma$ of the same geometry Γ, there is a unique isomorphism $h_{12} : \widehat{\Gamma}_2 \longrightarrow \widehat{\Gamma}_1$ such that $f_1 h_{12} = f_2$.

Thus, we may speak of 'the' universal m-cover of a given geometry, as if that m-cover were unique. Sometimes we also identify a universal m-cover $f : \widehat{\Gamma} \longrightarrow \Gamma$ with its domain $\widehat{\Gamma}$, saying that $\widehat{\Gamma}$ is the universal m-cover of Γ. This is actually an abuse, but it is quite harmless.

As every m-cover is also an r-cover for every $r = 1, 2, ..., m-1$, it is clear that the universal m-cover of Γ is a homomorphic image of the universal r-cover of Γ, for every $r = 1, 2, ..., m-1$ (provided these universal covers exist, of course)

We should now prove the existence of universal covers. That is, we should prove that, for every geometry Γ, there is a geometry $\widehat{\Gamma}$ (uniquely determined up to isomorphism) which is the universal m-cover of Γ. Actually, we can prove this only when $m = 1$ or $n-1$. When $1 < m < n-1$, we can only prove a weaker statement, namely that there is a coloured graph $\widehat{\mathcal{C}}$ which is the universal m-cover of the chamber system $\mathcal{C}(\Gamma)$ of Γ, in a sense to be defined later. Unfortunately, no proof is known that the coloured graph $\widehat{\mathcal{C}}$ is always the chamber system of some geometry $\widehat{\Gamma}$, although this is what happens in all examples we know.

In view of this, if we want to construct objects deserving to be called 'universal m-covers' we must forget about geometries for a while and switch to coloured graphs and chamber systems. However, before doing this, we will set up a general topological framework in which we can place different approaches to the construction of universal covers. We will do this in the next section. The terminology we will use there is mostly artificial, only suggested by some analogies with topology, on one side, and by the need to avoid homonymies with other concepts, on the other side.

12.2 Plate complexes

Let I_V and I_E be finite nonempty sets. A *plate complex* over the *sets of types* I_V, I_V is a 5-tuple $\mathcal{P} = (V, E, P, t_V, t_E)$ where (V, E) is a connected graph (with vertex set V and edge set E), t_V and t_E are surjective mappings from V to I_V and from E to I_E respectively, and P is a family of connected subgraphs of (V, E), called *plates*, such that every edge of (V, E) belongs to at least one plate and every plate has at least two vertices.

We call (V, E) the *underlying graph* of $\mathcal{P}$. The vertices and the edges of (V, E) are called *vertices* and *edges* of $\mathcal{P}$, respectively. The functions t_V and t_E will be called the *type functions* of $\mathcal{P}$. For every vertex $x \in V$ (for every edge $e \in E$) the element $t_V(x) \in I_V$ (the element $t_E(e) \in I_E$) is the *type* of x (of e).

We allow the possibility that a plate $X \in P$ contains both vertices of an edge $e \in E$ and nevertheless $e \notin X$. We also allow $|I_V|= 1$ or $|I_E| = 1$. When $|I_E|= 1$ (when $|I_E|= 1$), then all vertices (all edges) have the same type. In this case we may forget about types when speaking of vertices (of edges, respectively).

Given a vertex x of $\mathcal{P}$, the system of edges and plates of $\mathcal{P}$ containing x will we denoted by $\mathcal{P}_x$. We call it the *star* of x in $\mathcal{P}$.

12.2.1 Examples

Every connected simplicial complex $\mathcal{K}$ can be viewed as a plate complex, taking the skeleton of $\mathcal{K}$ as the underlying graph and the simplices of $\mathcal{K}$ as plates. We have $|I_V|= |I_E|= 1$ (types have no relevance here).

Let all simplices of $\mathcal{K}$ have dimension ≥ 2. We denote by $Tr_2(\mathcal{K})$ the simplicial complex obtained by deleting all faces of $\mathcal{K}$ of dimension ≥ 3, if any such face exists in $\mathcal{K}$. Needless to say, $Tr_2(\mathcal{K})$ is a plate complex with the same underlying graph as $\mathcal{K}$.

Thus, given a geometry Γ over the set of types I with type-function t, the flag-complex $\mathcal{K}(\Gamma)$ of Γ can be viewed as a plate complex. We can now give types to the vertices of this plate complex, taking t as the type function for vertices (hence $I_V = I$; we may forget about I_E here). The incidence graph of Γ is the underlying graph of the plate complex $\mathcal{K}(\Gamma)$. The chambers of Γ are the plates of $\mathcal{K}(\Gamma)$.

When Γ has rank $n \geq 3$, then we can also consider the plate complex $Tr_2(\mathcal{K}(\Gamma))$, taking the type function of Γ as type function for vertices. We call it the *2-dimensional flag-complex* of Γ. We write $\mathcal{K}_2(\Gamma)$ for $Tr_2(\mathcal{K}(\Gamma))$, for short.

Given a type $i \in I$, we can also define a plate complex for Γ inside the i-shadow space $\sigma_i(\Gamma)$ of Γ. We take the the collinearity graph of the i-Grassmann geometry $Gr_i(\Gamma)$ of Γ as the underlying graph of the plate complex we are going to define. We define plates starting from elements of $Gr_i(\Gamma)$ of type ≥ 1 in $Gr_i(\Gamma)$, as follows: if X is an element of $Gr_i(\Gamma)$ of type ≥ 2, then we take the collinearity graph of the lower residue of X as a plate; otherwise, if X is a line of $Gr_i(\Gamma)$, then we take the complete graph on the i-shadow $\sigma_i(X)$ of X in Γ as a plate. Types have no role in this plate complex (that is, we set $|I_V|=|I_E|= 1$). The plate complex defined above will be called the *i-shadow complex* of Γ. We denote it by the symbol $\Sigma_i(\Gamma)$.

Note that distinct elements of $Gr_i(\Gamma)$ might define the same plate of $\Sigma_i(\Gamma)$. On the other hand, two elements X, Y of $Gr_i(\Gamma)$ of type ≥ 1 with the same i-shadow in Γ might give rise to distinct plates of $\Sigma_i(\Gamma)$. For instance, if Γ is a flat C_3 geometry and 0, 2 are the types representing points and planes respectively, then all planes of Γ define the same plate of $\Sigma_0(\Gamma)$, whereas distinct points define distinct plates of $\Sigma_2(\Gamma)$, even if they have the same 2-shadow in Γ.

Another plate complex can be constructed on the i-shadow space $\sigma_i(\Gamma)$ of a geometry Γ, as follows. We take the elements of Γ of type i as vertices, as in $\Sigma_i(\Gamma)$. As plates we now take the shadows of the elements of $Gr_i(\Gamma)$ of type ≥ 1, viewed as complete graphs. The underlying graph of the plate complex is defined as follows: the vertices are the points of $Gr_i(\Gamma)$ and two distinct points of $Gr_i(\Gamma)$ are declared to be adjacent when there is some element of $Gr_i(\Gamma)$ incident to both of them. Types have no role, as in $\Sigma_i(\Gamma)$. We denote this plate complex by $\overline{\Sigma}_i(\Gamma)$ and we call it the *support* of $\Sigma_i(\Gamma)$.

Finally, a plate complex can also be defined on the chamber system $\mathcal{C}(\Gamma)$ of Γ for every $m = 1, 2, ..., n-1$ (where $n = |I|$). We know that $\mathcal{C}(\Gamma)$ can be viewed as a coloured graph. We choose $\mathcal{C}(\Gamma)$ as the underlying graph of the plate complex we are going to define and we set $I_E = I$ and $|I_V|= 1$. As plates we take the cells of the flags of Γ of corank m. The plate complex defined in this way will be called the *m-cell complex* of Γ. We denoted it by $\mathcal{C}_m(\Gamma)$.

It is worth noticing that there is some kind of duality between the $(n-1)$-cell complex $\mathcal{C}_{n-1}(\Gamma)$ and the flag-complex $\mathcal{K}(\Gamma)$. The vertices of $\mathcal{C}_{n-1}(\Gamma)$ are the plates of $\mathcal{K}(\Gamma)$ and the plates of $\mathcal{C}_{n-1}(\Gamma)$ are the (stars of the) vertices of $\mathcal{K}(\Gamma)$.

12.2.2 Covers

Let $\mathcal{P} = (V, E, P, t_V, t_E)$ and $\mathcal{P}' = (V', E', P', t'_V, t'_E)$ be plate complexes over the same sets of types $I_1 = I_V = I_{V'}$ and $I_2 = I_E = I_{E'}$. A *cover* from $\mathcal{P}'$ to $\mathcal{P}$ is a type preserving morphism of graphs $f : (V, E) \longrightarrow (V', E')$ such that:

(i) for every $x \in V'$, the morphism f induces an isomorphism from the star $\mathcal{P}'_x$ of x in $\mathcal{P}'$ to the star $\mathcal{P}_{f(x)}$ of $f(x)$ in $\mathcal{P}$;

(ii) for every $X \in P'$, the restriction of f to X in injective.

We write $f : \mathcal{P}' \longrightarrow \mathcal{P}$ to mean that f is a cover from $\mathcal{P}'$ to $\mathcal{P}$.

Let $f : \mathcal{P}' \longrightarrow \mathcal{P}$ be a cover. By the connectedness of (V', E') and by (i), the cover f maps the plates of P' onto plates of P. Hence, (ii) can also be restated as follows: for every plate $X \in P'$, the morphism f induces an isomorphism from X to the plate $f(X) \in P$.

Given a vertex a' of $\mathcal{P}'$ and a path $\alpha' = (a'_0, a'_1, ..., a'_h)$ of (V', E'), let $\alpha = f(\alpha') = (f(a'_0), f(a'_1), ..., f(a'_m))$. Then we say that α' is a *lifting* of α to $\mathcal{P}'$ at a'_0 *through* f. We also say that α *lifts* to α' *through* f.

Lemma 12.1 *Given a cover f from $\mathcal{P}'$ to $\mathcal{P}$, let a' be a vertex of $\mathcal{P}'$ and $a = f(a')$. Then every path α of (V, E) starting at a has a unique lifting to $\mathcal{P}'$ at a' through f.*

Proof. Let $\alpha = (a_0, a_1, ..., a_m)$, with $a_0 = a$. We work by induction on the length m of α. If $m = 1$, then the statement follows from (i). Let $m > 1$. By the inductive hypothesis, $(a_0, a_1, ..., a_{m-1})$ admits a unique lifting $(a'_0, a'_1, ..., a'_{m-1})$ with $a'_0 = a'$. This lifting can be extended to a lifting $(a'_0, a'_1, ..., a'_{m-1}, a'_m)$ of α in a unique way, by (i) in the star of a'_{m-1}. □

Note that Proposition 8.8 is just the special case of the above lemma for 1-cell complexes of geometries.

As an easy consequence of Lemma 12.1, of (i) and of the connectedness of (V, E), we obtain that a cover $f : \mathcal{P}' \longrightarrow \mathcal{P}$ is always surjective: it maps V' onto V, E' onto E and P' onto P (compare with Proposition 8.7).

An *automorphism* of a plate complex $\mathcal{P} = (V, E, P, t_V, t_E)$ is an automorphism g of the underlying graph (V, E) of $\mathcal{P}$ inducing a bijection from P to P and such that $t_V g = t_V$ and $t_E g = t_E$ (that is, g preverves types).

Deck transformations can be defined for covers of plate complexes as we have done for morphisms of geometries in Section 11.1. As we did there, we denote the group of deck transformations of a cover f by $Aut(f)$ and we call it the *deck group* of f. The identity element of $Aut(f)$ is called the *trivial* deck transformation of f.

Lemma 12.2 *The deck group of a cover $f : \mathcal{P}' \longrightarrow \mathcal{P}$ acts semi-regularly on the sets of vertices, on the set of edges and on the set of plates of $\mathcal{P}'$. Furthermore, for every non-trivial deck transformation g of f, we have $X \cap g(X) = \emptyset$ for every plate X of $\mathcal{P}'$ and $\mathcal{P}'_x \cap \mathcal{P}'_{g(x)} = \emptyset$ for every vertex x of $\mathcal{P}'$.*

Proof. Let $g \in Aut(f)$ fix a vertex x (an edge e; a plate X). By (i) and (ii), g also fixes all edges and plates in the star $\mathcal{P}'_x$ of x (both vertices of e and all plates containing e; all vertices and edges of X). By the connectedness of $\mathcal{P}'$ and iterating the previous argument, we obtain $g = 1$. The first claim is proved. The second one is a trivial consequence of (i) and (ii). □

When $Aut(f)$ is transitive on each of the fibres of f in one of the sets V', E' or P', it is also transitive on every fibre of f in the remaining two sets. In this case, the codomain $\mathcal{P}$ of f can be viewed as a *quotient* $\mathcal{P} = \mathcal{P}'/Aut(f)$ of $\mathcal{P}'$ by the group $Aut(f)$, as in the case of geometries (Section 11.1).

12.2.3 Homotopy groups

Given a plate complex $\mathcal{P} = (V, E, P, t_V, t_E)$, two paths $(a_0, a_1, ..., a_h)$ and $(b_0, b_1, ..., b_k)$ of (V, E) with $a_0 = b_0$ and $a_h = b_k$ are said to be *elementarily homotopic* if there is a plate containing both of them.

Let α, β, γ, δ, be paths such that both $\alpha\beta\delta$ and $\alpha\gamma\delta$ are defined. We extend the definition of elementary homotopy to this more general case, stating that $\alpha\beta\delta$ and $\alpha\gamma\delta$ are *elementarily homotopic* if β and γ are elementarily homotopic in the previous sense. Finally two paths α and β are said to be *homotopic* if there is a sequence $\alpha = \alpha_0, \alpha_1, ..., \alpha_m = \beta$ of paths such that α_{i-1} and α_i are elementarily homotopic for all $i = 1, 2, ..., m$ (needless to say, α and β must start at the same vertex and end at the same vertex). Trivially, the homotopy relation is an equivalence relation on the set of paths of $\mathcal{P}$. We will write $\alpha \equiv \beta$ to mean that α and β are homotopic paths.

Given a vertex a, let π_a be the set of all closed paths of $\mathcal{P}$ *based* at a (namely, starting and ending at a), including the null path (a). Trivially, π_a is a monoid with respect to the join of paths. The quotient of π_a by the homotopy relation is a group, called the *homotopy group* of $\mathcal{P}$ at a. We designate it by the symbol $\pi_a(\mathcal{P})$. The homotopy class of (a) is the null element of $\pi_a(\mathcal{P})$. We denote it by O_a.

More generally, given two vertices a and b, we write $\pi_{ab}(\mathcal{P})$ for the set of homotopy classes of paths from a to b. Trivially, given three vertices a, b, c and paths α, α' from a to b and β, β' from b to c, if $\alpha \equiv \alpha'$ and $\beta \equiv \beta'$, then $\alpha\beta \equiv \alpha'\beta'$. Therefore, if A, B are the homotopy classes of α and β respectively, then we can define the product AB stating that AB is the homotopy class of $\alpha\beta$. Trivially, if A' is the homotopy class of α^{-1}, then $AA' = O_a$ and $AA' = O_b$. That is, A' is the inverse of A. We denote it by A^{-1}. Given $A \in \pi_{ab}(\mathcal{P})$ and $B \in \pi_{bc}(\mathcal{P})$, we set

$$A \cdot \pi_{bc}(\mathcal{P}) = \{AX \mid X \in \pi_{bc}(\mathcal{P})\}$$

$$\pi_{ab}(\mathcal{P}) \cdot B = \{YB \mid Y \in \pi_{ab}(\mathcal{P})\}$$

Lemma 12.3 *Let a and b be distinct vertices and $A \in \pi_{ab}(\mathcal{P})$. Then $\pi_{ab}(\mathcal{P}) = A \cdot \pi_b(\mathcal{P}) = \pi_a(\mathcal{P}) \cdot A$ and $\pi_a(\mathcal{P}) = A \cdot \pi_b(\mathcal{P}) \cdot A^{-1}$.*

Proof. Trivially, $A \cdot \pi_b(\mathcal{P})$ and $\pi_a(\mathcal{P}) \cdot A$ are subsets of $\pi_{ab}(\mathcal{P})$. Conversely, let $X \in \pi_{ab}(\mathcal{P})$. Then $X = AA^{-1}X = XA^{-1}A$. This proves the first claim of the lemma. The second claim can be proved in a similar way. □

By Lemma 12.3 and by the connectedness of (V, E), we have $\pi_a(\mathcal{P}) \cong \pi_b(\mathcal{P})$ for any two vertices a, b of $\mathcal{P}$. We write $\pi(\mathcal{P})$ for $\pi_a(\mathcal{P})$, for short, calling $\pi(\mathcal{P})$ the *homotopy group* of $\mathcal{P}$.

12.2.4 Embeddings of homotopy groups via covers

Le $f : \mathcal{P}' \longrightarrow \mathcal{P}$ be a cover. Given a vertex b' of $\mathcal{P}'$, a path β' in (the underlying graph of) $\mathcal{P}'$ starting at b' and a plate X' of $\mathcal{P}'$ containing b', let us set $b = f(b')$, $\beta = f(\beta')$ and $X = f(X')$. Thus, β' is the unique lifting of β at b' through f (Lemma 12.1) and X' is the unique preimage of X containing b'.

Since f induces an isomorphism from X' to X, the path β' is also a path of the graph X' if and only if β is a path of the graph X. Therefore we have the following:

Lemma 12.4 *Given two paths α', β' of $\mathcal{P}$, we have $\alpha' \equiv \beta'$ if and only if $f(\alpha') \equiv f(\beta')$.*

Given a vertex a' of $\mathcal{P}'$, let $a = f(a')$. By the previous lemma, the cover f naturally defines an embedding $f_{a'} : \pi_{a'}(\mathcal{P}') \longrightarrow \pi_a(\mathcal{P})$. According to the conventions of Section 12.2.3, we can also write $f_{a'} : \pi(\mathcal{P}') \longrightarrow \pi(\mathcal{P})$, omitting the indices a' and a in the symbols denoting the homotopy groups. We call $f_{a'}$ the *embedding* of $\pi(\mathcal{P}')$ into $\pi(\mathcal{P})$ *induced* by f at the *base vertex* a'.

We should not omit the index a' in the symbol $f_{a'}$, since the embedding $f_{a'}$ in general depends on the choice of the base vertex a'. However, changing the base vertex does not change things too much. If α' is a path from a vertex b' to a' and A is the homotopy class of $f(\alpha')$, then we have $f_{b'}(\pi_{b'}(\mathcal{P}')) = A \cdot f_{a'}(\pi_{a'}(\mathcal{P}')) \cdot A^{-1}$.

Lemma 12.5 *The vertices of $\mathcal{P}'$ belonging to $f^{-1}(a)$ bijectively correspond to the left cosets of $f_{a'}(\pi(\mathcal{P}'))$ in $\pi(\mathcal{P})$.*

Proof. Given $b' \in f^{-1}(a)$, let α' be a path in $\mathcal{P}'$ from a' to b'. Then $\alpha = f(\alpha')$ is a closed path in $\mathcal{P}$ based at a. If β' is an another path in $\mathcal{P}'$ from a' to b', then $\beta' \equiv \beta'{\alpha'}^{-1}\alpha'$. Therefore, if A', B' are the homotopy classes of α' and β' respectively, their images $A = f(A')$ and $B = f(B')$ are in the same left coset of $f_{a'}(\pi(\mathcal{P}'))$. Let $\varphi(b')$ be that left coset. We have thus defined a mapping φ from $f^{-1}(a)$ to the set of left cosets of $f_{a'}(\pi(\mathcal{P}'))$ in $\pi(\mathcal{P})$.

Conversely, let α be a closed path of $\mathcal{P}$ based at a and let α' be its lifting to $\mathcal{P}'$ at a' through f (uniquely determined by Lemma 12.1). Let b' be the end vertex of α' and let A be the homotopy class of α. Then we have $\varphi(b') = f_{a'}(\pi(\mathcal{P}')) \cdot A$. Hence φ is surjective.

Finally, let b', c' be distinct elements of $f^{-1}(a)$ and β', γ' paths in $\mathcal{P}'$ from a' to b' and to c', respectively. If B, C are the homotopy classes of $f(\beta')$ and $f(\gamma')$, respectively, then we have $\varphi(b') = f_{a'}(\pi(\mathcal{P}')) \cdot B$ and $\varphi(c') = f_{a'}(\pi(\mathcal{P}')) \cdot C$. If $\varphi(b') = \varphi(c')$, then $BC^{-1} \in f_{a'}(\pi(\mathcal{P}'))$. Hence $C^{-1}B \in C^{-1} \cdot f_{a'}(\pi(\mathcal{P}')) \cdot C = f_{c'}(\pi(\mathcal{P}'))$. Therefore, every representative of $C^{-1}B$ lifts to a closed path of $\mathcal{P}'$ based at c'. However, $f({\gamma'}^{-1}\beta')$ is in fact a representative of $C^{-1}B$, the path ${\gamma'}^{-1}\beta'$ is the (unique) lifting of $f({\gamma'}^{-1}\beta')$ to $\mathcal{P}'$ at c', and b' is the end vertex of ${\gamma'}^{-1}\beta'$. Hence $b' = c'$. Therefore φ is injective. □

Corollary 12.6 *All fibres of f have the same size $|\pi(\mathcal{P})| \, / |f_{a'}(\pi(\mathcal{P}'))|$. In particular, f is an isomorphism if and only if $f_{a'}(\pi(\mathcal{P}')) = \pi(\mathcal{P})$.*

This is a trivial conequence of Lemma 12.5.

We now consider another cover $g : \mathcal{P}'' \longrightarrow \mathcal{P}'$ and a vertex $a'' \in g^{-1}(a)$ (we recall that $a = f(a')$).

Lemma 12.7 *Let $f_{a'}(\pi(\mathcal{P}'))$ be a subgroup of $g_{a''}(\pi(\mathcal{P}''))$. Then there is a unique cover $h : \mathcal{P}' \longrightarrow \mathcal{P}''$ such that $gh = f$.*

Proof. Given a vertex b' of $\mathcal{P}'$, let α' be a path from a' to b', let $\alpha = f(\alpha')$ and let α'' be the lifting of α to $\mathcal{P}''$ at a'' through g. Let b'' be the end vertex of α''. We set $h(b') = b''$.

We will prove that the above defines a mapping h, namely that b'' does not depend on the choice of α', but it only depends on b'. Once then this is proved, the rest of the statement can be obtained as a trivial consequence of the definition of covers and of Lemma 12.1.

Let β' be another path in $\mathcal{P}'$ from a' to b'. Trivially, if A and B are the homotopy classes of $\alpha = f(\alpha')$ and $\beta = f(\beta')$, then $BA^{-1} \in f_{a'}(\pi(\mathcal{P}')) \subseteq g_{a''}(\pi(\mathcal{P}''))$. Therefore $\beta\alpha^{-1}$ lifts to a closed path of $\mathcal{P}''$ based at a''. However, if β'' is the lifting of β at a'' and $\overline{\alpha}''$ is the lifting of α^{-1} at the end vertex of β'', then $\beta''\overline{\alpha}''$ is the lifting of $\beta\alpha^{-1}$ at a''. Therefore, as $\beta\alpha^{-1}$ lifts to a closed path based at a'', the path $\overline{\alpha}''$ ends at a''. This implies that $\overline{\alpha}''^{-1}$ is the (uniquely determined) lifting of α at a'', namely that $\overline{\alpha}''^{-1} = \alpha''$. Hence the end vertex of β'' is the end vertex b'' of α'' and the vertex $b'' = h(b')$ does not depend on the choice of the path α' from a' to b'. □

The next statement is an easy consequence of Corollary 12.6:

Corollary 12.8 *Let $f_{a'}(\pi(\mathcal{P}')) \leq g_{a''}(\pi(\mathcal{P}''))$ and let $h : \mathcal{P}' \longrightarrow \mathcal{P}''$ be the (unique) cover such that $gh = f$. Then h is an isomorphism if and only if $f_{a'}(\pi(\mathcal{P}')) = g_{a''}(\pi(\mathcal{P}''))$.*

Let $N(\pi(\mathcal{P}'))$ be the normalizer of $\pi(\mathcal{P}')$ in $\pi(\mathcal{P})$.

Proposition 12.9 *We have $Aut(f) \cong N(\pi(\mathcal{P}'))/\pi(\mathcal{P}')$.*

Proof. Let X be the orbit of a' in $Aut(f)$. Then X is a subset of $f^{-1}(a)$. By Lemma 12.5, the elements of $f^{-1}(a)$ bijectively correspond to the cosets of $f_{a'}(\pi(\mathcal{P}'))$ in $\pi(\mathcal{P})$. Therefore, as $Aut(f)$ acts semi-regularly on $f^{-1}(a)$, an injective mapping φ is defined from $Aut(f)$ to the set of cosets of $f_{a'}(\pi(\mathcal{P}'))$ in $\pi(\mathcal{P})$, as follows: if α' is a path from a' to $g(a')$ ($g \in Aut(f)$) and A is the homotopy class of $\alpha = f(\alpha')$ in $\mathcal{P}$, then $\varphi(g) = f_{a'}(\pi(\mathcal{P}')) \cdot A$. Trivially, if $b' = g(a')$, then

$$f_{b'}(\pi(\mathcal{P}')) = (fg^{-1})_{b'}(\pi(\mathcal{P}')) = A^{-1} \cdot f_{a'}(\pi(\mathcal{P}')) \cdot A.$$

However, g^{-1} is an isomorphism, $fg^{-1} = f$ and $g^{-1}(b') = a'$. Hence by Corollary 12.8 we have $f_{a'}(\pi(\mathcal{P}')) = (fg^{-1})_{b'}(\pi(\mathcal{P}'))$. It follows from this and the above that $f_{a'}(\pi(\mathcal{P}')) = A^{-1} \cdot f_{a'}(\pi(\mathcal{P}')) \cdot A$.

Let h be another element of $Aut(f)$, let $c' = h(b')$ and let β' a path from b' to c'. Let B be the homotopy class of $f(\beta')$. Then $f_{a'}(\pi(\mathcal{P}')) \cdot AB$ represents hg and, if γ' is a path from a' to $h(a')$ and C is the homotopy class of $f(\gamma')$ in $\mathcal{P}$, then $f_{a'}(\pi(\mathcal{P}')) \cdot C$ represents h. On the other hand, $\alpha'\beta'h(\alpha'^{-1})\gamma'^{-1}$ is a closed path. Hence, if D is the homotopy class of $f(h(\alpha'))$, then $ABDC \in f_{a'}(\pi(\mathcal{P}'))$. However, $D = A$ because $fh = f$. Therefore

$$f_{a'}(\pi(\mathcal{P}')) \cdot AB \; (= \varphi(hg)) \;=\; f_{a'}(\pi(\mathcal{P}')) \cdot CA \;=\; \varphi(h)\varphi(g).$$

Hence φ is an embedding of $Aut(f)$ into $N(\pi(\mathcal{P}'))/\pi(\mathcal{P}')$. We still must prove that φ is surjective. Let $A \in N(\pi(\mathcal{P}'))$, let α be a closed path in A and let α' be the lifting of α to $\mathcal{P}'$ through f at a'. Let b' be the end vertex of α'. Applying Lemma 12.7 and Corollary 12.8 with $\mathcal{P}'' = \mathcal{P}'$, $g = f$ and $a'' = b'$, we obtain that there is $h \in Aut(f)$ mapping a' onto b'. Trivially, $\varphi(h) = f_{a'}(\pi(\mathcal{P}')) \cdot A$. The surjectivity of φ is proved. □

Corollary 12.10 *We have $\mathcal{P} = \mathcal{P}'/Aut(f)$ if and only if $\pi(\mathcal{P}') \trianglelefteq \pi(\mathcal{P})$.*

(Easy by the previous proposition.)

12.2.5 Liftings of automorphisms

Liftings of automorphisms of $\mathcal{P}$ to $\mathcal{P}'$ through f can be defined as we did for geometries in Section 11.1. Let $L(f)$ be the set of all such liftings and let $N(Aut(f))$ be the normalizer of $Aut(f)$ in $Aut(\mathcal{P}')$.

Lemma 12.11 *We have $L(f) = N(Aut(f))$.*

Proof. Just rephrase the proof of Theorem 11.2. □

We now want to characterize those automorphisms of $\mathcal{P}$ that can be lifted to $\mathcal{P}'$ through f. In order to do that, we must preliminarily describe the embeddings of $\pi(\mathcal{P}')$ into $\pi(\mathcal{P})$ more carefully than we have done before.

Given a vertex a of $\mathcal{P}$ and a chosen element $a' \in f^{-1}(a)$, the cover f embeds $\pi_{a'}(\mathcal{P}')$ into $\pi_a(\mathcal{P})$ (this is in fact the embedding we have hitherto denoted by $f_{a'}$). However, if b' is another element of $f^{-1}(a)$, then the subgroups $f(\pi_{b'}(\mathcal{P}'))$ and $f(\pi_{a'}(\mathcal{P}'))$ of $\pi_a(\mathcal{P})$ may be different. Actually, if α' is a path from a' to b', if $\alpha = f(\alpha')$ and $A \in \pi_a(\mathcal{P})$ is the homotopy class of α, then $f(\pi_{b'}(\mathcal{P}')) = A \cdot f(\pi_{a'}(\mathcal{P}')) \cdot A^{-1}$. By Proposition 12.9, we have $f(\pi_{b'}(\mathcal{P}')) = f(\pi_{b'}(\mathcal{P}'))$ if and only if a' and b' belong to the same orbit of $Aut(f)$.

Conversely, let $A \in \pi_a(\mathcal{P})$, let α be an element of A and let α' be the lifting of α to $\mathcal{P}'$ at a'. Then $A \cdot f(\pi_{a'}(\mathcal{P}')) \cdot A^{-1} = f(\pi_{b'}(\mathcal{P}'))$, where $b' \in f^{-1}(a)$ is the end vertex of α'.

Therefore, the images of the homotopy groups $\pi_{a'}(\mathcal{P}')$, with a' ranging in $f^{-1}(a)$, form a conjugacy class in $\pi_a(\mathcal{P})$. Let $\overline{\pi}_a(f)$ designate this conjugacy class.

Theorem 12.12 *Let $g \in Aut(\mathcal{P})$. Then the following are equivalent:*

(i) the automorphism g lifts to an automorphism of $\mathcal{P}'$ through f;
(ii) for every vertex a of $\mathcal{P}$, the automorphism g induces a bijection from $\overline{\pi}_a(f)$ to $\overline{\pi}_{g(a)}(f)$;
(iii) we have $f(\pi) \in \overline{\pi}_{g(a)}(f)$ for some vertex a of $\mathcal{P}$ and some member π of $\overline{\pi}_a(f)$.

Proof. Trivially (ii) implies (iii). Let us prove that (iii) implies (i).

Given a, π as in (iii), let $a' \in f^{-1}(a)$ and $b' \in f^{-1}(g(a))$ be such that $f(\pi_{a'}(\mathcal{P}')) = \pi$ and $f(\pi_{b'}(\mathcal{P}')) = g(\pi)$. By Lemma 12.7 and Corollary 12.8, with $\mathcal{P}'$, $g^{-1}f$, b' in the roles of $\mathcal{P}''$, g and a'' respectively in Lemma 12.7, we obtain an automorphism h of $\mathcal{P}'$ such that $g^{-1}fh = f$. The automorphism h is the lifting of g we are looking for.

Finally, (i) implies (ii) because liftings of automorphisms of $\mathcal{P}$ through f permute the fibres of f. □

12.2.6 Universal covers and simple connectedness

Universal covers of plate complexes are defined as for geometries: a cover $f : \widehat{\mathcal{P}} \longrightarrow \mathcal{P}$ is *universal* if, for every cover $g : \mathcal{P}' \longrightarrow \mathcal{P}$, there is just one cover $h : \widehat{\mathcal{P}} \longrightarrow \mathcal{P}'$ such that $gh = f$.

We also say that a plate complex $\mathcal{P}$ is *simply connected* if $\pi(\mathcal{P}) = 1$.

Theorem 12.13 *Let $f : \widehat{\mathcal{P}} \longrightarrow \mathcal{P}$ be a cover and let $\widehat{\mathcal{P}}$ be simply connected. Then all the following hold:*

(i) the cover f is universal;
(ii) we have $Aut(f) \cong \pi(\mathcal{P})$ and $\mathcal{P} \cong \widehat{\mathcal{P}}/Aut(f)$;
(iii) every automorphism of $\mathcal{P}$ lifts to $\widehat{\mathcal{P}}$ through f.

Proof. The first statement follows from Lemma 12.7. The second statement is a consequence of Proposition 12.9 and Corollary 12.10. The third one follows from Proposition 12.12. □

Theorem 12.14 *Every plate complex $\mathcal{P}$ is covered by some simply connected plate complex $\widehat{\mathcal{P}}$. The simply connected plate complex $\widehat{\mathcal{P}}$ and the* *(universal) cover* $f : \widehat{\mathcal{P}} \longrightarrow \mathcal{P}$ *are uniquely determined by* $\mathcal{P}$*, up to isomorphism.*

Proof. The uniqueness claim is a consequence of Theorem 12.13(i). Note that f is universal by Theorem 12.3(i). Let us prove the existence statement.

Given $\mathcal{P} = (V, E, P, t_V, t_E)$, let us choose $a \in V$ and let $\widehat{V}$ be the set of homotopy classes of paths of $\mathcal{P}$ starting at a. Let us define $f : \widehat{V} \longrightarrow V$ stating that, for every $A \in \widehat{V}$, the vertex $f(A)$ is the common end vertex of all paths in A. Trivially, f is surjective and a type function $\widehat{t}_V$ can be defined on $\widehat{V}$ stating that $\widehat{t}_A = t_V(f(A))$. We also define a graph $(\widehat{V}, \widehat{E})$ on $\widehat{V}$ as follows: given $A, B \in \widehat{V}$, we have $\{A, B\} \in \widehat{E}$ if and only if $\{f(A), f(B)\} \in E$ and $\alpha \cdot (f(A), f(B)) \in B$ for some (equivalently, for every) element α of A (equivalently, $\beta \cdot (f(B), f(A)) \in A$ for $\beta \in B$).

The graph $(\widehat{V}, \widehat{E})$ defined in this way is connected. Indeed, given any vertex $A \in \widehat{V}$, let $(a_0, a_1, ..., a_h)$ be a member of A, with $a_0 = a$. For every $i = 0, 1, ..., h$, let A_i be the homotopy class of the path $(a_0, a_1, ..., a_i)$. In particular, A_0 is the homotopy class O_a of the null path (a). Then $(A_0, A_1, ..., A_h)$ is a path in $(\widehat{V}, \widehat{E})$ from O_a to A. Therefore every element of $\widehat{V}$ is joined by some path to O_a. Hence $(\widehat{V}, \widehat{E})$ is connected.

A type function $\widehat{t}_E$ can also be defined on $\widehat{E}$, stating that $\widehat{t}_E(\{A, B\}) = t_E(\{f(A), f(B)\})$ for every edge $\{A, B\} \in \widehat{E}$.

We still need to define an appropriate set $\widehat{P}$ of plates on $\widehat{V}$.

Given $X \in P$, and a vertex $b \in X$, let π_{ab} be the set of homotopy classes of paths from a to b, let $\pi_b(X)$ be the set of homotopy classes of paths in X starting at b and let us define $\pi_{ab} \cdot \pi_b(X) = \{AB \mid A \in \pi_{ab}, B \in \pi_b(X)\}$. Trivially, we have $f^{-1}(X) = \bigcup_{b \in X} \pi_{ab} \cdot \pi_b(X)$.

Let $f_b : \pi_b(X) \longrightarrow X$ be the function mapping every element B of $\pi_b(X)$ onto the common end vertex of its members. This function is a bijection. Indeed X is a connected subgraph of (V, E), hence f_b is surjective. Furthermore, two paths in X starting at b are (elementarily) homotopic if and only if they end at the same vertex. Therefore, we may identify $\pi_b(X)$ with X via f_b, writing X for $\pi_b(X)$, for short.

Moreover, if c is any other vertex of X and C is the homotopy class of a path in X from b to c, then $A \cdot X = AC \cdot X$ for every $A \in \pi_{ab}$. Therefore, we may keep b fixed and we have $f^{-1}(X) = \pi_{ab} \cdot X = \bigcap_{A \in \pi_{ab}} A \cdot X$.

Given two elements A, A' of π_{ab}, we have $A \cdot X \cap A' \cdot X \neq \emptyset$ if and only if $A \cdot X = A' \cdot X$. The latter equality holds if and only if $A = A'$. Therefore the sets $A \cdot X$ (with $A \in \pi_{ab}$) are pairwise disjoint and $f^{-1}(X)$ is their disjoint union. The graph structure of X can be copied on each of these sets via f. Indeed f isomorphically maps the graph induced by

$(\widehat{V}, \widehat{E})$ on $A \cdot X$ onto the graph induced by (V, E) on X. In order to imitate the structure of X in $A \cdot X$, we simply drop from the graph induced on $A \cdot X$ by $(\widehat{V}, \widehat{E})$ those edges (if any) that are mapped by f onto edges of (V, E) contained in the set X but not belonging to the graph structure of the plate X.

The sets $A \cdot X$ (with $A \in \widehat{V}$, $f(A) \in X$, $X \in P$), endowed with the graph stucture defined above, will be taken as plates. A plate complex $\widehat{\mathcal{P}}$ is obtained in this way. Checking that f is a cover from $\widehat{\mathcal{P}}$ to $\mathcal{P}$ is straightforward. To finish the proof, we need to show that $\widehat{\mathcal{P}}$ is simply connected.

Let $(A_0, A_1, ..., A_m)$ be a closed path in $\widehat{\mathcal{P}}$, with $A_0 = A_m = O_a$ and let $a_i = f(A_i)$ $(i = 0, 1, ..., m)$. Then $\alpha = (a_0, a_1, ..., a_m)$ is a closed path of $\mathcal{P}$, based at $a = a_0 = a_m$. For every $i = 0, 1, ..., m$, let us set $\alpha_i = (a_0, a_1, ..., a_i)$. In particular $\alpha_m = \alpha$. The path α_i is a representative of A_i. If $i = 0$, this claim is trivial. When $i > 0$ we can proceed by induction. The path α_{i-1} is a member of A_{i-1} by the inductive hypothesis. Then $\alpha_i \in A_i$ by the definition of $\widehat{E}$ and because $\{A_{i-1}, A_i\} \in \widehat{E}$. Therefore $\alpha \in A_m$. However, $A_m = O_a$, whence α is null homotopic. On the other hand, $(A_0, A_1, ..., A_m)$ is a lifting of α through the cover f. Hence $(A_0, A_1, ..., A_m)$ is null homotopic. Therefore $\pi(\widehat{\mathcal{P}}) = 1$. □

Corollary 12.15 *A cover $f : \widehat{\mathcal{P}} \longrightarrow \mathcal{P}$ is universal if and only if $\widehat{\mathcal{P}}$ is simply connected.*

Corollary 12.16 *A plate complex $\mathcal{P}$ is simply connected if and only if the identity automorphism of $\mathcal{P}$, viewed as a cover, is universal.*

Corollaries 12.15 and 12.16 follow from Theorems 12.13(i) and 12.14 and from the 'uniqueness' of universal covers. Following the linguistic abuse of identifying covers with their domains, Corollary 12.16 can be stated in the following concise way:

Corollary 12.17 *A plate complex is simply connected if and only if it is its own universal cover.*

Theorem 12.14 can be rephrased as follows, using Theorem 12.13(i).

Corollary 12.18 *Every plate complex admits a universal cover.*

Corollary 12.19 *The isomorphism classes of covers of a given plate complex $\mathcal{P}$ bijectively correspond to the subgroups of $\pi(\mathcal{P})$.*

Proof. The most appropriate way to (state and) prove this corollary would be to use the language of category theory. However, we prefer not to go so far. We use a rougher argument.

If $f : \widehat{\mathcal{P}} \longrightarrow \mathcal{P}$ is the universal cover of $\mathcal{P}$, then $Aut(f)$ and $\mathcal{P}$ can be identified with $\pi(\mathcal{P})$ and $\widehat{\mathcal{P}}/\pi(\mathcal{P})$, respectively (Theorem 12.13(ii)). Every subgroup K of $\pi(\mathcal{P})$ $(= Aut(f))$ gives us a qotient $\widehat{\mathcal{P}}/K$ and it is easily seen that f splits as $f = f_K p_K$ where p_K is the projection of $\widehat{\mathcal{P}}$ onto $\mathcal{P}/K$ and $f_K : \widehat{\mathcal{P}}/K \longrightarrow \mathcal{P}$ is a cover. On the other hand, since $\pi(\widehat{\mathcal{P}}) = 1$, every cover of $\mathcal{P}$ is covered by $\widehat{\mathcal{P}}$, by Lemma 12.7. Hence we can obtain it as a quotient of $\widehat{\mathcal{P}}$ by a subgroup of $\pi(\mathcal{P})$ (Theorem 12.13(ii)). Non-isomorphic covers of $\mathcal{P}$ correspond to distinct subgroups of $\pi(\mathcal{P})$, by Corollary 12.8. □

12.3 Chamber systems

In order to apply to m-covers of geometries the machinery set up in the previous section, we must associate a suitable plate complex with every geometry.

We choose m-cell complexes (Section 12.2.1). However, this choice will push us into a category larger than the category of geometries and their m-covers: the category of chamber systems and their m-covers.

12.3.1 Definitions

Keeping chamber systems of geometries in mind, we state the following general definition.

A *chamber system* over a (finite, nonempty) set of *types* I is a pair $\mathcal{C} = (C, (\Phi^i)_{i \in I})$ where C is a nonempty set, called the set of *chambers*, Φ^i (with $i \in I$) is an equivalence relation on C, called the *i-adjacency relation*, and the following hold:

($\mathbf{C}_1$) for every type $i \in I$, every equivalence class of Φ^i contains at least two chambers;

($\mathbf{C}_2$) for any two distinct types $i, j \in I$, we have $\Phi^i \cap \Phi^j = \mho$;

($\mathbf{C}_3$) we have $\bigvee_{i \in I} \Phi^i = \Omega$.

(More general definitions of chamber systems are often used in the literature, but the one we have state here is general enough for our purposes.)

As in Section 1.4, we define $\Phi^J = \bigvee_{j \in J} \Phi^j$ for every nonempty subset J of I. Thus, $\Phi^I = \Omega$, by $\mathbf{C}_3$. We set $\Phi^{\emptyset} = \mho$.

Given a chamber c and a set of types $J \subseteq I$, the *cell* of *type* J (of *cotype* $I - J$) containing c is the equivalence class $[c]\Phi^J$ of Φ^J containing c, endowed with the equivalence relations induced on it by the j-adjacency relations Φ^j ($j \in J$). Therefore, a cell of type $J \neq \emptyset$ is a chamber system over the set of types J.

The cells of type $\emptyset$ are just the chambers. $\mathcal{C}$ is the (unique) cell of type I. The *rank* (the *corank*) of a cell of type J is the number $|J|$ (respectively, $|I - J|$). In particular, $|I|$ is the *rank* of $\mathcal{C}$. The cells of rank 1 are called *panels* and the cells of corank 1 are called *vertices*.

When $\mathcal{C} = \mathcal{C}(\Gamma)$ for some geometry Γ, then the cells of $\mathcal{C}$ bijectively correspond to the flags of Γ (Lemma 1.22). Note that, if F is a flag of Γ and $X = \mathcal{C}_F$ is the cell of F in $\mathcal{C} = \mathcal{C}(\Gamma)$, then the type (the cotype, the rank, the corank) of X in $\mathcal{C}$, as defined above, is the cotype (the type, the corank, the rank respectively) of F in Γ. In particular, if F is a panel (a vertex) of Γ, then X is a panel (or a vertex) in $\mathcal{C}$, but it has rank 1 in $\mathcal{C}$ (rank $n - 1$, respectively).

The *adjacency graph* of a chamber $\mathcal{C}$ is the graph having the chambers of $\mathcal{C}$ as vertices, two distinct chambers being adjacent in this graph precisely when they are contained in the same panel (that is, when they are i-adjacent for some type i). We denote this adjacency relation by $\sim$.

The adjacency graph $(C, \sim)$ is a coloured graph, with I as set of colours for the edges of $(C, \sim)$, every edge having precisely one colour by $\mathbf{C}_2$. Cells are subgraphs of $(C, \sim)$ and they can be characterized as follows: a (nonempty) connected subgraph X of $(C, \sim)$ is a cell if and only if, for every colour $i \in I$, either none of the edges in X has colour i or, for every chamber $c \in X$, all edges of $(C, \sim)$ at c of colour i are in X. Trivially, the type of a cell X is the set of colours of the edges of X. It is also worth remarking that it may happen that two adjacent chambers c, d belong to a cell X but the edge $e = \{c, d\}$ is not in the graph X.

Adjacency graphs of chamber systems are easy to characterize. A connected coloured graph, with a finite set I of colours for its edges, is (the adjacency graph of) a chamber system over the set of types I if and only if both the following hold:

($\mathbf{C}_1'$) for every vertex x and every colour i, the vertex x belongs to some edges of colour i;

($\mathbf{C}_2'$) any two maximal monocromatic cliques of the same colour are disjoint.

Another graph can be associated with a chamber system $\mathcal{C}$. Let $V(\mathcal{C})$ be the set of vertices of $\mathcal{C}$. The *incidence graph* $\Gamma(\mathcal{C})$ of $\mathcal{C}$ is defined taking $V(\mathcal{C})$ as the set of vertices of the graph and stating that two vertices X, Y are adjacent in $\Gamma(\mathcal{C})$ precisely when $X \cap Y \neq \emptyset$. The adjacency relation of $\Gamma(\mathcal{C})$ is often called the *incidence relation* of $\mathcal{C}$. Trivially, $\Gamma(\mathcal{C})$ is n-partite, where n is the rank of $\mathcal{C}$. If $n \geq 2$, then $\Gamma(\mathcal{C})$ is connected, by $\mathbf{C}_3$. By Lemma 1.22, if $\mathcal{C} = \mathcal{C}(\Gamma)$ for some geometry Γ, then $\Gamma(\mathcal{C}) \cong \Gamma$ and $\mathcal{C}(\Gamma(\mathcal{C})) \cong \mathcal{C}$.

However, the graph $\Gamma(\mathcal{C})$ need not be a geometry in general (see Exercises 12.11–12.13). Furthermore, it may happen that $\Gamma(\mathcal{C})$ is a geometry and nevertheless $\mathcal{C}(\Gamma(\mathcal{C})) \not\cong \mathcal{C}$ (see Exercise 12.11). We shall turn back to this matter in Section 12.5.

12.3.2 Epimorphisms, isomorphisms and covers

Morphisms of chamber systems can be defined as morphisms of their adjacency graphs, as in Exercises 8.24, 8.25, or as morphisms of coloured graphs. A bit of general theory could also be developed for morphisms defined in one of these ways, but we are not going so far. We will only consider a fairly restricted class of morphisms.

Let $\mathcal{C}$ and $\mathcal{C}'$ be chamber systems over the same set of types I and with sets of chambers C and C' respectively. An *epimorphism* from $\mathcal{C}$ to $\mathcal{C}'$ is a mapping $f : C \longrightarrow C'$ such that, for every panel X of $\mathcal{C}$, the image $f(X)$ of X is a panel of $\mathcal{C}'$ of the same type as X.

As usual, we write $f : \mathcal{C} \longrightarrow \mathcal{C}'$ to mean that f is an epimorphism from $\mathcal{C}$ to $\mathcal{C}'$. Let $f : \mathcal{C} \longrightarrow \mathcal{C}'$ be an epimorphism. It is not hard to prove that $f(C) = C'$ (that is, f is surjective) and that, for every cell X of $\mathcal{C}$, the image $f(X)$ of X is a cell of $\mathcal{C}'$ of the same type as X (that is, f is 'residually surjective'; compare with Theorem 8.4).

When $\mathcal{C} = \mathcal{C}(\Gamma)$ and $\mathcal{C}' = \mathcal{C}(\Gamma')$ for geometries Γ and Γ', then the epimorphisms $f : \mathcal{C} \longrightarrow \mathcal{C}'$ are just those mappings $f : C \longrightarrow C'$ induced by type-preserving epimorphisms from Γ to Γ' (see Exercise 8.27 and use Lemma 1.22).

An *isomorphism* is an injective (hence, bijective) epimorphism. By Theorem 1.23, the isomorphisms between the chamber systems of two geometries are precisely the type-preserving isomorphisms of the corresponding geometries.

Let m be a positive integer less that $n = |I|$. An *m-cover* from $\mathcal{C}$ to $\mathcal{C}'$ is an epimorphism $f : \mathcal{C} \longrightarrow \mathcal{C}'$ such that, for every cell X of $\mathcal{C}$ of rank m, the restriction of f to X is an isomorphism from the chamber system induced by $\mathcal{C}$ on X to the chamber system induced by $\mathcal{C}'$ on $f(X)$.

Trivially, an m-cover is also a k-cover for every $k = m-1, m-2, ..., 1$.

By Theorem 1.23, an m-cover between chambers systems of geometries induces an m-cover between the corresponding geometries. Trivially, the converse also holds: every m-cover between two geometries induces an m-cover between the chamber systems of those geometries.

Universal m-covers of chamber systems are defined as for geometries (Section 12.1).

12.3.3 The existence of universal covers

We switch to plate complexes now, so as to take advantage of the results of Section 12.2.

Let $\mathcal{C}$ be a chamber system over the set of types I and let $n = |I|$ be the rank of $\mathcal{C}$. Let m be a positive integer less than n. The *m-cell complex* $\mathcal{C}_m$ of $\mathcal{C}$ is the plate complex having the adjacency graph of $\mathcal{C}$ as underlying graph, the set of types I as set of colours for the edges of this graph and the cells of $\mathcal{C}$ of rank m as plates.

Trivially, m-covers of chamber systems are just covers of their m-cell complexes. Furthermore, let $f : \mathcal{P} \longrightarrow \mathcal{C}_m$ be a cover from some plate complex $\mathcal{P}$ to the m-cell complex $\mathcal{C}_m$ of $\mathcal{C}$. It is easily seen that the properties $\mathbf{C}'_1$ and $\mathbf{C}'_2$ of Section 12.3.1 lift from $\mathcal{C}_m$ to $\mathcal{P}$ via f. Therefore, the underlying graph of $\mathcal{P}$ is the adjacency graph of a chamber system over the set of types I. Let us denote this chamber system by $\mathcal{C}(\mathcal{P})$. Furthermore, as f is a cover, the plates of $\mathcal{P}$ are isomorphically mapped onto the cells of $\mathcal{C}$ of rank m and the stars in $\mathcal{P}$ are isomorphic to stars in $\mathcal{C}_m$. It is now easy to check that the plates of $\mathcal{P}$ are just the cells of $\mathcal{C}(\mathcal{P})$ of rank m. Hence f is an m-cover from $\mathcal{C}(\mathcal{P})$ to $\mathcal{C}$.

Therefore, we can safely replace chamber systems and m-covers of chamber systems with m-cell complexes of chamber systems and covers of m-cell complexes. Thus, by Corollary 12.18 we obtain the following:

Theorem 12.20 (Existence theorem) *Every chamber system of rank n admits a universal m-cover, for every $m = 1, 2, ..., n-1$.*

We denote the (domain of the) universal m-cover of a chamber system $\mathcal{C}$ by $\widehat{\mathcal{C}}_m$. The homotopy group $\pi(\mathcal{C}_m)$ of the plate complex $\mathcal{C}_m$ is called the *m-homotopy group* of $\mathcal{C}$. We designated it by $\pi_m(\mathcal{C})$. We say that a chamber system $\mathcal{C}$ is *m-simply connected* if its m-cell complex $\mathcal{C}_m$ is simply connected, that is if $\pi_m(\mathcal{C}) = 1$.

By Corollary 12.15, a chamber system is m-simply connected if and only if it is its own universal m-cover. The proof of Theorem 12.14 also shows how to construct the universal m-cover $\widehat{\mathcal{C}}_m$ of a chamber system $\mathcal{C}$.

We will give another construction of $\widehat{\mathcal{C}}_m$ in Section 12.4.3, assuming that $\mathcal{C}$ admits a 'large' automorphism group.

Remark. The existence of universal m-covers of chamber systems has been discovered by Ronan [173] and Tits [227] (Section 5.1).

Exercise 12.1. Given a chamber system $\mathcal{C}$, let $\mathcal{P}(\mathcal{C})$ be the graph (possibly with multiple edges) having the panels and the chambers of $\mathcal{C}$ as vertices and edges, respectively. Prove that, if $\mathcal{C}$ is 1-simply connected, then $\mathcal{C}$ has no multiple edges and it is an infinite tree with no terminal nodes.

(Hint: we have $\pi_1(\mathcal{C}) = 1$ if and only if every closed path of the adjacency graph of $\mathcal{C}$ is contained in some clique.)

Exercise 12.2. Prove that a chamber system of rank 2 is 1-simply connected if and only if it is (the chamber system) of a generalized ∞-gon (if and only if its incidence graph is a tree).

Exercise 12.3. Let $\mathcal{C}$ be a chamber system of rank $n \geq 3$. Prove that, if $\mathcal{C}$ is $(n-1)$-simply connected and all cells of $\mathcal{C}$ of rank $m = 3, 4, ..., n-1$ are $(m-1)$-simply connected, then $\mathcal{C}$ is 2-simply connected.

12.4 Parabolic systems

An *automorphism* of a chamber system $\mathcal{C}$ is an isomorphism from $\mathcal{C}$ to $\mathcal{C}$. The automorphism group of $\mathcal{C}$ will be designated by $Aut(\mathcal{C})$. A subgroup $G \leq Aut(\mathcal{C})$ is said to be *transitive* (in $\mathcal{C}$) if it acts transitively on the set of chambers of $\mathcal{C}$. We say that $\mathcal{C}$ is *transitive* if $Aut(\mathcal{C})$ is transitive.

Let $\mathcal{C} = \mathcal{C}(\Gamma)$ for some geometry Γ. Then $Aut(\mathcal{C})$ can naturally be identified with $Aut_s(\Gamma)$. Thus, we take the symbols $Aut(\mathcal{C}(\Gamma))$ and $Aut_s(\Gamma)$ as synonymous, by a harmless abuse of notation. Trivially, $\mathcal{C}(\Gamma)$ is transitive if and only if Γ is flag-transitive.

We are considering only type-preserving automorphisms, according to the above definition of epimorphisms of chamber systems. Automorphisms of chamber systems could be defined in a more general way, so that to allow non-type-preserving automorphisms. These are interesting, too. However, the more restrictive definition we have given is sufficent for our purposes.

12.4.1 Chamber systems and parabolic systems

Let $\mathcal{C} = (C, (\Phi^i)_{i \in I})$ be a chamber system of rank $n = |I| \geq 2$.

Lemma 12.21 *Let $G \leq Aut(\mathcal{C})$ and $c \in C$. Then G is transitive if and only if, for every panel X containing c, the stabilizer G_X of X in G is transitive in X.*

Proof. Just rephrase the proof of Lemma 10.1. □

Henceforth G designates a given transitive subgroup of $Aut(\mathcal{C})$ and $c \in C$ is a given chamber of $\mathcal{C}$. The stabilizer of c in G will be denoted by B. We call it the *Borel* subgroup of G (relative to c). For every $J \subseteq I$, let X_J be the cell of $\mathcal{C}$ of type J containing c. We write G^J for the stabilizer of X_J in G. We call G^J the *parabolic* subgroup of G of *type* J (relative to c). The rank $|J|$ of X_J is called the *rank* of G^J. A parabolic subgroup G^J is said to be *proper* if $\emptyset \neq J \neq I$. Trivially, we have $G^{\emptyset} = B$ and $G^I = G$. That is, B and G are the only improper parabolic subgroups of G. The parabolic subgroups of rank 1 are the stabilizers of the panels containing c. We call them the *minimal parabolic* subgroups of G. As usual, we write G^i for $G^{\{i\}}$, for short. The *maximal parabolic* subgroups are the parabolic subgroups of rank $n-1$, that is the stabilizers of the vertices of $\mathcal{C}$ containing c.

Proposition 12.22 *The following hold:*

($\mathbf{C}_1''$) *we have* $B \neq G^i$ *for every* $i \in I$;
($\mathbf{C}_2''$) *we have* $G^i \cap G^j = B$ *for any two distinct types* $i, j \in I$;
($\mathbf{C}_3''$) *we have* $G = \langle G^i \rangle_{i \in I}$;
($\mathbf{C}_4''$) *we have* $\bigcap_{g \in G} gBg^{-1} = 1$.

Proof. Properties $\mathbf{C}_1''$ and $\mathbf{C}_2''$ are group-theoretic translations of $\mathbf{C}_1$ and $\mathbf{C}_2$, respectively. $\mathbf{C}_3''$ follows from $\mathbf{C}_3$ and from Lemma 12.21. Turning to $\mathbf{C}_4''$, the subgroup $\bigcap_{g \in G} gBg^{-1}$ fixes all chambers of $\mathcal{C}$, by the transitivity of G. Hence $\bigcap_{g \in G} gBg^{-1} = 1$. □

Actually, $\mathbf{C}_3''$ is a special case of the following proposition, which easily follows from Lemma 12.21 applied in cells containing c:

Proposition 12.23 *We have* $G^J = \langle G^j \rangle_{j \in J}$ *for every nonempty subset* $J \subseteq I$.

The following is a trivial consequence of $\mathbf{C}_2''$ and of the hypothesis $|I| \geq 2$:

Corollary 12.24 *We have* $B = \bigcap_{i \in I} G^i$.

It is worth remarking that $\mathbf{C}_1''$ and $\mathbf{C}_4''$ are just the same as $\mathbf{P}_1$ and $\mathbf{P}_1$ of Section 10.1.3. According to the definition of G^J as the stabilizer of the cell X_J, the statement of Proposition 12.23 resembles $\mathbf{P}_2$ of Section 10.1.3. However, something of $\mathbf{P}_2$ is missing in Proposition 12.23. Indeed, we can see now that $\mathbf{P}_2$ of Section 10.1.3 entailed two claims: that stabilizers of cells are intersections of maximal parabolics, and that we can also generate them by minimal parabolics. Proposition 12.23 only saves the second one of these two claims.

As we did in Section 10.1.2 for chamber systems of geometries, we can construct a model of $\mathcal{C}$ by means of cosets of B and cosets of the minimal parabolics G^i ($i \in I$). Furthermore, the system of minimal parabolics $(G^i)_{i\in I}$ uniquely determines B (by $\mathbf{C}''_2$). Therefore G and $(G^i)_{i\in I}$ uniquely determine $\mathcal{C}$.

We call $(G^i)_{i\in I}$ the *parabolic system* of $\mathcal{C}$ and G, relative to the chamber c and we denote it by $\mathcal{G}(\mathcal{C}, G)$.

Conversely, let $\mathcal{G} = (G^i)_{i\in I}$ be a family of subgroups of a group G, with $2 \leq |I| < \infty$. Let the properties $\mathbf{C}''_1$, $\mathbf{C}''_2$, $\mathbf{C}''_3$ and $\mathbf{C}''_4$ of Proposition 12.22 hold in $\mathcal{G}$, where we now write B for $\bigcap_{i\in I} G^i$.

Then we can define a chamber system $\mathcal{C}(\mathcal{G})$ over the set of types I, as follows. We take the right cosets of B as chambers. For every $i \in I$, the right cosets of G^i are taken as panels of type i (we are identifying a coset gG^i with the set of right cosets of B contained in it). Thus, two chambers gB, fB are i-adjacent in $\mathcal{C}(\mathcal{G})$ if and only if $f^{-1}g \in G^i$.

Given a nonempty subset $J \subseteq I$, we write G^J for $\langle G^j \rangle_{j\in J}$. The cells of $\mathcal{C}(\mathcal{G})$ of type J are the right cosets of G^J.

The group G, acting by left multiplication on right cosets of B, is a transitive subgroup of $Aut(\mathcal{C}(\mathcal{G}))$ and $\mathcal{G}$ is just the parabolic system of $\mathcal{C}(\mathcal{G})$ and G, with respect to the chamber B of $\mathcal{C}(\mathcal{G})$. The subgroup G^J ($\emptyset \neq J \subseteq I$) is the parabolic subgroup of type J. The subgroups G^i (respectively, G^{I-i}), with $i \in I$, are the minimal (the maximal) parabolics. The subgroup B is the Borel subgroup, namely the parabolic subgroup of type $\emptyset$. According to this, we often write $G^{\emptyset}$ for B.

It is clear that, if $\mathcal{G} = \mathcal{G}(\mathcal{C}, G)$ for some chamber system $\mathcal{C}$ and some transitive subgroup $G \leq Aut(\mathcal{C})$, then we have $\mathcal{C}(\mathcal{G}) \cong \mathcal{C}$ (actually, this is the precise meaning of what we said before, when we remarked that G and $\mathcal{G}(\mathcal{C}, G)$ uniquely determine $\mathcal{C}$).

We say that a family $\mathcal{G} = (G^i)_{i\in I}$ of subgroups of a group G (with $2 \leq |I| < \infty$) is a *generalized parabolic system* of *rank* $n = |I|$ over the *set of types* I if it satisfies $\mathbf{C}''_1$, $\mathbf{C}''_2$, $\mathbf{C}''_3$ and $\mathbf{C}''_4$.

We will always write 'parabolic system' for short, omitting the word 'generalized'. According to this convention, we change the terminology of Chapter 10 a bit: from now on, when we want to point out that a given parabolic system (in the above sense) defines a geometry (that is, it is a parabolic system in the sense of Chapter 10), then we call it a *geometric* parabolic system.

12.4.2 Morphisms of parabolic systems

Morphisms are defined for arbitrary parabolic systems as for geometric parabolic systems (Section 11.2.1), except that we now modify the notation a bit, substituting G^{I-J} $(= \langle G^i \rangle_{i\in I-J})$ for G_J $(= \bigcap_{j\in J} G_j)$ everywhere.

Indeed, Proposition 12.23 now only allows us to work with subgroups generated by sets of minimal parabolics: we have lost that part of $\mathbf{P}_2$ of Section 10.1.3 stating that the subgroups generated by sets of minimal parabolics are just those subgroups that can be obtained as intersections of sets of maximal parabolics. Hence we cannot now write G_J for G^{I-J} or G_{I-J} for G^J as in Chapter 10.

Apart from this change of notation, everything we have said in Section 11.2.1 remains true in the present more general context. In particular, we have the following generalization of Lemma 11.7:

Lemma 12.25 *Every morphism $\varphi : \mathcal{G} \longrightarrow \mathcal{G}'$ of parabolic systems uniquely determines an epimorphism $f_\varphi : \mathcal{C}(\mathcal{G}) \longrightarrow \mathcal{C}(\mathcal{G}')$ between the associated chamber systems.*

Proof. The proof is similar to that given for Lemma 11.7, except that we should now work with minimal parabolics and adjacency graphs instead of maximal parabolics and incidence graphs. □

Given a morphism $\varphi : \mathcal{G} \longrightarrow \mathcal{G}'$ of parabolic systems over the set of types I and given $J \subseteq I$, the symbol φ^J will designate the homomorphism induced by φ from G^J to G'^J. The next lemma is a straightforward generalization of Lemma 11.9:

Lemma 12.26 *Let $\varphi : \mathcal{G} \longrightarrow \mathcal{G}'$ be a morphism of parabolic systems over the set of types I and let m be a positive integer less than $n = |I|$. Then the epimorphism $f_\varphi : \mathcal{C}(\mathcal{G}) \longrightarrow \mathcal{C}(\mathcal{G}')$ is an m-cover if and only if φ^J is an isomorphism for every $J \subseteq I$ with $|J| = m$.*

12.4.3 Constructing universal covers as amalgams

Let $\mathcal{G} = (G^i)_{i \in I}$ be a parabolic system over the set of types I in a group G, let $\mathcal{C} = \mathcal{C}(\mathcal{G})$ be the chamber system defined by $\mathcal{G}$, and let m be a positive integer less than $n = |I|$.

Adjusting the notation of Section 11.2.2 to the present more general context, we denote by $\mathbf{G}_m$ the family of parabolic subgroups of G of rank m. As in Section 11.2.2, we designate by G_m the amalgamated product of the family $\mathbf{G}_m$, with amalgamation of the parabolic subgroups of rank $< m$. The canonical homomorphism from G_m to G will be denoted by φ_m.

For every $K \subseteq I$ with $|K| \leq m$, the homomorphism φ_m induces on $\varphi_m^{-1}(G^K)$ an isomorphism to G^K. Thus, we identify G^K with the subgroup $\varphi_m^{-1}(G^K)$ of G_m. In particular, $\mathbf{G}_m$ and $\mathcal{G}$ can also be viewed as systems of subgroups of G_m. According to the notation of Section 11.2.2, we write $\mathcal{G}_m$ for $\mathcal{G}$ when we regard $\mathcal{G}$ as a family of subgroups of G_m.

Lemma 12.27 *The system of subgroups $\mathcal{G}_m$ is a parabolic system in G_m and the homomorphism φ_m is a morphism from $\mathcal{G}_m$ to $\mathcal{G}$.*

Proof. According to the identification of parabolic subgroups of G of rank $\leq m$ with subgroups of G_m, we have $G_m = \langle G^J \rangle_{G^J \in \mathbf{G}_m}$. On the other hand, $G^J = \langle G^j \rangle_{j \in J}$. Hence $G_m = \langle G^i \rangle_{i \in I}$. Therefore, $\mathbf{C}_3''$ holds in $\mathcal{G}_m$. The properties $\mathbf{C}_1''$ and $\mathbf{C}_2''$ hold in $\mathcal{G}_m$ because they hold in $\mathcal{G}$. Let us prove that $\mathbf{C}_4''$ also holds in $\mathcal{G}_m$.

Let us write $N = \bigcap_{g \in G_m} gBg^{-1}$, for short (of course, we are regarding B as a subgroup of G_m). Trivially, $N \leq B$. On the other hand, $\varphi_m(N) = 1$ by $\mathbf{C}_4''$ in $\mathcal{G}$. Therefore $N \leq B \cap Ker(\varphi_m)$. However, the restriction of φ to B is injective. Therefore $N = 1$. Hence $\mathbf{C}_4''$ holds in $\mathcal{G}_m$.

Thus, we have proved that $\mathcal{G}_m$ is a parabolic system. Trivially, φ_m is a morphism of parabolic systems from $\mathcal{G}_m$ to $\mathcal{G}$. □

We call $\mathcal{G}_m$ the *m-amalgam* of $\mathcal{G}$ (as in Section 11.2.2).

Let $f_m : \mathcal{C}(\mathcal{G}_m) \longrightarrow \mathcal{C}$ be the epimorphism induced by $\varphi_m : \mathcal{G}_m \longrightarrow \mathcal{G}$ from the chamber system $\mathcal{C}(\mathcal{G}_m)$ defined by $\mathcal{G}_m$ to the chamber system $\mathcal{C} = \mathcal{C}(\mathcal{G})$ (Lemma 12.25). By Lemma 12.26, f_m is an m-cover.

Theorem 12.28 *The m-cover $f_m : \mathcal{C}(\mathcal{G}_m) \longrightarrow \mathcal{C}$ is universal.*

Proof. Let $\hat{f} : \widehat{\mathcal{C}}_m \longrightarrow \mathcal{C}$ be the universal cover of $\mathcal{C}$ (Theorem 12.20). By Theorem 12.13(iii), the group G in which $\mathcal{G}$ is given lifts to a transitive subgroup $\widehat{G}$ of $Aut(\widehat{\mathcal{C}}_m)$ through $\hat{f}$. We have $\widehat{G} \trianglerighteq Aut(\hat{f}) = \pi_m(\mathcal{C})$ (Lemma 12.11 and Theorem 12.13). Let $\hat{c}$ be a chamber of $\widehat{\mathcal{C}}_m$ such that $\hat{f}(\hat{c})$ is the chamber of $\mathcal{C}$ corresponding to the Borel subgroup B of $\mathcal{G}$. Let $\widehat{\mathcal{G}}$ be the parabolic system defined by $\hat{c}$ in $\widehat{G}$. According to Theorem 12.13(ii), we can identify $\mathcal{C}$ with $\widehat{\mathcal{C}}_m/Aut(\hat{f})$. We can also identify G with $\widehat{G}/Aut(\hat{f})$, as $Aut(\hat{f}) \trianglelefteq \widehat{G}$. The canonical projection $\hat{\varphi} : \widehat{G} \longrightarrow G = \widehat{G}/Aut(\hat{f})$ is a morphism from $\widehat{\mathcal{G}}$ to $\mathcal{G}$ and $\hat{f} : \widehat{\mathcal{C}} \longrightarrow \mathcal{C} = \widehat{\mathcal{C}}/Aut(\hat{f})$ is the epimorphism of chamber systems induced by $\hat{\varphi}$.

By Lemma 12.26 and since $\hat{f}$ is an m-cover, the homomorphism $\hat{\varphi}$ induces isomorphisms on each of the parabolic subgroups of $\widehat{\mathcal{G}}$ of rank $\leq m$. Therefore, we can identify each of these parabolic subgroups with the corresponding parabolic subgroups of $\mathcal{G}$, assuming that $\hat{\varphi}$ induces the identity isomorphism on each of them. Thus, the family $\mathbf{G}_m$ of subgroups of G can also viewed as a family of subgroups of $\widehat{G}$.

As G_m is the amalgamated product of the family $\mathbf{G}_m$, there is a unique surjective homomorphism $\varphi : G_m \longrightarrow \widehat{G}$ inducing the identity isomorphism on each of the members of $\mathbf{G}_m$. The homomorphism φ is also a morphism of parabolic systems from $\mathcal{G}_m$ to $\widehat{\mathcal{G}}$ and, by Lemmas 12.25 and 12.26, it induces an m-cover $f_\varphi : \mathcal{C}(\mathcal{G}_m) \longrightarrow \widehat{\mathcal{C}}_m$. The m-cover f_φ is an isomorphism, by Corollaries 12.15 and 12.16. □

Corollary 12.29 *We have* $Ker(\varphi_m) = \pi_m(\mathcal{C})$.

(Easy, by the above theorem and by Theorem 12.13(ii)).

Corollary 12.30 *The chamber system* $\mathcal{C} = \mathcal{C}(\mathcal{G})$ *is* m*-simply connected if and only if the group* G *is the amalgamated product of the family* $\mathbf{G}_m$.

It is worth recalling the construction of G_m, in order to understand the real meaning of Theorem 12.28. We take $X = \bigcup_{i \in I} G^i$ as set of generating symbols. For every $J \subseteq I$ with $|J| = m$, let $\mathbf{R}_J$ be the set of relations holding in G^J and involving only generators taken from $X_J = \bigcup_{j \in J} G^j$. We set

$$\mathbf{R}_m = \bigcup(\mathbf{R}_J \mid J \subseteq I, \ |J| = m).$$

Then G_m is the group presented by the set of relations $\mathbf{R}_m$. Needless to say, the set of relations $\mathbf{R}_m$ is often largely redundant. Nice refinements of it can be found in most cases. If we feed a computer with one of these refined sets of relations and apply coset enumeration, and if we are lucky, then we can concretely recognize the group G_m. Thus, we can check if $\mathcal{C}$ is m-simply connected or not (Corollary 12.30), provided that the group G_m is not too large for our computer.

An Example (Coxeter Complexes). It is now quite clear that Theorem 11.19 is just a special case of Theorem 12.28. We could now state it as follows: (chamber systems of) Coxeter complexes are the universal 2-covers of (chamber systems of) flag-transitive thin geometries. The only information contained in Theorem 11.19 that is missing in Theorem 12.28 is the following: the universal 2-cover of (the chamber system of) a flag-transitive thin geometry is (the chamber system of) a geometry.

Note that the set $\mathbf{M}$ of Coxeter relations of type $M = (m_{ij})_{i,j \in I}$ is a refinement of a larger set of relations, which we call $\mathbf{R}_2$ according to the above, and which contains all relations on generators s_i, s_j that hold in the dihedral group $D_{2m_{ij}}$ (or in $\mathbf{Z}$, when $m_{ij} = \infty$), for all $i, j \in I$ ($i \neq j$).

Revisiting Theorems 11.10 and 11.11. It is clear that the (chamber system of the) geometry $\Gamma(\mathcal{G}_m)$ of Theorem 11.10 is actually the universal m-cover of (the chamber system of) Γ. We assumed the tightness of $\mathcal{G}_m$

in that theorem, because we wanted $\mathcal{G}_m$ to be geometric (only geometric parabolic systems had been defined in Chapter 10).

We now see also that Theorem 11.12 can concisely be stated as follows: the parabolic system $\mathcal{G}_{n-1}$ is geometric. This is actually the group-theoretic translation of a more general result on universal $(n-1)$-covers of chamber systems of geometries (Theorem 12.39 of the next section).

Remark. Theorem 12.28 was proved almost simultaneously in [148] (pages 234–236), [230] (also [229]) and [189].

12.5 Back to geometries

12.5.1 Geometric chamber systems

Let $\mathcal{C}$ be a chamber system of rank $n = 1$ or 2. Then the incidence graph $\Gamma(\mathcal{C})$ of $\mathcal{C}$ is a geometry, by $\mathbf{C}_1$ and $\mathbf{C}_2$, and $\mathcal{C}$ can be identified with the chamber system of this geometry, dualizing chamber-vertex inclusions when $n = 2$.

The same is not true in general, when the chamber system $\mathcal{C}$ has rank $n \geq 3$. Let us spend a few words on this point.

Let γ be the function from the set of cells of $\mathcal{C}$ to the set of cliques of $\Gamma(\mathcal{C})$ mapping every cell X of $\mathcal{C}$ onto the set of vertices containing X. If $\mathcal{C}$ is the chamber system of some geometry Γ, then γ is bijective and $\Gamma(\mathcal{C}) \cong \Gamma$. On the other hand, it is easy to check that, if γ is bijective, then $\Gamma(\mathcal{C})$ is a geometry and $\mathcal{C}$ can be identified with the chamber system of $\Gamma(\mathcal{C})$.

We say that $\mathcal{C}$ is *geometric* if it is the chamber system of a geometry. We can now summarize the above as follow: a chamber system $\mathcal{C}$ is geometric if and only if γ is bijective. All chamber systems of rank 1 or 2 are geometric. However, chamber systems of rank $n \geq 3$ need not be geometric, in general (some examples of this kind will be given in Exercises 12.11–12.14; the reader may see [163] for more examples).

We will now examine some conditions characterizing geometric chamber systems. As in Section 1.5, we set $\Phi_i = \Phi^{I-i}$ and $\Phi_J = \bigcap_{j \in J} \Phi_j$ (for $i \in I$ and $J \subseteq I$, $J \neq \emptyset$). We also set $\Phi_\emptyset = \Phi^I = \Omega$.

Lemma 12.31 *The mapping γ defined above is injective if and only if the following holds:*

($\mathbf{GC}_1$) *we have $\Phi^J = \Phi_{I-J}$ for every $J \subseteq I$.*

Lemma 12.32 *The mapping γ is surjective if and only if the following holds:*

($\mathbf{GC}_2$) *given a subset $J \subseteq I$ and chambers c_j ($j \in J$), if the vertices $[c_j]\Phi_j$ ($j \in J$) pairwise have nonempty intersections, then $\bigcap_{j\in J}[c_j]\Phi_j \neq \emptyset$.*

The (easy) proof of these two lemmas is left for the reader. Gathering them in one statement, we obtain the next proposition:

Proposition 12.33 *A chamber system is geometric if and only if it satisfies both* $\mathbf{GC}_1$ *of Lemma 12.31 and* $\mathbf{GC}_2$ *of Lemma 12.32.*

It is not hard to prove that $\mathbf{GC}_1$ and $\mathbf{GC}_2$ are equivalent to the following conditions $\mathbf{GC}'_1$ and $\mathbf{GC}'_2$, respectively:

($\mathbf{GC}'_1$) we have $\Phi^J \cap \Phi^K = \Phi^{J\cap K}$, for all $J, K \subseteq I$.

($\mathbf{GC}'_2$) given a proper subset J of I and distinct types $i, j \in I - J$, we have $\Phi_J \cap (\Phi_i\Phi_j) = (\Phi_J \cap \Phi_i)(\Phi_J \cap \Phi_j)$.

We leave the proofs of these equivalences for the reader. Proposition 12.33 can now be restated as follows:

Proposition 12.34 *A chamber system $\mathcal{C}$ is geometric if and only if both* $\mathbf{GC}'_1$ *and* $\mathbf{GC}'_2$ *hold in it.*

Note that conditions $\mathbf{P}_2$ and $\mathbf{P}_3$ of Section 10.1.3 are just the group theoretic translations of $\mathbf{GC}_1$ and $\mathbf{GC}'_2$ respectively. Exercise 10.4 of Chapter 10 states the equivalence of $\mathbf{GC}_1$ and $\mathbf{GC}'_1$ in the case of parabolic systems. Lemma 10.11 and Exercise 10.1 state that $\mathbf{GC}_2$ and $\mathbf{GC}'_2$ are equivalent in parabolic systems. However, we now know that these equivalences hold for chamber systems in general. The following is also quite evident:

Corollary 12.35 *The chamber system $\mathcal{C}(\mathcal{G})$ of a parabolic system $\mathcal{G}$ is geometric if and only if $\mathcal{G}$ is geometric.*

If $\mathbf{GC}_1$ (equivalently, $\mathbf{GC}'_1$) is assumed, then $\mathbf{GC}'_2$ can be restated as follows:

($\mathbf{GC}''_2$) given a nonempty subset J of I and distinct types $i, j \in J$, we have $\Phi^J \cap (\Phi_i\Phi_j) = (\Phi^J \cap \Phi_i)(\Phi^J \cap \Phi_j)$.

Given $m = 2, 3, ..., n-1$, we may consider the following weaker versions of $\mathbf{G}'_1$ and $\mathbf{G}''_2$:

($\mathbf{GC}_{1,m}$) we have $\Phi^J \cap \Phi^K = \Phi^{J\cap K}$ for all $J, K \subseteq I$ with $|J| \leq m$;

($\mathbf{GC}_{2,m}$) given $J \subseteq I$ with $2 \leq |J| \leq m$ and distinct types $i, j \in J$, we have $\Phi^J \cap (\Phi_i\Phi_j) = (\Phi^J \cap \Phi_i)(\Phi^J \cap \Phi_j)$.

Trivially, $\mathbf{GC}_1^{(n-1)}$ and $\mathbf{GC}_2^{(n-1)}$ say just the same as $\mathbf{GC}_1'$ and $\mathbf{GC}_2''$ respectively. Therefore

Corollary 12.36 *A chamber system $\mathcal{C}$ of rank $n \geq 3$ is geometric if and only if it satisfies both $\mathbf{GC}_1^{(n-1)}$ and $\mathbf{GC}_2^{(n-1)}$.*

Lemma 12.37 *Let $\mathcal{C} = (C, (\Phi^i)_{i\in I})$ be a chamber system of rank n, let m be a positive integer $< n$ and let $f : \overline{\mathcal{C}} \longrightarrow \mathcal{C}$ be an m-cover. If the properties $\mathbf{GC}_1^{(m)}$ and $\mathbf{CG}_2^{(m)}$ hold in $\mathcal{C}$, then they also hold in $\overline{\mathcal{C}}$.*

Proof. Let $\mathbf{GC}_1^{(m)}$ and $\mathbf{GC}_2^{(m)}$ hold in $\mathcal{C}$. Let $\overline{C}$ and $\overline{\Phi}^i$ $(i \in I)$ be the set of chambers and the i-adjacency relation of $\overline{\mathcal{C}}$.

Let us prove that $\mathbf{GC}_1^{(m)}$ holds in $\overline{\mathcal{C}}$. Let $c \equiv d(\overline{\Phi}^J \cap \overline{\Phi}^K)$, with $c, d \in \overline{C}$, $J, K \subseteq I$ and $|J| \leq m$. Then $f(c) \equiv f(d)(\Phi^J \cap \Phi^K)$. By $\mathbf{GC}_1^{(m)}$ in $\mathcal{C}$, we have $f(c) \equiv f(d)(\Phi^{J\cap K})$. Therefore $c \equiv d(\overline{\Phi}^{J\cap K})$, because f isomorphically maps $[c]\overline{\Phi}^J$ onto $[f(c)]\Phi^J$ (indeed f is an m-cover and $|J| \leq m$), and because $d \in [c]\overline{\Phi}^J$.

By a similar argument we can prove that $\mathbf{GC}_1^{(m)}$ holds in $\overline{\mathcal{C}}$. We leave this proof as an exercise for the reader. □

Corollary 12.38 *Let $\mathcal{C}$ be a geometric chamber system of rank $n \geq 2$, let m be a positive integer $< n$ and let $\widehat{\mathcal{C}}_m$ be the universal m-cover of $\mathcal{C}$. Then both $\mathbf{GC}_1^{(m)}$ and $\mathbf{GC}_2^{(m)}$ hold in $\widehat{\mathcal{C}}_m$.*

Trivial, by the previous lemma. By that lemma and Corollary 12.36 we obtain the following, too:

Theorem 12.39 *The universal $(n-1)$-cover of a geometric chamber system of rank n is geometric.*

(compare with Theorem 11.12). The reader may see [134], [106] and [163] for more conditions sufficent for a chamber system to be geometric. We will only mention one of those conditions in Exercise 12.19, taken from [134].

Diagrams, diagram graphs and orders. All chamber systems of rank 2 are geometric, as we have remarked at the very beginning of this section. Therefore we can define *diagrams* and *diagram graphs* for chamber systems just as for geometries, considering rank 2 cells instead of rank 2 residues.

The diagram graph of a chamber system $\mathcal{C}$ will be denoted by $\mathcal{D}(\mathcal{C})$, according to the notation of Chapter 4. Trivially, diagrams and diagram graphs are preserved by m-covers, when $m \geq 2$.

Orders can also be defined for chamber systems as for geometries. Trivially, orders are also preserved by m-covers.

A chamber system $\mathcal{C}$ is said to be *thin* (respectively, *thick*) if all panels of $\mathcal{C}$ have precisely two chambers (at least three chambers, respectively).

Exercise 12.4. Prove that every 1-simply connected chamber system is geometric.

(Hint. Use Exercise 12.1.)

Exercise 12.5 Prove that all cells of rank 2 of a 1-simply connected chamber system of rank $n \geq 2$ are (chamber systems of) generalized ∞-gons.

(Compare with Exercise 12.3).

Exercise 12.6 Prove that the incidence graph of a 1-simply connected chamber system of rank $n \geq 2$ is an infinite tree with all vertices of valency at least $2(n-1)$.

Exercise 12.7 Prove that the adjacency graph of a thin 1-simply connected chamber system of rank n is a tree with all vertices of valency n.

Exercise 12.8 Let $(q_i)_{i \in I}$ be a finite family of non-null cardinal numbers. Prove that all chamber systems with set of types I and order $(q_i)_{i \in I}$ have the same universal 1-cover.

Exercise 12.9. Let $\mathcal{C}$ be a chamber system of rank $n \geq 3$ and let X be a vertex of $\Gamma(\mathcal{C})$. Prove that the neighbourhood of X in $\Gamma(\mathcal{C})$ is connected.

Exercise 12.10. Let $\mathcal{C}$ be a chamber system of rank 3. Prove that $\Gamma(\mathcal{C})$ is a geometry if and only if, for any two incident vertices X, Y of $\mathcal{C}$, the neighbourhood of $\{X, Y\}$ in $\Gamma(\mathcal{C})$ contains at least two vertices.

Exercise 12.11. Let G_1, G_2, G_3 be as in Exercise 10.24 and define $G^i = G_j \cap C_k$, for $\{i,j,k\} = \{1,2,3\}$. Prove that (G^1, G^2, G^3) is a parabolic system in $L_3(2)$, in the meaning of Section 12.4. The chamber system $\mathcal{C}$ defined by this parabolic system is non-geometric (see Exercise 10.24). Check that, nevertheless, $\Gamma(\mathcal{C})$ is a geometry.

(Remark. This example was discovered by Ronan [177].)

Exercise 12.12. Let G^1, G^2, G^3 be three distinct copies of $\mathbf{3}$ in $Frob_7^3$. Prove that (G^1, G^2, G^3) is a non-geometric parabolic system. Let $\mathcal{C}$ be the (non-geometric) chamber system defined by this parabolic system. Check that $\Gamma(\mathcal{C})$ is the complete graph on three vertices (hence it is not a geometry).

(Remark. This example is also due to Ronan [177].)

Exercise 12.13. Let G^1, G^2 be two distinct copies of $\mathbf{3}$ in $Frob_7^3$ and let G^3 be the Frobenius kernel $\mathbf{7}$ of $Frob_7^3$. Prove that (G^1, G^2, G^3) is a non-geometric parabolic system. Let $\mathcal{C}$ be the (non-geometric) chamber system defined by this parabolic system. Check that $\Gamma(\mathcal{C})$ is the complete graph on three vertices (hence it is not a geometry).

Exercise 12.14. Let G^1, G^2, G^3 be the three subgroups of order 2 in $\mathbf{2}^2$. Prove that (G^1, G^2, G^3) is a non-geometric parabolic system. Let $\mathcal{C}$ be the (non-geometric) chamber system defined by this parabolic system. Check that $\Gamma(\mathcal{C})$ is the complete graph on three vertices (hence it is not a geometry).

Exercise 12.15. Prove that the universal 2-cover of the chamber system of Exercise 12.13 is (the chamber system of) the direct sum of $PG(2,2)$ and of the geometry of rank 1 with seven elements (hence, it is geometric).

Exercise 12.16. Prove that the universal 2-cover of the chamber system of Exercise 12.14 is (the chamber system of) the direct sum of three copies of the geometry of rank 1 with two elements (hence it is geometric).

(Remark. The universal 2-covers of the chamber systems of Exercises 12.12 and 12.13 are geometric, too, by a result of Tits ([227], Corollary 3). We shall say something on this matter in the next chapter.)

Exercise 12.17. Prove that each of the following is equivalent to $\mathbf{GC}_2$:

(i) we have $\Phi_J(\Phi_i \cap \Phi_j) = (\Phi_J\Phi_i) \cap (\Phi_J\Phi_j)$ for every proper subset J of I and for any two distinct types $i, j \in I - J$;
(ii) we have $\Phi_J \cap (\Phi_K\Phi_H) = (\Phi_J \cap \Phi_K)(\Phi_J \cap \Phi_H)$ for every choice of subsets J, K, H of I;
(iii) we have $\Phi_J(\Phi_K \cap \Phi_H) = (\Phi_J\Phi_K) \cap (\Phi_J\Phi_H)$ for every choice of subsets J, K, H of I.

(Remark. Both $\mathbf{P}_{3,1}$ and $\mathbf{P}_{3,2}$ of Exercise 10.2 are group-theoretic paraphrases of (ii) above. $\mathbf{P}_{3,3}$ of Exercise 10.2 corresponds to (iii).)

Exercise 12.18. Let J, K be two nonempty disjoint subsets of the set of types I of a chamber system $\mathcal{C}$, such that $J \cup K = I$. Prove that the subsets J and K are not connected by any edge in the diagram graph $\mathcal{D}(\mathcal{C})$ of $\mathcal{C}$ if and only if $\Phi^j\Phi^k = \Phi^k\Phi^j$ for all $j \in J$, $k \in K$.

(Compare with Exercise 10.13.)

Exercise 12.19 (Meixner and Timmesfeld [134]). Let $\mathcal{D}(\mathcal{C})$ be a string or a disjoint union of strings and/or isolated vertices. Prove that $\mathcal{C}$ is geometric if and only if $\mathbf{GC}_1$ holds in it.

(Compare with Exercises 10.14 and 10.15.)

Exercise 12.20. Check that the chamber systems described in Exercises 12.11 and 12.12 belong to the Coxeter diagram $\widetilde{A}_2$.

Exercise 12.21. Check that the chamber systems of Exercises 12.13 and 12.14 belong to the following reducible Coxeter diagrams:

•———• • (Exercise 12.13)

• • • (Exercise 12.14)

12.5.2 Direct products and reducibility

Given a partition $\{I_h\}_{h=1}^r$ of a finite nonempty set I, let $\mathcal{C}_h = (C_h, (\Phi_h^i)_{i \in I_h})$ be a chamber system over the set of types I_h, for $h = 1, 2, ..., r$. We set $C = \prod_{h=1}^r C_h$ and, for every $h \in I$ and every $i \in I_h$, we define an equivalence relation Φ^i on C by the following clause: given elements $c = (c_h)_{h=1}^r$ and $d = (d_h)_{h=1}^r$ of C, we have $c \equiv d(\Phi^i)$ if $c_h \equiv d_h(\Phi_h^i)$ and $c_k = d_k$ for all $k = 1, 2, ..., h-1, h+1, ..., r$.

It is easily seen that $(C, (\Phi^i)_{i \in I})$ is a chamber system. We call it the *direct product* of the chamber systems $\mathcal{C}_1$, $\mathcal{C}_2$, ..., $\mathcal{C}_r$ and we denote it by $\mathcal{C}_1 \times \mathcal{C}_2 \times ... \times \mathcal{C}_r$; also by $\prod_{h=1}^r \mathcal{C}_h$, for short.

Trivially, the adjacency graph of $\prod_{h=1}^r \mathcal{C}_h$ is the product of the adjacency graphs of the factors $\mathcal{C}_1$, $\mathcal{C}_2$, ..., $\mathcal{C}_r$, in the sense of Exercise 4.1. It is also easily seen that, for every $k = 1, 2, ..., r$, the factor $\mathcal{C}_k$ of $\mathcal{C} = \prod_{h=1}^r \mathcal{C}_h$ is isomorphic to every cell of $\mathcal{C}$ of type I_k. The usual commutative and associative properties hold for direct products of chamber systems:

$$(\prod_{h=1}^{k} \mathcal{C}_h) \times (\prod_{h=k+1}^{r} \mathcal{C}_h) \cong \prod_{h=1}^{r} \mathcal{C}_h$$

for all $k = 1, 2, ..., r-1$, and

$$\prod_{h=1}^{r} \mathcal{C}_h \cong \prod_{h=1}^{r} \mathcal{C}_{p(h)}$$

for every permutation p of $\{1, 2, ..., r\}$.

Trivially, given geometries Γ_1, Γ_2, ..., Γ_r, we have

$$\prod_{h=1}^{r} \mathcal{C}(\Gamma_h) = \mathcal{C}(\bigoplus_{h=1}^{r} \Gamma_h)$$

Conversely, it is not difficult to prove that, if $\prod_{h=1}^{r} \mathcal{C}_h$ is geometric, then all factors $\mathcal{C}_h$ $(h = 1, 2, ..., r)$ are geometric. Therefore:

Proposition 12.40 *A direct product $\prod_{h=1}^{r} \mathcal{C}_h$ is geometric if and only if all of its factors are geometric.*

Given a chamber system $\mathcal{C} = (C, (\Phi^i)_{i \in I})$, let J be a proper subset of I such that both the following hold:

($\mathbf{T}_1$) we have $[c]\Phi^J \neq [c]\Phi^{J \cup i}$ for every $c \in C$ and for every $i \in I - J$;

($\mathbf{T}_2$) we have $\Phi^{J \cup i} \cap \Phi^{J \cup j} = \Phi^J$ for any two distinct types $i, j \in I - J$.

Then $(C/\Phi^J, (\Phi^{J \cup i}/\Phi^J)_{i \in I - J})$ is a chamber system with set of types $I - J$. We call it the *J-truncation* of $\mathcal{C}$, denoting it by $Tr_J(\mathcal{C})$. We say that $\mathcal{C}$ *admits truncation* of a proper subset J of I if both $\mathbf{T}_1$ and $\mathbf{T}_2$ hold for J.

Trivially, if $\mathcal{C} = \mathcal{C}(\Gamma)$ for some geometry Γ, then $\mathbf{T}_1$ and $\mathbf{T}_2$ hold for every proper subset J of I and we have $Tr_J(\mathcal{C}) = \mathcal{C}(Tr_{\bar{J}}^{-}(\Gamma))$.

It is also easily seen that, if $\{I_h\}_{h=1}^{r}$ is a partition of I in $r \geq 2$ subsets and $\mathcal{C} = \prod_{h=1}^{r} \mathcal{C}_h$ for chamber systems $\mathcal{C}_h = (C_h, (\Phi^i)_{i \in I_h})$ over the sets of types I_h $(h = 1, 2, ..., r)$, then $\mathcal{C}$ admits truncation of $I - I_h$ for every $h = 1, 2, ..., r$, and we have $\mathcal{C}_h \cong Tr_{I - I_h}(\mathcal{C})$.

A chamber system $\mathcal{C}$ over a set of types I is said to be *reducible* over a partition $\{I_h\}_{h=1}^{r}$ of I (with $r \geq 2$) if it admits truncation of $I - I_h$ for every $h = 1, 2, ..., r$ and if the function mapping every chamber $c \in C$ onto the sequence $([c]\Phi^{I - I_h})_{h=1}^{r}$ is an isomorphism of $\mathcal{C}$ onto $\prod_{h=1}^{r} Tr_{I - I_h}(\mathcal{C})$. In short, a chamber system is reducible if and only if it is the direct product of at least two chamber systems. If $\mathcal{C}$ is not reducible over any partition of its set of types, then we say that it is *irreducible*.

Lemma 12.41 *Let $\mathcal{C} = (C, (\Phi_i)_{i \in I})$ be a chamber system and let Π be a partition of I in at least two classes. The chamber system $\mathcal{C}$ is reducible over Π if and only if both the following hold:*

($\mathbf{R}_1$) *we have $\Phi^J \cap \Phi^{I-J} = \mho$ for every class $J \in \Pi$;*
($\mathbf{R}_2$) *we have $\Phi^i \Phi^j = \Phi^j \Phi^i$ for any two types $i, j \in I$ belonging to distinct classes of the partition Π.*

Proof. The 'only if' part is trivial. The first step in the proof of the 'if' part consists of showing that $\mathbf{R}_1$ and $\mathbf{R}_2$ imply $\mathbf{T}_1$ and $\mathbf{T}_2$ for $I - J$, for every class $J \in \Pi$.

Let $\mathbf{R}_1$ and $\mathbf{R}_2$ hold. Let us prove that $\mathbf{T}_1$ holds for every $J \in \Pi$. Given $J \in \Pi$ and $i \in J$, let $[c](\Phi^{(I-J) \cup i}) = [c]\Phi^{I-J}$ for some chamber c, if possible. Then we have $[c]\Phi^i \subseteq [c]\Phi^{I-J}$. Hence $[c]\Phi^i = c$, by $\mathbf{R}_1$. This contradicts $\mathbf{C}_1$ of Section 12.3.1. Therefore $\mathbf{T}_1$ must hold.

Turning to the proof of $\mathbf{T}_2$, let $J \in \Pi$ and let $i, j \in J$, with $i \neq j$. Let $c, d \in C$ correspond in $\Phi^{(I-J)\cup i} \cap \Phi^{(I-J)\cup j}$. By $\mathbf{R}_2$, there are chambers u, v such that $c \equiv u(\Phi^{I-J})$, $c \equiv v(\Phi^{I-J})$, $u \equiv d(\Phi^i)$ and $v \equiv d(\Phi^j)$. Therefore $u \equiv v(\Phi^{I-J} \cap (\Phi^i\Phi^j))$. We have $u = v$ by $\mathbf{R}_1$. Hence $u \equiv d(\Phi^i \cap \Phi^j)$. We obtain $u = d$ by $\mathbf{C}_2$ of Section 12.3.1. Therefore $c \equiv d(\Phi^{I-J})$. The property $\mathbf{T}_2$ is proved. Hence $\mathcal{C}$ admits truncation of $I - J$, for every $J \in \Pi$.

To finish the proof, we must show that the function f, mapping every chamber $c \in C$ onto the chamber $([c]\Phi^{I-J})_{J\in\Pi}$ of $\prod_{J\in\Pi} Tr_{I-J}(\mathcal{C})$ is a bijection. We prove this by induction on the number r of classes of Π. Let J_1, J_2, ..., J_r be the classes of Π. If $r \geq 3$, then we set $J_{1,2} = J_1 \cup J_2$.

To prove that f is injective we must show that $\bigcap_{h=1}^r \Phi^{I-J_h} = \mho$. If $r = 2$, then the statement we should prove is nothing but $\mathbf{R}_1$, which holds by assumption. Let $r \geq 3$ and let $c \equiv d\ (\bigcap_{h=1}^r \Phi^{I-J_h})$, for $c, d \in C$. Hence $c \equiv d\ (\Phi^{I-J_1} \cap \Phi^{I-J_2} \cap \bigcap_{h=3}^r \Phi^{I-J_h})$.

We have $\Phi^{I-J_1} = \Phi^{J_2}\Phi^{I-J_{1,2}}$ and $\Phi^{I-J_2} = \Phi^{J_1}\Phi^{I-J_{1,2}}$ by $\mathbf{R}_2$. Therefore there are chambers $u, v \in C$ such that $c \equiv u(\Phi^{J_1})$, $c \equiv u(\Phi^{J_2})$, $u \equiv d(\Phi^{I-J_{1,2}})$ and $v \equiv d(\Phi^{I-J_{1,2}})$. Applying $\mathbf{R}_1$ to J_1, c and u and to J_2, c and v, we obtain $c = u$ and $c = v$. Hence $c \equiv d(\Phi^{I-K})$. We can now apply the inductive hypothesis to the partition $\{J_{1,2}, J_3, ..., J_r\}$ of I (trivially, $\mathbf{R}_1$ and $\mathbf{R}_2$ remain true for this partition). We obtain $c = d$.

The injectivity of f is proved. To prove that f is surjective, we must show that $\bigcap_{h=1}^r [c_h]\Phi^{I-J_h} \neq \emptyset$ for every choice of chambers c_1, c_2, ..., c_r of $\mathcal{C}$.

When $r = 2$, we have $I - J_1 = J_2$, $I - J_2 = J_1$, and $\Phi^{J_1}\Phi^{J_2} = \Omega$ by $\mathbf{R}_2$ and $\mathbf{C}_3$. Hence $[c_1]\Phi^{I-J_1} \cap [c_2]\Phi^{I-J_2} \neq \emptyset$ for any two chambers $c_1, c_2 \in C$.

Let $r \geq 3$. Again $\Phi^{I-J_1}\Phi^{I-J_2} = \Omega$, by $\mathbf{R}_2$ and $\mathbf{C}_3$. Therefore there is a chamber $d \in C$ such that $c_1 \equiv d\ (\Phi^{I-J_1})$ and $d \equiv c_2\ (\Phi^{I-J_2})$. We have $[d]\Phi^{I-J_{1,2}} \subseteq [d]\Phi^{I-J_1} \cap [d]\Phi^{I-J_2} = [c_1]\Phi^{I-J_1} \cap [c_2]\Phi^{I-J_2}$.

By the inductive hypothesis applied to the partition $\{J_{1,2}, J_3, ..., J_r\}$ and to the sequence of chambers $(d, c_3, ..., c_r)$ we now obtain the following: $[d]\Phi^{I-J_{1,2}} \cap (\bigcap_{h=3}^r [c_h]\Phi^{I-J_h}) \neq \emptyset$.

Hence $\bigcap_{h=1}^r [c_h]\Phi^{I-J_h} \neq \emptyset$, as we wanted to prove. Therefore f is surjective. □

Trivially, if $\mathcal{C}$ is reducible over a partition Π of its set of types, then the connected components of the diagram graph $\mathcal{D}(\mathcal{C})$ of $\mathcal{C}$ form a (possibly improper) refinement of Π. Unfortunately, the converse need not hold in general (that is, the direct sum theorem fails to hold for arbitrary chamber system): it may happen that $\mathcal{D}(\mathcal{C})$ is disconnected and, nevertheless, $\mathcal{C}$

is irreducible. Examples of this kind have been given in Exercises 12.13 and 12.14 (see also Exercise 12.21). They are transitive and finite, but not 2-simply connected (see Exercises 12.5 and 12.16). Infinite 2-simply connected examples are mentioned by Tits ([227], Section 6.1.6(b)).

In general, if $\mathcal{D}(\mathcal{C})$ is disconnected and Π is the partition of $\mathcal{D}(\mathcal{C})$ in connected components, then we can only prove that $\mathbf{R}_2$ holds in Π (see Exercise 12.18). Thus, if $\widehat{\mathcal{C}}_m$ is the m-universal cover ($m \geq 2$) of a direct product $\mathcal{C}$ of irreducible chamber systems, we cannot appeal to its diagram graph to claim that $\widehat{\mathcal{C}}_m$ is the direct product of the universal m-covers of the factors of $\mathcal{C}$. Anyway, that claim is true, as we will see in a few lines.

Let $\mathcal{C}_1, \mathcal{C}_2, ..., \mathcal{C}_r$ be chamber systems of rank $n_1, n_2, ..., n_r$ respectively over pairwise disjoint sets of types $I_1, I_2, ..., I_r$ and $m = 2, 1, ...$ or $n-1$, where $n = \sum_{h=1}^{r}$ is the rank of $\prod_{h=1}^{r} \mathcal{C}_h$. Given $h = 1, 2, ..., r$, if $n_h > m$ then $f_h : \widehat{(\mathcal{C}_h)}_m \longrightarrow \mathcal{C}_h$ will designate the universal m-cover of $\mathcal{C}_h$. Otherwise, we set $\widehat{(\mathcal{C}_h)}_m = \mathcal{C}_h$ and f_h will be the identity automorphism of $\mathcal{C}_h$. For $h = 1, 2, ..., r$, we denote the set of chambers of $\mathcal{C}_h$ and $\widehat{(\mathcal{C}_h)}_m$ by C_h and $\widehat{C}_h$, respectively. Thus, $\prod_{h=1}^{r} C_h$ and $\prod_{h=1}^{r} \widehat{C}_h$ are the sets of chambers of $\prod_{h=1}^{r} \mathcal{C}_h$ and $\prod_{h=1}^{r} \widehat{(\mathcal{C}_h)}_m$.

Let $\prod_{h=1}^{r} f_h : \prod_{h=1}^{r} \widehat{C}_h \longrightarrow \prod_{h=1}^{r} C_h$ be the function mapping every element $(c_h)_{h=1}^{r} \in \prod_{h=1}^{r} \widehat{C}_h$ onto $(f_h(c_h))_{h=1}^{r} \in \prod_{h=1}^{r} C_h$. It is easy to check that $\prod_{h=1}^{r} f_h$ is an m-cover from $\prod_{h=1}^{r} \widehat{(\mathcal{C}_h)}_m$ to $\prod_{h=1}^{r} \mathcal{C}_h$.

Theorem 12.42 *The m-cover $\prod_{h=1}^{r} f_h$ is universal.*

Proof. Let us state the following notation:

$$I = \bigcup_{h=1}^{r} I_h, \qquad \Pi = \{I_h\}_{h=1}^{m}, \qquad f = \prod_{h=1}^{r} f_h$$

$$\overline{\mathcal{C}} = \prod_{h=1}^{r} \widehat{(\mathcal{C}_h)}_m, \qquad \overline{C} = \prod_{h=1}^{r} \widehat{C}_h$$

$$\mathcal{C} = \prod_{h=1}^{r} \mathcal{C}_h, \qquad C = \prod_{h=1}^{r} C_h$$

Let $\overline{\Phi}^i$ and Φ^i denote the i-adjacency relations of $\overline{\mathcal{C}}$ and $\mathcal{C}$ respectively, for $i \in I$. Let $\widehat{\mathcal{C}} = (\widehat{C}, (\widehat{\Phi}^i)_{i \in I})$ be the universal m-cover of $\mathcal{C}$.

As f is an m-cover, there is an m-cover $g : \widehat{\mathcal{C}} \longrightarrow \overline{\mathcal{C}}$ such that the m-cover fg is universal. We must prove that g is an isomorphism. According to the definition of f and the universality of fg, proving that g is an isomorphism is the same as proving that $\widehat{\mathcal{C}}$ is reducible over the partition Π of I. Therefore, all we have to do is to check that $\mathbf{R}_1$ and $\mathbf{R}_2$ of Lemma 12.41 hold in $\widehat{\mathcal{C}}$ with respect to Π.

Since $m \geq 2$, the chamber system $\widehat{\mathcal{C}}$ has the same diagram graph as $\mathcal{C}$. Hence $\mathbf{R}_2$ holds in $\widehat{\mathcal{C}}$ (see also Exercise 12.18). Let us prove that $\mathbf{R}_1$

holds, too. Let $c \equiv d\ (\widehat{\Phi}^{I_h} \cap \widehat{\Phi}^{I-I_h})$ for $c, d \in \widehat{\mathcal{C}}$ and $h = 1, 2, \ldots$ or r. Hence $g(c) \equiv g(d)\ (\overline{\Phi}^{I_h} \cap \overline{\Phi}^{I-I_h})$ in $\overline{\mathcal{C}}$. However, $\mathbf{R}_1$ holds in $\overline{\mathcal{C}}$, by Lemma 12.41. Therefore $g(c) = g(d)$. On the other hand, g induces an isomorphism from $[c]\widehat{\Phi}^{I_h} = [d]\widehat{\Phi}^{I_h}$ to $[g(c)]\overline{\Phi}^{I_h}$ because $[g(c)]\overline{\Phi}^{I_h} \cong \widehat{(\mathcal{C}_h)}_m$ is m-simply connected. Hence $c = d$. □

We have assumed the restriction $m \geq 2$ in the above. This is essential for the statement of Theorem 12.42. Indeed, the statement would be false when $m = 1$ (see Exercise 12.5).

The following are trivial consequences of Theorem 12.42:

Corollary 12.43 *Let $\mathcal{C}_1, \mathcal{C}_2, \ldots, \mathcal{C}_r$ be chamber systems over pairwise disjoint sets of types, of rank $n_1, n_2, \ldots, n_r$ respectively. Let $m = 2, 3, \ldots$ or $n-1$, where $n = \sum_{h=1}^{r} n_h$ and let us assume that, for every $h = 1, 2, \ldots, m$ such that $n_h > m$, the chamber system $\mathcal{C}_h$ is m-simply connected. Then $\prod_{h=1}^{m} \mathcal{C}_h$ is m-simply connected.*

Corollary 12.44 *Let Γ be a geometry of rank $n \geq 3$ with disconnected diagram graph and let $\Gamma_1, \Gamma_2, \ldots, \Gamma_r$ be the irreducible components of Γ. Let $m = 2, 3, \ldots$ or $n-1$. Given $h = 1, 2, \ldots, r$, if Γ_h has rank $> m$ then we denote the universal m-cover of $\mathcal{C}(\Gamma_h)$ by $\mathcal{C}_h$; otherwise, we set $\mathcal{C}_h = \mathcal{C}(\Gamma_h)$. Then $\prod_{h=1}^{r} \mathcal{C}_h$ is the universal m-cover of $\mathcal{C}(\Gamma)$.*

Exercise 12.22. Let $\mathcal{C}$ be a chamber system of rank $n \geq 3$, with set of types I. Prove that $\mathcal{C}$ is geometric if and only if, given any two subsets J, K of I with $J \subseteq K$ and $|J| = 3$, every cell X of $\mathcal{C}$ of type K admits truncation of $K - J$ and $Tr_{K-J}(X)$ is geometric.

Exercise 12.23. Let $\mathcal{C}$ be a chamber system over a set of types I, admitting truncation of a proper subset J of I. Let m be a positive integer less that $|I - J|$. Prove that, if $\mathcal{C}$ is m-simply connected, then $Tr_J(\mathcal{C})$ is m-simply connected.

(Hint. Let C be the set of chambers of $\mathcal{C}$. The natural projection of C onto C/Φ^J maps the set of closed paths of the m-cell complex $\mathcal{C}_m$ of $\mathcal{C}$ onto the set of closed paths of the m-cell complex $(Tr_J(\mathcal{C}))_m$ of $Tr_J(\mathcal{C})$ and the plates of $\mathcal{C}_m$ into plates of $(Tr_J(\mathcal{C}))_m$ ...)

Exercise 12.24. Keeping the notation and the hypotheses of the previous exercise, assume furthermore that, for every $K \subseteq I$ with $K \supset J$ and $|K| = |J| + m$, every cell of $\mathcal{C}$ of type K is m-simply connected. Prove that the universal m-cover $\widehat{\mathcal{C}}_m$ of $\mathcal{C}$ admits truncation of J and that $Tr_J(\widehat{\mathcal{C}}_m)$ is the universal m-cover of $Tr_J(\mathcal{C})$.

Exercise 12.25. Prove that the statement of the direct sum theorem holds in the class of chamber systems satisfying $\mathbf{GC}_1$.

Exercise 12.26. Find a category where direct products of chamber systems can be viewed as product objects.

(Warning: we are not claiming that direct products of chamber systems are product objects in the category of chamber systems and their epimorphisms, as defined in Sections 12.3.1 and 12.3.2. This is evidently false. The reader should look for a category a bit larger than that, with more objects or more morphisms. There are a number of different ways to solve this problem, and the reader may choose.)

12.5.3 Universal covers of geometries

Given a geometry Γ of rank $n \geq 2$ and a positive integer m less than n, we denote by $\widehat{\mathcal{C}}_m(\Gamma)$ the universal m-cover of the chamber system $\mathcal{C}(\Gamma)$ of Γ.

We cannot prove that $\widehat{\mathcal{C}}_m(\Gamma)$ is always geometric for every m (although no counterexample has been discovered till now). However, Corollary 12.38 gives us some information. It says that, the larger m is, the closer $\widehat{\mathcal{C}}_m(\Gamma)$ is to be geometric. In particular, $\widehat{\mathcal{C}}_{n-1}(\Gamma)$ is geometric (Theorem 12.39).

However, Corollary 12.38 does not say so much when m is small compared with n (for instance, when $m = 2$ and $n \geq 4$). Actually, Corollary 12.38 is not the only tool we could use for a proof that $\widehat{\mathcal{C}}_m(\Gamma)$ is geometric. For instance, the case of $m = 1$ can be settled without using that corollary (see Exercise 12.4); unfortunately, this case is not really interesting (see below).

When $\widehat{\mathcal{C}}_m(\Gamma)$ is geometric, then the geometry $\Gamma(\widehat{\mathcal{C}}_m(\Gamma))$ defined by $\widehat{\mathcal{C}}_m(\Gamma)$ is the universal m-cover of Γ, as defined in Section 12.1. We denote it by $\widehat{\Gamma}_m$.

The m-homotopy group of $\mathcal{C}(\Gamma)$ is called the *m-homotopy group* of Γ. We denote it by $\pi_m(\Gamma)$. If $\widehat{\mathcal{C}}_m(\Gamma) \cong \mathcal{C}(\Gamma)$ (that is, $\widehat{\mathcal{C}}_m(\Gamma)$ is geometric and $\widehat{\Gamma}_m \cong \Gamma$), then we say that Γ is *m-simply connected.* By Corollary 12.15, a geometry Γ is m-simply connected if and only if $\pi_m(\Gamma) = 1$.

The case of $m = n - 1$. The universal $(n-1)$-cover $\widehat{\mathcal{C}}_{n-1}(\Gamma)$ of $\mathcal{C}(\Gamma)$ is always geometric, as we have remarked above. That is, every geometry of rank n admits a universal $(n-1)$-cover $\widehat{\Gamma}_{n-1}$, in the sense of Section 12.1.

We will write $\widehat{\Gamma}$ and $\pi(\Gamma)$ for $\widehat{\Gamma}_{n-1}$ and $\pi_{n-1}(\Gamma)$ respectively, for short. We also write 'universal topological cover', 'topologically simply connected' and 'topological homotopy group' for 'universal $(n-1)$-cover', '$(n-1)$-simply connected' and '$(n-1)$-homotopy group', according to the conventions of Section 8. Section 12.6 will give us more motivation for this terminology.

The case of $m = 2$. This is the most interesting case. It is quite naturally related to the spirit of diagram geometry: the universal 2-cover $\widehat{\Gamma}_2$ of a geometry Γ (if it exists) is the largest geometry with the same local structure (the same rank 2 residues) as Γ. Thus, if we want to classify all geometries described by a diagram $\mathcal{D}$ and, possibly, by some additional 'local' properties, we can first try to classify all 2-simply connected chamber systems with that diagram and those properties, hoping to find that all those chamber systems are geometric. Then we should get control over their 2-quotients (see Section 12.5.5).

Unfortunately, we do not know if $\widehat{\mathcal{C}}_2(\Gamma)$ is always geometric when $n \geq 4$ (when $n = 3$, then 2-covers are topological covers; hence $\mathcal{C}_2(\Gamma)$ is geometric in this case). However, the devil is never so black as he is painted. When we consider an 'interesting' geometry Γ of rank $n \geq 4$, it often happens that all residues of elements of Γ are 2-simply connected. In that case, all 2-covers of Γ are forced to be topological covers (compare with Exercise 12.3). Hence $\widehat{\mathcal{C}}_2(\Gamma)$ is geometric. We will see many examples of this kind in Section 12.5.4. It may also happen that some truncations of Γ of rank 3 satisfy the hypotheses of Exercise 12.24, with $m = 2$. In that case, we can switch to topological covers of those rank 3 truncations, avoiding the obstacle in this way. The reader may see [149] for some examples where this trick is used.

The case of $m = 1$. The universal 1-cover $\widehat{\mathcal{C}}_1(\Gamma)$ of $\mathcal{C}(\Gamma)$ is always geometric (Exercise 12.4). However, there is no resemblance between Γ and $\Gamma(\widehat{\mathcal{C}}_1(\Gamma)$, in general; the only features of Γ that we can still recognize in $\widehat{\mathcal{C}}_1(\Gamma))$ are the sizes of the panels of Γ (see Exercises 12.5–12.8), which is of no great consequence. Thus, 1-covers do not seem to be very interesting in diagram geometry (but they are interesting in other contexts, of course).

A concise statement for the above could the following one: geometries of rank $n \geq 2$ admit too many 1-covers, in general. However, in spite of this, proper 1-covers may become quite rare if we prescribe a diagram to be preserved by them (see Section 8.4).

Remark. Many authors call $\widehat{\mathcal{C}}_m(\Gamma)$ the universal m-cover of Γ, whether or not it is geometric. I have preferred to avoid this terminology, as it might lead us to think that the existence problem for universal covers of geometries is completely solved in general, which is not true.

Problem. Prove that universal m-covers of geometric chamber systems of rank $n \geq m+2$ are always geometric, for every integer $m \geq 2$. That is, prove that every geometry of rank $n \geq m+2$ admits a universal m cover, for every $m \geq 2$.

The following is a subproblem of the above. Prove that, if a geometry Γ admits a universal m-cover $\overline{\Gamma}$, then the chamber system $\mathcal{C}(\overline{\Gamma})$ of $\overline{\Gamma}$ is the universal m-cover of $\mathcal{C}(\Gamma)$.

Exercise 12.27. Given a positive integer k, let Γ be a geometry of rank $n \geq k+2$, with set of types I and set of elements V. Let $F_k^+(\Gamma)$ be the graph defined as follows: the flags of Γ of rank k are the vertices of $F_k^+(\Gamma)$ and two flags F, G of rank k are adjacent in $F_k^+(\Gamma)$ if $F \cup G$ is a flag of rank $k+1$. The subsets of I of size k (of size $k+1$) can be taken as types for the vertices (the edges) of this graph. For every $x \in V$, we denote by $F_k^+(x)$ the graph induced by $F_k^+(\Gamma)$ on the set of flags of Γ_x of rank k. Let $\mathcal{F}_k^+(\Gamma)$ be the plate complex with underlying graph $F_k^+(\Gamma)$ and set of plates $\{F_k^+(x)\}_{x \in V}$.

Let $\widehat{\Gamma}$ be the universal topological cover of Γ. Prove that $\mathcal{F}_k^+(\widehat{\Gamma})$ is the universal cover of $\mathcal{F}_k^+(\Gamma)$.

(Remark: the graph $F_1^+(\Gamma)$ is just the incidence graph of Γ.)

Exercise 12.28. Let Γ, I, V be as in the previous exercise. Let $F_k^-(\Gamma)$ be the graph defined as follows: the flags of Γ of rank $k+1$ are the vertices of $F_k^-(\Gamma)$ and two flags F, G of rank $k+1$ are adjacent in $F_k^-(\Gamma)$ if $F \cap G$ is a flag of rank k. For every $x \in V$, let $F_k^-(x)$ be the graph induced by $F_k^-(\Gamma)$ on the set of flags of Γ_x of rank $k+1$. A plate complex $\mathcal{F}_k^-(\Gamma)$ can be defined as in the previous exercise, taking $F_k^-(\Gamma)$ as the underlying graph and $\{F_k^+(x)\}_{x \in V}$ as set of plates.

Let $\widehat{\Gamma}$ be the universal topological cover of Γ. Prove that $\mathcal{F}_k^-(\widehat{\Gamma})$ is the universal cover of $\mathcal{F}_k^-(\Gamma)$.

(Remark: the graph $F_1^-(\Gamma)$ is the dual of the incidence graph of Γ; if we allowed $n = k+1$, then $F_{n-1}^-(\Gamma) = \mathcal{C}(\Gamma)$.)

Exercise 12.29. Keep the notation of the previous exercise, but now allow $n = k+1$. Given $x \in V$, let $F_k^-[x]$ be the graph induced by $F_k^-(\Gamma)$ on the set of flags of rank $k+1$ containing x. Define a plate complex $\mathcal{F}_k^-[\Gamma]$ taking $F_k^-(\Gamma)$ as the underlying graph and $\{F_k^-[x]\}_{x \in V}$ as set of plates.

Let $\widehat{\Gamma}$ be the universal topological cover of Γ. Prove that $\mathcal{F}_k^-[\widehat{\Gamma}]$ is the universal cover of $\mathcal{F}_k^-[\Gamma]$.

(Remark: the plate complex $\mathcal{F}_{n-1}^-[\Gamma]$ is the $(n-1)$-cell complex of Γ.)

Exercise 12.30. Let $\mathcal{T}$ be a tessellation of a connected (possibly non-compact) topological surface $\mathcal{S}$ without borders and with no singularities. Prove that $\mathcal{T}$ is topologically simply connected in the meaning of this chapter (when $\mathcal{T}$ is viewed as a rank 3 geometry) if and only if the surface $\mathcal{S}$ is simply connected (in the usual topological meaning); that is, if and only if $\mathcal{S}$ is either the sphere or the affine plane.

Exercise 12.31. Let $\mathcal{T}$ be as above and assume furthermore that (the rank 3 geometry) $\mathcal{T}$ belongs to a Coxeter diagram. Prove that $\mathcal{T}$ is topologically simply connected if and only if it is a Coxeter complex (of type A_3, C_3, H_3, $\widetilde{C}_2$ or $\widetilde{G}_2$).

12.5.4 Some 2-simply connected geometries

We will now see that many geometries of rank $n \geq 3$ considered in previous chapters are 2-simply connected. Let us start with matroids.

Proposition 12.45 *All matroids of dimension $n \geq 3$ are 2-simply connected.*

Proof. The proof is by induction on n. The main step of the proof is showing that the universal 2-cover of the chamber system $\mathcal{C}(\Gamma)$ of an n-dimensional matroid Γ is geometric. Once we have proved this, the 2-simple connectedness of Γ is a trivial consequence of Theorem 7.6 and Proposition 8.14.

When $n = 3$, then $\mathcal{C}(\Gamma)$ is geometric by Theorem 12.39. Let $n \geq 4$. All residues of elements of Γ are 2-simply connected by the inductive hypothesis and by Theorem 12.42 (the latter must be used when the elements we consider are neither points nor hyperplanes). Therefore every 2-cover of $\mathcal{C}(\Gamma)$ is an $(n-1)$-cover. Hence $\widehat{\mathcal{C}}_2(\Gamma) = \widehat{\mathcal{C}}_{n-1}(\Gamma)$. The chamber system $\widehat{\mathcal{C}}_{n-1}(\Gamma)$ is geometric by Theorem 12.39. □

The next two corollaries are trivial consequences of the previous proposition:

Corollary 12.46 *All projective geometries of dimension $n \geq 3$ are 2-simply connected.*

Corollary 12.47 *All affine geometries of dimension $n \geq 3$ are 2-simply connected.*

Propositions 12.48–12.55 can be proved by the same argument used for Proposition 12.45, except that we must use other results instead of Proposition 8.14.

Proposition 12.48 *All polar spaces of rank $n \geq 3$ are 2-simply connected.*

(Use Proposition 8.18 instead of Proposition 8.14 in the proof of Proposition 12.45.)

Proposition 12.49 *All buildings of type D_n are 2-simply connected.*

(Use Proposition 8.20 instead of Proposition 8.14.)

Proposition 12.50 *All affine polar spaces are 2-simply connected.*

(Use Proposition 8.29 instead of Proposition 8.14.)

Proposition 12.51 *All geometries belonging to the diagram $Af.D_{n-1}$ are 2-simply connected.*

(Use Corollary 8.30 instead of Proposition 8.14.)

Proposition 12.52 *All geometries belonging to the Coxeter diagram E_6 are 2-simply connected.*

(Use Proposition 8.22.)

Proposition 12.53 *All geometries belonging to E_7 or E_8 and satisfying the intersection property* **IP** *are 2-simply connected.*

(Use Propositions 8.24 and 8.25.)

Proposition 12.54 *All geometries belonging to the Coxeter diagram F_4 and satisfying the intersection property* **IP** *are 2-simply connected.*

(use Proposition 8.27). By Theorems 11.19 and 12.28 we also have the following:

Proposition 12.55 *All Coxeter complexes of rank $n \geq 3$ are 2-simply connected.*

Exercise 12.32. Prove that bi-affine geometries are 2-simply connected. (Hint: use Exercise 8.62.)

Exercise 12.33. Prove that the geometry $\Gamma(HS)$ of Theorem 7.63(ii) is 2-simply connected.

(Hint: residues of points of $\Gamma(HS)$ are 2-simply connected by Proposition 12.45. Therefore, if Γ is the universal 2-cover of $\Gamma(HS)$, then Γ has a good system of lines, as this property holds in $\Gamma(HS)$. Hence $\Gamma \cong \Gamma(HS)$ by Theorem 7.63.)

Exercise 12.34. Prove that the $\{m-1, m, m+1\}_+$-truncation of an n-dimensional projective geometry ($n > m+1 > 1$) is 2-simply connected.

(Hint. The following statement is not hard to prove: given a topological cover $f : \Gamma' \longrightarrow \Gamma$, if $\mathbf{IP}_i$ holds in Γ with respect to some type i, then $\mathbf{IP}_i$ holds in Γ', too. Use this and Theorem 7.58.)

12.5.5 Quotients of simply connected geometries

In what follows $\widetilde{\Gamma}$ is an m-simply connected geometry of rank $n > m \geq 1$ and f is an m-cover from $\widetilde{\Gamma}$ to another geometry Γ. Hence, $\widetilde{\Gamma}$ is the universal m-cover $\widehat{\Gamma}_m$ of Γ.

Theorem 12.56 *The deck group $Aut(f)$ defines a quotient of $\widetilde{\Gamma}$ and we have $\widetilde{\Gamma}/Aut(f) \cong \Gamma$.*

Proof. We have $Aut(\mathcal{C}_m(\widetilde{\Gamma})) = Aut(\mathcal{C}(\widetilde{\Gamma})) = Aut_s(\widetilde{\Gamma})$ (the latter equality holds by Theorem 1.23). The conclusion follows from Theorem 12.13(ii). $\square$

Corollary 12.57 *If all residues of $\widetilde{\Gamma}$ of rank $> m$ are m-simply connected, then $Aut(f)$ residually defines quotients.*

Proof. We have $\Gamma \cong \widetilde{\Gamma}/Aut(f)$ by the previous theorem. Let F be a nonempty non-maximal flag of $\widetilde{\Gamma}$ and let $Aut(f)_F$ be the stabilizer of F in $Aut(f)$. The subgroup $Aut(f)_F$ acts faithfully in $\widetilde{\Gamma}_F$, by Lemma 11.4. We shall prove that $Aut(f)_F$ defines a quotient of $\widetilde{\Gamma}_F$ and that $\Gamma_{f(F)} \cong \widetilde{\Gamma}_F/Aut(f)_F$.

If F has corank $\leq m$, then f induces an isomorphism from $\widetilde{\Gamma}_F$ to $\Gamma_{f(F)}$ and we have $Aut(f)_F = 1$. The isomorphism $\Gamma_{f(F)} \cong \widetilde{\Gamma}_F/Aut(f)_F$ is quite evident in this case.

Let F have corank $> m$ and let f_F be the m-cover from $\widetilde{\Gamma}_F$ to $\Gamma_{f(F)}$ induced by f. By the previous theorem, $\Gamma_{f(F)} \cong \widetilde{\Gamma}_F/Aut(f_F)$. We also have $Aut(f_F) = \pi_m(\widetilde{\Gamma}_F)$ and $Aut(f) = \pi_m(\widetilde{\Gamma})$ by Theorem 12.9. The group $\pi_m(\widetilde{\Gamma}_F)$ can naturally be identified with a subgroup of $\pi_m(\widetilde{\Gamma})$. It is clear from the proof of Theorem 12.9 that the subgroup of $\pi_m(\widetilde{\Gamma}_F) = Aut(f)$ corresponding to $\pi_m(\widetilde{\Gamma}_F) = Aut(f_F)$ is the stabilizer $Aut(f)_F$ of F in $Aut(f)$. Therefore $Aut(f_F)$ is the stabilizer of F in $Aut(f)$. Hence $\Gamma_{f(F)} \cong \widetilde{\Gamma}_F/Aut(f)_F$, because $\Gamma_{f(F)} \cong \widetilde{\Gamma}_F/Aut(f_F)$. $\square$

Corollary 12.58 *Let $m = n-1$. Then $Aut(f)$ residually defines quotients.*

(Trivial, by Corollary 12.57; or by Theorem 12.56, using Exercise 11.4.)

Theorem 12.59 *Every flag-transitive m-quotient of $\widetilde{\Gamma}$ is obtained by factorizing by some subgroup G of $Aut_s(\widetilde{\Gamma})$, where G is normalized by a flag-transitive subgroup of $Aut_s(\widetilde{\Gamma})$ and residually defines quotients in $\widetilde{\Gamma}$.*

(Easy, by Theorem 12.28 and Corollary 11.3.)

We have anticipatively discussed some consequences of these theorems in Section 11.1.4, for quotients of polar spaces and affine polar spaces. We might also revisit Section 11.3 as an application of Theorem 12.59 and of Proposition 12.55. Theorem 12.56 and Corollary 12.57 now allow us to do a bit more than we did in Section 11.3. We remark that residues of flags of corank ≥ 3 of a Coxeter complex are Coxeter complexes, hence they are 2-simply connected by Proposition 12.55. Thus, we can use Corollary 12.57 to classify all (possibly non-flag-transitive) 2-quotients of a Coxeter complex $\Gamma(M)$. We must look for those subgroups of the Coxeter group $W(M)$ which residually define quotients in $\Gamma(M)$ and act semi-regularly on the set of flags of $\Gamma(M)$ of corank 2. The latter condition is equivalent to say that the subgroups we are speaking of have trivial intersections with $\langle s_i, s_j \rangle$, for any two distinct Coxeter involutions $s_i, s_j \in W(M)$.

Problem. The m-covers of Γ are m-quotients of $\widetilde{\Gamma}$, by the definition of universal m-covers. Furthermore, they are obtained factorizing $\widetilde{\Gamma}$ by subgroups of $\pi_m(\Gamma)$ (see Section 12.2.4 and Theorem 12.56). By Corollary 12.19, the subgroups of $\pi_m(\Gamma)$ bijectively correspond to the isomorphism classes of m-covers of $\mathcal{C}(\Gamma)$. However, when $m < n-1$ that corollary does not allow us to state a bijective correspondence between subgroups of $\pi_m(\Gamma)$ and isomorphism classes of m-covers of the geometry Γ. We could state such a correspondence if we knew that every m-cover of $\mathcal{C}(\Gamma)$ is geometric (this is true when $m = n-1$, by Lemma 12.37 and Corollary 12.36). On the other hand, we assumed that the geometry Γ admits an universal m-cover $\widetilde{\Gamma}$, namely that the universal m-cover of $\mathcal{C}(\Gamma)$ is geometric. Does this hypothesis imply that every m-cover of Γ is geometric ?

12.6 Topological covers

Henceforth Γ is a geometry of rank $n \geq 3$, with set of types I, $\mathcal{K}(\Gamma)$ is its flag-complex and $\mathcal{K}_2(\Gamma)$ is its 2-dimensional flag-complex, as in Section

12.2.1. According to the notation of Section 12.5.3, we designate the universal topological cover of Γ by $\widehat{\Gamma}$. Given a type $0 \in I$, $\Sigma_0(\Gamma)$ and $\overline{\Sigma}_0(\Gamma)$ are the 0-shadow complex of Γ and it support, respectively, as in Section 12.2.1.

We are interested in computing the topological homotopy group $\pi(\Gamma)$ of Γ. If Γ is reducible, then $\pi(\Gamma) = 1$ by Theorem 12.42 (and because $n \geq 3$). In this case, we have finished. Thus, from now on we assume that Γ is irreducible.

12.6.1 From $\pi(\Gamma)$ to $\pi(\mathcal{K}(\Gamma))$

Computing $\pi(\Gamma)$ inside the $(n-1)$-cell complex $\mathcal{C}_{n-1}(\Gamma)$ may be difficult (the plate complex $\mathcal{C}_{n-1}(\Gamma)$ is not handy at all to work with, in general). We want an easier way to compute $\pi(\Gamma)$. We consider $\mathcal{K}(\Gamma)$, to begin with.

We recall that $\mathcal{K}(\Gamma)$ can be viewed as a plate complex (Section 12.2.1), taking the chambers of Γ as plates, the incidence graph of Γ as the underlying graph of $\mathcal{K}(\Gamma)$ and the type function of Γ as the type function for the vertices of $\mathcal{K}(\Gamma)$.

Let $f : \mathcal{K} \longrightarrow \mathcal{K}(\Gamma)$ be a cover of the plate complex $\mathcal{K}(\Gamma)$. It is straightforward to check that $\mathcal{K}$ is the flag-complex of a (uniquely determined) geometry, let us call it Γ'. The cover f induces a topological cover from Γ' to Γ. Conversely, every topological cover $f : \Gamma' \longrightarrow \Gamma$ induces a cover of plate complexes from $\mathcal{K}(\Gamma')$ to $\mathcal{K}(\Gamma)$. Therefore, the flag-complex $\mathcal{K}(\widehat{\Gamma})$ of the universal topological cover $\widehat{\Gamma}$ of Γ is the universal cover of $\mathcal{K}(\Gamma)$. By Theorem 12.13(ii) and by the above we obtain the following:

Lemma 12.60 *We have* $\pi(\mathcal{K}(\Gamma)) \cong \pi(\Gamma)$.

We have given types to the vertices of the plate complex $\mathcal{K}(\Gamma)$, according to their types as elements of Γ. If we forget types, then we are left with the simplicial complex $\mathcal{K}(\Gamma)$, which is still a plate complex (Section 12.2.1). Forgetting types is just choosing the usual topological point of view. However, types have no role in the definition of homotopy groups of plate complexes. Therefore, the homotopy group $\pi(\mathcal{K}(\Gamma))$ of the plate complex $\mathcal{K}(\Gamma)$ is just the homotopy group of the simplicial complex $\mathcal{K}(\Gamma)$, in the usual topological meaning, and $\mathcal{K}(\widehat{\Gamma})$ is the universal cover of the simplicial complex $\mathcal{K}(\Gamma)$, in the usual topological meaning.

It is straightforward to check that $\pi(\mathcal{K}_2(\Gamma)) = \pi(\mathcal{K}(\Gamma))$. Hence Lemma 12.60 can be rephrased as follows:

Corollary 12.61 *We have* $\pi(\Gamma) \cong \pi(\mathcal{K}_2(\Gamma))$.

The previous corollary allows us to compute $\pi(\Gamma)$ inside $\mathcal{K}_2(\Gamma)$. This is certainly better than working in $\mathcal{C}_{n-1}(\Gamma)$. However, that computation may still be difficult. We will show that some information on $\pi(\mathcal{K}_2(\Gamma)$ can be obtained from the homotopy groups $\pi(\Sigma_0(\Gamma))$ and $\pi(\overline{\Sigma}_0(\Gamma))$ of the 0-shadow complex $\Sigma_0(\Gamma)$ and of its support $\overline{\Sigma}_0(\Gamma)$. These groups are sometimes easier to compute than $\pi(\mathcal{K}_2(\Gamma))$.

12.6.2 From $\pi(\Gamma)$ to $\pi(\Sigma_0(\Gamma))$

Theorem 12.62 *The group $\pi(\Sigma_0(\Gamma))$ is a homomorphic image of $\pi(\Gamma)$.*

Proof. As $\pi(\mathcal{K}_2(\Gamma)) \cong \pi(\Gamma)$ (Corollary 12.61), we can work with $\pi(\mathcal{K}_2(\Gamma))$ instead of $\pi(\Gamma)$. Let a be an element of Γ of type 0.

Given a closed path $\alpha = (a_0, a_1, ..., a_m)$ in $\mathcal{K}_2(\Gamma)$ with $a_0 = a_m = a$, for every $i = 1, 2, ..., m$ we choose an element a'_i of Γ of type 0 incident to both a_{i-1} and a_i. Trivially, if a_{i-1} (if a_i) has type 0, then $a'_i = a_{i-1}$ (respectively, $a'_i = a_i$). In particular, $a'_m = a_m$ $(= a_0)$. We set $a'_0 = a_0$.

For every $i = 1, 2, ..., m$, the elements a'_{i-1} and a'_i are joined by a path α'_i in the plate $\sigma_0(a_{i-1})$ of $\Sigma_0(\Gamma)$. Needless to say, if $a'_{i-1} = a'_i$, then we can take the null path for α'_i (actually, this is the only possible choice for α'_i when $a'_{i-1} = a'_i = a_i$). The path $\alpha' = \alpha'_0\alpha'_1...\alpha'_m$ of $\Sigma_0(\Gamma)$ is closed and based at a. Let $p(\alpha)$ be the set of the closed paths of $\Sigma_0(\Gamma)$ constructed as above starting from α. We will now prove that $p(\alpha)$ is contained in one homotopy class of $\Sigma_0(\Gamma)$.

Let $\alpha' = \alpha'_0\alpha'_1...\alpha'_m$ and $\alpha'' = \alpha''_0\alpha''_1...\alpha''_m$ be members of $p(\alpha)$. For every $i = 1, 2, ..., m-1$, we can take a path τ_i in the plate $\sigma_0(a_{i-1}, a_i)$ from the terminal node a'_i of α'_i to the terminal node a''_i of α''_i. By τ_0 and τ_m we denote the null path (a). Since τ_{i-1}, τ_i, α'_i and α''_i are contained in the plate $\sigma_0(a_i)$, we have $\alpha'_i \equiv \tau_{i-1}\alpha''_i\tau_i^{-1}$. Therefore:

$$\alpha' \equiv \tau_0\alpha''_1\tau_1^{-1}\tau_1\alpha''_2\tau_2^{-1}...\tau_{m-1}\alpha''_m\tau_m^{-1} \equiv \alpha''$$

Let β be another closed path in $\mathcal{K}_2(\Gamma)$ based at a. We will now prove that, if $\alpha \equiv \beta$ in $\mathcal{K}_2(\Gamma)$, then $p(\alpha)$ and $p(\beta)$ are contained in the same homotopy class of $\Sigma_0(\Gamma)$.

We can assume without loss of generality that α and β are elementarily homotopic in $\mathcal{K}_2(\Gamma)$ and that $\beta = (a_0, a_1, ..., a_{h-1}, b, a_h, a_{h+1}, ..., a_m)$ with $(a_0, a_1, ..., a_m) = \alpha$. The elements a_{h-1}, b and a_h form a triangle in the underlying graph of $\mathcal{K}_2(\Gamma)$. That is, they form a flag of Γ. Hence there is an element b' of Γ of type 0 incident to all of them and we can choose $\alpha' \in p(\alpha)$ as follows: $\alpha' = \alpha'_1\alpha'_2...\alpha'_{h-1}\alpha'_{h+1}...\alpha'_m$ with α'_{h-1} ending at b' and α'_h starting at b'. We can now obtain an element β' of $p(\beta)$ inserting the null path (b) between α'_{h-1} and α'_h. Trivially, $\beta' \equiv \alpha'$ (actually, these two paths are practically the same). Therefore, $p(\alpha)$ and $p(\beta)$ are contained in the same homotopy class of $\Sigma_0(\Gamma)$.

By the above, we have a homomorphism $p_0 : \pi(\mathcal{K}_2(\Gamma)) \longrightarrow \pi(\Sigma_0(\Gamma))$ mapping the homotopy class of a closed path α of $\mathcal{K}_2(\Gamma)$ onto the homotopy class of $\Sigma_0(\Gamma)$ containing $p(\alpha)$. We will now prove that p_0 is surjective.

Let $\alpha' = (a'_0, a'_1, ..., a'_h)$ be a closed path of $\Sigma_0(\Gamma)$ based at $a = a'_0 = a'_h$. For every $i = 0, 1, ..., h-1$, let a_i be an element of Γ of type $\neq 0$ incident to both a'_i and a'_{i+1} and let $\alpha = (a'_0, a_0, a'_1, a_1, a'_2, ..., a'_{h-1}, a_{h-1}, a'_h)$. It is easily seen that $\alpha' \in p(\alpha)$. The surjectivity of p_0 is proved. □

Corollary 12.63 *The geometry Γ is topologically simply connected only if the plate complex $\Sigma_0(\Gamma)$ is simply connected.*

(Trivial, by the previous theorem.)

The surjective morphism $p_0 : \pi(\Gamma) \longrightarrow \pi(\Sigma_0(\Gamma))$ constructed in the proof of Theorem 12.62 will be called the *projection* of $\pi(\Gamma)$ onto $\pi(\Sigma_0(\Gamma))$ (*based* at the vertex a).

Theorem 12.64 *Let $Gr_0(\Gamma)$ have a good system of lines. Then we have $\pi(\Gamma) \cong \pi(\Sigma_0(\Gamma))$.*

Proof. We work with $\pi(\mathcal{K}_2(\Gamma))$ instead of $\pi(\Gamma)$, exploiting the isomorphism $\pi(\mathcal{K}_2(\Gamma)) \cong \pi(\Gamma)$. Let K be the type of those flags of Γ which are lines of $Gr_0(\Gamma)$ and let us choose a type $k \in K$.

Given an element a of Γ of type 0, let $\alpha = (a_0, a_1, ..., a_m)$ be a closed path in $\Sigma_0(\Gamma)$ based at a. For every $i = 1, 2, ..., m$, let X_i be the line of $Gr_0(\Gamma)$ through the points a_{i-1} and a_i and let a'_i be the element of type k in the flag X_i of Γ (the line X_i is unique because $Gr_0(\Gamma)$ has a good system of lines). Let us set $q(\alpha) = (a_0, a'_1, a_1, a'_2, a_2, ..., a'_m, a_m)$. Trivially, $q(\alpha)$ is a closed path of $\mathcal{K}_2(\Gamma)$ based at a. We shall prove that the homotopy class of $q(\alpha)$ in $\mathcal{K}_2(\Gamma)$ only depends on the homotopy class of α in $\Sigma_0(\Gamma)$.

Thus, let $\beta = (b_0, b_1, ..., b_h)$ be another closed path of $\Sigma_0(\Gamma)$ based at $a = b_0 = b_h$, homotopic to α. We want to prove that $q(\beta) \equiv q(\alpha)$. We may assume without loss of generality that α and β are elementarily homotopic. That is, there are nonnegative integers u, v, w with $0 \leq u$, $u < v$, $u < w$, $v \leq m$, $w \leq h$ and $m - v = h - w$ and there is a plate X of $\Sigma_0(\Gamma)$ such that all the following hold: $a_i = b_i$ for $i = 0, 1, ..., u$, we have $a_{v+j} = b_{w+j}$ for $j = 0, 1, ..., m - v$ $(= h - w)$ and both $(a_u, a_{u+1}, ..., a_v)$ and $(b_u, b_{u+1}, ..., b_w)$ are paths in the graph X.

We may always assume that X is represented by an element c of Γ of type > 1 in $Gr_0(\Gamma)$. Hence $(a_u, a_{u+1}, ..., a_v)$ and $(b_u, b_{u+1}, ..., b_w)$ are paths in the collinearity graph of the lower residue of c in $Gr_0(\Gamma)$. For every $t = u+1, u+2, ..., v$, let X_t be the flag of Γ representing the (unique) line of $Gr_0(\Gamma)$ through the points a_{t-1} and a_t and let a'_t be the element of X_t of type k. Similarly, for every $s = u+1, u+2, ..., w$, let b'_s be the element of type k in the flag Y_s of Γ representing the (unique) line of $Gr_0(\Gamma)$ through b_{s-1} and b_s. Let us set $\alpha' = (a_u, a'_{u+1}, a_{u+1}, a'_{u+2}, a_{u+2}, ..., a'_v, a_v)$ and $\beta' = (b_u, b'_{u+1}, b_{u+1}, b'_{u+2}, b_{u+2}, ..., b'_w, b_w)$. Since a_{t-1} and a_t are collinear in the lower residue of c in $Gr_0(\Gamma)$, we have $X_t * c$ in Γ for every $t = u+1, u+2, ..., v$. Hence $a'_t * c$ for $t = u+1, u+2, ..., v$. Similarly, $b'_s * c$ for every $s = u+1, u+2, ..., w$. Considering the triangles (c, a_{t-1}, a'_t, c), (c, a'_t, a_t, c) and (c, b_{s-1}, b'_s, c), (c, b'_s, b_s, c) (for $t = u+1, u+2, ..., v$ and $s = u+1, u+2, ..., w$) we easily see that $\alpha' \equiv \beta'$ in $\mathcal{K}_2(\Gamma)$. Since $q(\alpha) = \gamma'\alpha'\delta'$ and $q(\beta) = \gamma'\beta'\delta'$ for suitable paths γ' and δ' of $\mathcal{K}_2(\Gamma)$, we have $q(\alpha) \equiv q(\beta)$, as we wanted.

Therefore, we have a homomorphism $q_0 : \pi(\Sigma_0(\Gamma)) \longrightarrow \pi(\mathcal{K}_2(\Gamma))$, mapping every element $A \in \pi(\Sigma_0(\Gamma))$ onto the homotopy class of $q(\alpha)$, with $\alpha \in A$.

Let us prove that q_0 is surjective. That is, we must show that every closed path of $\mathcal{K}_2(\Gamma)$ based at a is homotopic to $q(\alpha)$ for some closed path α of $\Sigma_0(\Gamma)$ based at a.

Let $\alpha' = (a'_0, a'_1, ..., a'_m)$ be a closed path of $\mathcal{K}_2(\Gamma)$ based at $a = a'_0 = a'_m$. For every $i = 0, 1, ..., m-1$, let c_i be an element of Γ of type 0 incident to both a'_i and a'_{i+1}. Trivially, we have $c_0 = c_{m-1} = a$ $(= a'_0 = a'_m)$. Let $i = 1, 2, ..., m-1$. If a'_i has type > 1 in $Gr_0(\Gamma)$, then we choose a path γ_i from c_{i-1} to c_i in the lower residue of a'_i in $Gr_0(\Gamma)$. If a'_i is a line of $Gr_0(\Gamma)$, then we define $\gamma_i = (c_{i-1}, c_i)$. If a'_i has type 0, then $c_{i-1} = c_i = a'_i$. In this case γ_i will denote the null path at a'_i. Let us set $\alpha = \gamma_1\gamma_2...\gamma_{m-1}$. It is not difficult to prove that $q(\alpha)$ is homotopic to the following path of $\mathcal{K}_2(\Gamma)$:

$$(a'_0, a'_1, c_1, a'_2, c_2, a'_3, ..., a'_{m-1}, c_{m-1}, a'_m)$$

(by abuse of notation, we are now allowing repetitions of consecutive terms in sequences describing paths). On the other hand, this path is homotopic to α'. Therefore, $q(\alpha) \equiv \alpha'$ in $\mathcal{K}_2(\Gamma)$. Thus, we have constructed an epimorphism from $\pi(\Sigma_0(\Gamma))$ to $\pi(\Gamma)$. It is easily seen that this epimorphism is actually the inverse of the projection p_0 of $\pi(\Gamma)$ onto $\pi(\Sigma_0(\Gamma))$. Therefore, $\pi(\Gamma) \cong \pi(\Sigma_0(\Gamma))$. □

Note that, if we drop the assumption that $Gr_0(\Gamma)$ has a good system of lines, then the conclusion of the above theorem is no longer valid. For instance, let Γ be the halved octahedron and let 0 be the type representing the points of Γ. We know that Γ does not have a good system of lines.

It is straightforward to check that $\pi(\Sigma_0(\Gamma)) = 1$. Nevertheless, the octahedron is a 2-fold topological cover of Γ. The octahedron is topologically simply connected (Proposition 12.48, or Proposition 12.55), hence it is the universal topological cover of Γ (Corollary 12.15). Therefore $\pi(\Gamma) = \mathbf{2}$.

On the other hand, the halved octahedron has a good dual system of lines (Lemma 7.29). Hence $\pi(\Sigma_2(\Gamma)) = \pi(\Gamma) = \mathbf{2}$. Nevertheless, by the flatness of Γ we obtain $\pi(\overline{\Sigma}_2(\Gamma)) = 1$.

12.6.3 From $\pi(\Gamma)$ to $\pi(\overline{\Sigma}_0(\Gamma))$

We turn back to the general theory of plate complexes for a few lines, to better understand the relations between the homotopy groups of the plate complexes $\Sigma_0(\Gamma)$ and $\overline{\Sigma}_0(\Gamma)$.

Given a plate complex $\mathcal{P} = (V, E, P, t_V, t_E)$, let $\overline{\mathcal{P}}$ be the plate complex defined as follows. For every plate X of $\mathcal{P}$, we forget the graph structure of X and we replace it by the complete graph on the set of vertices of X. The plates of $\overline{\mathcal{P}}$ are the complete graphs obtained in this way. According to this, $\overline{\mathcal{P}}$ has just the same vertices as $\mathcal{P}$, but any two distinct vertices contained in some common plate of $\mathcal{P}$ now form an edge of the underlying graph of $\overline{\mathcal{P}}$. As for types, the vertices of $\overline{\mathcal{P}}$ keep the types they had in $\mathcal{P}$. On the other hand, we take the trivial type function (one type for all edges) instead of t_E. Note that it might happen that distinct plates of $\mathcal{P}$ have the same set of vertices, thus giving us the same plate of $\overline{\mathcal{P}}$.

We call $\overline{\mathcal{P}}$ the *support* of $\mathcal{P}$. Trivially, all paths in the underlying graph (V, E) of $\mathcal{P}$ are also paths in the underlying graph of the support $\overline{\mathcal{P}}$ of $\mathcal{P}$. Hence a homomorphism $\varphi_{\mathcal{P}} : \pi(\mathcal{P}) \longrightarrow \pi(\overline{\mathcal{P}})$ is naturally defined, mapping the homotopy class in $\mathcal{P}$ of a closed path α of $\mathcal{P}$ onto the homotopy class of α in $\overline{\mathcal{P}}$. Furthermore, since the plates of $\mathcal{P}$ are connected subgraphs of the underlying graph (V, E) of $\mathcal{P}$, every closed path in the underlying graph of $\overline{\mathcal{P}}$ is homotopic in $\overline{\mathcal{P}}$ to some closed path of (V, E). Therefore $\varphi_{\mathcal{P}}$ is surjective. We call it the *projection* of $\pi(\mathcal{P})$ onto $\pi(\overline{\mathcal{P}})$.

We say that $\mathcal{P}$ satisfies the *intersection property* for plate complexes (that it satisfies **IPP**, for short) if both the following hold:

(**IPP**$_1$) the set $P \cup V \cup \{\emptyset\}$ is a semilattice with respect to $\cap$.

(**IPP**$_2$) for every plate X of $\mathcal{P}$, the graph structure of X is the graph induced on it by the underlying graph of $\mathcal{P}$.

Lemma 12.65 *Let* **IPP** *hold in* $\mathcal{P}$. *Then the projection* $\varphi_{\mathcal{P}}$ *of* $\pi(\mathcal{P})$ *onto* $\pi(\overline{\mathcal{P}})$ *is an isomorphism.*

Proof. We need to show that $\varphi_{\mathcal{P}}$ is invertible.

Given a closed path $\alpha = (a_0, a_1, ..., a_m)$ in $\overline{\mathcal{P}}$, for every $i = 1, 2, ..., m$ we choose a plate X_i of $\mathcal{P}$ containing both a_{i-1} and a_i and a path β_i in X_i from a_{i-1} to a_i. We set $\beta = \beta_1\beta_2...\beta_m$. If A and B_α are the homotopy classes of α and β in $\mathcal{P}$ and $\overline{\mathcal{P}}$ respectively, then $\varphi_{\mathcal{P}}(B_\alpha) = A$. The lemma will be proved if we show that B_α only depends on A.

We first prove that B_α does not depend on the choice of the plates X_1, X_2, ..., X_m and of the paths β_1, β_2, ..., β_m, once α is given. Let $\beta' = \beta'_1\beta'_2...\beta'_m$ be obtained as β from α, but from another choice of plates X'_1, X'_2, ..., X'_m and of paths β'_1, β'_2, ..., β'_m. By $\mathbf{IPP}_1$, $Y_i = X_i \cap X'_i$ is a plate, for every $i = 1, 2, ..., m$. Therefore we can choose a path γ_i from a_{i-1} to a_i inside Y_i, for every $i = 1, 2, ..., m$. By $\mathbf{IPP}_2$, the path γ_i belongs to both graphs X_i and X'_i. Hence $\beta_i \equiv \gamma_i \equiv \beta'_i$, for $i = 1, 2, ..., m$. Therefore, $\beta' \equiv \beta$. That is, $\beta' \in B_\alpha$. Hence B_α does not depend on the particular choice of X_1, X_2, ..., X_m, β_1, β_2, ..., β_m.

We still need to prove that B_α does not depend on the choice of the representative α of A. Let $\alpha' = (a'_0, a'_1, ..., a'_h)$ be another closed path in $\overline{\mathcal{P}}$, based at $a'_0 = a_0$ and homotopic to α. We only consider the case where α and α' are elementarily homotopic (when this is done, all the rest becomes trivial). Thus, let α and α' be elementarily homotopic. There are integers u, v, w with $0 \leq u$, $u < v$, $u < w$, $v \leq m$, $w \leq h$ and $m - v = h - w$ and there is a plate $\overline{X}$ of $\overline{\mathcal{P}}$ such that the following hold: $a_i = a'_i$ for $i = 0, 1, ..., u$, we have $a_{v+j} = a'_{w+j}$ for $j = 0, 1, ..., m - v$ $(= h - w)$, and $\overline{X}$ contains all of $a_u, a_{u+1}, ..., a_v$ and $a'_u, a'_{u+1}, ..., a'_w$. Let us write r for $m - v$ $(= h - w)$.

For $i = 1, 2, ..., u$, let X_i be a plate of $\mathcal{P}$ containing a_{i-1} $(= a'_{i-1})$ and a_i $(= a'_i)$ and let β_i be a path in X_i from a_{i-1} to a_i. For $j = 1, 2, ..., r$, let Y_j be a plate of $\mathcal{P}$ containing both a_{v+j-1} $(= a'_{w+j-1})$ and a_{v+j} $(= a'_{w+j})$ and let γ_j be a path in Y_i from a_{v+j-1} to a_{v+j}. By $\mathbf{IPP}_2$, distinct plates of $\mathcal{P}$ have distinct sets of vertices. Therefore, there is a unique plate Z of $\mathcal{P}$ having $\overline{X}$ as set of vertices. There are paths δ and δ' of Z from a_u $(= a'_u)$ to a_v $(= a'_w)$ such that δ and δ' run through $a_{u+1}, a_{u+2}, ..., a_{v-1}$ and through $a'_{u+1}, a'_{u+2}, ..., a'_{w-1}$, respectively. Let us set $\beta = \beta_1\beta_2...\beta_u\delta\gamma_1\gamma_2...\gamma_r$ and $\beta' = \beta_1\beta_2...\beta_u\delta'\gamma_1\gamma_2...\gamma_r$. We have $\beta \in B_\alpha$ and $\beta' \in B_{\alpha'}$. Moreover, $\beta \equiv \beta'$ because both δ and δ' belong to the plate Z. Therefore $B_\alpha = B_{\alpha'}$. $\square$

We now turn back to $\Sigma_0(\Gamma)$ and $\overline{\Sigma}_0(\Gamma)$. Trivially, $\overline{\Sigma}_0(\Gamma)$ is just the support of $\Sigma_0(\Gamma)$ as defined above. It is quite evident that $\mathbf{IPP}_1$ holds in $\Sigma_0(\Gamma)$ if and only if the weak intersection property of Section 6.2 holds in Γ, with respect to the type 0. Therefore, if the weak intersection property holds in Γ with respect to 0 and if $\mathbf{IPP}_2$ holds in $\Sigma_0(\Gamma)$, then $\mathbf{IPP}$ holds in $\Sigma_0(\Gamma)$ and we have $\pi(\Sigma_0(\Gamma)) \cong \pi(\overline{\Sigma}_0(\Gamma))$, by Lemma 12.65.

The property $\mathbf{IPP}_2$ states that, for every element X of $Gr_0(\Gamma)$ of type

≥ 2, the graph induced on $\sigma_0(X)$ by the collinearity graph of $Gr_0(\Gamma)$ is the collinearity graph of the lower residue of X in $Gr_0(\Gamma)$.

We also remark that, if the property $\mathbf{IP}_0^1$ of Section 6.7.2 holds in Γ, then $\mathbf{IPP}_2$ holds in $\Sigma_0(\Gamma)$. Therefore, if Γ satisfies the intersection property $\mathbf{IP}_0$ with respect to 0, then **IPP** holds in $\Sigma_0(\Gamma)$ and we have $\pi(\Sigma_0(\Gamma)) \cong \pi(\overline{\Sigma}_0(\Gamma))$.

Theorem 12.66 *Let Γ satisfy the intersection property $\mathbf{IP}_0$, with respect to the type* 0. *Then $\pi(\Gamma) \cong \pi(\overline{\Sigma}_0(\Gamma))$.*

Proof. Since $\mathbf{IP}_0$ holds, the 0-Grassmann geometry $Gr_0(\Gamma)$ of Γ has a good system of lines. Hence $\pi(\Sigma_0(\Gamma)) \cong \pi(\Gamma)$, by Theorem 12.64. On the other hand, **IPP** holds in $\Sigma_0(\Gamma)$, because $\mathbf{IP}_0$ holds in Γ. Therefore $\pi(\overline{\Sigma}_0(\Gamma)) \cong \pi(\Sigma_0(\Gamma))$. □

Actually, a property a bit weaker than $\mathbf{IP}_0$ is sufficent to obtain the isomorphisms $\pi(\Gamma) \cong \pi(\overline{\Sigma}_0(\Gamma))$. Indeed, if $Gr_0(\Gamma)$ satisfies the property $\mathbf{IP}_0^1$ of Section 6.7.2 and if the weak intersection property of Section 6.2 holds in Γ with respect to 0, then $Gr_0(\Gamma)$ has a good system of lines and **IPP** holds in $\Sigma_0(\Gamma)$. Hence Theorem 12.64 and Lemma 12.65 can be applied and we obtain the conclusion as in the proof of Theorem 12.66.

However, if we do not assume any kind of 'intersection property', then the isomorphisms stated in Lemma 12.65 and Theorem 12.66 may fail to hold, as we can see by considering the halved octahedron (Section 12.6.2, final remarks).

12.6.4 Examples

The *Alt*(7)-geometry. Let Γ be the *Alt*(7)-geometry. We will prove that Γ is topologically simply connected (hence 2-simply connected, as it has rank 3). Let $\pi(\Gamma) \neq 1$, if possible. Then the universal topological cover $f : \widehat{\Gamma} \longrightarrow \Gamma$ of Γ is a proper cover.

The geometry Γ has a good dual system of lines (Lemma 7.29). Hence $\pi(\Gamma) \cong \pi(\Sigma_2(\Gamma))$, where 2 is the type representing planes. The dual point–line system of Γ is the point–line system of $PG(3,2)$ (see Section 6.4.2 and the model of $PG(3,2)$ described at the end of Section 9.3.4). Therefore every closed path of $\Sigma_2(\Gamma)$ splits in triangles. Furthermore, A_7 acts in the dual point-line system of Γ just as it acts in the point-line system of $PG(3,2)$. The stabilizer in A_7 of a plane of $PG(3,2)$ is $L_3(2)$ (see Section 9.3.4). Therefore A_7 has two orbits on the set of triangles of the collinearity graph of $\Sigma_2(\Gamma)$: one orbit consists of the 35 triangles corresponding to the 35 lines of $PG(3,2)$ and the other one consists of the 420 proper triangles of $PG(3,2)$. Trivially, the 35 triangles corresponding to lines of $PG(3,2)$ are null-homotopic in $\Sigma_2(\Gamma)$. Since we have assumed that $\pi(\Sigma_2(\Gamma)) \neq 1$ and all closed paths of $\pi(\Sigma_2(\Gamma))$ split into triangles, some proper triangle of $PG(3,2)$ is not null-homotopic in $\Sigma_2(\Gamma)$ and $\pi(\Sigma_2(\Gamma))$ is generated by some proper triangles of $PG(3,2)$. As A_7 transitively permutes the proper triangles of $PG(3,2)$, none of them is null-homotopic in $\Sigma_2(\Gamma)$.

The universal cover $\widehat{\Gamma}$ of Γ is not a polar space, by Proposition 8.18. By Corollary 7.39, Γ has not a good system of lines. Therefore, there are three distinct planes of $\widehat{\Gamma}$ forming a triangle T in the dual collinearity graph of $\widehat{\Gamma}$ but not on the same line of Γ (see Exercise 7.13). Since $\widehat{\Gamma}$ is topologically simply connected, T is null-homotopic in $\Sigma_2(\widehat{\Gamma})$. Hence $f(T)$ is null-homotopic in $\Sigma_2(\Gamma)$. Therefore $f(T)$ consists of the three planes on a line of Γ. As f is a topological cover, the above forces T to consist of the three planes on a line of $\widehat{\Gamma}$, contrary to our choice of T. We have reached a contradiction. Therefore Γ is topologically simply connected.

C_3 geometries thin on top. Let Γ be a top-thin C_3 geometry, with types 0, 1, 2 denoting points, lines and planes, respectively. We have $\pi(\Gamma) \cong \pi(\Sigma_2(\Gamma))$ by Theorem 12.64 and Lemma 7.29. The plates of $\Sigma_2(\Gamma)$ are collinearity graphs of dual grids, hence they are bipartite graphs. Therefore, if α and β are homotopic closed paths in $\Sigma_2(\Gamma)$ with no repetitions of consecutive vertices and of length m and r respectively, then $m \equiv r$ $(mod\ 2)$. Hence every null-homotopic closed path of $\Sigma_2(\Gamma)$ (with no repetition of consecutive vertices) has even length. It easily follows from this that, if $\pi(\Sigma_2(\Gamma)) = 1$, then the underlying graph of $\Sigma_2(\Gamma)$ (namely, the dual collinearity graph of Γ) is bipartite. We can now prove the following:

Proposition 12.67 *A top-thin C_3 geometry is 2-simply connected if and only if it is a polar space.*

Proof. The 'if' claim is just a special case of Proposition 12.48. Turning to the 'only if' part, let Γ be a 2-simply connected top-thin C_3 geometry. If Γ has a good system of lines, then it is a polar space by Corollary 7.39. On the other hand, if Γ has not a good system of lines, then we can construct a triangle in the dual collinearity graph of Γ (Exercise 7.13). However, the dual collinearity graph of Γ is bipartite, hence there are no triangles in it. Therefore Γ is a polar space. □

Corollary 12.68 *Every finite thick-lined top-thin C_3 geometry is a polar space.*

Easy, by the previous proposition and Corollary 8.17. Turning to the general case, let Γ be a non-2-simply connected top-thin C_3 geometry and let

$\mathcal{P}$ be its universal 2-cover. Since $\mathcal{P}$ is a polar space by Proposition 12.67 and it is top-thin, we can consider the projective geometry $\Gamma(\mathcal{P})$ obtained unfolding $\mathcal{P}$.

Corollary 12.69 *Given Γ and $\mathcal{P}$ as above, we have $\pi(\Gamma) = \mathbf{2}$ and $\Gamma = \mathcal{P}/\langle\delta\rangle$ for some polarity δ of $\Gamma(\mathcal{P})$. Furthermore, δ has no absolute points.*

Proof. We have $\Gamma = \mathcal{P}/G$ for some subgroup G of $Aut_s(\mathcal{P})$, by Theorem 12.56. Since the diagram A_3 admits recovery from Grassmann geometries with respect to its central node, we have $Aut_s(\mathcal{P}) = Aut(\Gamma(\mathcal{P}))$. Let $g \in G \cap Aut_s(\Gamma(\mathcal{P}))$. For every plane u of $\mathcal{P}$, u and $g(u)$ meet in at least one point of $\mathcal{P}$. Therefore $u = g(u)$, because g residually defines quotients of $\mathcal{P}$ (Corollary 12.57) and $\mathcal{P}/G$ is a 2-quotient of $\mathcal{P}$. Therefore g fixes all planes of $\mathcal{P}$, that is $g = 1$. Hence $G \cap Aut_s(\Gamma(\mathcal{P})) = 1$. Therefore $G = \langle\delta\rangle$ for some polarity δ of $\Gamma(\mathcal{P})$. Hence $\pi(\Gamma) = \mathbf{2}$, by Theorem 12.13(ii).

Let $a \in \delta(a)$ for some point a of $\Gamma(\mathcal{P})$, if possible. Then δ fixes the line $\{a, \delta(a)\}$ of $\mathcal{P}$. However, this is impossible, by Lemma 11.4. Therefore δ has no absolute points. □

The geometries $\Gamma(U_5(2))$ and $\Gamma(O_6^-(3))$. We now consider the geometries $\Gamma(U_5(2))$ and $\Gamma(O_6^-(3))$ of Section 10.2.3. We write Γ_U and Γ_O for $\Gamma(U_5(2))$ and $\Gamma(O_6^-(3))$ respectively, for short.

The geometry Γ_U has a good system of lines. By Theorem 12.64, we have $\pi(\Gamma_U) \cong \pi(\Sigma_0(\Gamma_U))$. It is not difficult to check that the triangular property holds in Γ_U, that the collinearity graph of Γ_U has diameter 2 and that the following hold in it:

(i) given any two points a, b of Γ at distance 2 in the collinearity graph of Γ_U, the graph induced on $a^\perp \cap b^\perp$ by the collinearity graph of Γ_U is connected;

(ii) given points a, b, c of Γ_U with $a \perp b$ and c at distance 2 from both a and b in the collinearity graph of Γ_U, we have $a^\perp \cap b^\perp \cap c^\perp \neq \emptyset$.

As the collinearity graph of Γ_U has diameter 2 and satisfies (i) and (ii), every closed path of $\Sigma_0(\Gamma_U)$ splits in triangles. Furthermore, every triangle of $\Sigma_0(\Gamma_U)$ is null-homotopic, because the triangular property holds in Γ_U. Hence $\pi(\Sigma_0(\Gamma_U)) = 1$ and Γ_U is 2-simply conected.

Let us turn to Γ_O. This geometry satisfies the triangular property and its collinearity graph has diameter 2. However, (i) now fails to hold in the collinearity graph of Γ_O. Indeed, given points a, b at distance 2 in the collinearity graph of Γ_O, the graph induced on $a^\perp \cap b^\perp$ has three connected components. Thus, contrary to the previous case, we cannot prove that Γ_O is 2-simply connected. Actually, Γ_O is not 2-simply connected: it admits a 3-fold 2-cover, called $3 \cdot \Gamma(O_6^-(3))$ in Section 10.2.3. Let us write Γ' for $3 \cdot \Gamma(O_6^-(3))$, for short. We will now prove that Γ' is the universal 2-cover of Γ_O.

It is not difficult to prove that every 2-cover of Γ_O has a good system of lines and satisfies the triangular property, as these properties hold in Γ_O. Therefore Γ' has a good system of lines and it satisfies the triangular property. Hence $\pi(\Gamma') \cong \pi(\Sigma_0(\Gamma'))$ (Theorem 12.64) and every triangle in the collinearity graph of Γ' is null-homotopic in $\Sigma_0(\Gamma')$. The collinearity graph of Γ' is the graph considered in Section 13.2.C of [13]. It follows from the information given in [13] on that graph that the above property (i) holds in it. Therefore, every quadrangle in the collinearity graph of Γ' is null-homotopic.

The following result has been obtained in [34] (Corollary 3.5 and Theorem 3.15):

Proposition 12.70 *Let Γ be a geometry with diagram and orders as follows:*

$$(c.C_2) \qquad \overset{c}{\bullet\!\!-\!\!-\!\!-\!\!-\!\!-\!\!\bullet}\!\!=\!\!=\!\!=\!\!=\!\!\bullet \qquad s \quad t$$

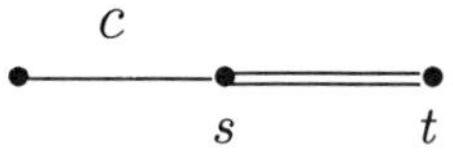

where $s, t < \infty$ and $2 \leq t$. Then the collinearity graph of Γ has diameter $d \leq s$.

The reader is referred to [34] for the proof (we warn the reader that only geometries with a good system of lines are considered in [34]; however that assumption plays no role in the proof of the above proposition).

Let $f : \widehat{\Gamma} \longrightarrow \Gamma'$ be the universal 2-cover of Γ'. By Proposition 12.70, $\widehat{\Gamma}$ has diameter $d \leq 4$. Let α be a non-null-homotopic closed path in $\Sigma_0(\Gamma')$, if possible, and let a be the base point of α. Given $b \in f^{-1}(a)$, let β be the lifting of α through f at b and let c be the end-point of β. As $\widehat{\Gamma}$ is 2-simply connected and α is not null-homotopic, we have $c \neq b$. On the other hand, b and c have distance ≤ 4 in the collinearity graph of $\widehat{\Gamma}$. Therefore there is a path γ of $\Sigma_0(\widehat{\Gamma})$ from c to b of length ≤ 4. As $\widehat{\Gamma}$ is 2-simply connected, the closed path $\beta\gamma^{-1}$ is null-homotopic in $\Sigma_0(\widehat{\Gamma})$. Therefore α and $f(\gamma)$ are homotopic in $\Sigma_0(\Gamma')$. However, the closed path $f(\gamma)$ has length ≤ 4, hence it is null-homotopic. Hence α is null-homotopic, contrary to our assumptions on it.

Therefore Γ' is 2-simply connected. Hence $\pi(\Gamma_O) = \mathbf{3}$.

12.6.5 Exercises

Exercise 12.35. Give an explicit construction of an isomorphism between

$\pi(\Gamma)$ and $\pi(\mathcal{K}(\Gamma))$.

(Hint: given a closed gallery $(C_0, C_1, ..., C_m)$ of Γ based at a chamber $C = C_0 = C_m$ and an element $a \in C$, choose an element $a_i \in C_{i-1} \cap C_i$ for every $i = 1, 2, ..., m$ ...)

Exercise 12.36. Let $\widehat{\Gamma}$ be a bi-affine geometry of rank $n \geq 3$ obtained from $PG(n, K)$ removing a hyperplane H and the star of a point p. Let Φ_p be the equivalence relation defined in Exercise 8.54 and let Γ be the quotient of $\widehat{\Gamma}$ by Φ_p. Prove the following statements:

(i) the geometry $\widehat{\Gamma}$ is the universal 2-cover of Γ;
(ii) the group $\pi(\Gamma)$ is isomorphic to the additive group of K or to the multiplicative group of K, according to whether $p \in K$ or $p \notin K$;
(iii) we have $\pi(\Sigma_0(\Gamma)) = \pi(\overline{\Sigma}_0(\Gamma)) = 1$.

(Hint: see Section 11.1.4(4) and use Exercise 12.32.)

Exercise 12.37. Let Γ be the flat $Af.C_2$ geometry of Exercise 7.20 and let the type 0 represent the points of Γ. Check that $\pi(\Sigma_0(\Gamma)) = \pi(\overline{\Sigma}_0(\Gamma)) = 1$ but $\pi(\Gamma) = \mathbf{2}$.

Exercise 12.38. Prove that the $Alt(8)$-geometry (Exercise 7.21) is 2-simply connected.

(Hint: residues of elements of the $Alt(8)$-geometry are 2-simply connected, by Proposition 12.45 and Corollary 12.43. Hence, we only need to prove that the $Alt(8)$-geometry is topologically simply connected. The triangular property holds in it and its collinearity graph is a complete graph. Therefore ...)

Exercise 12.39. Let $f : \Gamma' \longrightarrow \Gamma$ be a topological cover of irreducible geometries. Prove that f induces a homomorphism $\pi_0(f)$ from $\pi(\Sigma_0(\Gamma'))$ to $\pi(\Sigma_0(\Gamma))$ and a homomorphism $\overline{\pi}_0(f)$ from $\pi(\overline{\Sigma}_0(\Gamma')$ to $\pi(\overline{\Sigma}_0(\Gamma))$ (as usual, these homorphisms are determined up to isomorphisms, depending on the choice of a 'base vertex').

Exercise 12.40. Let $f : \Gamma' \longrightarrow \Gamma$ be a topological cover of irreducible geometries and let $Gr_0(\Gamma)$ have a good system of lines. Prove that f induces a cover $\Sigma_0(f)$ from $\Sigma_0(\Gamma')$ to $\Sigma_0(\Gamma)$).

Exercise 12.41. Keep the hypotheses of the previous exercise and let $\pi_0(f)$ be the homomorphism from $\pi(\Sigma_0(\Gamma'))$ to $\pi(\Sigma_0(\Gamma))$ induced by f (Exercise 12.39). Prove that $\pi_0(f)$ is injective.

Exercise 12.42. Let $f : \Gamma' \longrightarrow \Gamma$ be a topological cover of irreducible geometries and let $\mathbf{IP}_0^1$ of Section 6.7.2 hold in $Gr_0(\Gamma)$. Prove that $\mathbf{IP}_0^1$ also holds in $Gr_0(\Gamma')$ and that f induces a cover $\overline{\Sigma}_0(f)$ from $\overline{\Sigma}_0(\Gamma')$ to

$\overline{\Sigma}_0(\Gamma))$.

Exercise 12.43. Keep the hypotheses of the previous exercise and let $\overline{\pi}_0(f)$ be the homomorphism from $\overline{\pi}(\Sigma_0(\Gamma'))$ to $\overline{\pi}(\Sigma_0(\Gamma))$ induced by f (Exercise 12.39). Prove that $\overline{\pi}_0(f)$ is injective.

Exercise 12.44. Keep the notation of Exercise 12.39 and let $\pi(f)$ be the embedding of $\pi(\Gamma')$ into $\pi(\Gamma)$ induced by f (see Section 12.2.4). Let p_0 and p'_0 be the projections of $\pi(\Gamma)$ and $\pi(\Gamma')$ onto $\pi(\Sigma_0(\Gamma))$ and $\pi(\Sigma_0(\Gamma'))$, respectively (Theorem 12.62). Prove that we can choose the base vertices for $\pi(f)$, $\pi_0(f)$, p_0 and p'_0 in such a way that $\pi_0(f) \cdot p'_0 = p_0 \cdot \pi(f)$.

Exercise 12.45. Prove that, if $Gr_0(\Gamma)$ has a good system of lines, then $\pi(\Gamma) \cong \pi(Gr_0(\Gamma))$.

Exercise 12.46. Let $Gr_0(\Gamma)$ have a good system of lines. Prove that the universal topological cover of $Gr_0(\Gamma)$ is the 0-Grassmann geometry of the universal topological cover of Γ.

Exercise 12.47. Let $Gr_0(\Gamma)$ have a good system of lines. Prove that the topological covers of $Gr_0(\Gamma)$ are the 0-Grassmann geometries of the topological covers of Γ.

Exercise 12.48. Let J be a nonempty subset of the set I of types of an irreducible geometry Γ of rank $n > m \geq 1$. Prove that the J-Grassmann geometries of the m-covers of Γ are m-covers of $Gr_J(\Gamma)$ (hence, if $Gr_J(\Gamma)$ is m-simply connected, then Γ is m-simply connected).

Problem. A homomorphism from $\pi(Gr_0(\Gamma))$ to $\pi(\Gamma)$ can be constructed as follows. Given a chamber $C = (F_1, F_2, ..., F_n)$ of $Gr_0(\Gamma)$, let us set $f(C) = \bigcup_{i=1}^n F_i$. It is easily seen that $f(C)$ is a chamber of Γ. The function f mapping a chamber C of $Gr_0(\Gamma)$ onto $f(C)$ is a surjective (possibly non-injective) morphism from the graph $\mathcal{C}(Gr_0(\Gamma))$ to the graph $\mathcal{C}(\Gamma)$. This morphism induces a homomorphism φ from $\pi(Gr_0(\Gamma))$ to $\pi(\Gamma)$. It is not difficult to prove that f maps the set of closed galleries of $\mathcal{C}(Gr_0(\Gamma))$ based at a given chamber C_0 of $Gr_0(\Gamma)$ onto the set of closed galleries of $\mathcal{C}(\Gamma)$ based at $f(C_0)$. Therefore the homomorphism φ is surjective. Is it always an isomorphism ?

In other words: does the statement of Exercise 12.45 remain valid even if we do not assume that $Gr_0(\Gamma)$ has a good system of lines ?

Chapter 13

Coxeter Diagrams

We are now ready to come back to the main theme of Chapter 7: recovering geometries from diagrams. We are not going to offer an exhaustive account of all theorems where some description is achieved for some class of geometries with a given diagram; this would be too long. We only want to show how the tools prepared in chapters 8-12 can be used to continue the work started in Chapter 7, with no aim to finish it.

Actually, important progresses with respect to Chapter 7 have already been done in Section 9.6, where finite flag-transitive matroids have been classified, and in 11.3.3, where something like a classification has been obtained for flag-transitive thin geometries.

In this chapter we consider Coxeter diagrams, focusing on the thin case (Section 13.2) and on Dynkin diagrams (Section 13.3). The investigation of geometries belonging to the Dynkin diagrams C_n and F_4 will be continued in Chapter 14. Finally, we will consider the diagram $L.C_{n-1}$ (Chapter 15).

Buildings are the most important geometries belonging to Coxeter diagrams. We give a brief introduction to buildings in the next section. Many important concepts and results will be passed under silence. The reader wanting to know more on this subject should consult [222] and [227], or [14], [180] or [186].

13.1 Buildings and BN-pairs

Many propositions and theorems of this section are taken from [222] and [227]. We will state them without proofs, except when a few lines will be sufficent for a proof.

13.1.1 Buildings

We need a general definition of subgeometries before coming to buildings. Given a geometry $\Gamma = (V, *)$ and a nonempty set X of the set of vertices V of Γ, we say that X defines a *subgeometry* of Γ if the graph $\mathcal{X} = (X, *)$ induced on X by the incidence graph of Γ is a geometry. If I and $t : V \longrightarrow I$ are the set of types and the type function of Γ, then $t(X)$ and the restriction of t to X are taken as set of types and type function respectively for the geometry $\mathcal{X} = (X, *)$. We can define buildings now.

Given a Coxeter diagram M, we denote by $\Gamma(M)$ the Coxeter complex of type M, as in Section 11.3.2. Let Γ be a geometry belonging to M. We say that Γ is a *building* of *type* M if it admits a family $\mathcal{A}$ of thin subgeometries satisfying the following:

($\mathbf{B}_1$) all members of $\mathcal{A}$ are isomorphic to $\Gamma(M)$;

($\mathbf{B}_2$) given any two chambers C, D of Γ, there is a member $\mathcal{X}$ of $\mathcal{A}$ such that both C and D are chambers of $\mathcal{X}$;

($\mathbf{B}_3$) given a chamber C and a flag F of Γ and members $\mathcal{X}$, $\mathcal{Y}$ of $\mathcal{A}$ containing both C and F, there is a type-preserving isomorphism α from $\mathcal{X}$ to $\mathcal{Y}$ such that $\alpha(C) = C$ and $\alpha(F) = F$.

(Note that we do not assume in $\mathbf{B}_3$ that α is induced by some automorphism of Γ.) The members of $\mathcal{A}$ are called *apartments* and $\mathcal{A}$ is called an *apartment system* for Γ. Note that Γ might be thin. In that case Γ is its own unique apartment.

Given a chamber C and a flag F of Γ, a gallery γ from C to a chamber containing F is said to be *shortest* if no gallery from C to a chamber containing F is shorter than γ. The *convex hull* $[C, F]_\Gamma$ of C and F in Γ is the set of chambers of Γ belonging to some shortest gallery from C to a chamber containing F. The following can be obtained as a consequence of $\mathbf{B}_1$, $\mathbf{B}_2$ and $\mathbf{B}_3$ (see [222], 3.4):

($\mathbf{B}_4$) for every apartment $\mathcal{X} \in \mathcal{A}$, for every chamber C of $\mathcal{X}$ and every flag F of $\mathcal{X}$, the convex hull $[C, F]_\Gamma$ is contained in the set of chambers of $\mathcal{X}$.

Note that a building might admit more than one apartment system. However, the following holds:

Proposition 13.1 *Let* $(\mathcal{A}_k)_{k \in K}$ *be a family of apartment systems for a building* Γ *and let* $\mathcal{A} = \bigcup_{k \in K} \mathcal{A}_k$. *Then* $\mathcal{A}$ *is an apartment system for* Γ.

Proof. This proposition is stated as a remark in [222], without proof. However, its proof is not completely trivial. We therefore give that proof here.

Trivially, $\mathbf{B}_1$ and $\mathbf{B}_2$ hold for the family $\mathcal{A}$ defined as above. To prove $\mathbf{B}_3$ we need the following general property of Coxeter complexes:

(1) Let $\mathcal{X}$ and $\mathcal{Y}$ be distinct models of the same Coxeter complex of type M and let γ and δ be galleries of $\mathcal{X}$ and $\mathcal{Y}$ respectively, of the same type. Then there is a (unique) type-preserving isomorphism from $\mathcal{X}$ to $\mathcal{Y}$ mapping γ onto γ'.

Let us prove (1). Let C, D be the initial chambers of the galleries γ and δ. We recall that we have defined $\Gamma(M)$ in Section 11.3.2 as a coset geometry in $W(M)$. It is clear from Section 10.1.2 that there are (uniquely determined) isomorphisms f and g from $\Gamma(M)$ to $\mathcal{X}$ and $\mathcal{Y}$ respectively, mapping $1 \in W(M)$ onto C and D respectively. Then gf^{-1} is an isomorphism from $\mathcal{X}$ to $\mathcal{Y}$. It maps γ onto δ.

We can now prove $\mathbf{B}_3$ for $\mathcal{A}$. Let C, F, $\mathcal{X}$ and $\mathcal{Y}$ be as in the hypotheses of $\mathbf{B}_3$ and let γ be a shortest gallery from C to F. Then γ belongs to both $\mathcal{X}$ and $\mathcal{Y}$, by $\mathbf{B}_4$. By (1), there is a type-preserving isomorphism $\alpha : \mathcal{X} \longrightarrow \mathcal{Y}$ mapping γ onto itself. In particular, $\alpha(C) = C$ and $\alpha(F) = F$. □

By the above proposition, the join of all apartment systems of a building Γ is still an apartment system for Γ, the *maximal* one. Thus, we can speak of apartments of Γ without saying which apartment system we are thinking of; we may always assume to have chosen the maximal one.

A building is said to be of *spherical type* or of *affine type* if its Coxeter diagram is spherical or of affine type, respectively. It follows from $\mathbf{B}_2$ that the chamber system of a building Γ has finite diameter if and only if Γ is of spherical type. Therefore, a building is finite if and only if it is of spherical type and locally finite.

Proposition 13.2 *The apartments of a building of spherical type are the convex hulls of pairs of chambers at maximal distance.*

Proof. The following holds in every Coxeter complex $\mathcal{X}$ of spherical type:

(2) for every chamber C of $\mathcal{X}$, there is a (unique) chamber D at maximal distance from C and we have $\mathcal{X} = [C, D]_{\mathcal{X}}$.

The reader can find a proof of (2) in [222] (2.34 and 2.36). Otherwise, he can obtain it from properties of Coxeter groups of spherical type [11] (we recall that we have defined Coxeter complexes as coset geometries in Coxeter groups).

Let Γ be a building of spherical type. Given two chambers C, D at maximal distance in Γ and an apartment $\mathcal{X}$ containing them, we have $\mathcal{X} = [C, D]_\Gamma$, by $\mathbf{B}_4$ and (2). On the other hand, given an apartment $\mathcal{X}$ and a pair of chambers C, D of $\mathcal{X}$ at maximal distance in $\mathcal{X}$, the chambers C, D have maximal distance in Γ too, by $\mathbf{B}_4$. Therefore the apartments of Γ are the convex hulls of pairs of chambers of Γ at maximal distance. □

Corollary 13.3 *A building of spherical type admits just one apartment system.*

(Trivial, by the previous proposition.) We have already mentioned the following property of buildings in Section 6.6:

Proposition 13.4 (Tits [222]) *The strong intersection property holds in all buildings.*

Note that there are many geometries belonging to Coxeter diagrams and satisfying **IP** that are not buildings. For instance, **IP** holds in the halved icosahedron, which is not a Coxeter complex (whence it is not a building, as it is thin). Indeed, it is a 2-fold quotient of the Coxeter complex of type H_3 (namely, of the icosahedron). More generally, **IP** holds in all tessellations of surfaces, but only eight of them are models of Coxeter complexes: the tetrahedron, the octahedron and its dual (the cube), the icosahedron and its dual (the dodecahedron), the tessellation of the Euclidean plane into squares, the tessellation of the plane into equilateral triangles and its dual (the tessellation of the plane into regular hexagons).

Let us turn to residues. Let Γ be a building of type M over the set of types I and let $\mathcal{A}$ be an apartment system of Γ. Given a flag F of Γ of cotype $J \neq \emptyset$, let $\mathcal{A}(F)$ be the set of apartments $\mathcal{X} \in \mathcal{A}$ containing F. For every $\mathcal{X} \in \mathcal{A}$, the residue $\mathcal{X}_F$ of F in $\mathcal{X}$ is a Coxeter complex of type M_J, where M_J is the diagram induced by M on J, as in Section 11.3.1.

We set $\mathcal{A}_F = \{\mathcal{X}_F \mid \mathcal{X} \in \mathcal{A}(F)\}$

Proposition 13.5 *The residue Γ_F of F in Γ is a building of type M_J, with $\mathcal{A}_F$ as an apartment system.*

(The proof is straightforward; we leave it for the reader.)

Given buildings Γ_1, Γ_2, ..., Γ_m over mutually disjoint sets of types I_1, I_2, ..., I_m, let $\mathcal{A}_i$ be an apartment system of Γ_i (for $i = 1, 2, ..., m$). We set $\Gamma = \bigoplus_{i=1}^m \Gamma_i$ and $\bigoplus_{i=1}^m \mathcal{A}_i = \{\bigoplus_{i=1}^m \mathcal{X}_i \mid \mathcal{X}_i \in \mathcal{A}_i,\ i = 1, 2, ..., m\}$.

The (easy) proof of the following proposition is left for the reader.

Proposition 13.6 *The direct sum* $\bigoplus_{i=1}^{m} \Gamma_i$ *is a building and* $\bigoplus_{i=1}^{m} \mathcal{A}_i$ *is an apartment system for it.*

Conversely, let Γ be a building of type M with apartment system $\mathcal{A}$ and let M be reducible with irreducible components M_1, M_2, ..., M_m. For every $h = 1, 2, ..., m$, let Γ_h be the irreducible component of Γ relative to the component M_h of M and let I_h be the set of types of M_h. Using Proposition 13.5, the reader can easily prove the following:

Proposition 13.7 *For every* $h = 1, 2, ..., m$, *the geometry* Γ_h *is a building and, for every flag* F *of* Γ *of cotype* I_h, *the apartment system* $\mathcal{A}_F$ *of* Γ_F *is (isomorphic to) an apartment system of* Γ_h $(\cong \Gamma_F)$.

Remark. The definition of buildings stated in this section is that of [227], where Tits drops the thickness hypothesis formerly made in [222]. On the other hand, Coxeter diagrams where not mentioned at all in the definition of buildings of [222]. The thickness assumption was needed in [222] to prove that apartments of thick buildings as defined in [222] are Coxeter complexes. However, if we assume in the definition of buildings that apartments are Coxeter complexes, as we have done here (and as Tits does in [227]), then we no longer need the thickness assumption and most of the properties proved in [222] in the thick case remain true in the non-thick case as well. In particular, $\mathbf{B}_4$ remains true. It will be understood that, whenever we refer to [222] for properties of buildings, we will be considering properties that also hold in the non-thick case.

13.1.2 Examples

Thin buildings are just Coxeter complexes. Thus, **IP** holds in all Coxeter complexes, by Proposition 13.4.

Trivially, every geometry of rank 1 is a building, with pairs of distinct elements as apartments.

Generalized m-gons are buildings. When $m < \infty$, then the apartments are the circuits of length $2m$ in the incidence graph. When $m = \infty$, then the apartments are the two-sided infinite paths of the incidence graph with mutually distinct vertices (see Exercises 1.5 and 1.6).

It is clear from Section 9.3.2 that n-dimensional projective geometries are buildings of type A_n. As all geometries belonging to the Coxeter diagram A_n are projective geometries (Corollary 7.7), buildings of type A_n and n-dimensional projective geometries are just the same things.

Polar spaces of rank n are buildings of type C_n (see Section 9.4.2). On the other hand, all buildings of type C_n are polar spaces, by Proposition

13.4 and Theorem 7.38. Therefore, buildings of type C_n and polar spaces of rank n are the same things.

Let $\Gamma = \Gamma(\mathcal{P})$ be the unfolding of a top-thin polar space $\mathcal{P}$ of rank $n \geq 4$. The unfolding construction transforms the apartments of $\mathcal{P}$ into thin subgeometries of Γ of type D_n isomorphic to the Coxeter complex of type D_n. It is easily seen that $\mathbf{B}_2$ and $\mathbf{B}_3$ hold for these subgeometries of Γ. Hence Γ is a building. Thus, the terminology of Sections 1.3.7 and 5.3.4 is consistent with the general definition of buildings stated in this chapter.

All geometries belonging to the Coxeter diagram D_n are unfoldings of top-thin polar spaces (Theorem 7.40). Therefore, all D_n geometries are buildings (and **IP** holds in them, by Proposition 13.4).

Thick buildings of type E_6, E_7, E_8 and F_4 have been classified by Tits [222]. We are not going to report on that classification here. The reader should consult [222] (Theorem 6.13 and Chapter 10) for it. Theorem 1.3 is a classification of thick projective geometries of dimension $n \geq 3$. Thick lined polar spaces of rank $n \geq 3$ have also been classified by Tits, as we remarked in Section 1.3.6 (Theorem 1.5 and remarks following it). By Theorem 7.40, the classification of finite thick-lined polar spaces of rank $n \geq 3$ also entails the classification of thick buildings of type D_n. Thus, all thick buildings belonging to Dynkin diagrams of rank ≥ 3 are classified.

A reducible geometry Γ is the direct sum of its irreducible components (Corollary 4.8) and these are buildings if and only if Γ is a building (Propositions 13.6 and 13.7). Thus, all thick buildings belonging to Coxeter diagrams where all connected components are Dynkin diagrams of rank $\neq 2$ are classified, too.

Let us turn to the two irreducible spherical diagrams of rank ≥ 3 that are not Dynkin diagrams, namely H_3 and H_4. No locally finite thick generalized pentagons exist, by Proposition 3.3 and Theorem 3.6. Therefore, no locally finite thick buildings exist for H_3 and H_4. But more than this is known (see [225]; also [222], Addenda):

Proposition 13.8 *There are no thick buildings of type H_3 or H_4.*

Corollary 13.9 *The Coxeter complexes of type H_3 and H_4 are the unique flag-transitive buildings of type H_3 or H_4.*

Proof. The flag-transitivity implies the existence of orders. Furthermore, if an ordinary pentagon admits order, then it admits uniform order, by Proposition 3.3. The conclusion follows from Proposition 13.8. □

13.1.3 BN-pairs

Given a Coxeter diagram M over a set of types I and a building Γ of type M, let $\mathcal{A}$ be an apartment system of Γ and let G be a subgroup of $Aut_s(\Gamma)$ stabilizing $\mathcal{A}$. We say that G is *chamber-apartment* transitive in Γ with respect to $\mathcal{A}$ if it is transitive on the set of pairs $(C, \mathcal{X})$ with $\mathcal{X} \in \mathcal{A}$ and C a chamber of $\mathcal{X}$ (trivially, the chamber-apartment transitivity implies the flag-transitivity, by $\mathbf{B}_2$). For instance, the groups considered in Section 9.3.3 for projective geometries, in Section 9.4.3 for polar spaces and in Section 9.4.4 for D_n-buildings are chamber-apartment transitive.

Given a chamber C_0 and an apartment $\mathcal{X}_0$ containing it, let B and N be the stabilizers in G of C_0 and $\mathcal{X}_0$ respectively. By the flag-transitivity of G, we have $\bigcap_{g\in G} gBg^{-1} = 1$ (Corollary 10.8). By the hypotheses made on G, the subgroup N acts flag-transitively in $\mathcal{X}_0$. Let H denote the kernel of that action; then the factor group N/H is the special automorphism group of the Coxeter complex $\mathcal{X}_0$ ($\cong \Gamma(M)$), by Corollary 9.2. That is, N/H is the Coxeter group $W(M)$ of type M. It acts chamber regularly in $\mathcal{X}_0$ (Corollary 9.3). Therefore, $H = B \cap N = \bigcap_{n\in N} nBn^{-1}$.

Let $(s_i)_{i\in I}$ be the system of Coxeter involutions of $W(M) = N/H$ relative to the chamber C_0 (Theorem 9.4) and, for every $i \in I$, let $\bar{s}_i$ be a representative of s_i in N. For every $i \in I$ and every $n \in N$ we have $\bar{s}_i Bn \subseteq B\bar{s}_i nB \cup BnB$ (see [222], Chapter 3). Therefore $\langle \bar{s}_i, B\rangle = B \cup B\bar{s}_i B$.

The subgroup $\langle \bar{s}_i, B\rangle$ is the stabilizer of the panel of C_0 of cotype i. Therefore, Γ is thin at i if and only if B has index 2 in $\langle \bar{s}_i, B\rangle$. This happens if and only if $\bar{s}_i$ normalizes B.

The argument used to prove Proposition 9.7 can easily be rephrased in general. Thus, we obtain that $G = BNB$ (hence $G = \langle B, N\rangle$). This is called the *Bruhat decomposition* of G.

According to the next definition, we can summarize the above saying that (B, N) is a BN-pair in G. Indeed, given a group G and a Coxeter group $W(M)$ of type M with set of Coxeter involutions $(s_i)_{i\in I}$, two subgroups B and N of G are said to form a *BN-pair* of *type* M if all the following hold:

($\mathbf{BN}_1$) $G = \langle B, N\rangle$;

($\mathbf{BN}_2$) $B \cap N = \bigcap_{n\in N} nBn^{-1}$ (hence $B \cap N \trianglelefteq N$);

($\mathbf{BN}_3$) there is a surjective homomorphism $\nu : N \longrightarrow W(M)$ such that $Ker(\nu) = B \cap N$ and $\bar{s}_i Bn \subseteq BnB \cup B\bar{s}_i nB$ for every $n \in N$, for every $i \in I$ and for every $\bar{s}_i \in \nu^{-1}(s_i)$;

($\mathbf{BN}_4$) $\bigcap_{g\in G} gBg^{-1} = 1$.

The factor group $N/(B \cap N)$ is called the *Weyl group* of the BN-pair (B, N) and a homomorphism ν as in $\mathbf{BN}_3$ is called a *distinguished homomorphism* from N to $W(M)$.

The set of types I of M and the rank $|I|$ of M are called the *set of types* and the *rank* of the BN-pair (B, N).

In what follows, (B, N) is a BN-pair of type M in a group G, over a set of types I with $|I| \geq 2$. We write H for $B \cap N$, for short. We set $S = \{\bar{s}_i\}_{i \in I}$, where the elements $\bar{s}_i$ of N $(i \in I)$ are as in $\mathbf{BN}_3$. Trivially, $\langle H \cup S \rangle = N$. For every subset J of I, we set $N^J = \langle \{\bar{s}_j\}_{j \in J}, H \rangle$. In particular, $N^I = N$ and $N^\emptyset = H$. Trivially, $N^J = \nu^{-1}(W^J)$ where ν is as in $\mathbf{BN}_3$ and $W^J = \langle s_j \rangle_{j \in J} \leq W(M)$.

We set $G^i = \langle B, \bar{s}_i \rangle$ for every $i \in I$, $G^J = \langle G^j \rangle_{j \in J}$ for every nonempty subset J of I and $G^\emptyset = B$.

For every element $w \in W(M)$, we choose a word $\sigma(w)$ of minimal length in the letters $\bar{s}_i$ $(i \in I)$ such that $\nu(\sigma(w)) = w$. As $\bar{s}_i^2 \in H$ for every $i \in I$, we can assume to have chosen those words in such a way that none of the letters $\bar{s}_i$ appears with exponent -1 in any of them.

Lemma 13.10 *Given $a_1, b_1, a_2, b_2 \in B$ and $n_1, n_2 \in N$, if $a_1 n_1 b_1 = a_2 n_2 b_2$, then $n_1^{-1} n_2 \in H$.*

Proof. Let us write w_i for $\sigma(\nu(n_i))$ $(i = 1, 2)$, for short. Let $a_1 n_1 b_1 = a_2 n_2 b_2$. Hence $b_2 b_1^{-1} = n_2^{-1} a_2^{-1} a_1 n_1$. Therefore $w_2^{-1} b w_1 \in B$ for some $b \in B$. Iteratively applying $\mathbf{BN}_3$, we get $w_3^{-1} w_1 \in B$ for a suitable subword w_3 of w_2. Trivially, $w_3^{-1} w_1 \in B \cap N = H$. Therefore w_1 and w_3 represent the same element of N/H. Since w_1 has minimal length with respect to the property of representing $\nu(n_1)$, the word w_3 is not shorter than w_1. On the other hand, w_3 is a subword of w_2. Therefore w_2 is not shorter than w_1. Repeating the above argument starting from the equality $a_1^{-1} a_2 = n_1 b_1 b_2^{-1} n_2^{-1}$, we obtain that w_1 is not shorter than w_2. Therefore w_1 and w_2 have the same length and $w_3 = w_1$, as w_3 is a subword of w_1 not shorter than w_2. Therefore $w_1^{-1} w_2 \in H$. That is, $n_1^{-1} n_2 \in H$ □

Lemma 13.11 (Bruhat decomposition) *We have $G^J = BN^J B$ for all $J \subseteq I$. In particular, $G = BNB$.*

(This is a trivial consequence of $\mathbf{BN}_3$.)

Lemma 13.12 *We have $G^J \cap G^K = G^{J \cap K}$ for all $J, K \subseteq I$.*

Proof. We have $G^J \cap G^K = B(N^J \cap N^K)B$ by Lemmas 13.11 and 13.10. On the other hand, $N^J \cap N^K = \nu^{-1}(W^J \cap W^K)$, $N^{J \cap K} = \nu^{-1}(W^{J \cap K})$ and $W^J \cap W^K = W^{J \cap K}$ (see $\mathbf{P}'_2$ of Exercise 10.3 and Section 11.3.2). Therefore $G^J \cap G^K = G^{J \cap K}$, by Lemma 8. □

Lemma 13.13 *We have* $(G^J G^K) \cap (G^J G^L) = G^J(G^K \cap G^L)$ *for every choice of subsets* J, K, L *of* I.

Proof. By $\mathbf{BN}_3$ and Lemma 13.11 we have $G^J G^K = BN^J N^K B$, $G^J G^L = BN^J N^L B$ and $G^J G^{K\cap L} = BN^J N^{K\cap L} B$. Furthermore, $G^K \cap G^L = G^{K\cap L}$, by Lemma 13.12. Therefore, the equality we want to prove is equivalent to the following one: $(N^J N^K) \cap (N^J N^L) = N^J N^{K\cap L}$, which is in turn equivalent to the relation $(W^J W^K) \cap (W^J W^L) = W^J W^{K\cap L}$. The latter holds, by properties of Coxeter systems (Section 11.3.2 and $\mathbf{P}_{3,3}$ of Exercise 10.2). □

Lemma 13.14 *The family* $\mathcal{G} = (G^i)_{i\in I}$ *of subgroups of* G *is a geometric parabolic system in* G *with Borel subgroup* B *and minimal parabolics* G^i $(i \in I)$.

Proof. We have defined $\mathcal{G}$ by means of minimal parabolics, as in Section 12.4.1. The property $\mathbf{C}''_4$ of 12.4.1 is just $\mathbf{BN}_4$. $\mathbf{C}''_1$ is trivial and $\mathbf{C}''_3$ is an easy consequence of $\mathbf{BN}_1$ and $\mathbf{BN}_3$. The property $\mathbf{C}''_2$ of Section 12.4.1 is a special case of Lemma 13.12. Therefore, $\mathcal{G}$ is a parabolic system with Borel subgroup B. Let us prove that $\mathcal{G}$ is geometric. Let us set $G_i = G^{I-i}$ for every $i \in I$, $G_J = \bigcap_{j\in J} G_j$ for every nonempty subset J of I and $G_\emptyset = G$. It is easily seen that $G^J = G_{I-J}$ for every $J \subseteq I$, by Lemma 13.12. Therefore $G^i = G_{I-i}$ for every $i \in I$ and $\mathbf{P}_2$ of Section 10.1.3 holds in $\mathcal{G}$ (see also Exercise 10.3). The statement of Lemma 13.13 is nothing but $\mathbf{P}_{3,3}$ of Exercise 10.2, which is equivalent to $\mathbf{P}_3$ of Section 10.1.3. Therefore $\mathbf{P}_3$ also holds in $\mathcal{G}$. Hence $\mathcal{G}$ is geometric. □

Given $\mathcal{G}$ as in the previous lemma, let $\Gamma(\mathcal{G})$ be the geometry (of the geometric chamber system) defined by $\mathcal{G}$ in G. Given $g \in G$, we set $\mathcal{N}_g = \{gnB\}_{n\in N}$. As the chambers of $\Gamma(\mathcal{G})$ are represented by right cosets of B, the set $\mathcal{N}_g$ represents a set of chambers of $\Gamma(\mathcal{G})$. Hence we can identify $\mathcal{N}_g$ with the subgraph of $\mathcal{C}(\Gamma(\mathcal{G}))$ induced on that set of chambers.

We recall that $W(M)$ represents the set of chambers of the Coxeter complex $\Gamma(M)$. Given $w \in W(M)$, we set $\mu_g(w) = g\sigma(w)B$. This clause defines a mapping μ_g from the chamber system of $\Gamma(M)$ to the subgraph $\mathcal{N}_g$ of the chamber system $\mathcal{C}(\Gamma(\mathcal{G}))$ of $\Gamma(\mathcal{G})$. Chamber systems can also be viewed as coloured graphs, with types as colours.

Lemma 13.15 *The mapping* μ_g *is an isomorphism of coloured graphs.*

Proof. As $\mathcal{N}_g$ is obtained by left-multiplying $\mathcal{N}_1$ by g, proving the above statement in the case of $g = 1$ is sufficent to prove it in the general case as well.

The mapping μ_1 is evidently surjective. Let $\sigma(u)B = \sigma(v)B$ for $u, v \in W(M)$. Then $\sigma(v)^{-1}\sigma(u) \in B \cap N = H$, whence $u = v$. Therefore μ_1 is also injective. Note that $G^i = B \cup B\bar{s}_i B = \nu^{-1}(\langle s_i \rangle)$ by $\mathbf{BN}_3$ and Lemma 13.11. It is clear from this remark that, given $u, v \in W(M)$ and $i \in I$, we have $v^{-1}u \in \langle s_i \rangle$ if and only if $\sigma(v)^{-1}\sigma(u) \in G^i$. The lemma is proved. □

Since chamber systems uniquely determine their geometries when they are viewed as coloured graphs (Theorem 1.23), the isomorphism of coloured graphs μ_g uniquely determines an isomorphism from the Coxeter complex $\Gamma(M)$ to a thin subgeometry $\mathcal{X}_g$ of $\Gamma(\mathcal{G})$ with $\mathcal{N}_g$ as set of chambers. We set $\mathcal{A} = \{\mathcal{X}_g\}_{g \in G}$.

Theorem 13.16 *The geometry $\Gamma(\mathcal{G})$ is a building of type M and $\mathcal{A}$ is an apartment system for it.*

The group G (acting by left multiplication on the right cosets of B) is a chamber–apartment transitive subgroup of $Aut_s(\Gamma(\mathcal{G}))$.

The subgroup N is the stabilizer in G of the apartment $\mathcal{X}_1$ and H is the kernel of the action of N in $\mathcal{X}_1$.

Proof. The property $\mathbf{B}_1$ holds, as the subgeometries $\mathcal{X}_g$ ($g \in G$) are isomorphic to $\Gamma(M)$.

Let C_0 be the chamber represented by B. Given any other chamber C of $\Gamma(\mathcal{G})$, we have $C = g(C_0)$ for some $g \in G$, by the flag-transitivity of G in $\Gamma(\mathcal{G})$. By Lemma 13.11, we have $gB = bnB$ for some $b \in B$ and some $n \in N$. Therefore $C = bn(C_0)$ and both C and C_0 are chambers of $\mathcal{X}_b$. Let $C = g(C_0)$ and $D = f(C_0)$ be any two chambers of $\Gamma(\mathcal{G})$. By the above, there is a subgeometry $\mathcal{X}_b$ ($b \in B$) containing both $C_0 = g^{-1}(C)$ and $g^{-1}f(C_0) = g^{-1}(D)$. Trivially, both C and D belong to $\mathcal{X}_{gb}$. Therefore $\mathbf{B}_2$ holds in $\Gamma(\mathcal{G})$.

Let us prove that $\mathbf{B}_3$ also holds in $\Gamma(\mathcal{G})$. Let $\mathcal{X}$, $\mathcal{Y}$, C and F be as in the hypotheses of $\mathbf{B}_3$. By the flag-transitivity of G we can assume that $C = C_0$. As $C = C_0$ is a chamber of $\mathcal{X}$, we have $\mathcal{X} = \mathcal{X}_g$ with $g \in G$ such that $gnB = B$ for some $n \in N$. Hence $g = bn^{-1}$ for some $b \in B$. Multiplying on the left by b^{-1}, we map $\mathcal{X}_g$ onto $\mathcal{X}_1$, fixing C_0. Thus, we can also assume that $\mathcal{X} = \mathcal{X}_1$. By Lemma 13.11, we have $\mathcal{Y} = \mathcal{X}_{bnb'}$ for suitable $b, b' \in B$ and $n \in N$. We can assume $b' = 1$, as C_0 belongs to $\mathcal{Y}$. Therefore $\mathcal{Y} = \mathcal{X}_{bn} = \mathcal{X}_b$. The flag F belongs to $\mathcal{X} = \mathcal{X}_1$. Therefore F is represented by a right coset uG^J with $u \in N$, where J is the cotype of F. On the other hand, F also belongs to $\mathcal{Y} = \mathcal{X}_b$. Therefore we have $bvG^J = uG^J$ for some $v \in N$. We have $G^J = BN^JB$ by Lemma 13.11. Therefore $b_1\sigma(w)b_2u^{-1} = v^{-1}b^{-1}$ for suitable $b_1, b_2 \in B$ and $w \in W^J = \langle s_j \rangle_{j \in J}$. As $\sigma(w)$ has minimal length with respect to the property of representing w, it only involves letters $\bar{s}_j$ with $j \in J$. Indeed the analogous statement holds for words of minimal length representing w in $W(M)$ with $(s_i)_{i\in I}$ as set of letters (see [11], Chapter 4). Applying $\mathbf{BN}_3$, we obtain $b_1\sigma(w)b_2u^{-1} = b_1'\bar{w}u^{-1}b_2'$ for suitable $b_1', b_2' \in B$ and $\bar{w} \in N^J$. Therefore $b_1'\bar{w}u^{-1}b_2' = v^{-1}b^{-1}$. By Lemma 13.10 we have $v\bar{w}H = uH$. Therefore, $vG^J = uG^J$ and $buG^J = bvG^J = uG^J$. Hence the left-multiplication by b maps $\mathcal{X}_1$ $(= \mathcal{X})$ onto $\mathcal{X}_b$ $(= \mathcal{Y})$ and it fixes C_0 $(= C)$ and the flag F (represented by uG^J). The property $\mathbf{B}_3$ is proved. Therefore $\Gamma(\mathcal{G})$ is a building.

Trivially, G is chamber-apartment transitive in $\Gamma(\mathcal{G})$. It remains to prove that N and H are the setwise and the elementwise stabilizers of $\mathcal{X}_1$, respectively. Clearly, H fixes all elements of $\mathcal{X}_1$ and N setwise fixes $\mathcal{X}_1$. Let $g \in G$ be such that $gnB = nB$ for all $n \in N$. Then $n^{-1}gn \in B$ for all $n \in N$. That is, $g \in \bigcap_{n\in N} nBn^{-1}$. Hence $g \in H$ by $\mathbf{BN}_2$. Let now $f \in G$ be such that $fnB \in \mathcal{N}_1$ for every $n \in N$. Then $fB = uB$ for some $u \in N$, whence $f = ub$ for some $b \in B$ such that $fnB \in \mathcal{N}_1$ for all $n \in N$. Therefore, for every $n \in N$ there is an element $n' \in N$ such that $bnB = n'B$. By Lemma 13.10 we have $nH = n'H$. Hence $bnB = nB$ for every $n \in N$. By the above, $b \in H$. Therefore $g \in N$. □

Proposition 13.17 *The geometry $\Gamma(\mathcal{G})$ is thin at i if and only if $\bar{s}_i$ normalizes B.*

Proof. Let F be the panel represented by G^i. Let $\bar{s}_i$ normalize B. Then $G^i = B \cup \bar{s}_i B$, whence F is contained in precisely two chambers, represented by B and $\bar{s}_i B$ respectively. Hence $\Gamma(\mathcal{G})$ is thin at i. Conversely, let $\Gamma(\mathcal{G})$ be thin at i. Then F is contained in just two chambers. On the other hand, $b\bar{s}_i B \subseteq G^i$ for every $b \in B$, by Lemma 13.11. As F is contained in precisely two chambers and $\bar{s}_i \notin B$, we have $b\bar{s}_i B = \bar{s}_i B$ for every $b \in B$. That is, $\bar{s}_i^{-1} b \bar{s}_i \in B$ for all $b \in B$. Hence $\bar{s}_i$ normalizes B. □

The next two corollaries are easy consequences of Proposition 13.17:

Corollary 13.18 *The following are equivalent:*

(i) for every $i \in I$, $\bar{s}_i$ normalizes B;
(ii) we have $B = 1$ and $N = G \cong W(M)$;
(iii) $\Gamma(\mathcal{G})$ is thin (hence $\Gamma(\mathcal{G}) \cong \Gamma(M)$).

Corollary 13.19 *The geometry $\Gamma(\mathcal{G})$ is thick if and only if the following holds:*

($\mathbf{BN}_5$) $\bar{s}_i B \bar{s}_i^{-1} \neq B$ *for all $i \in I$.*

A BN-pair satisfying (i) (equivalently, (ii)) of Corollary 13.18 is said to be *trivial.*

We have only considered BN-pairs of rank ≥ 2 in the above, as we have not defined parabolic systems of rank 1 in sections 10.1 and 12.4. A BN-pair of rank 1 defines a building of rank 1. A group G is chamber-apartment transitive in a building Γ of rank 1 if and only if it is 2-transitive on the set of elements of the rank 1 geometry Γ.

Remark. It is implicit in the proof of Theorem 13.16 that, if a building Γ admits a chamber-apartment transitive automorphism group G, then we can choose the isomorphism α in $\mathbf{B}_3$ in such a way that it is induced by some element of G. Such a choice of α in $\mathbf{B}_3$ may be impossible if the special automorphism group of Γ is not chamber-apartment transitive. We may think of non-Desarguesian finite projective planes here, or of non-flag-transitive (hence non-classical) finite generalized quadrangles, or of a non-thick and non-thin projective geometry, or of a product of thick polar spaces ...

Remark. The definition of BN-pairs stated here is different from that of [222], mainly because our definition of buildings is not the same as in [222]. Coxeter groups are not mentioned at all in the definition of BN-pairs of [222] (just as $\mathbf{B}_1$ is absent from the definition of buildings of [222]) and the elements s_i $(i \in I)$ of $\mathbf{BN}_3$ are only assumed to be generators of $N/(B\cap N)$ in [222]. On the other hand, Tits assumes $\mathbf{BN}_5$ of Corollary 13.19 in his definition of BN-pairs [222], as he assumes buildings to be thick in [222]. Using $\mathbf{BN}_5$ and the relation $\bar{s}_i BnB \subseteq BnB \cup B\bar{s}_i nB$ of $\mathbf{BN}_3$, it is possible to prove that $N/(B \cap N)$ is a Coxeter group with $(s_i)_{i\in I}$ as Coxeter set of involutions (see [222]). However, we have dropped the thickness hypothesis in our definition of buildings. Thus, we have no reason to insert $\mathbf{BN}_5$ in our definition of BN-pairs. Therefore, we must assume in our definition of BN-pairs that $N/(B\cap N) \cong W(M)$ with $(s_i)_{i\in I}$ as a Coxeter set of involution, as we cannot obtain this as a theorem (similarly, we have inserted $\mathbf{B}_1$ in the definition of buildings). We also remark that, if we assume $\mathbf{BN}_5$, then the surjective morphism ν of $\mathbf{BN}_3$ is uniquely determined, modulo automorphisms of $W(M)$ stabilizing the set $\{s_i\}_{i\in I}$ of Coxeter involutions [222]. On the other hand, this might be false when $\mathbf{BN}_5$ fails to hold. For instance, if $G = N = W(M)$ and $B = 1$, then ν can be any automorphism of $W(M)$.

The property $\mathbf{BN}_2$ is stated in [222] in the weaker form $B \cap N \trianglelefteq N$. The stronger requirement $B \cap N = \bigcap_{n\in N} nBn^{-1}$ is called the *saturation condition* in [222]. We need it to obtain the third statement of Theorem

13.16.

The property $\mathbf{BN}_4$ is not mentioned at all in the definition of BN-pairs of [222]. This property is the same as $\mathbf{P}_4$ of 10.1.3 and $\mathbf{C}_4''$ of Section 12.4.1 in the definition of parabolic systems. If we drop it, then we should factorize G by $\bigcap_{g\in G} gBg^{-1}$ when interpreting G as an automorphism group of a building of type M.

13.1.4 Examples

BN-Pairs of Spherical Type. Non-trivial BN-pairs of type A_n, C_n and D_n have been considered in Sections 9.3.3, 9.4.3 and 9.4.4.

Let Γ be a thick building of rank ≥ 3 and of irreducible spherical type. It follows from the classification of thick buildings of irreducible spherical type and rank ≥ 3 that Γ arises from a BN-pair ([222], Chapter 11). Furthermore, if all projective planes occurring as rank 2 residues in Γ are coordinatized by a field K (in particular, if Γ is finite), then Γ arises from a BN-pair in the group of K-rational elements of a K-defined subgroup G of a quasi-simple algebraic group $\overline{G}$ over the algebraic closure $\overline{K}$ of K, except when Γ is a polar space defined by a non-degenerate sesquilinear or non-singular quadratic form in an infinite-dimensional vector space ([222]; see [8] and [9] for the theory of BN-pairs in K-defined subgroups of algebraic groups; also Chapter 5 of [222] or Chapter 12 of [92]). BN-pairs as in [8] are called *algebraic* BN-pairs. We are not going to summarize the theory of [8] here. We only remark that the BN-pairs for $PGL(n+1,K)$ (K a field), $PGO(\varphi)$ and $P\Omega^+(2n,K)$ examined in Sections 9.3.3, 9.4.3 and 9.4.4 actually arise from algebraic BN-pairs.

The following is a by-product of results stated in Sections 9.3.4 and 9.4.5: let Γ be a finite thick building of rank $n \geq 3$ and type A_n, C_n or D_n and let G be a flag-transitive subgroup of $Aut_s(\Gamma)$. Then G is chamber-apartment transitive in Γ (hence it admits a BN-pair), except when $\Gamma = PG(3,2)$ and $G = A_7$. Note that, if a BN-pair (B,N) of type A_3 existed in A_7, we should have $B = D_8$ (Sylow 2-subgroup of A_7) because $PG(3,2)$ has 315 chambers and $N \cap B = 1$ because there are 105 apartments in $PG(3,2)$ and each apartment has 24 chambers (and A_7 should be chamber-apartment transitive in $PG(3,2)$). Therefore, $N = S_4$. However, $B \cap N$ contains at least one involution, for every copy B of D_8 and every copy N of S_4 in A_7. Therefore, no BN-pair of type A_3 exists in A_7 (the same conclusion could be obtained from a more general result of [222], 11.7).

The special automorphism group of a direct sum of geometries is the direct product of the special automorphism groups of the summands. Therefore, all thick buildings of spherical type with no irreducible components of rank 2 arise from BN-pairs.

One example of affine type. Let $G = PGL(3, Q_p)$, where Q_p is the field of p-adic numbers, for some prime $p > 1$. By R_p and (p) we denote the ring of p-adic integers and the ideal of R_p generated by p, respectively. Thus, $R_p - (p)$ is the set of invertible elements of R_p. Let T be the group of 3×3 upper triangular matrices with entries in R_p and invertible elements of R_p on the main diagonal. Let P be the set of 3×3 matrices with all entries in (p) and let B be the subgroup of G consisting of the elements represented by matrices of the form $M_T + M_P$ with $M_T \in T$ and $M_P \in P$, with respect to a given ordered basis (e_1, e_2, e_3) of $V(3, Q_p)$. Let H be the subgroup of those elements of B represented by matrices $M_T + M_P$ with M_T diagonal.

For $i = 1, 2$, let s_i be the element of G represented by the linear transformation permuting e_i with e_{i+1} and fixing the remaining vector of the basis. Let s_0 be the element of G represented by the linear transformation mapping e_1 onto pe_3, e_3 onto $p^{-1}e_1$ and fixing e_2. Let W be the subgroup of G generated by s_0, s_1, s_2. The reader can check that s_0, s_1, s_2 are involutions satisfying the Coxeter set of relations of type $\widetilde{A}_2$ and that $W \approx \mathbf{Z}^2 : D_6$. Hence W is the Coxeter group of type $\widetilde{A}_2$, with Coxeter involutions s_0, s_1, s_3 (see Exercise 9.7). Furthermore, $W \cap H = 1$ and W normalizes H. Let us set $N = HW$. It is straightforward to check that B and N form a BN-pair in G, of type $\widetilde{A}_2$.

Let Γ be the building (of affine type $\widetilde{A}_2$) defined by the above BN-pair (Theorem 13.16). We take 0, 1, 2 as types for Γ. Let x_0 be the element of Γ of type 0, represented by the maximal parabolic $G_0 = \langle B, s_1, s_2 \rangle$. It is not difficult to check that the elements of G_0 are just those represented by matrices with entries in R_p and determinant in $R_p - (p)$. The kernel $K_0 = \bigcap_{g \in G_0} gBg^{-1}$ of G_0 in the residue Γ_{x_0} of x_0 consists of those elements of G_0 represented by matrices of the form $aI + M_P$ with $M_P \in P$ and $a \in R_p - (p)$. Therefore $G_0/K_0 \cong PGL(3, p)$. Hence $\Gamma_{x_0} \cong PG(2, p)$ and Γ has uniform order p. Actually, $\Gamma_x \cong PG(2, p)$ for every element x of Γ. When x has type 0, this claim is evident by the above and by the flag-transitivity of G in Γ. The cases with x of type 1 or 2 are left for the reader.

The diagram $\widetilde{A}_2$ is the rank 3 case of the affine Coxeter diagram $\widetilde{A}_{n-1}$, which looks like a cycle of length n (see [11]). It is not difficult to generalize the previous construction to the case of $PGL(n, Q_p)$ with $n \geq 2$, thus obtaining BN-pairs of affine type $\widetilde{A}_{n-1}$ for every $n \geq 2$. Many other BN-pairs of affine type exist and it can be shown that those of rank $n \geq 4$ always arise from groups defined over local fields ([228] and [226]; the reader wanting to know more on this topic may also consult [15]).

The BN-pair constructed above in $PGL(3, Q_p)$ does not arise from any algebraic BN-pair, since algebraic BN-pairs define buildings of spherical type. This makes it clear that there are groups admitting quite different BN-pairs, of different types.

13.1.5 Buildings and universal 2-covers of geometries belonging to Coxeter diagrams

Let $M = (m_{ij})_{i,j \in I}$ be a Coxeter diagram of rank $n \geq 3$ and let Γ be a geometry belonging to M (a geometry of *type* M, for short).

Theorem 13.20 (Tits [227]) *Let the following hold:*

(∗) for every subset J of I of size 3 such that the diagram M_J induced by M on J is C_3 or H_3, all residues of Γ of type J are 2-covered by buildings;

Then the universal 2-cover of the (chamber system of) Γ is (the chamber system of) a building.

That is, if (∗) holds, then the universal 2-cover of the geometry Γ exists and it is a building.

The proof of Theorem 13.20 is neither easy nor short. The reader can find it in [227]. The next corollary is a trivial consequence of Theorem 13.20:

Corollary 13.21 *If M does not contain any induced subdiagram of type C_3 or H_3, then the universal 2-cover of Γ (exists and it) is a building.*

Theorem 13.22 (Tits [227]) *All buildings of rank $n \geq 3$ are 2-simply connected.*

(See [227] for the proof.) Corollary 12.46 and Propositions 12.48, 12.49 and 12.55 are special cases of Theorem 13.22. It will turn out from results of Section 13.3 (Theorems 13.25, 13.26 and 13.29) that Propositions 12.52, 12.53 and 12.54 are also special cases of Theorem 13.22.

Residues in buildings are buildings (Proposition 13.5). Hence (∗) of Theorem 13.20 holds in every geometry 2-covered by a building. Therefore, by Theorems 13.20 and 13.22, buildings of rank $n \geq 3$ are precisely the universal 2-covers of geometries belonging to Coxeter diagrams of rank n and satisfying (∗). By Theorem 12.56, these geometries can be obtained as 2-quotients of buildings by suitable groups. Thus, classifying the geometries of type M that satisfy (∗) is the same as classifying all buildings of type M and their automorphism groups that define 2-quotients. Needless to say, none of these two parts of the classification programme are easy to accomplish in general. However, important results have been obtained for some families of diagrams (see Sections 13.1.2, 13.1.4 for the classification of thick buildings of irreducible spherical type and rank $n \geq 3$; see Section 13.3 for some results on their quotients; see [228] and [226] (also Chapter 10 of [180]) for a classification of thick buildings of affine type and rank $n \geq 4$; see [106], [112], [109], [110] for finite quotients of buildings of affine type).

As residues in buildings are buildings (Proposition 13.5), every residue of rank ≥ 3 in a building is 2-simply connected, by Theorem 13.22. Hence a group defining a 2-quotient of a building residually defines quotients, by Corollary 12.57.

Condition $(*)$ is essential in Theorem 13.20. Non-building 2-simply connected geometries of type C_3 and H_3 are mentioned in [227] (1.6), obtained by some kind of free construction. They are not locally finite. Non-building 2-simply connected finite C_3 geometries also exist. Some of them have thin lines (see [150], Example 3), but we also know a thick example, namely the $Alt(7)$-geometry (we proved in Section 12.6.4 that the $Alt(7)$-geometry is 2-simply connected). The $Alt(7)$-geometry appears as a rank 3 residue in some geometries belonging to Coxeter diagrams of rank ≥ 4 (see [3], [201], [203]). Needless to say, the universal 2-covers of these geometries are not buildings.

13.2 Thin geometries

Theorem 11.19 almost gives us a classification of flag-transitive thin geometries. We will now use Theorem 13.20 to generalize that result to all (possibly non flag-transitive) thin geometries belonging to Coxeter diagrams. In order to apply Theorem 13.20, we need to get control over thin geometries of type C_3 and H_3.

Lemma 13.23 *The octahedron and the halved octahedron are the only thin geometries of type C_3. The icosahedron and the halved icosahedron are the only thin geometries of type H_3.*

Proof. The statement for C_3 is straightforward. Let us turn to H_3. Let Γ be a thin geometry of type H_3

(5)

(H_3) •———•———•

Having drawn the diagram in this way, Γ has a good system of lines (Exercise 4.5). Let N_0, N_1, N_2 be the numbers of points, lines and planes of Γ, respectively. Counting point-line flags, point–plane flags and line–planes flags we see that $3N_0 = 2N_1 = 5N_2$. Hence there is a positive integer k such that $N_0 = 10k$, $N_1 = 15k$ and $N_2 = 6k$.

Given a point a, we say that a plane u has *distance* i from a if it contains some point at distance i from a in the collinearity graph of Γ, but no points at distance $< i$ from a. Trivially, a plane u has distance 0 from a if and only if it is incident to a.

Let u be a plane and let a_0, a_1, a_2, a_3, a_4 be a path of length 4 in the collinearity graph of Γ from $a = a_0$ to a point $a_4 * u$. Let x_i be the line joining a_{i-1} with a_i $(i = 1, 2, 3, 4)$. Stars of points of Γ are ordinary triangles. Therefore there are planes v, w in Γ_{a_1} and Γ_{a_3} respectively incident to x_1 and x_2 and to x_3 and x_4, respectively. In Γ_{a_2} we find a line x incident to both v and w. The residue Γ_w of w is an ordinary pentagon. Hence there are points b_1, b_2 in it such that $x * b_1 \perp b_2 * x_4$. Trivially, $b_2 * u$ and either $b_1 \perp a_1$ or $b_1 = a_1$. Thus, we have obtained a path of length ≤ 3 from a to a point $b_2 * u$. Therefore, every plane has distance ≤ 3 from a.

Trivially, there are precisely three planes at distance 0 from a. If b is a point at distance 1 from a, two of the three planes on b are incident to the line $\{a, b\}$, hence to a. Therefore, just one of those planes has distance 1 from a. As there are three lines on a, there are at most three points at distance 1 from a. Hence there are at most three planes at distance 1 from a.

Let u be a plane at distance 2 from a and let a_0, a_1, a_2 be a path in the collinearity graph of Γ from $a = a_0$ to a point $a_2 * u$. Let x_i be the line joining a_{i-1} with a_i $(i = 1, 2)$. In the triangle Γ_{a_1} we find a plane v incident to both x_1 and x_2. In the triangle Γ_{a_2} we find a line x incident to both v and u. If b is the point of x other than a_2, then b has distance 2 from a. Therefore there are at least two points in u at distance 2 from a. In a similar way we can prove that, if u is a plane at distance three from a, then u is incident to at least two points at distance 3 from a.

Using the above, it is not difficult to prove that there are at most three planes at distance 2 from a and at most three planes at distance 3 from a. Hence there are at most 12 planes in Γ, as every plane of Γ has distance ≤ 3 from a. Therefore $k = 1$ or 2 and Γ has either 10 or 20 points.

It is straightforward to check that, if $k = 2$, then the collinearity graph of Γ is isomorphic to the collinearity graph of the dodecahedron. The planes of Γ are pentagons of that graph. On the other hand, Γ has $6k = 12$ planes and there are 12 pentagons in that graph. Therefore the planes of Γ are the 12 pentagons of the collinearity graph of the dodecahedron. Hence Γ is the dodecahedron (dually, the icosahedron).

It is not difficult to prove that, if $k = 1$, then the collinearity graph of Γ is the Petersen graph, which is the collinearity graph of the halved dodecahedron. There are $6k = 6$ planes in Γ and they appear as pentagons in the collinearity graph of Γ. On the other hand, there are 12 pentagons in the Petersen graph. However, there are only two ways of selecting six pentagons in the Petersen graph so as to construct an H_3 geometry. These two ways give us two models of the halved dodecahedron. Therefore Γ is the halved dodecahedron (dually, the halved icosahedron). □

By the previous lemma and by Theorem 13.20 we obtain the following:

Theorem 13.24 *All thin geometries belonging to Coxeter diagrams of rank ≥ 3 are 2-quotients of Coxeter complexes.*

13.3 Dynkin diagrams

We will now apply the results of Section 13.1.5 to the investigation of geometries belonging to Dynkin diagrams of rank $n \geq 3$.

13.3.1 The general case

We have already remarked in Section 13.1.2 that all geometries of type A_n and D_n are buildings and that a geometry of type C_n is a building if and only if **IP** holds in it (that is, if and only if it is a polar space). Combining Theorem 13.20 with Corollary 12.69 we also obtain the following:

Proposition 13.25 *Every top-thin geometry of type C_n ($n \geq 3$) is a 2-quotient of a polar space*

Only E_6, E_7, E_8 and F_4 remain to examine. The universal 2-cover of a geometry of type E_6, E_7 or E_8 (exists and it) is a building, by Corollary 13.21. On the other hand, all geometries of type E_6 are 2-simply connected (Proposition 12.52). Therefore

Theorem 13.26 *All geometries of type E_6 are buildings.*

All buildings satisfy **IP** (Proposition 13.4). On the other hand, all geometries of type E_7 or E_8 satisfying **IP** are 2-simply connected (Proposition 12.53). Therefore

Theorem 13.27 *A geometry of type E_7 or E_8 is a building if and only if* **IP** *holds in it.*

By this theorem and by Propositions 7.47 and 7.51 we obtain the following corollaries (where points, dual points and lines are defined as in Sections 7.6.2 and 7.6.3):

Corollary 13.28 *A geometry Γ of type E_7 is a building if and only if every line x of Γ is incident to all dual points of Γ that are incident to at least two points of x.*

Corollary 13.29 *A geometry Γ of type E_8 is a building if and only if both the following hold in it:*

(i) every line x is incident to all dual points that are incident to at least two points of x;
(ii) given any two distinct points a, b, if there are distinct dual points incident to both a and b, then a and b are collinear.

Theorem 13.30 *A geometry of type F_4 is a building if and only if* **IP** *holds in it.*

Proof. The 'only if' claim is contained in Proposition 13.4. Turning to the 'if' part, let Γ be a geometry of type F_4 satisfying **IP**. Then all residues of Γ of type C_3 satisfy **IP**, as **IP** is preserved when taking residues. Hence they are polar spaces and Γ is a 2-quotient of a building by Theorem 13.20. On the other hand, Γ is 2-simply connected (Proposition 12.54). Therefore Γ is a building. □

By this theorem and by Proposition 7.53 we obtain the following:

Corollary 13.31 *A geometry Γ of type F_4 is a building if and only it has a good system of lines and both the following hold in it:*

(i) every line x is incident to all dual points that are incident to at least two points of x;
(ii) given any two distinct points a, b, if there are distinct dual points incident to both a and b, then a and b are collinear.

13.3.2 The finite thick case

Theorem 13.32 *All finite thick geometries of type E_7 or E_8 are buildings.*

Proof. The universal 2-cover of a geometry of type E_7 or E_8 is a building by Corollary 13.21 and **IP** holds in it by Proposition 13.4. On the other hand, a finite thick geometry of type E_7 or E_8 satisfying **IP** does not admit any proper 2-quotients, by Proposition 8.26. Therefore, all finite thick geometries of type E_7 or E_8 are buildings. □

We remark that an infinite or non-thick building of type E_7 or E_8 may admit proper 2-quotients (an infinite example is given in [12], for instance).

Finite thick-lined polar spaces of rank $n \geq 3$ do not admit any proper 2-quotients (Corollary 8.17). By this and by Theorem 13.20 we obtain the following:

Proposition 13.33 *A finite thick-lined geometry Γ of type C_n ($n \geq 3$) is a polar space if and only if all residues of Γ of type C_3 are polar spaces.*

Proposition 13.34 *A finite thick geometry Γ of type F_4 is a building if and only if all residues of Γ of type C_3 are polar spaces.*

Proof. The 'only if' part of the statement is a trivial consequence of Proposition 13.5. Let us prove the 'if' claim. Let all residues of Γ of type C_3 be polar spaces. Then the universal 2-cover of Γ is a building by Theorem 13.20 and **IP** holds in it, by Proposition 13.4. On the other hand, a finite thick geometry of type F_4 satisfying **IP** does not admit any proper 2-quotients (Proposition 8.28). Therefore Γ is a building. □

The next corollary is a trivial consequence of Proposition 13.29 and Corollary 12.68:

Corollary 13.35 *All finite thick-lined top-thin C_n geometries are polar spaces.*

Of course, if we aim to a classification of finite thick geometries of type C_n or F_4, then we cannot be satisfied with the above. In order to achieve such a classification we need to know which non-buildings finite thick C_3-geometries exist and if any of them can occur as a rank 3 residue in a C_n geometry with $n \geq 4$ or in a geometry of type F_4. The next chapter will be devoted to this.

Chapter 14

C_n and F_4

This chapter is a continuation of the last section of Chapter 13. We will focus on finite thick-lined geometries of type C_n ($n \geq 3$) and on flag-transitive finite thick geometries of type F_4.

14.1 The diagram C_3

14.1.1 The Ott–Liebler index

In this section Γ is a geometry of type C_3. We do not assume that Γ is finite or thick-lined, for the moment.

Given two distinct dually collinear planes u, v of Γ, the line incident to u and v (unique by Lemma 7.29) will be denoted by $u \wedge v$. The dual collinearity relation will be denoted by $\top$, as in Section 7.4. When we say that $u \top v$ in Γ_a for some point a, then we mean that $u \wedge v \in \Gamma_a$.

We designate the shadow operators of Γ by σ_0, σ_1 and σ_2, as usual, where 0, 1, 2 represent points, lines and planes, respectively. Given a flag $\{x, y\}$, we write $\sigma_i(x, y)$ for $\sigma_i(\{x, y\})$, for short ($i = 0, 1$ or 2).

Given a point–plane flag (a, u) of Γ, let $X(a, u)$ be the set of planes $v \neq u$ such that $a * v \top u$ and $a \notin \sigma_0(u \wedge v)$. We set $\alpha(a, u) = |X(a, u)|$.

Lemma 14.1 *The number $\alpha(a, u)$ does not depend on the particular choice of the point–plane flag (a, u).*

Proof. Given a line x incident to both a and u, let u' be a plane incident to x and distinct from u. Given $v \in X(a, u)$, we have $x \neq u \wedge v$. Let b be the meet point of x and $u \wedge v$ in the projective plane Γ_u. Trivially, $b \neq a$. Let y be the line through a and b in the projective plane Γ_v. As

$x \neq u \wedge v$, we have $y \neq x$. Hence $y \notin \sigma_1(u)$, otherwise x and y would be distinct lines of the projective plane Γ_u meeting in the two distinct points a and b. Therefore, there is a unique plane $v' * b$ such that $y * v' \top u$ in the generalized quadrangle Γ_b. It is easily seen that $v' \wedge u \notin \sigma_1(a)$. Therefore $v' \in X(a, u')$.

We have thus defined a function $f : X(a,u) \longrightarrow X(a,u')$, mapping $v \in X(a,u)$ onto the plane $v' \in X(a,u')$ constructed as above. However, we can reread that construction in the reverse sense, starting from v' and going back to v. Therefore f admits an inverse $f^{-1} : X(a,u') \longrightarrow X(a,u)$. That is, f is a bijection. Hence:

$$(1) \quad \alpha(a,u) = \alpha(a,u')$$

Given x as above, let a' be a point of x other than a and let $v \in X(a,u)$. There is just one plane w in Γ_a such that $x * w \top v$. Trivially, $w \wedge v \neq u \wedge v$. Let b be the meet point of $w \wedge v$ and $u \wedge v$ in the projective plane Γ_v. We have $b \neq a'$, otherwise $u \wedge v$, $w \wedge v$, x, v, w, u would form a proper triangle in the generalized quadrangle $\Gamma_{a'}$. Let y be the line through b and a' in the projective plane Γ_w. It is easily seen that $y \notin \sigma_1(u)$. There is just one plane v' in Γ_b such that $y * v' \top u$. It is easy to check that $v' \in X(a', u)$.

We have thus defined a function $g : X(a,u) \longrightarrow X(a',u)$, mapping $v \in X(a,u)$ onto the plane $v' \in X(a',u)$ constructed as above. It is clear from that construction that g admits an inverse. That is, g is a bijection. Hence:

$$(2) \quad \alpha(a,u) = \alpha(a',u)$$

Let (a', u') be another point–plane flag with $a' \perp a$. Given a line x incident to both a and a', there are planes w, w' on x such that $w \top u$ in the generalized quadrangle Γ_a and $w' \top u'$ in the generalized quadrangle $\Gamma_{a'}$. We have $\alpha(a,u) = \alpha(a,w)$, $\alpha(a,w) = \alpha(a,w')$, $\alpha(a',w') = \alpha(a',u')$ by (1) and $\alpha(a,w') = \alpha(a',w')$ by (2). Therefore $\alpha(a,u) = \alpha(a',u')$.

By the above and by the connectedness of the collinearity graph of Γ we obtain that $\alpha(a',u') = \alpha(a,u)$ for every point–plane flag (a',u'). □

As $\alpha(a,u)$ does not depend on the particular choice of the point–plane flag (a,u), we write α for $\alpha(a,u)$. We call α the *Ott–Liebler index* of Γ, as Ott [140] and Liebler [119] were the first to exploit this constant in the investigation of C_3 geometries.

Given a non-incident point–plane pair (a,w), let $Y(a,w)$ be the set of planes on a dually collinear with w.

Lemma 14.2 *We have* $|Y(a,w)| = \alpha + 1$.

Proof. We have $Y(a,w) \neq \emptyset$ by Lemma 7.30. Given $u \in Y(a,w)$, we can define a mapping $f : Y(a,w) \longrightarrow X(a,u) \cup u$ as follows.

Let $v \in Y(a,w)$. If $v \wedge w = u \wedge w$, then either $v = u$ or $v \in X(a,u)$. In any case, we set $f(v) = v$. Let $v \wedge w \neq u \wedge w$ and let b be the meet point of $v \wedge w$ and $u \wedge w$ in the projective plane Γ_w. Let x be the line through a and b in Γ_v. We have $x \notin \sigma_1(u)$, otherwise x, $v \wedge w$, $u \wedge w$, v, w, u would form a proper triangle in the generalized quadrangle Γ_b. Therefore there is just one plane $v' * b$ such that $x * v' \top u$ in Γ_b. It is easily seen that $v' \in X(a,u)$. We set $f(v) = v'$.

It is clear from the above construction that f admits an inverse. Hence f is a bijection. Therefore $|Y(a,w)| = |X(a,u) \cup u| = \alpha + 1$. □

Given distinct points a, b of Γ, let $\lambda_1(a,b)+1$ be the number of lines incident to both a and b. Note that $a \not\perp b$ if and only if $\lambda_1(a,b) = -1$. Note also that Γ is a polar space if and only if $\lambda_1(a,b) = 0$ for any two distinct collinear points a, b (Corollary 7.39).

Lemma 14.3 *For every point–plane flag (a,u), we have*

$$(3) \qquad \sum_{b \in \sigma_0(u) - a} \lambda_1(a,b) \quad = \sum_{v \in X(a,u)} |\, \sigma_0(u \wedge v) \,|$$

Proof. Given a point $b * u$ distinct from a and a line x through a and b not in Γ_u, there is just one plane $v * b$ such that $x * v \top u$ in the generalized quadrangle Γ_b. It is clear that $v \in X(a,u)$. On the other hand, given any plane $v \in X(a,u)$ and any point $b * u \wedge v$, the line joining a with b in Γ_v is not incident to u. The equality (3) is now quite evident. □

Corollary 14.4 *We have $\alpha = 0$ if and only if Γ is a polar space.*

Proof. We have $\alpha = 0$ if and only if $X(a,u) = \emptyset$ for any point–plane flag (a,u). Hence $\alpha = 0$ if and only if Γ has a good system of lines, by (3). That is, $\alpha = 0$ if and only if Γ is a polar space, by Corollary 7.39. □

Corollary 14.5 *Let Γ be non-flat. Then Γ is a polar space if and only if, for any non-incident point-plane pair (a,w), the plane v such that $a * v \top u$ (existing by Lemma 7.30) is unique.*

(Easy, by Lemma 14.2 and Corollary 14.4.) Given a non-incident point–line pair (a,x), let $\lambda_2(a,x)+1$ be the number of planes incident to both a and x.

Lemma 14.6 *For every point–plane flag (a,u), we have*

$$(4) \qquad \alpha = \sum_{x \in \sigma_1(u) - \sigma_1(a)} \lambda_2(a, x)$$

Proof. If $v \in X(a,u)$, then $u \wedge v \in \sigma_1(u) - \sigma_1(a)$. On the other hand, if $x \in \sigma_1(u) - \sigma_1(a)$, then u is one of the $\lambda_2(a,x) + 1$ planes on a and x. The remaining $\lambda_2(a,x)$ planes on a and x are in $X(a,u)$. □

Lemma 14.7 *Given a non-incident point–plane pair* (a, w) *and a point* $b \in a^{\perp} \cap \sigma_0(w)$, *we have*

$$(5) \qquad \lambda_1(a,b) + 1 = \sum_{x \in \sigma_1(b,w)} (\lambda_2(a,x) + 1)$$

Proof. For every line y through a and b, there is just one plane $v * b$ such that $y * v \top u$ in Γ_b. On the other hand, if $x \in \sigma_1(b,w)$ is coplanar with a and v is a plane on a and x, then a line joining a with b is uniquely determined in Γ_v. □

14.1.2 Lemmas on C_3 geometries with finite orders

We now assume that Γ admits finite orders q, q, t:

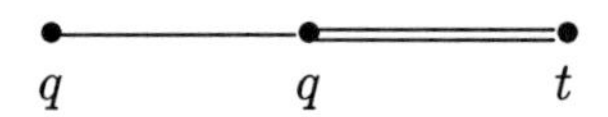

Hence Γ is finite (Corollary 7.32). We do not assume that q is a prime power. The equality (3) can now be restated as follows:

$$(3') \qquad \sum_{b \in \sigma_0(u) - a} \lambda_1(a,b) = (q+1)\alpha$$

Let n_0, n_1, n_2 be the number of points, lines and planes of Γ, respectively. We can compute n_0 as follows.

Given a plane u, there are $q^2 + q + 1$ points in u. The points of Γ not in u, if any, are in planes dually collinear with u. As Γ has a good dual system of lines, there are precisely $(q^2 + q + 1)t$ such planes. If v is a plane dually collinear with u, there are q^2 points in $\sigma_0(v) - \sigma_0(u \wedge v)$. Let a be one of those points. If $a * u$, then $v \in X(a,u)$. Otherwise, $v \in Y(a,u)$. By the definition of α and by Lemma 14.2 we obtain the following relation:

$$(q^2 + q + 1)tq^2 = (q^2 + q + 1)\alpha + [n_0 - (q^2 + q + 1)](\alpha + 1)$$

which is equivalent to the following one:

$$(6) \qquad n_0 = \frac{(q^2 + q + 1)(q^2 t + 1)}{\alpha + 1}$$

Every point is incident with $(qt+1)(q+1)$ lines and $(qt+1)(t+1)$ planes (Section 1.2.6(4)). Every line has $q+1$ points and every plane has q^2+q+1 points. Therefore:

$$(7) \quad n_1 = \frac{(q^2+q+1)(q^2t+1)(qt+1)}{\alpha+1}$$

$$(8) \quad n_2 = \frac{(q^2t+1)(qt+1)(t+1)}{\alpha+1}$$

(The reader may compare the above relations with the inequalities stated in Exercise 7.8.) By (6) and (8) we obtain the following divisibility condition:

$$(9) \quad \alpha \mid (q^2t+1)(q^2+q+1,(qt+1)(t+1))$$

Every plane has q^2+q+1 points. Hence $n_0 \geq q^2+q+1$. By this inequality and by (6) we obtain the following

$$(10) \quad \alpha \leq q^2t$$

Lemma 14.8 *We have $\alpha = q^2t$ if and only if Γ is flat.*

Proof. Trivially, Γ is flat if and only if $n_0 = q^2+q+1$. By (6), this holds if and only if $\alpha = q^2t$. □

A *k-arc* of a generalized quadrangle $\mathcal{Q}$ is a set of k mutually non-collinear points ([166], Section 2.7). If $\mathcal{Q}$ has order (q,t) and K is a k-arc of $\mathcal{Q}$, then precisely $k(t+1)$ lines of $\mathcal{Q}$ meet K, as there are $t+1$ lines through each point of K and none of these lines meets K in more than 1 point. Therefore $k \leq qt+1$, as there are just $(qt+1)(t+1)$ lines in $\mathcal{Q}$ (see Section 1.2.6(4)). It $k = qt+1$, then K is called an *ovoid.*

Lemma 14.9 *Given two distinct collinear points a, b of Γ, the $\lambda_1(a,b)+1$ lines through a and b form a $(\lambda_1(a,b)+1)$-arc in the generalized quadrangle Γ_a.*

(Indeed, distinct lines through a and b are never coplanar.) Therefore, the following holds for any two distinct points a and b:

$$(11) \quad \lambda_1(a,b) \leq qt$$

Note that (10) could also be obtained by (11) and (3′).

Lemma 14.10 *The geometry Γ is flat if and only if, for any two distinct collinear points a, b of Γ, the lines joining a with b form an ovoid in the generalized quadrangle Γ_a.*

(Easy, by (11), (3′) and Lemmas 14.8 and 14.9.)

Lemma 14.11 *let $q \geq 2$ and let λ_1, λ_2 be nonnegative integers such that $\lambda_1(a,b) = \lambda_1$ for any two distinct collinear points a, b and $\lambda_2(a,x) = \lambda_2$ for every non-incident but coplanar point–line pair (a,x). Then Γ is either a polar space or flat.*

Proof. The relations (3′) and (4) can now be rewritten as follows:

$$(3'') \qquad \alpha = q\lambda_1$$

$$(4') \qquad \alpha = q^2\lambda_2$$

Therefore

$$(12) \qquad \lambda_1 = q\lambda_2$$

Also, $\lambda_2 \leq t$ by (10) and (4′). If $\lambda_2 = t$, then Γ is flat (Lemma 14.8). If $\lambda_2 = 0$, then Γ is a polar space (Corollary 14.4). Let $1 \leq \lambda_2 < t$, if possible. Then Γ is non-flat (Lemma 14.8). Hence, there is a non-incident point–plane pair (a,w). Given $b \in a^\perp \cap \sigma_0(w)$, we can rewrite (5) as follows:

$$(5') \qquad \lambda_1 + 1 = \nu(\lambda_2 + 1)$$

where ν is the number of lines of w on b coplanar with a. Hence this number does not depend on the choice of the triplet (a,w,b) as above. By (12) and (5′) we get

$$(13) \qquad q\lambda_2 + 1 = \nu(\lambda_2 + 1)$$

Threfore $\lambda_2 + 1 \mid q - 1$. That is, $q = r\lambda_2 + r + 1$ for some positive integer r (indeed we have assumed $q \geq 2$). Hence $\nu = r\lambda_2 + 1$, by (5′).

The lines coplanar with a in the projective plane Γ_w form a dual linear space on the set of points $a^\perp \cap \sigma_0(w)$ with orders $q = r\lambda_2 + r + 1$ and $\nu - 1 = r\lambda_2$. We have $\nu \mid q(q+1)$ by a divisibility condition on the orders of a finite linear space (Section 1.2.6(4)). Hence

$$r\lambda_2 + 1 \quad | \quad (r\lambda_2 + r + 1)(r\lambda_2 + r + 2)$$

and this implies $r\lambda_2 + 1 \mid r + 1$, which holds only if $\lambda_2 = 1$ (we recall that we have assumed $\lambda_2 \geq 1$). Therefore $q = 2r + 1$, $\nu = r + 1$ and there are $2(r+1)r + 1$ lines in w coplanar with a (see Section 1.2.6(4); also Lemma 7.16). Since $q = 2r + 1 > r$, there are lines in w that are not coplanar with a. Let x be one of those lines. The line x meets each of the $2(r+1)r+1$ lines of w coplanar with a in precisely one point and every point in $a^\perp \cap \sigma_0(w)$ is in precisely $\nu = r + 1$ lines of w coplanar with a. Therefore

$$| a^{\perp} \cap \sigma_0(x) | = \frac{2(r+1)r+1}{r+1}$$

and $r+1 \mid 2(r+1)r+1$. However, this forces $r = 0$, which contradicts the assumption $q \geq 2$. □

Lemma 14.12 (Liebler) *We have* $\alpha \leq t^3$. *Furthermore, the following divisibility conditions hold:*

(13.*a*) $(q^2+t)(q+t)(\alpha+1) \mid (1+qt)(1+q^2t)(q^6+\alpha)$

(13.*b*) $(q^2+t)(q+t)(\alpha+1) \mid (1+qt)(1+q^2t)(t^3-\alpha)$

(13.*c*) $q(q^2+t)(\alpha+1) \mid (1+q^2t)(1+q+q^2)(q^4t-\alpha)$

(13.*d*) $q(q^2+t)(\alpha+1) \mid (1+q^2t)(1+q+q^2)(q^2t^2+\alpha)$

(13.*e*) $q(q+t)(\alpha+1) \mid (1+qt)(1+q^2t)(q^3+\alpha)$

(13.*f*) $q(q+t)(\alpha+1) \mid (1+qt)(1+q^2t)(q^3t^3-\alpha)$

(13.*g*) $q(q+t)(\alpha+1) \mid (1+qt)(1+q+q^2)(q^2t-\alpha)$

(13.*h*) $q(q+t)(\alpha+1) \mid (1+qt)(1+q+q^2)(q^2t^2+\alpha)$

(13.*i*) $\alpha+1 \mid (q^6t^3-\alpha)$

These conditions are due to Liebler, who obtained them reinterpreting for C_3 geometries results by Hoefsmit [83] on multiplicities of irreducible representations of Hecke algebras of groups of Lie type. Unfortunately, Liebler never published his discovery. Anyway, the reader can find a proof of the above lemma in [121] (Section 3.2).

Note that, if $q = 1$, then the above conditions are equivalent to the following one: $\alpha+1 \mid (t+1)(3, t+1)$ (and $\alpha+1 \leq t+1$ by (10)). It has been proved by Rees [171] that all finite C_3 geometries of order $(1,1,t)$ can be obtained from suitable sets of Latin squares. By that construction we can get examples with $\alpha+1 = d$ for every divisor d of $t+1$. In particular, when $d \neq 1, t+1$, then we obtain (finite, thin-lined) C_3 geometries that are neither polar spaces nor flat (by Corollary 14.4 and Lemma 14.8). We can even choose the Latin squares in such a way as to obtain (non-building and non-flat) 2-simply connected examples ([150], Example 3).

14.1.3 Finite thick-lined flag-transitive geometries of type C_3

Theorem 14.13 *Let Γ be a finite flag-transitive C_3 geometry with thick lines. Then Γ is one of the following:*

(i) a polar space;
(ii) the $Alt(7)$-geometry;
(iii) an unknown geometry that is neither a polar space nor flat and where residues of planes are non-Desarguesian projective planes with sharply flag-transitive automorphism groups, as in (iii) of Theorem 9.9.

We state some notation and we prove a couple of Lemmas, before coming the proof of the above theorem.

In what follows Γ is a flag-transitive finite C_3 geometry with thick lines, as in the hypotheses of Theorem 14.13. Given a plane u of Γ, we denote the stabilizer of u in $Aut(\Gamma)$ by G_u. The kernel of the action of G_u in the residue Γ_u of u will be denoted by K_u. Trivially, Γ admits order (q, q, t), with $q \geq 2$ because Γ is thick-lined.

Lemma 14.14 *Let q be a prime power. Given a plane u of Γ, let $\Gamma_u = PG(2, q)$ and $G_u/K_u \geq L_3(q)$. Then Γ is either a polar space or flat.*

Proof. The group $L_3(q)$ ($\leq G_u/K_u$) is 2-transitive on the set of points of $PG(2, q)$ and transitive on the set of non-incident point-line pairs of $PG(2, q)$. Furthermore, $Aut(\Gamma)$ is (flag-transitive, hence it is) transitive on the set of planes of Γ. Therefore $Aut(\Gamma)$ is transitive on the set of pairs of distinct collinear points of Γ and on the set of non-incident coplanar point–line pairs of Γ. Hence the hypotheses of Lemma 14.11 hold in Γ. By that lemma, Γ is either a polar space or flat. □

Lemma 14.15 *The $Alt(7)$-geometry is the only flag-transitive finite thick-lined flat geometry of type C_3.*

Outline of the Proof. The reader can find the proof of this lemma in [120] (also [121], 5.3). We give only a sketch of it here. We can use Theorem 9.9 to get control over the action G_u/K_u in Γ_u of the stabilizer of a plane u of Γ. The Frobenius case ((ii) and (iii) of Theorem 9.9) can be ruled out. Thus, we are left with the classical case ((i) of Theorem 9.9). In this case $Aut(\Gamma)$ acts 2-transitively on the q^2+q+1 points of Γ and $G_u/K_u \geq L_3(q)$. Note that G_u/K_u is a subgroup of the action of $Aut(\Gamma)$ on the set of points of Γ, as Γ is flat.

It is straightforward to prove that, if $q = 2$, then Γ is the $Alt(7)$-geometry.

Let $q \geq 3$, if possible. Let S be the set of points of Γ. For every plane u of Γ, the 0-shadows of the lines of u form a model of $PG(2,q)$ on the set of points S. Using some results on 2-transitive groups of degree q^2+q+1 containing $L_3(q)$, it is possible to prove that all planes of Γ determine the same model of $PG(2,q)$ on S. That is, Γ arises from one model Π of $PG(2,q)$ on S, counting each line $qt+1$ times and Π itself $(qt+1)(t+1)$ times, in some (unknown) suitable way (note that the assumption $q \geq 3$ is essential to obtain this description for Γ; indeed, the $Alt(7)$-geometry would be a counterexample if we allowed $q = 2$). Using this information and with some more work on certain configurations of lines and planes of Γ, it is possible to prove that the 1-Grassmann point–line system of Γ is the point–line system of $\mathcal{Q}_5^+(q)$ and that Γ arises from q^2+q+1 points of $PG(5,q)$ forming an exterior set for $\mathcal{Q}_5^+(q)$, in the same way as the $Alt(7)$-geometry is obtained in [170] from seven points of $PG(5,2)$ forming an exterior set for $\mathcal{Q}_5^+(2)$. However, q^2+q+1 points of $PG(5,q)$ can form a set exterior to $\mathcal{Q}_5^+(q)$ only if $q = 2$, by a result of Thas [209]. We have reached a contradiction. □

Proof of Theorem 14.13. We can now prove Theorem 14.13. By Theorem 9.9, given a plane u of Γ, either $G_u/K_u \geq L_3(q)$ (classical case) or G_u/K_u is a Frobenius group sharply flag-transitive in Γ_u (Frobenius case). In the classical case, (i) and (ii) of Theorem 14.13 are the only possibilities, by Lemmas 14.14 and 14.15.

In the Frobenius case, Γ is neither flat (by Lemma 14.15) nor a polar space. Hence $0 < \alpha < q^2t$ (Corollary 14.4 and Lemma 14.8). If $q = 2$, then $t = 1, 2$ or 4 (see Section 3.2.4). Substituting in Lemma 14.12, we see that any of these values of t together with $q = 2$ force $\alpha = 0$ or $q^2t = 8$, contradicting the above. Therefore $q \geq 3$. If $q = 8$, then the following are the feasible values for t, by Section 3.2.4 and Appendix II of Chapter 1:

1, 4, 6, 7, 8, 9, 12, 14, 16, 18, 21, 24, 28, 32, 36, 42, 56, 64.

Substituting in Lemma 14.12, we see that $\alpha = 0$ or $q^2t = 64t$ are the only surviving possibilities. However, we have $0 < \alpha < q^2t$, as we have remarked above. Therefore $q \neq 8$. By Theorem 9.9, Γ_u is not Desarguesian. □

The computations by which we have ruled out the cases of $q = 2$ and $q = 8$ in the above proof mainly depend on the fact that q^2+q+1 is prime in those cases. However, q^2+q+1 is always prime in case (iii) of Theorem 14.13. Exploiting this information in Lemma 14.12, we obtain the following improvement of Theorem 14.13:

Proposition 14.16 *Let Γ be as in (iii) of Theorem 14.13. Then the orders q, t and the Ott–Liebler index α of Γ satisfy the following relations, where* *d designates the greatest common divisor of* q^2 *and* t*:*

$$q > d^2,$$
$$q^2 - q > t \geq (q-1)d^2 + d,$$
$$qd \mid \alpha,$$
$$(\alpha+1)qd \mid q^2t - \alpha,$$
$$(\alpha+1)(q+t)q \mid (1+qt)(q^2t-\alpha),$$
$$(\alpha+1)(q^2+t)q \mid (1+q^2t)(q^4t-\alpha);$$

Furthermore, $q \equiv 0 \ (mod\ 8)$, q is not a prime power and $q+1 \equiv 0 \ (mod\ 3)$ in case (iii) of Theorem 14.13 (see Theorem 9.9 (iii)).

Also, $q+t \mid qt(q+1)(t+1)$ (Chapter 1, Appendix II), and $q > 14,400,008$ (see the remark after Theorem 9.9).

Problem. Prove that (iii) of Theorem 14.13 is impossible. A partial result on this problem is obtained in [243]: if Γ is as in (iii) of Theorem 14.13, then t is odd and $Aut(\Gamma)$ is non-solvable.

Remark. The following result by Aschbacher [2] is the 'ancestor' of Theorem 14.13: let Γ be a finite flag-transitive C_3 geometry where residues of planes are Desarguesian projective planes and stars of points are classical generalized quadrangles; then Γ is either a polar space or the $Alt(7)$-geometry. It is clear from Theorem 14.13 that, in order to obtain this conclusion, we only need the hypothesis that the planes of Γ are Desarguesian.

14.1.4 Finite thick-lined C_3 geometries with orders of known type

In this section Γ is a finite C_3 geometry admitting order (q,q,t) with $q \geq 2$. We say that (q,q,t) is of *known type* if the pair (q,t) is as in one of the following cases (compare with (1)–(5) of Section 3.2.4):

(1) $t = 1$ (top-thin case);
(2) $q = t$ (uniform order);
(3) $q^2 = t$;
(4) $q = t^2$;
(5) $q^3 = t^2$;
(6) $q^2 = t^3$;
(7) $q + 2 = t$;
(8) $q = t + 2$.

Theorem 14.17 *Let (q, q, t) be of known type. Then Γ is one of the following:*

(i) a polar space;
(ii) a flat C_3 geometry with uniform order;
(iii) a flat C_3 geometry with $q^3 = t^2$;
(iv) a C_3 geometry with $q = t^2$ and Ott-Liebler index $\alpha = t^3$.

Proof. Finite thick-lined top-thin geometries are polar spaces (Corollary 12.68). Thus, case (1) is settled. Using Lemma 14.12, Corollary 14.4 and Lemma 14.9, it is not difficult to check that the following are the only possiblities for Γ in each of the above cases (2)–(8):

Case (2): either a polar space or flat;
Case (3): either a polar space or flat;
Case (4): either a polar space or $\alpha = t^3$;
Case (5): polar space;
Cases (6), (7) and (8): impossible.

It is known that generalized quadrangles of order (q, q^2) do not admit any ovoids ([166], 1.8.3). Therefore Γ cannot be flat in case (3), by Lemma 14.10. □

The *Alt*(7)-geometry is the only example known for (ii). No examples are known for (iii) and (iv). Note that Γ is neither a polar space nor flat in (iv), by Corollary 14.4 and Lemma 14.9. Note also that stars of points in (iii) (if any example existed for that case) cannot be isomorphic to any of the known generalized quadrangles. Indeed $\mathcal{H}_4(q^2)$ is the only generalized quadrangle of order (q^2, q^3), but it has no ovoids ([166], 3.4.1(ii)), whereas stars of points of flat C_3 geometries admit ovoids, by Lemma 14.10.

Problem. Prove that (iii) and (iv) of Theorem 14.17 are impossible. Prove that the *Alt*(7)-geometry is the only possibility in (ii).

14.1.5 Exercises

Exercise 14.1. Let $f : \Gamma' \longrightarrow \Gamma$ be a 2-cover of a C_3 geometry Γ and let α, α' be the Ott–Liebler indices of Γ and Γ', respectively. Prove that $\alpha + 1$ divides $\alpha' + 1$ and that $(\alpha' + 1)/(\alpha + 1)$ is the multiplicity of f.

(Remark: in particular, if $\Gamma = \Gamma'/G$ for some $G \leq Aut_s(\Gamma')$, then $(\alpha' + 1)/(\alpha + 1) = |G|$; see Section 11.1.3.)

Exercise 14.2. Prove that the Ott–Liebler index of a C_3 geometry Γ is equal to the number of closed galleries of Γ of type 012012012 based at a given chamber.

(Remark: this is the way in which α has been defined in [140].)

Exercise 14.3. Let (a, w) be a non-incident point–plane pair of C_3 geometry Γ. Prove the following:

$$\sum_{v \in Y(a,w)} \mid \sigma_0(v \wedge w) \mid \; = \sum_{b \in a^{\perp} \cap \sigma_0(w)} (\lambda_1(a, b) + 1)$$

Exercise 14.4. Let Γ be a C_3 geometry admitting (possibly infinite) orders q, q, t. Given a line-plane flag (x, u) of Γ, let $\tau(x, u)$ be the number of oriented proper triangles in the dual point-line system of Γ containing the flag (u, x). Prove that $\tau(u, x) = (q + 1)\alpha$.

(Compare with the statement of Exercise 7.13.)

Exercise 14.5. Let Γ be a flat C_3 geometry with finite orders q, q, t. Prove that we have $q = t$ if and only if the dual point-line system of Γ is a linear space.

(Hint: see the hint given for Exercise 7.12.)

Exercise 14.6. Let Γ be a flat C_3 geometry with finite orders q, q, t, with $q < t$. Prove that the dual collinearity graph of Γ is strongly regular.

Exercise 14.7. Let Γ be a flat C_3 geometry with finite orders q, q, t. Prove that $q^2 + t \mid t(t - q)(qt + 1)$.

(Hint: note that $q \leq t$ by Exercise 7.12 (or by the inequality $\alpha \leq t^3$ of Lemma 14.12). If $q < t$, then use Exercise 14.6 and Appendix II of Chapter 1.)

Exercise 14.8. Let Γ be as in (iii) of Theorem 14.13. Prove that $q < t$ and that the order of Γ is not of known type.

Exercise 14.9. (Ott–Liebler Indices of arbitrary rank [168]). Let Γ be a C_n geometry, $n \geq 2$. We can recursively define an index $\beta(\Gamma)$ as follows. We set $\beta(\Gamma) = 1$ when $n = 2$. If $n \geq 3$, then we set

$$(*) \quad \beta(\Gamma) = (1 + \alpha(a, u))\beta(\Gamma_a)$$

where a is a point of Γ (hence $\beta(\Gamma_a)$ has been defined at the $(n-1)$th step), u is a dual point on a and $\alpha(a, u)$ is the number of dual points v distinct from u and such that $a * v \top u$ and $a \notin u \wedge v$ (where $u \wedge v$ is the dual line incident to u and v). Prove that the previous definition, namely that $\beta(\Gamma)$ does not depend on the choice of the point-dual-point flag (a, u) in $(*)$, is consistent.

(Hint: given a point–dual-point flag (a, u), we have $\alpha(a, u) = \alpha(a, v)$ for every dual point $v * a$ dually collinear with u in Γ_a and $(1 + \alpha(a, u))\beta(\Gamma_a) = (1 + \alpha(b, u))\beta(\Gamma_b)$ for every point $b * u$.)

Note that, if $n = 3$, then $\beta(\Gamma) = \alpha + 1$, where α is the Ott–Liebler index of Γ. Thus, we call $\beta(\Gamma) - 1$ the *Ott–Liebler index* of Γ. We denote it by $\alpha(\Gamma)$.

Exercise 14.10. Prove that a C_n geometry Γ ($n \geq 3$) is a polar space if and only if $\alpha(\Gamma) = 0$, where $\alpha(\Gamma)$ is the Ott–Liebler index of Γ (defined in the previous exercise).

(Hint: work by induction on n, using Corollary 7.28 and Theorem 7.38.)

Exercise 14.11. Let $\mathcal{P}$ be a polar space of rank $n \geq 3$ and let $G \leq Aut(\mathcal{P})$ define a 2-quotient of $\mathcal{P}$. Prove that $|G| = \beta(\mathcal{P}/G) + 1$ (notation as in Exercise 14.9).

(Hint: the group G residually defines quotients of $\mathcal{P}$, by Corollary 12.57. Hence we can work by induction on n, using Exercise 14.1 to start.)

Exercise 14.12. Show by some examples that the factors $\beta(\Gamma_a)$ and $1 + \alpha(a, u)$ of $\beta(\Gamma)$ in $(*)$ of Exercise 14.9 may vary if we let a vary (although their product $\beta(\Gamma)$ is constant).

(Hint: take a topological quotient $\Gamma = \mathcal{P}/G$ of a polar space $\mathcal{P}$ by a subgroup $G \leq Aut(\mathcal{P})$, choosing G in such a way that there are points of $\mathcal{P}$ with stabilizers of different orders in G. For instance, G might fix some points of $\mathcal{P}$ and act semi-regularly on the other ones. The group G residually defines quotients of $\mathcal{P}$ (Exercise 11.4). Therefore, by Exercise 14.11 ...)

Exercise 14.13. Let Γ be a C_4 geometry. Prove that, given a dual point u of Γ and any two distinct points a, b of u, we have $\alpha(\Gamma_a) \leq \alpha(b, u)$ (notation as in Exercise 14.9).

Exercise 14.14. Given a C_4 geometry Γ, let a be a point of Γ with $\alpha(a, u) = 0$ for some (hence every) dual point $u * a$ (notation as in Exercise 14.9) and let $b \perp a$. Prove that Γ_b is a polar space and that b is joined with a by precisely one line.

(Hint: use Exercise 14.13.)

Exercise 14.15. Let Γ be a geometry with diagram and orders as follows:

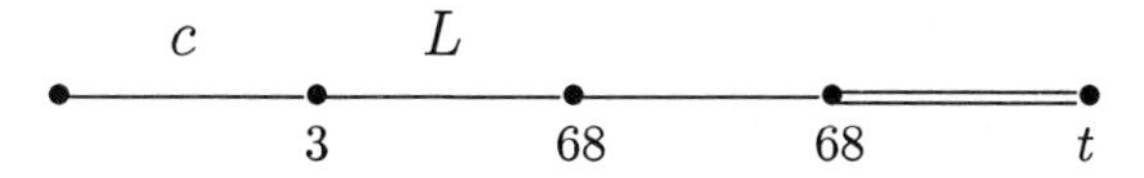

(note that residues of dual points of Γ are as in Theorem 7.23(vi); of course, we are not claiming that such a geometry exists). Let x be a line of Γ. Prove that the star Γ_x^+ of x is flat.

(Hint: by Lemma 14.12 we get that either $\alpha = 0$ or $\alpha = 68^2 t$. As 68 is not a prime power, Γ_x^+ is not a polar space, by Theorem 1.6. Hence ...)

14.2 The diagram C_n with $n \geq 4$

In this section Γ is a C_n geometry of rank $n \geq 4$. The nonnegative integers 0, 1, ..., $n-1$ are taken as types, as in Section 7.4. Thus, the residues of Γ of type $\{n-3, n-2, n-1\}$ are those belonging to the diagram C_3. We call them C_3 *residues*, for short.

We begin with a lemma on covers. Then we will turn to finite thick-lined C_n geometries, focusing on the flag-transitive case and on the case with order of known type.

Lemma 14.18 *If every C_3 residue of Γ is either a polar space or flat, then Γ is either a 2-quotient of a polar space or flat.*

Outline of the Proof. We give only a sketch of the proof, assuming $n = 4$ for ease of exposition. The reader should consult [164] for the complete proof.

Let $n = 4$. If all stars of points are flat, then Γ is flat (see Exercise 4.11). Thus, we can assume that the star of some point of Γ is a polar space. Let a be a point with flat residue.

We say that an element x of Γ is *far* from a if there is no dual point incident to both a and x. Note that the dual points far from a are those that are not incident to a. The points far from a are those at distance 2 from a in the collinearity graph of Γ. The lines far from a are those that are neither incident nor coplanar with a.

If x is an element of Γ of type $i \geq 1$ far from a, then there is a flag (y, z) of type $(i-1, i)$ such that $x * y$ and $a * z$ (see Exercise 7.7). This flag is uniquely determined by a and x ([164], Lemma 8). We call z the *projection* of x *onto* a and we denote it by $p_a(x)$.

It is possible to prove that there is a point b far from a and such that Γ_b is a polar space ([164], Lemma 10). Trivially, all elements of Γ_b are far from a. It is not hard to check that the function mapping every element x of Γ_b onto its projection $p_a(x) \in \Gamma_a$ is a 2-cover from Γ_b to Γ_a. Therefore Γ_a is a 2-quotient of a polar space. This holds for every flat C_3 residue in Γ. All non-flat C_3 residues are polar spaces by assumption. Therefore Γ is a 2-quotient of a polar space, by Theorem 13.20. □

Corollary 14.19 *Let Γ be thick-lined and finite. If every C_3 residue of Γ is either a polar space or flat, then Γ is a polar space (and all C_3 residues of Γ are polar spaces).*

Proof. Under the previous hypotheses, Γ is either flat or 2-covered by a polar space, by Lemma 14.18. However, there are no finite thick-lined flat C_n geometries of rank $n \geq 4$ (Exercise 7.11) and finite thick-lined polar spaces do not admit any proper 2-quotients (Corollary 8.17). Therefore Γ is a polar space. □

Theorem 14.20 *All finite thick-lined flag-transitive C_n geometries of rank $n \geq 4$ are polar spaces.*

Proof. Let Γ be flag-transitive, thick-lined and finite. Given a flag F of Γ of cotype $\{n-3, n-2, n-1\}$, let u be a dual point incident to F. We have $\Gamma_u = PG(n-1, q)$ for some prime power q (Theorem 1.3). Hence $\Gamma_{F\cup u} = PG(2, q)$. Therefore Γ_F is either a polar space or the $Alt(7)$-geometry, by Theorem 14.13 (we recall that planes are non-Desarguesian in (iii) of Theorem 14.13). By Corollary 14.19, Γ is a polar space. □

Orders of *known type* can be defined for finite thick-lined C_n geometries of rank $n \geq 4$ just as for finite thick-lined C_3 geometries (Section 14.1.4). Note that, since $n \geq 4$, q is now a prime power.

Theorem 14.21 *All finite thick-lined C_n geometries ($n \geq 4$) admitting order of known type are polar spaces.*

Outline of the Proof. Let Γ be a finite thick-lined C_n geometry with $n \geq 4$ and order $(q, q, ..., q, t)$ of known type. By Theorem 14.17 and Corollary 14.19, Γ is a polar space, except possibly when $q = t^2$ and C_3 residues have Ott-Liebler index $\alpha = t^3$. However, the latter case can be ruled out by some computations on the Ott–Liebler index of Γ (defined in Exercise 14.9). The reader should consult [153] for the details. □

We also mention the following result, proved in [121] (Theorem 4):

Theorem 14.22 *Let Γ be a finite thick-lined C_n geometry, with $n \geq 4$ and order $(q, q, ..., q, t)$ where q and t are powers of the same prime. Then Γ is a polar space.*

Problem. Prove that all finite thick-lined C_n geometries with $n \geq 4$ are polar spaces.

14.3 The diagram F_4

In this section Γ is a finite thick geometry of type F_4. By Proposition 3.3, Γ admits orders s, s, t, t

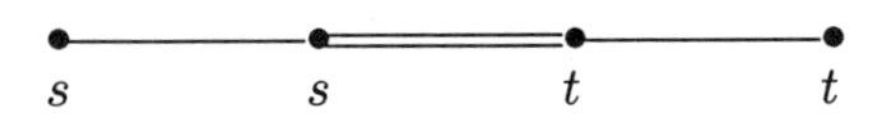

and we have $2 \leq s, t < \infty$, as Γ is thick and finite. We begin with a few lemmas on the case of uniform order. Then we will focus on the flag-transitive case.

Lemma 14.23 *Let $s = t$ and let a be a point of Γ such that Γ_a is flat. Then every point collinear with a is joined to a by just one line.*

Proof. Given $b \perp a$, let x, y be lines incident to both a and b. Since Γ_a is flat and $s = t$, the lines x and y are coplanar (see Exercise 14.5). Therefore $x = y$. □

Lemma 14.24 *Let $s = t$ and let a be a point of Γ such that Γ_a is flat. Then Γ_u is a polar space, for every dual point $u * a$.*

Proof. Let u be a dual point incident to a. Then Γ_u has a good system of lines by the previous lemma. Hence it is a polar space. □

Lemma 14.25 *Let $s = t$ and let a, b be distinct points of Γ with flat residues. Then $a \not\perp b$.*

Proof. Let $a \perp b$ if possible and let x be the line joining a with b (unique by Lemma 14.23). Let u be a dual point incident to x. Every line $y * a$ is coplanar with x (Exercise 14.5) and incident to u. Therefore, $a^\perp \subseteq b^\perp \cap \sigma_0(u)$. Similarly, $b^\perp \subseteq a^\perp \cap \sigma_0(u)$. Hence $a^\perp = b^\perp \subseteq \sigma_0(u)$.

As Γ_a is flat, there are just $(s^2+1)(s+1)$ lines on a (indeed, there are just $(s^2+1)(s+1)$ lines in $\Gamma_{\{a,u\}}$; see Section 1.2.6(4)). By this and by Lemma 14.23, we obtain that $|a^\perp| = s(s^2+1)(s+1)+1$. Therefore, and because $a^\perp = b^\perp$, we can see at most $s(s^2+1)(s+1)+1$ points collinear with either a or b inside Γ_u. On the other hand, Γ_u is a polar space by Lemma 14.24 and it is not difficult to check that, given two distinct collinear points a, b in a polar space $\mathcal{P}$ of rank 3 and uniform order s, there are just $2s^4 + (s^2+1)(s+1)$ points of $\mathcal{P}$ collinear with either a or b. Trivially, $2s^4 + (s^2+1)(s+1) > s(s^2+1)(s+1)+1$. We have reached a contradiction. □

Theorem 14.26 *All finite thick flag-transitive geometries of type F_4 are buildings.*

Proof. Let Γ be a finite thick flag-transitive F_4 geometry. If all C_3 residues of Γ are polar spaces, then Γ is a building by Proposition 13.30.

Let some C_3 residue be not a polar space, if possible. For instance, let stars of points be not polar spaces. Then they are either isomorphic to the $Alt(7)$-geometry or as in (iii) of Theorem 14.13. On the other hand, they cannot be isomorphic to the $Alt(7)$-geometry, by Lemma 14.25. Therefore they are as in (iii) of Theorem 14.13 and s and t satisfy the conditions of Proposition 14.16. Hence (s, s, t) is not of known type and $s < t$ (Exercise 14.8). Therefore the pair (t, s) does not satisfy any of the relations (1)–(8) of Section 14.1.4.

Let us turn to residues of dual points now. As (t, s) does not satisfy any of the relations (1)-(8) of 14.1.4, residues of dual points are neither polar spaces nor isomorphic to the $Alt(7)$-geometry. On the other hand, they cannot be as in (iii) of Theorem 14.13, since $s < t$, whereas we should have $t < s$ in that case (Exercise 14.8). The final contradiction is reached. □

Problem. Is it true that every flag-transitive top-thin F_4 geometry with thick lines is a building ?

Remark. We have assumed that Γ is finite. However, by Proposition 7.52 and Corollary 7.32, a geometry of type F_4 is finite if and only if it is locally finite. Thus, we could substitute the finiteness assumption with the seemingly weaker hypothesis that Γ is locally finite. The same can be said in Section 13.3.2 for E_7 and E_8, by Proposition 7.45 and by Corollary 7.50 (note that the finiteness and the local finiteness are equivalent in E_6 geometries, by Proposition 7.41).

Chapter 15

$L.C_{n-1}$ and more diagrams

As in Chapter 7, we designate the following diagram by $L.C_{n-1}$:

$$(L.C_{n-1}) \quad \overset{L}{\bullet\!\!-\!\!\!-\!\!\!-\!\!\bullet}\!\!-\!\!\!-\!\!\!-\!\!\bullet\!\!-\ \cdots\cdots\ -\!\!\bullet\!\!-\!\!\!-\!\!\!-\!\!\bullet\!\!=\!\!=\!\!=\!\!\bullet$$

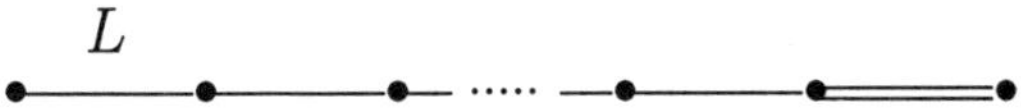

The diagram C_n (examined in the previous chapter) is a special case of $L.C_{n-1}$. Another special case of $L.C_{n-1}$ is $Af.C_{n-1}$:

$$(Af.C_{n-1}) \quad \overset{Af}{\bullet\!\!-\!\!\!-\!\!\!-\!\!\bullet}\!\!-\!\!\!-\!\!\!-\!\!\bullet\!\!-\ \cdots\cdots\ -\!\!\bullet\!\!-\!\!\!-\!\!\!-\!\!\bullet\!\!=\!\!=\!\!=\!\!\bullet$$

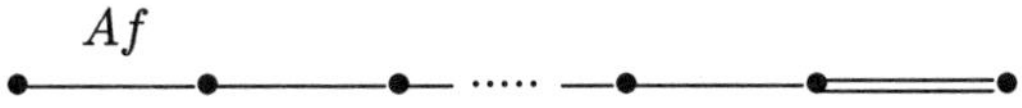

We have already studied geometries belonging to $Af.C_{n-1}$ in Section 7.7.1. We will come back to them in the next section.

15.1 The diagram $Af.C_{n-1}$

The geometries belonging to $Af.C_{n-1}$ will be called $Af.C_{n-1}$ *geometries*, for short. Affine polar spaces of rank n, their 2-quotients (in particular, standard quotients, defined in Section 8.4.7) and the $Alt(8)$-geometry (Exercise 7.22) are the only $Af.C_{n-1}$ geometries we know. Note that an affine polar space can be viewed as an (improper) standard quotient of itself. Thus, when we speak of standard quotients of affine polar spaces, it will be understood that we include affine polar spaces among them.

15.1.1 $Af.C_{n-1}$ geometries satisfying IP

We remarked in Section 8.4.7 that the intersection property **IP** holds in standard quotients of affine polar spaces. In particular, it holds in affine polar spaces.

Theorem 15.1 ([49]) *All $Af.C_{n-1}$ geometries of rank $n \geq 4$ satisfying* **IP** *are standard quotients of affine polar spaces.*

Theorem 15.2 ([46]) *All finite $Af.C_2$ geometries satisfying* **IP** *are standard quotients of affine polar spaces.*

Sketch of the Proof of Theorem 15.1. The basic idea of the proof is quite easy: just an application of the definition of universal topological covers, as the reader will see. However, we need a long analysis of properties of $Af.C_{n-1}$ geometries with **IP** before being ready for the main (and final) step of the proof. We will mention those properties without proving them. The reader should consult [49] for the proofs.

Let Γ be an $Af.C_{n-1}$ geometry ($n \geq 4$), satisfying **IP**. As $n \geq 4$, residues of dual points of Γ are isomorphic to $AG(n-1, K)$ for some division ring K (Theorem 1.4). Therefore, stars of points are classical polar spaces (Theorem 1.5).

The collinearity graph of Γ has diameter $d \leq 3$ (see Exercise 7.19). Let Φ_Γ be the reflexive closure of the relation 'being at distance 3' in the collinearity graph of Γ. It can be proved that Φ_Γ is an equivalence relation on the set of points of Γ and that it can be extended to an equivalence relation on the set of elements of Γ, as we have done for affine polar spaces in Section 8.4.7. Furthermore, Φ_Γ defines a topological quotient of Γ and Γ/Φ_Γ has a good system of lines (whence **IP** holds in it, by Corollary 7.28).

A parallelism relation $\|$ can also be defined between lines, planes, ... dual points of Γ, extending the parallelism relations of the (affine) planes of Γ. We can take the parallelism classes of $\|$ as new elements, defining an incidence relation between them in the natural way: two classes X, Y are declared to be incident if $x * y$ for some $x \in X$ and $y \in Y$. Thus we obtain a 'geometry at infinity' Γ^∞ with string diagram graph of rank n or $n-1$ and where planes are isomorphic to $PG(2, K)$. We have two cases, which we call the *tangent* and *secant* cases respectively (motivations for this terminology will become evident later).

The tangent case. The geometry Γ^∞ is obtained from a degenerate sesquilinear (or singular quadratic) form f of Witt index n in a vector space V over K, with $dim(Rad(f)) = 1$ (the Witt index of a sesquilinear or quadratic form can be defined in degenerate (singular) cases just as in the non-degenerate (non-singular) case). The elements of Γ^∞ are the non-trivial linear subspaces of V totally isotropic (respectively, totally singular) for f and not containing $Rad(f)$. The incidence relation of Γ^∞ is the natural one, defined as symmetrized inclusion. The rank of Γ^∞ is $n-1$. The system of all linear subspaces of V totally isotropic (totally singular) for f is a 'classical degenerate polar space', in the meaning of [25]. Let us denote it by $\mathcal{P}^\infty$. Note that $\mathcal{P}^\infty$ is obtained from the geometry Γ^∞ taking as additional elements the totally isotropic (totally singular) non-trivial linear subspaces of V containing $Rad(f)$. Note that $\mathcal{P}^\infty$ is not a geometry, because every dual point of Γ^∞ is contained in just one maximal element of $\mathcal{P}^\infty$. Hence it is not a polar space in our meaning.

For every hyperplane H of V such that $Rad(f) \not\subseteq H$, the set of elements of Γ^∞ contained in H is called a *secant hyperplane* of $\mathcal{P}^\infty$. We identify this set with H, even if doing so is an abuse.

The secant case. The geometry Γ^∞ is a polar space of rank $n-1$ or n, with planes isomorphic to $PG(2, K)$. Hence the polar space Γ^∞ is classical (Theorem 1.5). That is, it arises from a non-degenerate sesquilinear form (or a non-singular quadratic form) f in a vector space over K. We now take $\mathcal{P}^\infty$ as a synonym of Γ^∞. Secant hyperplanes of polar spaces have been defined in Section 2.2.2. As $\mathcal{P}^\infty$ is now a polar space, we refer to that definition when speaking of secant hyperplanes of $\mathcal{P}^\infty$.

Both in the tangent case and in the secant case, it is possible to prove that the set of (parallelism classes of) lines through a given point a of Γ is a secant hyperplane of $\mathcal{P}^\infty$. Let us denote it by H_a. It can also be proved that, given any two points a, b of Γ, we have $H_a = H_b$ if and only if $a \equiv b$ (Φ_Γ). Let $\mathcal{H}$ be the family of secant hyperplanes of $\mathcal{P}^\infty$ of the form H_a, with a a point of Γ. The points of the quotient geometry Γ/Φ_Γ bijectively correspond to the members of $\mathcal{H}$. Furthermore, if x is an element of Γ of type $m > 0$ and a_0, a_1, ..., a_m are points of Γ forming a base of the m-dimensional affine geometry Γ_x, then the points of Γ/Φ_Γ in $[x]\Phi_\Gamma$ bijectively correspond to the members of $\mathcal{H}$ containing $\bigcap_{i=0}^m H_{a_i}$. Thus, we can build a model $\Gamma(\mathcal{H})$ of Γ/Φ_Γ taking the members of $\mathcal{H}$ as points and defining the remaining elements as suitable subspaces of members of $\mathcal{H}$.

The sesquilinear (or quadratic) form f defining $\mathcal{P}^\infty$ can be extended to a non-degenerate sesquilinear form (a non-singular quadratic form) $\hat{f}$ defining a classical polar space $\mathcal{P}$ of rank n such that $\mathcal{P}^\infty$ is a hyperplane of $\mathcal{P}$ (of tangent or secant type according to whether we are in the tangent case or in the secant case). Let $\widehat{\Gamma}$ be the affine polar space obtained by dropping the hyperplane $\mathcal{P}^\infty$ from $\mathcal{P}$. Let φ be the function mapping every point p of $\widehat{\Gamma}$ onto the hyperplane $p^\perp \cap \mathcal{P}^\infty$ of $\mathcal{P}^\infty$ (needless to say, $p^\perp$ is defined in $\mathcal{P}$). The partition $Ker(\varphi)$ of the set of points of $\widehat{\Gamma}$ is just the relation 'being at distance 3' in $\widehat{\Gamma}$, denoted by Φ_0 in Section 8.4.7 and defining the minimal standard quotient of $\widehat{\Gamma}$. Furthermore, we can choose (the extension $\hat{f}$ of f defining) $\mathcal{P}$ in such a way that $\varphi(p) \in \mathcal{H}$ for some point p of $\widehat{\Gamma}$. With this choice of $\mathcal{P}$ (that is, of $\hat{f}$) the function φ induces an isomorphism from the minimal standard quotient $\widehat{\Gamma}/\Phi_0$ to the geometry $\Gamma(\mathcal{H})$ (hence φ is a topological cover from $\widehat{\Gamma}$ to $\Gamma(\mathcal{H})$). Since $\Gamma(\mathcal{H})$ is a model of Γ/Φ_Γ, we obtain $\Gamma/\Phi_\Gamma \cong \widehat{\Gamma}/\Phi_0$.

We can now finish the proof. The geometry $\widehat{\Gamma}$ is topologically simply connected (Proposition 12.50) and $\widehat{\Gamma}/\Phi_0$ is a topological quotient. Therefore $\widehat{\Gamma}$ is the universal topological cover of $\widehat{\Gamma}/\Phi_0$, that is of Γ/Φ_Γ ($\cong \widehat{\Gamma}/\Phi_0$). By the definition of universal topological covers, there is a topological cover $f : \widehat{\Gamma} \longrightarrow \Gamma$ such that the composition of f with the projection of Γ onto Γ/Φ_Γ is (isomorphic to) the projection of $\widehat{\Gamma}$ onto its minimal standard quotient $\widehat{\Gamma}/\Phi_0$. Therefore $Ker(f)$ is a refinement of Φ_0. Hence Γ is a standard quotient of $\widehat{\Gamma}$. □

Sketch of the Proof of Theorem 15.2. Theorem 15.2 can be proved by an argument quite similar to that used for Theorem 15.1, except for more work needed to obtain for Γ^∞ a description as in the previous proof. Indeed we cannot appeal to Theorem 1.4 now to prove that the planes of Γ are Desarguesian, and yet this information is essential to describe Γ^∞ in the proof of Theorem 15.1. On the other hand, we are now assuming that Γ is finite. Thus, we can overstep the gap using some combinatorial characterizations of finite classical generalized quadrangles (Chapter 5 of [166]); but this needs some work (see [46]). □

15.1.2 Finite flag-transitive $Af.C_{n-1}$ geometries

Theorem 15.3 *Let Γ be a flag-transitive $Af.C_{n-1}$ geometry of rank $n \geq 4$. Then Γ is one of the following:*

(i) a standard quotient of an affine polar space;
(ii) the Alt(8)-geometry.

Proof. The geometry Γ has order $(q-1, q, ..., q, t)$, as it is flag-transitive (or by Propositions 3.3 and 3.4 and because all affine planes admit orders). By Theorem 1.4, the order q is a prime power and residues of dual points of Γ are isomorphic to $AG(n-1, q)$. Hence case (iii) of Theorem 14.13 cannot occur for stars of points (when $n = 4$). Therefore stars of points of Γ are either polar spaces or the *Alt*(7)-geometry, by Theorems 14.13 and 14.20.

Let stars of points of Γ be isomorphic to the *Alt*(7)-geometry. Then Γ is flat (Exercise 4.11) and it is straightforward to prove that Γ is the

$Alt(8)$-geometry.

Thus, from now on we assume that stars of points of Γ are polar spaces. They are classical polar spaces, by Theorem 1.6. Let us prove that Γ has a good system of lines.

Let (a, x) be a point–line flag of Γ and let G_a and $G_{a,x}$ be the stabilizer of a and (a, x) respectively in $Aut(\Gamma)$. Given a point $b \neq a$ in x, let L_b be the set of lines of Γ through a and b other than x and let $L = \bigcup_{b \in \sigma_0(x)-a} Y_b$. If $L = \emptyset$, then Γ has a good system of lines, hence **IP** holds in it by Corollary 7.28. By Theorem 15.1, the geometry Γ is a standard quotient of an affine polar space.

Let $L \neq \emptyset$, if possible. By Theorem 9.15 and Corollaries 9.18 and 9.19, either the action G_a/K_a of G_a in Γ_a is classical or $\Gamma_a = \mathcal{Q}_5^+(2)$ and $G_a/K_a = S_7$.

Let G_a/K_a be classical. Then $G_{a,x}$ is transitive on the set L' of lines through a non-coplanar with x (this can be proved by straightforward computations, or using Propositions 9.12 and 9.13). Trivially, $L \subseteq L'$ and $G_{a,x}$ fixes L. Therefore $L = L'$, as L' is an orbit of $G_{a,x}$. We will now obtain a contradiction computing $|L'|$ and an upper bound for $|L|$.

There are just $(q^{n-2}t+1)(q^{n-1}-1)/(q-1)$ lines on a and each of them is in precisely $(q^{n-3}t+1)(q^{n-2}-1)/(q-1)$ planes (see Exercises 7.8 and 7.9). Every plane on x contains q lines through a other than x. Therefore

$$(1) \qquad |\, L' \,| \; = \; q^{2n-4}t$$

Furthermore, there are just

$$\frac{(q^{n-2}t+1)(q^{n-3}t+1)(q^{n-1}-1)(q^{n-2}-1)}{(q-1)^2(q+1)}$$

planes on a (see also Exercises 7.8 and 7.9). Given $b \in \sigma_0(x)$ distinct from a, the lines in $L_b \cup x$ are mutually non-coplanar. Comparing with the number of planes on a, we see that there at most $(q^{n-2}t+1)(q^{n-1}-1)/(q^2-1)$ lines in $L_b \cup x$. Therefore

$$(2) \qquad |\, L \,| \; \leq \; (q^{2n-3}t + q^{n-1} - q^{n-2}t - q^2)/(q+1)$$

Comparing (1) with (2), we see that $|L|<|L'|$. This contradicts the equality $L = L'$ obtained above. Therefore G_a/K_a cannot be classical.

Let $G_a/K_a = S_7$ with $\Gamma_a = \mathcal{Q}_5^+(2)$. Let us write $x \equiv y$ to mean that two lines x, y through a meet in another point $b \neq a$. That is, $x \equiv y$ if and only if x and y have the same points, since Γ now has order $(1, 2, 2, 1)$. Trivially, $\equiv$ is an equivalence relation and G_a/K_a ($= S_7$) preserves it. However, it is clear from the description of the actions of A_7 and S_7 in $PG(3, 2)$ (see Section 9.3.4) that S_7 acts primitively on the set of points of $\mathcal{Q}_5^+(2)$ (that is, on the set of lines of $PG(3,2)$). Therefore $\equiv$ is either the identity relation or the trivial relation. In the first case, Γ has a good system of lines, as we wanted to prove. In the second case all lines on a should have the same points, which is evidently absurd. □

Theorem 15.4 (Del Fra [55]) *Let Γ be a flag-transitive $Af.C_2$ geometry with stars of points isomorphic to classical generalized quadrangles. Then Γ is a standard quotient of an affine polar space.*

Sketch of the Proof. We can use Theorem 9.16 to get control over the action G_a/K_a of G_a in Γ_a. When that action is classical, then we can prove that Γ has a good system of lines, as in the proof of Theorem 15.3. Hence **IP** holds in Γ by Corollary 7.28 and Γ is a standard quotient of an affine polar space by Theorem 15.2.

If case (i) of Theorem 9.16 occurs for G_a/K_a, then we can prove that Γ has a good system of lines by an argument similar to that used in the proof of Theorem 15.3 for the case of $G_a/K_a = S_7$. Therefore Γ is a standard quotient of an affine polar space.

Cases (ii) and (iii) of Theorem 9.16 require considerably more work. Given a chamber (a, x, u), we can compare a feasible action in the affine plane Γ_u of the stabilizer G_u of u in $Aut(\Gamma)$ with orbits in Γ_a of the stabilizer $G_{a,x}$ of the flag (a, x) and of the stabilizer $G_{a,u}$ of (a, u). In this way, Del Fra [55] has proved that Γ has a good system of lines (hence it is a standard quotient of an affine polar space, by Theorem 15.2). □

Problems. We have assumed that stars of points are classical generalized quadrangles in Theorem 15.4. What about the case where residues of points are grids ? We recall that the flat $Af.C_2$ geometry of order $(1,2,1)$ constructed in Exercise 7.21 is a non-standard 2-quotient of the affine polar space obtained from $\mathcal{Q}_5^+(2)$ by removing a tangent hyperplane (see Exercise 8.57). It is clear from the construction given in Exercise 7.21 that this geometry is flag-transitive, with automorphism group $S_4 \times S_3$. Is this the only example of this kind ?

Only two finite thick non-classical flag-transitive generalized quadrangles are known (see Section 9.4.4). Can we construct any flag-transitive $Af.C_2$ geometry where stars of points are isomorphic or dually isomorphic to one of those non-classical generalized quadrangles ?

Remark. We have assumed the finiteness in Theorems 15.2, 15.3 and 15.4, for ease of exposition. However, we could assume the local finiteness as well. Indeed, the collinearity graph of a locally finite $Af.C_{n-1}$ geometry with stars of points satisfying **IP** has diameter $d \leq 3$ (see [154]; note that, if we also assume that the geometry has a good system of lines, then the same conclusion can be obtained in the infinite case too (Exercise 7.19)). Furthermore, a C_{n-1} geometry is finite if it is locally finite (Corollary 7.32). Therefore all locally finite $Af.C_{n-1}$ with stars of points satisfying **IP** are finite. We have not assumed **IP** on stars of points in Theorem 15.3; however, the first step of the proof of that theorem is just to prove that **IP** holds in stars of points.

15.2 The diagram $L.C_{n-1}$

In this section Γ is a flag-transitive locally finite geometry belonging to the diagram $L.C_{n-1}$. As Γ is flag-transitive and locally finite, it admits finite orders s, q, ..., q, t:

($L.C_{n-1}$) [diagram: nodes s —L— q — q ····· q — q = t]

Theorem 15.5 *Let $n \geq 4$. Then Γ is one of the following:*

(i) a C_n geometry with thin lines;
(ii) a polar space;
(iii) a standard quotient of an affine polar space;
(iv) the $Alt(8)$-geometry;
(v) the geometry $\Gamma(Fi_{22})$ mentioned in Section 10.2.3.

Sketch of the Proof. The proof is divided in two parts. A great deal of work is done in the first part, which is fairly easy. Then we must cope with one rather difficult 'exceptional' case, to finish; we will only give references for this part of the proof.

By Theorem 9.22, given a dual point u of Γ, the residue of u is one of the following: the $(n-1)$-dimensional simplex, $PG(n-1,q)$, $AG(n-1,q)$ or $S(M_{22})$. If Γ_u is the $(n-1)$-dimensional simplex, then Γ is a C_n geometry with thin lines. If $\Gamma_u = PG(n-1,q)$, then Γ is a polar space by Theorem 14.20. If $\Gamma_u = AG(n-1,q)$ then Γ is as in (iii) or (iv), by Theorem 15.3.

Finally, let $\Gamma_u = S(M_{22})$. Given a point a of Γ, let G_a be the stabilizer of a in $Aut(\Gamma)$. The star Γ_a of a is a polar space by Theorem 14.13 and G_a acts a classical group in it, by Theorem 9.15. Therefore G_a is transitive on the ordered pairs of non-coplanar lines on a. Using this information it is easy to show that Γ has a good system of lines, as in the proof of Theorem 14.3. Furthermore, the triangular property holds in Γ. Indeed, suppose the contrary. Then there are two collinear points b, c collinear with a but such that the lines x, y joining a with b and c respectively are not coplanar. As G_a is transitive on the ordered pairs of non-coplanar lines through a, the set $a^\perp$ is a clique in the collinearity graph of Γ. By the connectedness of this graph and the flag-transitivity of $Aut(\Gamma)$, the collinearity graph of Γ is a complete graph and $Aut_s(\Gamma)$ is 2-transitive on the set of points of Γ. A contradiction can now be obtained, either by using results on 2-transitive groups, as in [17], or by purely combinatorial arguments [68]. Therefore the triangular property holds in Γ.

Buekenhout and Hubaut [26] have proved that $\Gamma(Fi_{22})$ is the unique geometry with the above properties. The proof by Buekenhout and Hubaut is a good example of interaction between geometry and group theory. We are not going to report on it here, as it is a bit long and not trivial at all. We only mention the two main steps of that proof.

We have not yet determined the isomorphism type of residues of points of Γ. We have only remarked that they are polar spaces of rank 3 and, as residues of dual points are isomorphic to $S(M_{22})$, stars of points have order $(4, 4, t)$. However, there are five mutually non-isomorphic polar spaces of rank 3 and order $(4, 4, t)$, corresponding to each of the values $t = 1, 2, 4, 8, 16$. Thus, the first thing to do is to prove that we have $t = 2$, as in $\Gamma(Fi_{22})$; that is, that stars of points are isomorphic to $\mathcal{H}_5(2^2)$). The proof of this claim is mainly combinatorial; however, a bit of group theory is needed at some points.

Once the isomorphism type of stars of points is determined, the local structure of Γ and $Aut(\Gamma)$ is under control. It is now possible to recover certain subspaces inside Γ whose stabilizers in $Aut(\Gamma)$ generate $Aut(\Gamma)$ and have properties that allow us to recognize $Aut(\Gamma)$ as $Aut(Fi_{22})$, keeping in mind the construction of Fi_{22} as a group generated by 3-transpositions [67].

A purely group-theoretic proof exploiting generators and relations in the spirit of Theorem 11.28 is given by Van Bon and Weiss in [232]. The basic idea of that proof is quite simple. It is worth explaining it. Exploiting the local information we have collected above on Γ (hence on $Aut(\Gamma)$), we can find a suitable nice refinement $\mathbf{R}$ of the set $\mathbf{R}_3$ of all relations holding in maximal parabolics of $Aut(\Gamma)$ (see Section 12.4.3). We need not know at this stage if $\mathbf{R}$ and $\mathbf{R}_3$ are really equivalent. We need not even know if the local information we have on $Aut(\Gamma)$ is sufficient to completely determine $\mathbf{R}_3$. We only know that $\mathbf{R}$ is certainly a refinement of $\mathbf{R}_3$. Therefore the group $Aut(\Gamma)$ is a homomorphic image of the group presented by $\mathbf{R}$. On the other hand, it turns out that $\mathbf{R}$ is a presentation of $Aut(Fi_{22})$. Hence $Aut(\Gamma)$ is a homomorphic image of $Aut(Fi_{22})$. On the other hand, $Aut(Fi_{22}) \approx Fi_{22} \cdot \mathbf{2}$. Therefore $Aut(\Gamma) = Aut(Fi_{22})$.

Another proof, more in the spirit of [26], is given by Meixner in [129].
□

The method chosen by Van Bon and Weiss (described above) is perhaps not very elegant, but it is often quite efficient. We call it the *generators-and-relations method.*

Theorem 15.6 *Let* $n = 3$ *and let stars of points of* Γ *be classical generalized quadrangles. Then* Γ *is one of the following:*

(i) a polar space;
(ii) the $Alt(7)$*-geometry;*
(iii) a standard quotient of an affine polar space;
(iv) one of the geometries $\Gamma(U_4(2))$, $\Gamma(O_6^-(3))$, $3 \cdot \Gamma(O_6^-(3))$, $\Gamma(McL)$, $\Gamma(Suz)$ *of Section 10.2.3;*
(v) the Yoshiara geometry for $Aut(HS)$ *(see 10.2.3).*

Sketch of the Proof. As in the previous theorem, the first part of the proof runs quite smoothly. We will expose it in detail. The second part deals with a few exceptional cases and it is rather more difficult. We will not go into details there, but we will give references.

Let $F = (a, u)$ be a point–plane flag of Γ. As Γ_a is a classical generalized quadrangle, q is a prime power. By this and by Theorem 9.21, Γ_u is one of the following:

(a) a projective plane of order q;
(b) an affine plane of order q;
(c) the circular space on $q + 2$ points;
(d) the Hermitian unital $U_H(3)$ (order $(3, 8)$);
(e) the Ree unital $U_R(3)$ (order $(3, 8)$);
(f) the Witt–Bose–Shrikande space $W(q)$, with $q \geq 8$;
(g) a linear space as in (iii) of Theorem 9.21, with at least five lines on every point.

Note that $q \leq 3$ implies $s = 1, q - 1$ or q, as $s \leq q$ (Section 1.2.6(4)). When $s = 1, q - 1$ or q, we are in case (c), (b) or (a), respectively. Thus, we have assumed $q \geq 4$ in (g) and we have not explicitly mentioned the Hermitian unital $U_H(2)$ and the Witt–Bose–Shrikande space $W(4)$. Let us prove that (d), (e), (f) and (g) do not occur. We can do this by comparison of induced actions (see Section 9.6.2).

Let $\Gamma_u = U_H(3)$. Then $G_u/K_u = U_4(3)$ or $U_4(3) \cdot \mathbf{2}$ $(= G_2(3))$ and the stabilizer $(G_u/K_u)_a$ of a in G_u/K_u is either $\mathbf{3}^{1+2} : \mathbf{8}$ or $\mathbf{3}^{1+2} : \mathbf{8} : \mathbf{2}$ (see [41]), acting in Γ_F as $\mathbf{3}^2 : \mathbf{8}$ or $\mathbf{3}^2 : \mathbf{8} : \mathbf{2}$, respectively. On the other hand, G_a/K_a is a classical group (Theorem 9.16), whence the action of $(G_a/K_a)_u$

in Γ_F contains $L_2(8)$. This does not fit with what we have above obtained for $(G_u/K_u)_a$. Therefore $\Gamma_u \neq U_H(3)$.

Let $\Gamma_u = U_R(3)$. Then G_u/K_u is a subgroup of the Ree group ${}^2G_2(3)$. Hence we have $(G_u/K_u)_a \leq \mathbf{9} : \mathbf{6}$ (see [41]). On the other hand, $(G_a/K_a) \geq L_2(8)$, as above. Again, we have a contradiction. Therefore $\Gamma_u \neq U_R(3)$.

Let $\Gamma_u = W(q)$, $q \geq 8$. We now have $L_2(q) \leq G_u/K_u \leq P\Gamma L(2,q)$ and $(G_u/K_u)_a$ has index $q(q-1)/2$ in G_u/K_u. On the other hand, the action of $(G_a/K_a)_u$ in Γ_F contains $L_2(q)$, as G_a/K_a is classical by Theorem 9.16. This does not fit with the above. Hence $\Gamma_u \neq W(q)$.

Finally, let Γ_u be as in (iii) of Theorem 9.21. As $s \leq q$ (Section 1.2.6(4)), we have $s = 1, q-1$ or q if $q \leq 3$. Hence, if $q \leq 3$, then we are in one of the cases (a), (b) or (c). Thus, we assume $q \geq 4$.

The linear space Γ_u has p^m points for some prime p and some positive integer m. Furthermore, $G_u/K_u \leq A\Gamma L(1,p^m)$ and $(G_u/K_u)_a = A \cdot B$ with $A \leq Z_{q-1}$ and $B \leq Z_m$. Therefore, the action of $(G_u/K_u)_a$ in Γ_F has the form $\overline{A} \cdot \overline{B}$ for some $\overline{A} \leq Z_a$ and $\overline{B} \leq Z_b$, where a, b are (possibly improper) divisors of $q-1$ and m, respectively. If G_a/K_a is classical, then the action of $(G_a/K_a)_u$ in Γ_F contains $L_2(q)$, which is not of the form $\overline{A} \cdot \overline{B}$ with $\overline{A}$ and $\overline{B}$ cyclic (because we have assumed $q \geq 4$). Therefore G_a/K_a cannot be classical. As $q \geq 4$, by Theorem 9.16 the following is the only surviving possibility: $q = 9$, $\Gamma_u = \mathcal{H}_3(3^2)$ and $G_a/K_a = L_3(4) \cdot \mathbf{2}_2$, $L_3(4) \cdot \mathbf{2}_3$ or $L_3(4) \cdot (\mathbf{2}^2)$ (notation as in [41]). The action induced by $(G_a/K_a)_u$ in Γ_F is A_6 $(= L_2(9))$ in the first two cases and M_{10} $(\approx L_2(9) \cdot \mathbf{2})$ in the third case. However, this evidently does not fit with the above description $\overline{A} \cdot \overline{B}$ of the action of $(G_u/K_u)_a$ in Γ_F. Thus, (g) does not occur. The linear space Γ_u is a projective plane, an affine plane or a circular space.

If Γ_u is a projective plane, then Γ is either a polar space or the $Alt(7)$-geometry, by Theorem 14.13. Indeed, stars of points cannot be classical generalized quadrangles in case (iii) of Theorem 14.13 (see Exercise 14.8). If Γ_u is an affine plane, then Γ is a standard quotient of an affine polar space (Theorem 15.4).

We are left with the case where Γ_u is a circular space. We have $q \geq 2$ since stars of points are classical (hence thick) generalized quadrangles. The circular space on four points is $AG(2,2)$ (case (b)). Thus, we may assume that $q \geq 3$. By the flag-transitivity and since Γ_u is a circular space, G_u acts 2-transitively on the $q+2$ points of Γ_u.

If G_a/K_a is not classical, then $q = 3$ or 9, by Theorem 9.16. That Theorem also tells us which are the possibilities for G_a/K_a. On the other hand, the feasible actions of G_u on the $q+2$ $(= 5$ or $9)$ points of Γ_u are known, by the classification of 2-transitive groups ([32]; also [40]). By comparison of induced actions we can prove that $q = 9$ with $G_a/K_a = L_3(4) \cdot \mathbf{2}^2$ is the only surviving possibility [60]. By the generators-and-relations method it is possible to prove that the Yoshiara geometry (Section 10.2.3) is the unique example for this case (see [238]; also [159]).

Let G_a/K_a be classical. Then the stabilizer of a in the 2-transitive action of G_u/K_u on the $q+2$ points of Γ_u contains $L_2(q)$ as a normal subgroup. This situation has been studied by Suzuki [205] and by Tits [221]. They have independently proved that the above situation occurs only if $q \leq 4$ or $q = 9$ with $G_u/K_u = M_{11}$. Therefore $q = 3$, 4 or 9 and the last case occurs only if $G_u/K_u = M_{11}$.

It seems we are close to the end of the proof; however, the most difficult part of the proof begins here. To finish, we still need to prove that Γ is one of the geometries mentioned in (iv). This can be done either by combining geometric and group-theoretic arguments (see [26], [60] and [131]) or by the generators-and-relations method (see [241] and [238]) or by some modification of that method [158]. We omit the details. □

15.2.1 Some 'counterexamples'

The hypothesis that stars of points are classical is essential in Theorem 15.6. Indeed, if we drop it, a number of examples can be constructed that do not fall into any of the cases (i)–(v) of that theorem. We mention some of them here.

(1) As we remarked in Section 9.4.4, two non-classical finite thick flag-transitive generalized quadrangles are known (there are four if we distinguish between dual cases). One of them is the affine expansion $T_2^*(O)$ of a hyperoval O of $PG(2,4)$. It has order $(3,5)$. Yoshiara [242] has constructed three flag-transitive geometries belonging to the diagram $c.C_2$,

$(c.C_2)$ •—c—•═══•

with stars of points isomorphic to $T_2^*(O)$ and to the dual of $T_2^*(O)$, with O as above. We have $T_2^*(O)$ as isomorphism type of stars of points in one of those examples. In the other two examples, stars of points are dually isomorphic to $T_2^*(O)$. The examples found by Yoshiara are 2-simply connected.

(2) Let Γ be the $\{m-2, m-1, m\}_+$-truncation of $PG(2m-1,q)$ $(m \geq 3)$ and let Γ' be the Grassmann geometry of Γ with respect to the central node of the diagram of Γ, representing $(m-1)$-dimensional elements of $PG(2m-1,q)$. The geometry Γ' belongs to $L.C_2$, residues of planes of Γ' are isomorphic to the point–line system of $PG(m,q)$ and stars of points of

Γ' are grids. The geometry $PG(2m-1,q)$ admits a polarity, which acts as a type-preserving automorphism in Γ'. It is clear from this that Γ' is flag-transitive, with $Aut(\Gamma') = Aut(PG(2m-1,q))$.

If we repeat this construction starting from the $(2m-1)$-dimensional simplex $\mathcal{S}$, then we obtain a flag-transitive geometry Γ' belonging to the diagram $c.C_2$ where stars of points are grids and residues of planes are circular spaces with $m+1$ points. We have $Aut(\Gamma') = Aut(\mathcal{S}) = S_{2m} \times \langle\delta\rangle$, with δ an involutory duality of $\mathcal{S}$ (see Section 9.2.2(2)). The duality δ defines a 2-fold 2-quotient of Γ'. By Corollary 11.3, the quotient $\Gamma'/\langle\delta\rangle$ is flag-transitive and $Aut(\Gamma'/\langle\delta\rangle) = S_{2m}$.

In particular, when $m = 3$ and $q = 2$, then Γ' is the affine polar space obtained by dropping a secant hyperplane from $\mathcal{Q}_5^+(2)$ and $\Gamma'/\langle\delta\rangle$ is the (unique) proper standard quotient of that affine polar space (the reader can prove these claims using Theorem 15.2, then counting the number of points in order to determine the right hyperplane to consider).

We have assumed $m \geq 3$. If $m = 2$, then the above is just the construction of a top-thin rank 3 polar space from a projective geometry of rank 3.

(3) Given an m-dimensional subspace U of $PG(2m-1,q)$ $(m \geq 3)$, let Γ be the geometry consisting of all subspaces X of $PG(2m-1,q)$ of dimension $m-1$, m or $m+1$ such that either $U \cap X = \emptyset$ or $U \cup X$ spans $PG(2m-1,q)$. The geometry Γ is a truncation of an amalgam of attenuated spaces (Section 2.2.1). It belongs to the diagram $L^*.L$ (Section 7.8) and residues of planes (stars of points) of Γ are isomorphic (dually isomorphic) to the point–line system of $AG(m,q)$. Let Γ' be the Grassmann geometry of Γ with respect to the central node of the diagram of Γ. The geometry Γ' belongs to $L.C_2$, stars of points of Γ' are grids and the planes of Γ' are isomorphic to the point–line system of $AG(m,q)$. We can always define a nondegenerate symplectic form f on $V(2m,q)$ in such a way that U is totally isotropic for f. The polarity defined by f in $PG(2m-1,q)$ (mapping a subspace X of $PG(2m-1,q)$ onto its orthogonal $X^\perp$ with respect to f) is a dual automorphism of Γ, hence it is a type-preserving automorphism of Γ'. It is now clear that Γ' is flag-transitive. Note that, when $q = 2$, then Γ' belongs to $c.C_2$.

We have above assumed $m \geq 3$. If we repeat the previous construction in the case of $m = 2$, then we get the affine polar space obtained by dropping a tangent hyperplane from $\mathcal{Q}_5^+(q)$ (see Exercise 5.15; also Section 7.8.2).

(4) Let Γ be the geometry of Exercise 7.29 and let Γ' be the Grassmann geometry of Γ with respect to the central node of the diagram of Γ. The

geometry Γ' belongs to $L.C_2$, stars of points of Γ' are grids and the planes of Γ' are isomorphic to the Witt–Bose–Shrikande space $W(q)$. We recall that Γ is obtained from $\mathcal{Q}_3^+(q)$, $q = 2^h$, $h \geq 2$. Let π be the polarity defined by $\mathcal{Q}_3^+(q)$ in $PG(3,q)$, mapping a subspace of $PG(3,q)$ onto its orthogonal with respect to the quadratic form defining $\mathcal{Q}_3^+(q)$. It is not difficult to prove that π is a dual automorphism of Γ. Hence it is a type-preserving automorphism of Γ'. It is now evident that Γ' is flag-transitive.

Note that, if $h = 2$, then Γ' belongs to $c.C_2$.

(5) Let Γ be the geometry of Exercise 7.30 and let Γ' be its Grassmann geometry with respect to the central node of the diagram. Γ' belongs to $L.C_2$, stars of points of Γ' are grids and the planes of Γ' are isomorphic to the Hermitian unital $\mathcal{H}(q^2)$. The polarity π defined by $\mathcal{H}_4(q^2)$ in $PG(4,q^2)$ is a dual automorphism of Γ, hence it is a type-preserving automorphism of Γ'. Clearly, Γ' is flag-transitive.

When $q = 2$, Γ' belongs to $Af.C_2$

Many flag-transitive finite geometries exist belonging to the diagram $c.C_2$ and with grids as stars of points. Some examples of this kind have been mentioned above. Two examples are also implicit in the constructions of Exercises 7.27 and 7.28 (the geometries constructed there belong to $c^*.c$; in order to pass to $c.C_2$, we should take Grassmann geometries with respect to the central node of the $c^*.c$ diagram). More examples are given in [132].

15.2.2 Problems

(1) As we have remarked in Section 9.4.4, another flag-transitive non-classical generalized quadrangle is known besides the one obtained from a hyperoval of $PG(2,4)$. It has order $(15,17)$. Are there any flag-transitive $c.C_2$ geometries admitting that generalized quandrangle (or its dual) as isomorphism type of stars of points ?

(1) The only linear spaces that can occur as residues of planes in Theorem 15.6 are projective planes, affine planes and circular spaces (the latter only with 4, 5, 6 or 11 points). The above examples show that, if we allow stars of points to be grids, then point–line systems of projective or affine geometries of dimension $m \geq 3$, Witt–Bose–Shrikande spaces and all circular spaces can also appear as residues of planes. Are there any other linear spaces that can occur as isomorphism types of residues of points in finite flag-transitive $L.C_2$ geometries ?

(2) Try to classify finite flag-transitive $c.C_2$ geometries where stars of points

are grids. Some partial results on this problem have been obtained in [157] and [133].

(3) Try to classify all finite flag-transitive C_n geometries with thin lines (case (i) of Theorem 15.5).

15.3 The diagram $L_m.C_k$

By $L_m.C_k$ we designate the following diagram of rank $n \geq 3$:

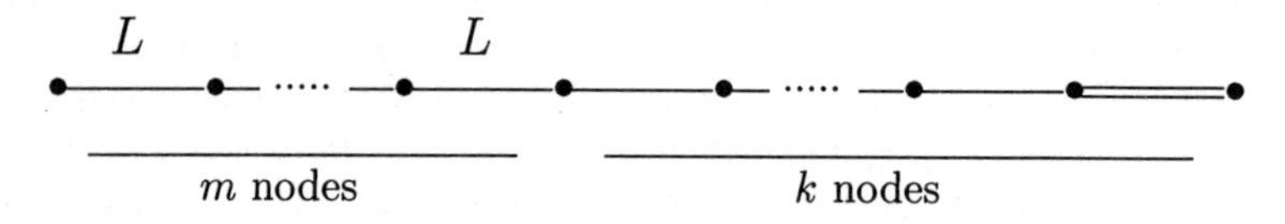

where $m + k = n + 1$, $m \geq 2$ and $k \geq 2$. Note that, if $m \geq 3$, then $L_m.C_k$ includes $L_{m-1}.C_{k+1}$ as a special case. Note also that the rank 3 case $L_2.C_2$ of $L_m.C_k$ is just the diagram $L.C_2$ considered in Theorem 15.6. Thus, we will consider only cases of rank $n \geq 4$ in this section.

We take the non-negative integers 0, 1, ..., $n-1$ as types, labelling the nodes of the diagram from left to right, as usual.

Throughout this section, Γ is a locally finite flag-transitive geometry belonging to the diagram $L_m.C_k$. As Γ is finite and flag-transitive, it admits finite orders s_0, s_1, ..., s_{m-2}, q, ..., q, t

L L

s_0 s_1 s_{m-2} q q q q t

with $s_0 \leq s_1 \leq ... \leq s_{m-2} \leq q$ (see Section 1.2.6(4)).

Theorem 15.7 *Let $k \geq 3$ (hence $n \geq 4$). Then Γ is one of the following:*

(i) a C_n geometry with thin lines;
(ii) a polar space;
(iii) a standard quotient of an affine polar space;
(iv) the $Alt(8)$-geometry;
(v) one of the geometries $\Gamma(Fi_{22})$, $\Gamma(Fi_{23})$, $\Gamma(Fi_{24})$ or $3{\cdot}\Gamma(Fi_{24})$ mentioned in Section 10.2.3.

Sketch of the Proof. The case of $m = 2$ has already been examined in Theorem 15.5. Thus, we can assume $m \geq 3$ and $s_{m-2} < q$. Given a dual point u of Γ, let a be a point incident to u. Let G_u and G_a be the stabilizers in $Aut(\Gamma)$ of u and a, respectively.

By Theorem 9.22, we have either $\Gamma_u = S(M_{23})$ ($m = 3$, $n = 5$ and $k = 3$) or $\Gamma_u = S(M_{24})$ ($m = 4$, $n = 6$, $k = 3$).

Let $\Gamma_u = S(M_{23})$ (hence G_u acts as M_{23} in Γ_u). We have $\Gamma_a = \Gamma(Fi_{22})$ by Theorem 15.5. Hence G_a acts as Fi_{22} or $Aut(Fi_{22})$ in Γ_a. The kernels K_a and K_u of G_a and G_u can be determined [232]. Thus, we get on G_a and G_u enough information to use the generators-and-relation method. By that method, Van Bon and Weiss [232] have proved that $Aut(\Gamma) = Fi_{23}$. Therefore $\Gamma = \Gamma(Fi_{23})$.

Let $\Gamma_u = S(M_{24})$ (hence G_u acts as M_{24} in Γ_u). The star Γ_a of a and the action of G_a in Γ_a is known by the previous step of the proof. As in that step, the generators-and-relations method can be used. It turns out that $Aut(\Gamma)$ is a homomorphic image of the central non-split extension $\mathbf{3} \cdot Fi_{24}$ of Fi_{24} (see [232]). Hence $Aut(\Gamma) = Fi_{24}$ or $\mathbf{3} \cdot Fi_{24}$, as the commutator subgroup Fi'_{24} of Fi_{24} is simple and has index 2 in Fi_{24}. Therefore $\Gamma = \Gamma(Fi_{24})$ or $3 \cdot \Gamma(Fi_{24})$.

We have referred to [232] in the above because we wished to mention one more application of the generators-and-relation method. The reader wanting to see a more geometric proof may consult [129] (also [143]). □

We finish with a result on the case of $k = 2$:

$(L_{n-1}.C_2)$

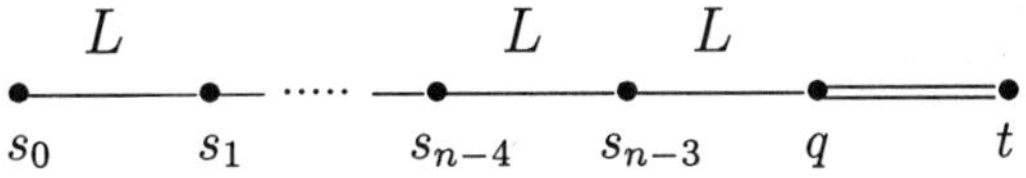

We assume $s_{n-3} < q$ and $n \geq 4$, to keep this diagram distinct from the one considered in Theorem 15.7 and from $L.C_2$, considered in Theorem 15.6. Furthermore, we assume that stars of elements of Γ of type $n-3$ are classical generalized quadrangles.

Proposition 15.8 *Under the previous hypotheses, one of the following holds:*

(i) the geometry Γ *belongs the following diagram:*

$(c^{n-2}.C_2)$

(that is, $s_0 = s_1 = ... = s_{n-3} = 1$*);*

(ii) the geometry Γ *has rank* 4 *and order* $(1, 2, 3, t)$*, with* $t = 3$ *or* 9 *and residues of dual points of* Γ *are isomorphic to the enrichment of the Mőbius plane arising from* $Q_3^-(3)$*.*

Proof. If $s_{n-3} = 1$, then we are in case (i). Let $s_{n-3} > 1$ and let $F = (a, u)$

be a flag of type $(n-4, n-1)$. Feasible structures for Γ_u are known by Theorem 9.22 and feasible structures for the star of Γ_a^+ are known by Theorem 15.6. The information obtained in this way on Γ_u fits with that got on Γ_a^+ (and with the hypothesis $1 < s_{n-3} < q$) only if Γ_u is the enrichment of a (classical) Mőbius plane (see Section 9.6.2(5)). In this case, we have $n = 4$, a is a point, Γ belongs to $Af.C_2$ and Γ has order $(1, q-1, q, t)$ (with $q \geq 3$ because $q - 1 = s_{n-3} > 1$). We have $q = p^m$ for some prime p and some positive integer m.

Thus, let Γ_u be the enrichment of a Mőbius plane $\mathcal{M}$. As we have remarked in Section 9.6.2(5), $\mathcal{M}$ is the Mőbius plane obtained from $\mathcal{Q}_3^-(q)$.

We can now proceed by comparison of induced actions. We denote the stabilizers in $Aut(\Gamma)$ of u and F by G_u and G_F respectively and their kernels by K_u and K_F, as usual. It is not hard to prove that the action induced by $(G_u/K_u)_a$ in Γ_F is a subgroup of $\mathbf{p}^{2m} : (Z_{q^2-1} \times Z_2)$ (we can prove this either by using the information given in Section 9.6.2(5) or by straightforward computations in the automorphism group of the Mőbius plane $\mathcal{M}$).

Given a line x incident to F, let $G_{a,x}$ and $G_{a,x,u}$ be the stabilizers in $Aut(\Gamma)$ of the flags (a, x) and $F \cup x = (a, x, u)$, respectively. Using Theorem 9.16, it is not difficult to prove that the action $G_{a,x}/K_{a,x}$ of $G_{a,x}$ in $\Gamma_{\{a,x\}}$ is a classical group, except perhaps when $q = 3$ and $t = 3$ or 9. If $G_{a,x}/K_{a,x}$ is classical, then $G_{a,x,u}/K_{a,x,u} \geq L_2(q)$ and $G_F/K_F \geq \mathbf{p}^{2m} : SL(2, q)$. This does not fit with the information we have previously obtained on the action of $(G_u/K_u)_a$ in Γ_F. Therefore $q = 3$ and $t = 3$ or 9. □

Problem. Prove that case (ii) of the above proposition cannot occur.

Meixner [129] obtained an almost complete classification for case (i) of Proposition 15.8. The reader should consult [129] (also [165]) for the details of that classification.

Many top-thin geometries exist belonging to the diagram $L^{n-2}.C_2$ ($n \geq 4$). For instance, we can start from a building Γ belonging to a Coxeter diagram of the following form:

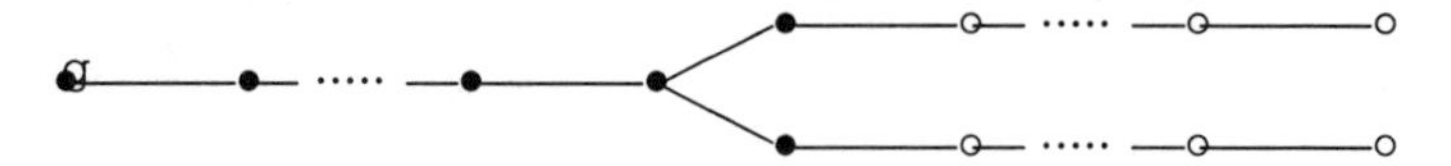

and we can take a truncation of Γ, dropping the elements corresponding to the types represented by empty discs in the above picture. Then we can form the Grassmann geometry of that truncation with respect to the initial node of the diagram, circled in the above picture; we may also take a proper

2-quotient of that Grassmann geometry, if we like (and if it admits proper 2-quotients). A geometry obtained in this way belongs to $L_{n-1}.C_2$ (where n is the number of nodes represented by full discs in the above picture) and it is top-thin (stars of elements of type $n-3$ are grids). In particular, if Γ is a Coxeter complex, then we get top-thin geometries belonging to $c^{n-2}.C_2$. Note that (2) of Section 15.2 is a special case of the previous construction.

The above geometries are not included in (i) of Proposition 15.8, since the hypotheses of Proposition 15.8 do not allow grids as C_2 residues.

Most flag-transitive top-thin geometries belonging to $c^{n-2}.C_2$ ($n \geq 4$) arise from Coxeter complexes by the construction described above. However, some examples exist that cannot be obtained in that way (see [132]; also [165]). A complete classification of these geometries does not seem to be easy. Nevertheless, some results have been obtained on this problem; for instance, 2-simply connected flag-transitive top-thin $c^{n-2}.C_2$ geometries of rank $n \geq 7$ have been classified in [133].

15.4 More diagrams

A great amount of work has been done on diagrams obtained by replacing some strokes representing projective planes in a spherical Coxeter diagram of rank ≥ 3 with strokes labelled by c or c^* or even by L or L^*, or adding some strokes of that kind at some nodes of a spherical Coxeter diagram of rank ≥ 2.

Many diagrams considered in Chapter 7 and those examined in the previous sections of this chapter are just of this kind. A survey of results on this topic is given in [27] (Section 4). We only mention a few of them here.

We have considered the diagram $L.A_{n-2}.L^*$ ($n \geq 4$) in Theorem 7.63, obtaining a quasi-classification of the geometries belonging to that diagram, with a good system of lines and admitting finite orders:

$$(L.A_{n-2}.L^*) \qquad \overset{L}{\underset{s}{\bullet}\!\!-\!\!\underset{q}{\bullet}}\!-\!\underset{q}{\bullet}\cdots\underset{q}{\bullet}\!-\!\overset{L^*}{\underset{q}{\bullet}\!\!-\!\!\underset{t}{\bullet}}$$

Assuming flag-transitivity and $s = t = 1 < q$, but dropping the assumption of a good system of lines, it is not difficult to prove that the following are the only examples that can arise ([165], Theorem 8.4):

(i) a bi-affine geometry defined over $GF(2)$;
(ii) the almost flat quotient of a bi-affine geometry defined by a point–hyperplane flag in $PG(n, 2)$ (see Exercises 8.59, 8.60 and 8.61);
(iii) the geometry $\Gamma(HS)$.

A classification of the rank 3 case of the above diagram is much harder, even if we assume the flag-transitivity and $s = t = 1$, as above. The reader may find some interesting results on this problem in [4].

The following diagram of rank $n \geq 3$ can be viewed as a 'dual' of $L.C_{n-1}$:

$(C_{n-1}.L)$ $\overline{L}$

where $\overline{L}$ designates the class of those linear spaces where the full set of points can be spanned by some finite set of points.

Ronan has investigated geometries belonging to this diagram in [176] and [178]. He has proved that a geometry Γ belonging $C_{n-1}.L$ ($n \geq 4$) is a (possibly improper) 2-quotient of a truncation of a polar space if and only if all C_3 residues of Γ are 2-covered by polar spaces.

In the rank 3 case he has proved that a $C_2.L$ geometry satisfying IP and with stars of points isomorphic to the point–line system of a 3-dimensional projective geometry is either a (possiby improper) 2-quotient of a truncated polar space or one of two 'sporadic' examples, arising from the groups M_{24} and $L_3(2) \cdot \mathbf{2}$, respectively. However, many examples exist for $C_2.L$ where either IP does not hold or stars of points are not point–line systems of 3-dimensional projective geometries (see [141]).

A classification for the following diagram of rank $n \geq 4$ is also implicit in the above:

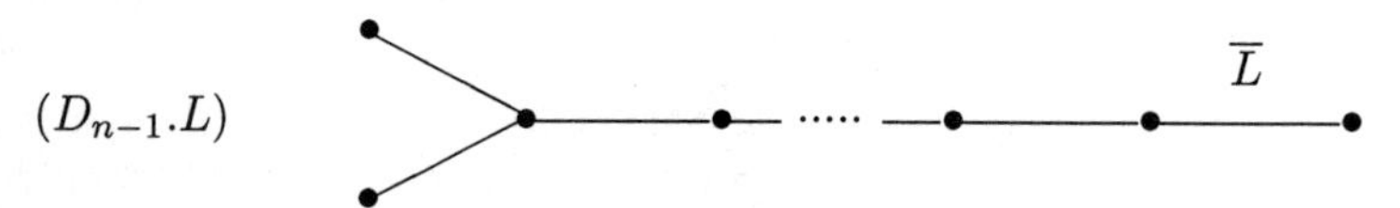

Applying the previous result on $C_{n-1}.L$ to Grassmann geometries with respect to the last node of the $D_{n-1}.L$ diagram, we get that every geometry belonging to $D_{n-1}.L$ with $n \geq 4$ is a (possibly improper) 2-quotient of a truncation of a D_m building, for some $m \geq n$.

Similarly, a theorem on $L.D_{n-1}$ is implicit in every theorem on $L.C_{n-1}$.

The following diagram has been investigated by Ivanov and Shpectorov ([93], [96]–[101], [188], [190]):

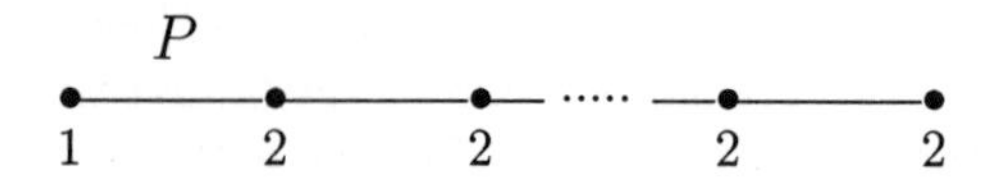

where P denotes the Petersen graph (that is, the point–line system of the halved dodecahedron). They have proved that only eight flag-transitive examples exist for this diagram. Two of them have rank 3 and arise from the groups M_{22} and $\mathbf{3} \cdot M_{22}$ (the example for $\mathbf{3} \cdot M_{22}$ is the universal 2-cover of that for M_{22}). There are four examples of rank 4, related to the following groups: M_{23}, J_4, Co_2 and $\mathbf{3}^{23} \cdot Co_2$ (the last one is the universal 2-cover of the third one). Two examples have rank 5. One of them arises from Fischer's Baby Monster F_2. It admits just one proper 2-cover, which arises from a non-split extension $\mathbf{3}^{4371} \cdot F_2$ of F_2.

The classification achieved by Ivanov and Sphectorov for the above diagram has stimulated an impressive amount of work on geometries involving the Petersen graph as a rank 2 residue. The reader should consult Section 5 of [27] for a survey of results on this topic.

Flag-transitive geometries belonging to Coxeter diagrams where rank 2 residues other than generalized digons are finite classical generalized m-gons ($m = 3, 4, 6$ or 8) have been systematically investigated by Timmsfeld, Stroth and Meixner, assuming the finiteness of Borel subgroups. A paper by Aschbacher [2] and a paper by Niles [137] are the starting points of this inquiry. The work begun in [2] and [137] has been continued by Timmesfeld ([212]–[218]) and by Stroth ([198]–[202], [204]) and finished by Meixner ([130], [128]). The reader can find a survey of that classification in [27] (Section 3).

Classification theorems have also been obtained for other diagrams, often with some additional hypotheses such as flag-transitivity, local finiteness, thickness at certain nodes of the diagram (see [27]). As it happens in many cases mentioned before, it often turns out that we have a class of well-known 'standard' examples for the diagram we are considering and a few exceptional geometries in addition to them, often related to some sporadic simple groups or to some 'sporadic' property of a small simple group. Sometimes no 'standard' class exists and we have only a short list of sporadic geometries. In other cases we also get a class of 'extreme' examples besides 'standard' and 'sporadic' ones, thin at almost every node of their diagram; 'discard-examples', some might say. However, these 'bad' geometries are sometimes fairly interesting, too.

One might now be tempted to dream of a 'Mendeleev Table' of geometries, perhaps a geometric counterpart of the classification of finite simple groups; but we are still very far from that.

Bibliography

[1] M. Aschbacher, *Flag structures in Tits geometries*, Geom. Dedicata **14** (1983), 21–32.

[2] M. Aschbacher, *Finite geometries of type C_3 with flag-transitive automorphism group*, Geom. Dedicata **16** (1984), 195–200.

[3] M. Aschbacher and S. Smith, *Tits geometries over $GF(2)$ defined by groups over $GF(3)$*, Comm. Alg. **11** (1983), 1675–1684.

[4] B. Baumeister, *Fahnentransitive rang 3 Geometrien, die lokal vollstãndige Graphen sind*, Diplomarbeit, Freie Universitaet Berlin, Summer 1992.

[5] M. Biliotti and A. Pasini, *Intersection properties in geometries*, Geom. Dedicata **13** (1982), 257–275.

[6] G. Birkhoff, Lattice Theory, A.M.S. Math. Colloq. Publ. **25** (3*rd* edition), N.Y. 1967.

[7] R. Bose, *Strongly regular graphs, partial geometries and balanced designs*, Pacific J. Math. **13** (1963), 389–419.

[8] A. Borel and J. Tits, *Groupes réductifs*, Publ. Math. I.H.E.S. **27** (1965), 55–150.

[9] A. Borel and J. Tits, *Complément à l' article 'Groupes réductifs'*, Publ. Math. I.H.E.S. **41** (1972), 253–276.

[10] N. Bourbaki, Algèbre, chap. 9: *Formes sesquilinèaires et formes quadratiques*, Hermann, Paris 1959.

[11] N. Bourbaki, Groupes et Algèbres de Lie, chaps. 4, 5 and 6, Hermann, Paris 1968.

[12] A. Brouwer and A. Cohen, *Some remarks on Tits's geometries*, Indagationes Math. **45** (1983), 393–402.

[13] A. Brouwer, A. Cohen and A. Neumaier, Distance Regular Graphs, Springer, Berlin 1989.

[14] K. Brown, Buildings, Springer, Berlin 1989.

[15] F. Bruhat and J. Tits, *Groupes réductifs sur un corps local. I. Données radicielles*, Publ. Math. I.H.E.S. **41** (1972), 5–251.

[16] F. Buekenhout, *Une caractérisation des espaces affins basées sur la notion de droite*, Math. Zeit. **3** (1969), 367–371.

[17] F. Buekenhout, *Extensions of polar spaces and the doubly transitive symplectic groups*, Geom. Dedicata **6** (1977), 13–21.

[18] F. Buekenhout, *Diagrams for geometries and groups*, J. Comb. Th. A **27** (1979), 121–151.

[19] F. Buekenhout, *The basic diagram of a geometry*, in Geometries and Groups, L.N. **893**, Springer, Berlin (1981), 1–29.

[20] F. Buekenhout, *Le plans de Benz: une approche unifiée des plans de Mőbius, Laguerre et Minkowski*, J. Geometry **17** (1981), 61–68.

[21] F. Buekenhout, *An approach to buildings based on points, lines and convexity*, European J. Comb. **3** (1982), 103–118.

[22] F. Buekenhout, *Diagram geometries for sporadic groups*, in Finite Groups Coming of Age, A.M.S series Contemp. Math. **45** (1985), 1–32.

[23] F. Buekenhout, *The geometry of finite simple groups*, in Buildings and the Geometry of Diagrams (L.A. Rosati ed.), L.N. **1181**, Springer, Berlin (1986), 1–78.

[24] F. Buekenhout and D. Buset, *On the foundations of incidence geometry*, Geom. Dedicata **25** (1988), 269–296.

[25] F. Buekenhout and A. Cohen, Diagram Geometry, to appear.

[26] F. Buekenhout and X. Hubaut, *Locally polar spaces and related rank 3 groups*, J. Algebra **45** (1977), 391–434.

[27] F. Buekenhout and A. Pasini, *Finite geometries extending buildings*, Chapter 22 of Handbook of Incidence Geometry (F. Buekenhout ed.), to appear.

[28] F. Buekenhout and E. Shult, *On the foundation of polar geometry*, Geom. Dedicata **3** (1974), 155–170.

[29] F. Buekenhout and Sprague, *Polar spaces having some lines of cardinality two*, J. Comb. Th. A **33** (1982), 223–228.

[30] F. Buekenhout, A. Delandtsheer and J. Doyen, *Finite linear spaces with flag-transitive groups*, J. Comb. Th. A, **49** (1988), 268–293.

[31] F. Buekenhout, A. Delandtsheer, J. Doyen, P. Kleidman, M. Liebeck and J. Saxl, *Linear spaces with flag-transitive automorphism groups*, Geom. Dedicata, **36** (1990), 89–94.

[32] P. Cameron, *Finite permutation groups and finite simple groups*, Bull. London Math. Soc. **13** (1981), 1–22.

[33] P. Cameron, *Dual polar spaces*, Geom. Dedicata **12** (1982), 75–85.

[34] P. Cameron, D. Hughes and A. Pasini, *Extended generalized quadrangles*, Geom. Dedicata **35** (1990), 193–228.

[35] R. Carter, Simple Groups of Lie Type, Wiley, London 1972.

[36] A. Cohen, *Point-line characterizations of buildings*, in Buildings and the Geometry of Diagrams (L.A. Rosati ed.), L.N. **1181**, Springer, Berlin (1986), 191–206.

[37] A. Cohen, *Point-line spaces related to buildings*, Chapter 12 of Handbook of Incidence Geometry (F. Buekenhout ed.), to appear.

[38] A. Cohen and B. Cooperstein, *A characterization of some geometries of exceptional Lie type*, Geom. Dedicata **15** (1983), 73–105.

[39] A. Cohen and E. Shult, *Affine polar spaces*, Geom. Dedicata **35** (1990), 43–76.

[40] A. Cohen and H. Zantema, *A computation concerning doubly transitive permutation groups*, J. Reine und Angewandte Math. **374** (1984), 196–211.

[41] J. Conway, R. Curtis, S. Norton, R. Parker and R. Wilson, Atlas of Finite Groups, Clarendon, Oxford 1985.

[42] B. Cooperstein, *A characterization of some Lie incidence structures*, Geom. Dedicata **6** (1977), 205–258.

[43] H. Coxeter and W. Moser, Generators and Relations for Discrete Groups, Springer, Berlin 1980.

[44] H. Crapo and G.C. Rota, On the Foundations of Combinatorial Theory II: Combinatorial Geometries, M.I.T. Press, Cambridge 1970.

[45] H. Cuypers, *The affine unitary space* (unpublished).

[46] H. Cuypers, *Finite locally generalized quadrangles with affine planes*, European J. Comb., **13** (1992), 439–453.

[47] H. Cuypers, *Symplectic geometries, transvection groups and modules*, J. Comb. Th. A, to appear.

[48] H. Cuypers, *Affine Grassmannians*, to appear.

[49] H. Cuypers and A. Pasini, *Locally polar geometries with affine planes*, European J. Comb. **13** (1992), 39–57.

[50] F. De Clerck and J. Thas, *Partial geometries in finite projective spaces*, Arch. Math. **30** (1978), 537–540.

[51] F. De Clerck and H. Van Maldeghem, *Some classes of rank 2 geometries*, Chapter 10 of Handbook of Incidence Geometry (F. Buekenhout ed.), to appear.

[52] A. Delandtsheer, *Finite (line, plane)-flag transitive linear spaces*, Geom. Dedicata, **41** (1992), 145–153.

[53] A. Delandtsheer, *Dimensional linear spaces whose automorphism group is (line, hyperplane)-flag transitive*, Designs Codes and Cryptography, **1** (1992), 237–245.

[54] A. Delandtsheer, *Dimensional linear spaces*, Chapter 6 of Handbook of Incidence Geometry (F. Buekenhout ed.), to appear.

[55] A. Del Fra, *Flag-transitive* $Af.C_2$ *geometries*, to appear.

[56] A. Del Fra and D. Ghinelli, $Af^*.Af$ *geometries, the Klein quadric and* $\mathcal{H}_q^n$, to appear.

[57] A. del Fra and A. Pasini, $C_2.c$ *geometries and generalized quadrangles of order* $(s-1, s+1)$, Results in Maths., **22** (1992), 489–508.

[58] A. Del Fra, D. Ghinelli and A. Pasini, *One diagram for many geometries*, in Advances in Finite Geometries and Designs (J. Hirschfeld, D. Hughes and J. Thas eds.), Oxford University Press, Oxford (1991), 111–140.

[59] A. Del Fra, D. Ghinelli and A. Pasini, *Affine attenuated spaces*, J. Comb. Th. A, **63** (1993), 175–194.

[60] A. Del Fra, D. Ghinelli, T. Meixner and A. Pasini, *Flag transitive extensions of C_n geometries*, Geom. Dedicata **37** (1990), 253–273.

[61] P. Dembowski, Finite Geometries, Springer, Berlin 1968.

[62] J. Dieudonné, Sur les Groupes Classiques, Actualités scientifiques et industrielles 1040, Hermann, Paris, 1948.

[63] J. Dieudonné, La Géometrie des Groupes Classiques, Springer, Berlin 1963.

[64] J. Doyen and X. Hubaut, *Finite regular locally projective spaces*, Math. Zeit. **119** (1971), 83–88.

[65] W. Feit, *Finite projective planes and a question about primes*, Proc. A.M.S. **108** (1990), 561–564.

[66] W. Feit and D. Higman, *The nonexistence of certain generalized polygons*, J. Algebra **1** (1964), 114–131.

[67] B. Fischer, *Finite groups generated by* 3*-transpositions*, Invent. Math. **13** (1971), 232–246.

[68] P. Fisher and T. Penttila, *One-point extensions of polar spaces*, European J. Comb. **11** (1990), 535–540.

[69] O. Frink, *Complemented modular lattices and projective spaces of infinite dimension*, Bull. A. M. S. **52** (1946), 452–467.

[70] H. Gevaert and D. Keppens, *Epimorphisms of partial geometries onto generalized quadrangles*, Arch. Math. **53** (1989), 95–98.

[71] J. Hall, *Steiner triple systems with minimally generated subsystems*, Quart. J. Math. Oxford Ser.(2), **25** (1974), 41–50.

[72] J. Hall, *Graphs, geometry,* 3*-transpositions and symplectic F_2-transvection groups*, Proc. London Math. Soc., **58** (1989), 89–111.

[73] J. Hall, *Some* 3*-transposition groups with normal* 2*-subgroups*, Proc. London Math. Soc., **58** (1989), 112–136.

[74] J. Hall and E. Shult, *Geometric hyperplanes of non-embeddable Grassmannians*, European J. Comb., **14** (1993), 29–36.

[75] M. Hall, The Theory of Groups, McMillan, New York, 1959.

[76] M. Hall, *Automorphisms of Steiner triple systems*, IBM J. Res. Develop. **4** (1960), 460–472.

[77] M. Hall, *Incidence axioms for affine geometry*, J. Algebra, **21** (1972), 535–547.

[78] G. Hanssens, *A characterization of point-line geometries for finite buildings*, Geom. Dedicata **25** (1988), 297–315.

[79] C. Hering, *Two new sporadic doubly transitive linear spaces*, in Finite Geometries (C. Baker and L. Batten eds.), Dekker, New York (1985), 127–129.

[80] D. Higman, *Flag-transitive collineation groups of finite projectice spaces*, Illinois J. Math. **6** (1962), 434–446.

[81] J. Hirschfeld, Projective Geometries over Finite Fields, Clarendon, Oxford 1979.

[82] S. Hobart and D. Hughes, *Extended partial geometries: nets and dual nets*, European J. Comb. **11** (1990), 357–372.

[83] P. Hoefsmit, *Representations of Hecke algebras of finite groups with BN-pairs of classical type*, Ph.D. Thesis, University of British Columbia, August 1974.

[84] D. Hughes, *On homomorphisms of projective planes*, A. M. S. Proc. Symp. Appl. Math. **10** (1960), 45–52.

[85] D. Hughes, *Extensions of designs and groups: projective, symplectic and certain affine groups*, Math. Zeit. **89** (1965), 199–205.

[86] D. Hughes, *On designs*, in Geometries and Groups, L. N. 893, Springer, Berlin (1981), 43–67.

[87] D. Hughes, *Semi-symmetric* 3*-designs*, in Finite Geometries (N. Johnson, M. Kallaher and C. Long eds.), L. N. in Pure and Applied Math. **82**, Dekker, New York (1982), 223–235.

[88] D. Hughes, *On the non-existence of a semi-symmetric* 3*-design with* 78 *points*, Ann. Discr. Math. **18** (1983), 473–480.

[89] D. Hughes, *Extended partial geometries: dual* 2*-deisgns*, European J. Comb., **11** (1990), 459–471.

[90] D. Hughes and F. Piper, Projective Planes, Springer, Berlin 1973.

[91] D. Hughes and F. Piper, Design Theory, Cambridge University Press, Cambridge 1985.

[92] J. Humphreys, Linear Algebraic Groups, Springer, Berlin 1975.

[93] A.A. Ivanov, *Graphs of girth* 5 *and diagram geometries related to the Petersen graph*, Soviet Mat. Doklady **295** (1987), 529–533 (in Russian).

[94] A.A Ivanov, *Geometric presentations of groups with an application to the Monster*, Proceedings of the International Congress of Mathematicians Kyoto 1990, Springer, Berlin (1991), 1443–1453.

[95] A.A Ivanov, *The minimal parabolic geometry of the Comway groups* Co_1 *is simply connected*, in Combinatorics '90 (A. Barlotti et al. eds.), North-Holland (1992), 259–273.

[96] A.A. Ivanov, *A geometric characterization of Fischer's Baby Monster*, J. Alg. Comb., **1** (1992), 45–69.

[97] A.A. Ivanov and S. Shpectorov, *Geometries for sporadic groups related to the Petersen graph. I*, Comm. Alg. **16** (1988), 925–953.

[98] A.A. Ivanov and S. Shpectorov, *Geometries for sporadic groups related to the Petersen graph. II*, European J. Comb. **10** (1989), 347–362.

[99] A.A. Ivanov and S. Shpectorov, P*-geometries of* J_4*-type have no natural representations*, Bull. Soc. Math. Belgique **42** (1990), 547–560.

[100] A.A. Ivanov and S. Shpectorov, *Natural representations of the* P*-geometries of* Co_2*-type*, J. Algebra, to appear.

[101] A.A. Ivanov and S. Shpectorov, *The last flag-transitive* P*-geometry*, Israel J. Math., to appear.

[102] B. Jónsson, *Lattice-theoretic approach to projective and affine geometry*, in The Axiomatic Method (L. Henkin, P. Suppes and A. Tarski eds.), North Holland, Amsterdam (1959), 188–203.

[103] W. Kantor, *Dimension and embedding theorems for geometric lattices*, J. Comb. Th. A **17** (1974), 173–195.

[104] W. Kantor, *Some geometries that are almost buildings*, European J. Comb. **2** (1981), 239–247.

[105] W. Kantor, *Some exceptional* 2*-adic buildings*, J. Algebra **92** (1985), 208–233.

[106] W. Kantor, *Generalized polygons, SCABs abd GABs*, in Buildings and The Geometry of Diagrams (L.A Rosati ed.), L.N. **1181**, Springer, Berlin (1986), 79–158.

[107] W. Kantor, *Some locally finite flag-transitive buildings*, European J. Comb. **8** (1987), 429–436.

[108] W. Kantor, *Primitive groups of odd degree and an application to finite projective planes*, J. Algebra **106** (1987), 14–45.

[109] W. Kantor, *Reflections on concrete buildings*, Geom. Dedicata **25** (1988), 121–145.

[110] W. Kantor, *Finite geometries via algebraic affine buildings*, Finite Geometries, Buildings and Related Topics (W. Kantor, R. Liebler, S. Payne and E. Shult eds.), Oxford University Press, Oxford (1990), 37–44.

[111] W. Kantor, 2*-Transitive and flag-transitive designs*, in Proceedings of M. Hall Conference (Vermont, September 1990), to appear.

[112] W. Kantor, R. Liebler and J. Tits, *On discrete chamber-transitive automorphism groups of affine buildings*, Bull. A.M.S. **16** (1987), 129–133.

[113] W. Kantor, T. Meixner and M. Wester, *Two exceptional* 3*-adic affine buildings*, Geom. Dedicata **33** (1990), 1–11.

[114] W. Klingenberg, *Projektive Geometrien mit Homomorphismus*, Math. Ann. **132** (1956), 190–200.

[115] C. Lam, *The search for a finite projective plane of order* 10, Am. Math. Mon. **98** (1991), 305–318.

[116] C. Lefèvre-Percsy, *Infinite* (Af, Af^*) *geometries*, J. Comb. Th. A **55** (1990), 133–139.

[117] H. Lenz, *Zur Begrunding der Analityschen Geometrie*, Sitzungsber. Bayer. Akad. Wiss. Math. Natur. K1 (1954), 17–72.

[118] H. Lenz, Grundlagen der Elementarmathematik, Deutscher Verlag d. Wiss., Berlin, 1961.

[119] R. Liebler, *A representation theoretic approach to finite geometries of spherical type*, 1985 (unpublished).

[120] G. Lunardon and A. Pasini, *A result on C_3 geometries*, European J. Comb. **10** (1989), 265–271.

[121] G. Lunardon and A. Pasini, *Finite C_n geometries (a survey)*, Note di Matematica **10** (1990), 1–35.

[122] H. Luneburg, Translation Planes, Springer, Berlin 1980.

[123] T. Meixner, *Chamber systems with extended diagram*, Coxeter Festschrift (III), Mitt. Math. Sem. Giessen **165** (1984), 93–104.

[124] T. Meixner, *Construction of chamber systems of type M with transitive automorphism group*, in Buildings and the Geometry of Diagrams (L.A. Rosati ed.), L.N. **1181**, Springer, Berlin (1986), 207–217.

[125] T. Meixner, *Klassische Tits Kammersysteme mit einer transitiven Automorphismengruppen*, Mitteilungen Math. Sem. Giessen, **174** (1986).

[126] T. Meixner, *Folding down classical Tits chamber systems*, Geom. Dedicata **25** (1988), 148–157.

[127] T. Meixner, *Groups acting transitively on locally finite Tits chamber systems*, in Finite Geometries, Buildings and Related Topics (W. Kantor, R. Liebler, S. Payne and E. Shult eds.), Oxford University Press, Oxford (1990), 45–65.

[128] T. Meixner, *Locally finite Tits chamber systems with transitive group of automorphism in characteristic* 3, Geom. Dedicata **35** (1990), 13–30.

[129] T. Meixner, *Some polar towers*, European J. Comb. **12** (1991), 397–415.

[130] T. Meixner, *Parabolic systems: the $GF(3)$-case*, Trans. A. M. S., to appear.

[131] T. Meixner, *A computer free proof of a theorem of Weiss and Yoshiara*, Abhandlungen aus dem Math. Sem. Hamburg, to appear.

[132] T. Meixner and A. Pasini, *A census of known flag transitive extended grids*, in Finite Geometries and Conbinatorics, (F. De Clerck et al. eds.), Cambridge University Press, Cambridge (1993), 249–268.

[133] T. Meixner and A. Pasini, *On flag-transitive extended grids*, to appear in Atti Sem. Mat. Fis. Univ. Modena.

[134] T. Meixner and F. Timmesfeld, *Chamber systems with string diagrams*, Geom. Dedicata, **15** (1983), 115–123.

[135] B. Mortimer, *A geometric proof of a theorem of Hughes on homomorphisms of projective planes*, Bull. London Math. Soc. **7** (1975), 267–268.

[136] A. Neumaier, *Some sporadic geometries related to* $PG(3,2)$, Arch. Math. **42** (1984), 89–96.

[137] R. Niles, *Finite groups with parabolic type subgroups must have a* BN*-pair*, J. Algebra **75** (1982), 484–494.

[138] U. Ott, *Bericht uber Hecke Algebren un Coxeter Algebren endlicher Geometrien*, in Finite Geometries and Designs (P. Cameron, J. Hirschfeld and D. Hughes eds.), Cambridge University Press, Cambridge (1981), 260–271.

[139] U. Ott, *Some remarks on representation theory in finite geometries*, in Geometries and Groups, L.N. **893**, Springer, Berlin (1981), 68–110.

[140] U. Ott, *On finite geometries of type* B_3, J. Comb. Th. A **39** (1985), 209–221.

[141] D. Pasechnik, *Dual linear extensions of generalised quadrangles*, European J. Comb. **12** (1991), 541–548.

[142] D. Pasechnik, *Geometric characterization of graphs from the Suzuki chain*, European J. Comb., **14** (1993), 491–499.

[143] D. Pasechnik, *Geometric characterization of sporadic groups* Fi_{22}, Fi_{23} *and* Fi_{24}, J. Comb. Th.A, to appear.

[144] A. Pasini, *Diagrams and incidence structures*, J. Comb. Th. A, **33** (1982), 186–194.

[145] A. Pasini, *On the number of elements in finite projective cyclic geometries*, Boll. Un. Mat. Italiana (6) **2**-B (1983), 791–802.

[146] A. Pasini, *Canonical linearizations of pure geometries*, J. Comb. Th. A, **35** (1983), 10–32.

[147] A. Pasini, *The non-existence of proper epimorphisms of finite thick generalized polygons*, Geom. Dedicata **15** (1984), 389–397.

[148] A. Pasini, *Some remarks on covers and apartments*, in Finite Geometries (C. Baker and L. Batten eds.), Dekker (1985), 223–250.

[149] A. Pasini, *Covers of finite geometries with non-spherical minimal circuit diagram*, in Buildings and the Geometry of Diagrams (L.A. Rosati ed.), L.N. **1181**, Springer, Berlin (1986), 218–241.

[150] A. Pasini, *On geometries of type C_3 that are either buildings or flat*, Bull. Soc. Math. Belgique **38** (1986), 75–99.

[151] A. Pasini, *On finite geometries of type C_3 with thick lines*, Note di Matematica **6** (1986), 205–236.

[152] A. Pasini, *Geometries of type C_n and F_4 with flag transitive automorphism group*, Geom. Dedicata **25** (1988), 317–337.

[153] A. Pasini, *On the classification of finite C_n geometries with thick lines*, A.M.S. series Contemp. Math. **111** (1990), 153–169.

[154] A. Pasini, *A bound for the collinearity graph of certain locally polar geometries*, J. Comb. Th. A **58** (1991), 127–130.

[155] A. Pasini, *On locally polar geometries whose planes are affine*, Geom. Dedicata **34** (1990), 35–56.

[156] A. Pasini, *Quotients of affine polar spaces*, Bull. Soc. Math. Belgique **42** (1990), 643–658.

[157] A. Pasini, *On extended grids with large automorphism groups*, Ars Combinatoria **29** (1990), 643–658.

[158] A. Pasini, *A classification of a class of Buekenhout geometries exploiting amalgams and simple connectedness (groups for geometries in given diagrams III)*, Atti Sem. Mat. Fis. Univ. Modena **39** (1990), 247–283.

[159] A. Pasini, *Groups for geometries in given diagrams II: on a characterization of the group $Aut(HS)$*, European J. Comb. **12** (1991), 147–158.

[160] A. Pasini, *Diagram geometries for sharply n-transitive sets of permutations or of Mappings*, Designs, Codes and Cryptography, **1** (1992), 275–297.

[161] A. Pasini, *Flag-transitive C_3 geometries*, Discrete Math. **117** (1993), 169–182.

[162] A. Pasini, *Shadow geometries and simple connectedness*, European J. Comb., **15** (1994), 17–34.

[163] A. Pasini, *Reducibility of chamber systems and factorization of parabolic systems*, Proceedings of the International Conference of Algebra, Barnaul, August 1991, to appear.

[164] A. Pasini and S. Rees, *A theorem on Tits geometries of type C_n*, J. Geometry **30** (1987), 123–143.

[165] A. Pasini and S. Yoshiara, *Flag transitive Buekenhout geometries*, Combinatorics '90 (A. Barlotti et al. eds.), North-Holland, Amsterdam (1992), 403–447.

[166] S. Payne and J. Thas, Finite Generalized Quadrangles, Pitman, Boston 1984.

[167] T. Penttila, *One-point extensions of inversive planes*, Australasian J. Combinatorics, to appear.

[168] G. Pica, *Ott-Liebler numbers*, Geom. Dedicata **39** (1991), 307–320.

[169] G. Pica and A. Pasini, *Flag-transitivity in shadow geometries*, European J. Comb., **15** (1994), 35–44.

[170] S. Rees, *C_3 geometries arising from the Klein quadric*, Geom. Dedicata **18** (1985), 67–85.

[171] S. Rees, *Finite C_3 geometries in which all lines are thin*, Math. Zeit. **189** (1985), 263–271.

[172] R. Robinson, *The structure of certain triple systems*, Math. Comp. **29** (1975), 223–241.

[173] M. Ronan, *Coverings and automorphisms of chamber systems*, European J. Comb. **1** (1980), 259–269.

[174] M. Ronan, *Coverings of certain finite geometries*, in Finite Geometries and Designs (P. Cameron, J. Hirschfeld and D. Hughes eds.), Cambridge University Press, Cambridge (1981), 316–331.

[175] M. Ronan, *On the second homotopy group of certain simplicial complexes and some combinatorial applications*, Quart. J. Math. Oxford, **32** (1981), 225–233.

[176] M. Ronan, *Locally truncated buildings and M_{24}*, Math. Zeit. **180** (1982), 489–501.

[177] M. Ronan, *Triangle geometries*, J. Comb. Th. A, **37** (1984), 294–319.

[178] M. Ronan, *Extending locally truncated buildings and chamber systems*, Proc. London Math. Soc. **53** (1986), 385–406.

[179] M. Ronan, *A construction of buildings with no rank* 3 *residues of spherical type*, Buildings and the Geometry of Diagrams (L.A. Rosati ed.) , L.N. **1181**, Springer, Berlin (1986), 242–248.

[180] M. Ronan, Lectures on Buildings, Academic Press (Perspectives in Mathematics **7**), 1989.

[181] M. Ronan, *Classification and construction of buildings*, Finite Geometries, Buildings and Related Topics (W. Kantor, R. Liebler, S. Payne and E. Shult eds.), Oxford University Press, Oxford (1990), 1–15.

[182] M. Ronan and S. Smith, 2-*Local geometries for some sporadic groups*, A.M.S. Proc. Symp. Pure Math. **37** (1980), 283–289.

[183] M. Ronan and G. Stroth, *Minimal parabolic geometries for the sporadic groups*, European J. Comb., **5** (1984), 59–91.

[184] M. Ronan and J. Tits, *Building buildings*, Math. Ann. **278** (1987), 291–306.

[185] R. Scharlau, *Geometric realizations of shadow geometries*, Proc. London Math. Soc. **61** (1990), 615–656.

[186] R. Scharlau, *Buildings*, Chapter 11 of Handbook of Incidence Geometry (F. Buekenhout ed.), to appear.

[187] G. Seitz, *Flag-transitive subgroups of Chevalley groups*, Ann. of Math. **97** (1973), 27–56. Correction (unpublished).

[188] S. Shpectorov, *A geometric characterization of the group* M_{22}, in Investigations in the Algebraic Theory of Combinatorial Objects, VNI-ISI, Moscow (1985), 112–123 (in Russian).

[189] S. Shpectorov, unpublished (1986).

[190] S. Shpectorov, *The universal* 2-*cover of the* P-*geometry* $\mathcal{G}(Co_2)$, European J. Comb. **13** (1992), 291–312.

[191] E. Shult, *Remarks on geometries of type* C_n, Geom. Dedicata **25** (1988), 223–268.

[192] E. Shult, *Geometric hyperplanes in embeddable Grassmannians*, J. Algebra, **145** (1992), 55–82.

[193] L. A. Skornjakov, *On homomorphisms of projective planes and T-homomorphisms of ternary rings* (in Russian), Math. Sbornik **43** (1957), 285–294.

[194] L. Soicher, *On simplicial complexes related to the Suzuki sequence graphs*, in Groups, Combinatorics and Geometry (M. Liebeck and J. Saxl eds.), Cambridge University Press, Cambridge (1992), 240–248.

[195] A. Sprague, *A characterization of* 3*-nets*, J. Comb. Th. A **27** (1979), 223–353.

[196] A. Sprague, *Incidence structures whose planes are nets*, European J. Comb. **2** (1981), 193–204.

[197] A. Sprague, *Rank* 3 *incidence structures admitting dual-linear linear diagram*, J. Comb. Th. A **38** (1985), 254–259.

[198] G. Stroth, *Geometries of type M related to A_7*, Geom. Dedicata **20** (1986), 265–293.

[199] G. Stroth, *Some sporadic geometries having C_3-residues of type A_7*, Geom. Dedicata **23** (1987), 215–219.

[200] G. Stroth, *One node extensions of buildings*, Geom. Dedicata, **25** (1988), 71–120.

[201] G. Stroth, *A local classification of some finite classical Tits geometries and chamber systems of characteristic* $\neq 3$, Geom. Dedicata **28** (1988), 93–106.

[202] G. Stroth, *The nonexistence of certain Tits geometries with affine diagrams*, Geom. Dedicata **28** (1988), 278–319.

[203] G. Stroth, *Some geometry for McL*, Comm. Alg. **17** (1989), 2825–2833.

[204] G. Stroth, *Chamber systems, geometries and parabolic systems whose diagram contains only bonds of strength* 1 *and* 2, Invent. Math. **102** (1990).

[205] M. Suzuki, *Transitive extensions of a class of doubly transitive groups*, Nagoya Math. J. **27** (1966), 159–169.

[206] G. Tallini, *Partial line spaces and algebraic varieties*, Symp. Math. **28**, Academic Press (1986), 203–217.

[207] J. Thas, *Partial geometries in affine spaces*, Math. Zeit. **158** (1978), 1–13.

[208] J. Thas, *Flocks, maximal exterior sets and inversive planes*, A.M.S. series Contemp. Math. **111** (1990), 187–218.

[209] J. Thas, *Maximal exterior sets of hyperbolic quadrics: the complete classification*, J. Comb. Th. A **56** (1991), 303–308.

[210] J. Thas, *Generalized polygons*, Chapter 9 of Handbook of Incidence Geometry (F. Buekenhout ed.), to appear.

[211] J. Thas, private communication.

[212] F. Timmesfeld, *Tits geometries and parabolic systems in finitely generated groups, I, II*, Math. Zeit. **1984** (1983), 377–396 and 449–487.

[213] F. Timmesfeld, *Tits chamber systems and finite group theory*, in Buildings and the Geometry of Diagrams (L.A. Rosati ed.), L.N. **1181**, Springer, Berlin (1986), 249–269.

[214] F. Timmesfeld, *Tits geometries and revisionism of the classification of finite simple groups of characteristic* 2 *type*, Proc. Rutgers Group Theoretic Year, Cambridge University Press, Cambridge (1984), 229–242.

[215] F. Timmesfeld, *On the 'non-existence' of flag-transitive triangle geometries*, Coxeter Festschrift (I), Mitt. Math. Sem. Giessen **163** (1984), 19–44.

[216] F. Timmesfeld, *Tits geometries and parabolic systems of rank* 3, J. Algebra **96** (1985), 442–478.

[217] F. Timmesfeld, *Locally finite classical Tits chamber systems of large order*, Invent. Math. **87** (1987), 603–641.

[218] F. Timmesfeld, *Classical locally finite Tits chamber systems of rank* 3, J. Algebra **124** (1989), 9–59.

[219] J. Tits, *Les groupes de Lie exceptionnels et leur interpretation géometrique*, Bull. Soc. Math. Belgique **8** (1956), 48–81.

[220] J. Tits, *Classification of algebraic semisimple groups*, Proc. Symp. Pure Math. **9**, A.M.S. (1966), 33–62.

[221] J. Tits, *Non-existence de certaines extensions transitives I. Groups projectifs à une dimension*, Bull. Soc. Math. Belgique **23** (1971), 481–492.

[222] J. Tits, Buildings of Spherical Type and Finite BN-pairs, L.N. **386**, Springer, Berlin 1974.

[223] J. Tits, *Non-existence de certains polygones generalisés, I*, Invent. Math. **36** (1976), 275–284.

[224] J. Tits, *Non-existence de certains polygones generalisés, II*, Invent. Math. **51** (1979), 267–269.

[225] J. Tits, *Endliche Spiegelungsgruppen, die als Weylgruppen auftreten*, Invent. Math. **43** (1977), 283–295.

[226] J. Tits, *Reductive groups over local fields*, Proc. Symp. Pure Math. **33** (1979), 29–69.

[227] J. Tits, *A local approach to buildings*, in The Geometric Vein (C. Davis et al. eds.), Springer, Berlin (1981), 519–547.

[228] J. Tits, *Immeubles de type affine*, in Buildings and the Geometry of Diagrams (L. A. Rosati ed.), L.N. **1181**, Sringer (1986), 157–190.

[229] J. Tits, *Ensembles ordonnes, immeubles et sommes amalgamées*, Bull. Soc. Math. Belgique A, **38** (1986), 367–387.

[230] J. Tits, *Buildings and groups amalgamation*, London Math. Soc. Lect. Notes, **121** (1986), 110–127.

[231] F. Toth, Regular Figures, Pergamon Press, Oxford, 1964.

[232] J. Van Bon and R. Weiss, *A characterization of the groups* Fi_{22}, Fi_{23} *and* Fi_{24}, Forum Mathematicum, **4** (1992), 425–432.

[233] L. Van Nypelseer, *Rank n geometries with affine hyperplanes and dual affine point residues*, European J. Comb. **12** (1991), 561–566.

[234] O. Veblen and J. Young, Projective Geometry, Ginn, Boston, 1916.

[235] F. Veldkamp, *Polar geometry*, Proc. Kon. Akad. Wet., A**62** (1959), 512–551 and A**63** (1959), 207–212.

[236] R. Weiss, *Extended generalized hexagons*, Proc. Cambridge Phil. Soc., **108** (1990), 7–19.

[237] R. Weiss, *A geometric characterization of the groups M_{12}, He and Ru*, J. London Math. Soc. (2), **44** (1991), 261–269.

[238] R. Weiss and S. Yoshiara, *A geometric characterization of the groups Suz and HS*, J. Algebra **1333** (1990), 182–196.

[239] D. Welsh, Matroid Theory, Academic Press, London 1974.

[240] S. Yoshiara, *A locally polar geometry associated with the group HS*, European J. Comb. **11** (1990), 81–93.

[241] S. Yoshiara, *A classification of flag-transitive classical cC_2 geometries by means of generators and relations*, European J. Comb. **12** (1991), 159–181.

[242] S. Yoshiara, *On some flag-transitive non-classical c.C_2 geometries*, European J. Comb., **14** (1993), 59–77.

[243] S. Yoshiara and A. Pasini, *On flag-transitive anomalous C_3-geometries*, to appear in Contributions to Alg. and Geom.

Index